CW00937723

Tolley's Handbook of Disaster and Emergency Management

Tolley's Handbook of Disaster and Emergency Management

Third edition

Edited by

Tony Moore
Cranfield University

and

Raj Lakha
Safety Solutions (UK) Ltd

AMSTERDAM • BOSTON • HEIDELBERG • LONDON
NEW YORK • OXFORD • PARIS • SAN DIEGO
SAN FRANCISCO • SINGAPORE • SYDNEY • TOKYO

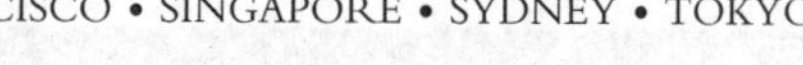

Newnes is an imprint of Elsevier

Newnes is an imprint of Elsevier
Linacre House, Jordan Hill, Oxford OX2 8DP, UK
30 Corporate Drive, Suite 400, Burlington, MA 01803, USA

First edition 2002
Second edition 2003
Third edition 2006

British Library Cataloguing in Publication Data
A catalogue record for this book is available from the British Library

Library of Congress Cataloging in Publication Data
A catalogue record for this book is available from the Library of Congress

ISBN-13: 978-0-75-066990-0
ISBN-10: 0-75-066990-X

For information on all Newnes publications visit our website at http://books.elsevier.com

Printed and bound in Great Britain
06 07 08 09 10 10 9 8 7 6 5 4 3 2 1

A book dedicated to the victims of and responders to September the 11th 2001 and all human-made disasters

Contents

Foreword

Disasters are as old as mankind. In ancient Egypt the Nile flooded at regular intervals, bringing in its wake a swathe of devastation involving great loss of life and ruined crops. The Roman cities of Pompeii and Herculaneum were destroyed, and countless lives lost, when Mount Vesuvius erupted in AD 79, whilst many cities in the middle East have been ravaged by earthquakes. The list is endless, but to confine such events to history would be foolhardy: the volcanic eruption on the island of Montserrat, as well as the recent flooding of China's Yangtse river, serve as a timely reminder that the devastating forces of nature are forever with us.

However nature is not alone in generating disasters. Man's ingenuity has seemingly no bounds and ambitious feats of engineering have led for example to faster travel by land, sea and air; the use of nuclear power; projects such as the Channel Tunnel and the World Trade Centre as well as bridges of prodigious span and height. All have been designed or developed to improve the quality of life and in most cases they have. Sometimes, though, things go badly wrong and when they do the effects can be no less devastating than what insurers and lawyers euphemistically describe as an 'Act of God'.

Occasionally our daily routine is disturbed by a major disaster such as Chernobyl, when millions across Europe were affected by the explosion at the nuclear power plant and the subsequent radiation fallout. Etched, too, on the minds of many people are the King's Cross underground fire, the explosions on the Piper Alpha platform and at Buncefield and the recent train crashes at Ladbroke Grove, Hatfield and Potters Bar. All such disasters, whether natural or man made, ancient or modern, share certain characteristics: sudden and overwhelming destruction combined with extensive death and serious injury for all in the vicinity.

If we can be certain of one thing it is that other disasters are waiting to happen. The questions are what?, where?, and when? What devastation would result if for any reason a plane faltered on its approach to a major airport, flying over the densely populated city it serves? What if we suffer another Chernobyl-type explosion? Could there be another September the 11th and further terrorist outrages such as those that occurred in Spain, Jordan, England and Egypt? These are questions many dare not contemplate, preferring instead to hope that such events will never happen.

Some, however, do not enjoy that luxury. Those entrusted with the vital roles of disaster management and emergency response must constantly plan to mitigate, so far as possible, the carnage and damage which inevitably follow a disaster and restore normality as soon as possible. There is no place today for the fatalism of earlier times when the view prevailed that little could be done, either before or afterwards, either to prevent or alleviate disaster. Notwithstanding that Noah's Ark may be regarded as one of the earliest examples of disaster management, it is only comparatively recently that serious attempts have been made to effectively plan, integrate and co-ordinate the efforts of services – public and private – as they address the effects of a major incident.

Disasters do not come much larger than the 11th September atrocity in New York, but the speed and efficiency displayed by the emergency services to firstly save life and, subsequently, clear the site could not have been achieved without an emergency response plan. Furthermore, the speed with which London got back to normal business following the explosions on 7 July 2005 was the result of careful planning for such an emergency. Both events remind us that we all have to be prepared for such incidents, particularly businesses for which major interruption could be fatal.

Matters such as emergency planning, business continuity, crisis management, disaster recovery and medical response need to be considered not only by the emergency services but also within organisations vulnerable to disaster. This book, wide in its scope, carefully pulls together all these issues. Looking at case studies, and the important areas of insurance and loss prevention, the book also reviews the modern scourge of terrorism. It is a timely publication and will be a valuable source of reference for those with responsibility for Disaster and Emergency Planning as well as those studying for the NEBOSH Module.

John Barrell
Leicester, March 2006

Preface

Tolley's Handbook of Disaster and Emergency Management: Principles and Practice has been produced to assist those who are either directly or peripatetically engaged in Disaster or Emergency Management issues. In the broadest context the Handbook will be of use to those in Occupational Safety and Health, Emergency Planning, COMAH operations, Civil Defence, Business Continuity and Insurance, Loss Prevention, Risk Management, Fire Safety, Public Health and Environmental Management. It is relevant to both the private and the public sectors.

The publication will also be of interest to practitioners in countries which have a similar approach to that of the UK because of the inclusion of information and case studies relevant to all those with responsibility for Crisis, Disaster and Emergency Management.

The Handbook's remit is to focus on 'human-made' and industrial disasters. The field of Natural Disaster Management is likewise a comprehensive body of knowledge with its own methods and theories that deserves a separate treatment from this Handbook.

As with its previous editions, the third edition emphasises common generic themes that cut across industrial sectors. Topics such as crisis planning, business continuity, insurance, fire safety, construction and design, training to name a few are themes that affect all organisations. Where appropriate, chapters have been updated and the book now incorporates the Civil Contingencies Act of 2004, the COMAH (Amendment) Regulations of 2005, and discusses the draft Corporate Manslaughter Bill for England and Wales. The third edition also has new publishers in Elsevier. Nevertheless, it follows the well-known and respected style of clarity and accuracy that is synonymous with the Handbooks developed by its previous publishers, LexisNexis Tolley.

Essentially, this publication is a reference source written by practitioners in their specialist fields of operation. It forms the starting point for further reading and research rather than a complete treatise on what is a very vast and developing body of knowledge.

Apart from its value as a practitioner's Handbook, it is also a source for academic and vocational courses in Disaster and Emergency Management such as the new NEBOSH Specialist Diploma in 'Disaster and Emergency Management', courses offered by the Emergency Planning College, those of the Emergency Planning Society, Business Continuity Institute, the Institute of Civil Defence and Disaster Studies, the International Institute of Security and Safety Management, the College of Teachers and other professional bodies.

This Handbook is one of many other useful sources in the field of Disaster and Emergency Management. However, it is probably the first compilation by practitioners in the UK of a generic and practical source of information in the field of Industrial Disaster Management.

We would like to thank Elsevier, in particular Doris Funke, for their co-operation and patience in the updating of this Handbook, as well as all the original contributors who generously gave of their time to produce their respective chapters. That having been said, the third edition has been brought up to date solely by ourselves, and any omission or error is solely the responsibility of the co-editors and not that of the original contributor.

We hope that the Handbook continues to provide information and guidance to all those readers who are pursuing academic careers but, more importantly, perhaps, assists those who are tasked with the responsibility of making their organisations safer and more efficient in terms of the emergency management, in its widest context, by providing relevant information and guidance.

Tony Moore and Raj Lakha
March 2006

Contributors

CO-EDITORS

Raj Lakha

Raj Lakha co-founded in 1994 Safety Solutions UK Limited, one of the UK's leading safety related training companies. He has advised and trained many hundreds of blue-chip organisations and local and national government bodies worldwide on topics ranging from Safety, Risk, Disaster and Business Continuity Management. He has also developed DEMSoft, the new checklist software. He conceived and developed for NEBOSH the new Specialist Diploma in Disaster and Emergency Management Systems (DEMS). He holds degrees in Economics and Government, an MBA in International Business and is a Member of IOSH and the IIRSM.

Tony Moore

Tony Moore spent 28 years in the London Metropolitan Police, retiring in the rank of Chief Superintendent in 1986. Since then he has specialised in the management of human-made disasters, including civil disorder and terrorism. He is currently a Visiting Fellow within the Disaster Management Centre, a part of Cranfield University based at the UK Defence Academy. He holds a Master of Philosophy (M Phil) degree from the University of Southampton, is the Chairman of the Institute of Civil Defence and Disaster Studies and a Member of the Society of Industrial Emergency Services Officers.

AUTHORS

Michael Appleby

Michael Appleby is a solicitor. In September 2003 he joined Fisher Scoggins LLP who have vast experience in 'disaster' litigation both from the civil and criminal perspective.

Michael defended the two train drivers at the Watford and Southall train crashes in 1996 and 1997 charged with manslaughter and health and safety offences. Both were acquitted. He also acted in the inquiries that followed the Southall and the 1999 Ladbroke Grove train crashes.

He frequently writes and lectures on health and safety issues. He writes a regular column for *Safety and Health Practitioner*. He is joint editor of *Corporate Liability: Work Related Deaths and Criminal Prosecutions* (*2003*) with Gerard Forlin.

Gary Beckley MIOSH NdipM
Since leaving the Army (where he was employed for over 15 years as a military engineer and Bomb Disposal Officer) Gary Beckley has worked as a full-time Safety, Health and Environmental (SHE) Professional. He initially joined a civil engineering company as a regional safety advisor and then moved to a construction health and safety company as a consultant and trainer. In 1999, Gary became an independent consultant and regular specialist lecturer on a range of courses, including bespoke, NEBOSH and IOSH accredited, as well as for the prestigious Roben's Institute on their MSc.

Gary is now the owner and director of SHE Management (UK) Ltd, a successful company providing practical and cost effective safety, health and environmental business solutions to a wide-ranging client base.

Toby Clark BSc FIOSH AIEMA SpDipEM CertEd
Toby Clark spent 12 years working in industrial chemistry (metal finishing, plastics and rubber, polyether foam production). He was a lead lecturer to London Fire Brigade and lectures on NEBOSH Environmental and DEMS Diploma courses as well as the IOSH course Managing the Environment. He is an author of Rapid Results College NEBOSH Safety Technology module and a NEBOSH examiner. Toby is a Director of Safety Solutions UK Limited.

Peter Dawkins BA MSc DMS MCIEH Sp DipEN
Peter qualified as an Environmental Health Officer in 1984 and has spent the majority of his professional career working for local government in inner London. Since qualification Peter has specialised in health and safety, food safety enforcement and communicable disease control and has gained wide experience in related enforcement issues. Peter is also an associate lecturer with Safety Solutions UK Ltd where he teaches on various health and safety and environment qualification courses.

Christopher Eskell TD LLB MCMI, Partner, Bond Pearce
Christopher Eskell qualified as a solicitor in 1975 and for the past 27 years has specialised in acting for claimants and defendants in a wide range of personal injury actions, including employers' liability, public liability, occupational and industrial disease claims. Christopher has also represented clients at numerous inquests, HSE and Environment Agency prosecutions and a major public inquiry to investigate the causes of an outbreak of Legionnaires Disease at Stafford Hospital. Christopher lectures and has contributed chapters on Employers' Liability Insurance and the Control of Major Accident Hazards to *Tolley's Health and Safety at Work Handbook*. He is also the legal editor of *Tolley's Fire Safety Management Handbook*.

Gerry Forlin
Gerard Forlin is a barrister at 2–3 Gray's Inn Square, and is widely accepted as a leading expert in the field of regulatory crime. He has undertaken disaster cases all over the world, including in this country the Southall and Ladbroke Grove train

crashes. He is standing Counsel to a number of plcs, government departments and trade unions, both in the UK and abroad. He has a vast amount of experience in representing doctors, nurses, directors, individuals and companies charged with manslaughter and other regulatory offences. He lectures to lawyers, companies and other safety professionals and regularly appears on television and radio as an expert. He is recommended as a leader in Health and Safety in *Chambers and Partners Directory 2003–4* and the *Legal 500*, where he is described as a natural Advocate and an excellent orator and named *'The Manslaughter Guy'*. He is general editor of *Corporate Liability: Work Related Deaths and Criminal Prosecutions – December 2003* published by LexisNexis and has many other publications to his name.

Ken Golding MSc
Ken holds a MSc in Safety Management, is a partner of Heritage Safety Management Services and senior consultant with GsD Safety, which are members of the Seraph Consulting Group of companies. Prior to establishing Heritage Safety Management in 1999, Ken trained as a production/quality controller in the printing industry before joining the National Trust as a member of their conservation management team. He designed the conservation emergency procedures training for the Trust and has advised numerous heritage sites in methods of integrating safety and emergency planning as part of the daily management of the business. He has addressed national fire safety conferences, published articles on the subject and is a guest lecturer at the Ironbridge Institute. Ken is a member of the IOSH Healthcare Specialist Group and has been involved with a number of NHS Trusts as an advisor in the risk management strategies necessary when construction works are taking place.

Ira Gupta MBA
Ira has over six years of working experience in the field of Information Security. She has been responsible for and carried out software project management activities with NewGen Software Technologies, a document management solutions company. She has been a part of KPMG's Information Risk Management practice for over three years and carried out IT consulting and advisory assignments for various clients in the area of Information Security. Currently, Ira works as an independent information security consultant.

Ira holds an MBA in Systems and is also a Certified Information Systems Security Professional (CISSP) from International Information Systems Security Certification Consortium (ISC^2), USA and a Certified Information Systems Auditor (CISA) from Information Systems Audit and Control Association (ISACA), USA.

Jim Munday
Jim Munday joined the Metropolitan Police Forensic Science Laboratory in 1972 and moved into full-time fire and explosion investigation in 1979. He left the Laboratory in 1998 to set up a private consultancy providing fire and explosion investigation services. He has investigated around 1,700 fire scenes including over 300 fatal incidents. Jim is heavily involved in devising and delivering training programmes and is a member of the Arson Control Forum group creating a national standard of competence for investigators. He is a Member of the Institution of Fire Engineers and holds the Diploma of the Forensic Science Society in fire investigation.

Dr Brian Robertson OStJ TD MIEM
Dr Brian Robertson has a background in General Practice and the Armed Forces along with lifelong interest in the delivery of pre-hospital emergency care. He is National Chairman of the British Association for Immediate Care (BASICS). His special interests are major incident management and mass gathering medicine. He was involved with the Hungerford multiple shooting (1987) the Clapham (1988) and Purley (1989) railway accidents and the loss of the *Marchioness* on the River Thames (1989). Brian lectures widely on pre-hospital care, mass gathering medicine and major incident management. He has been Medical Director of the Farnborough International Air Show since 1996 and is the owner and director of The Event Medicine Company Ltd delivering mass gathering medicine to all types of events.

David Schofield
David is a senior lecturer and course director for the Master of Science degree in Safety Management at Bournemouth University and until recently was the University safety adviser. He started his career as an Environmental Health Officer and for twelve years enforced standards in food, pollution, disease, housing and safety, giving him much enforcement experience. Subsequently, he became senior lecturer at Highbury College, Portsmouth and was the NEBOSH Diploma course tutor for six years. He has a considerable consultancy record. His professional roles include chief examiner for the NEBOSH Diploma and previously as a full NEBOSH Board member for three years. Professional membership includes IOSH and the American Society of Safety Engineers. He has published over 30 case studies in health and safety.

Charles van Oppen
Charles van Oppen MA AIRM FICDDS is Managing Director of van Oppen & Co, risk and insurance information managers based in the City of London (see www.vanoppen.co.uk). Charles has written a number of papers on disaster risk insurance themes and edits two specialist journals. He is the Vice Chairman of the Institute of Civil Defence and Disaster Studies (ICDDS). His career started as an insurance broker at Lloyd's in 1988. He has a BA in Risk Management, an MA in Development Studies and is an Affiliate Member of the Benfield Greig Hazard Research Centre, University College London (UCL).

Sam Webb
Sam Webb is an architect with 40 years experience of the building industry. Elected to the Council of the RIBA he served as Chairman of the Building Legislation and Technology Committee. In 1983 he presented evidence to the Chairman of the London Borough of Newham Council's Housing Committee and the tenants of Ronan Point. The Minister released all the Inquiry papers to him in 1985 following a campaign in *The Times* and Parliament. Ronan Point was dismantled in 1986, followed by the other eight blocks on the site. He was a consultant for the Channel 4 TV series: *Collapse: Why buildings fall down* shown in 2000.

Acknowledgements

The Code of Practice in Chapter 12 is reproduced with permission from the International Labour Organization. The original edition of the Code of Practice was published by the International Labour Office, Geneva, under the title, *Prevention of major industrial accidents: An ILO code of practice*. Copyright © 1991 International Labour Organization.

The Appendices in Chapter 13 are reproduced with permissions as follows: Appendix 1 by agreement with the British Insurance Brokers Association; Appendices 2 and 3 by agreement with the Association of British Insurers.

The 'Cruciform' triage card shown in Chapter 15 is reproduced with the permission of Carl Wallin of CWC Services.

Acknowledgements

Abbreviations

ABI	Association of British Insurers
ADR	Alternative Dispute Resolution
ACC	American Chemistry Council
ACOP	Approved Code of Practice
ACPO	Association of Chief Police Officers
AFR	Accident Frequency Rate
AN	Ammonium Nitrate
ANAO	Australian National Audit Office
APRIA	Asia-Pacific Risk and Insurance Association
ASC	Assured Safe Catering
ASCE	American Society of Civil Engineers
ASF	Anti-shatter film
BAT	Best Available Techniques
BBNC	Bomb Blast Curtains
BATNEEC	Best Available Techniques Not Entailing Excessive Cost
BCI	Business Continuity Institute
BCM	Business Continuity Management
BCP	Business Continuity Planning
BI	Business Impacts
BIBA	British Insurance Brokers' Association
BLEVE	Boiling Liquid Expanding Vapour Explosion
BRE	Building Research Establishment
BRO	Business Recovery Organization
BSI	The British Standards Institute
CA	Competent Authority
CAD	Computer Aided Design
CAD	Civil Aviation Division
CAP	Community Action Panel
CB	chemical or biological (weapon)
CBA	Cost Benefit Analysis
CBRN	Chemical Biological Radiological and Nuclear
CCDC	Consultants in Communicable Disease Control

CCTA	Computer & Telecommunications Agency
CDM	Construction (Design and Management) Regulations 1994
CEO	Chief Executive Officer
CILA	Chartered Institute of Loss Adjusters
CIMAH	Control of Industrial Major Accident Hazards Regulations 1984
CMP	Codes of Management Practices
COMAH	Control of Major Accident Hazards Regulations 1999
COMAT	Company Material
COP	Code of Practice
CPS	Crown Prosecution Service
CRAMM	Risk Analysis and Management Model
CRT	Coastguard Rescue Team
EMS	Environmental Management System
EPIRB	Emergency Position Indicating Radio Beacons
EPN	Emergency Prohibition Notice
EPO	Emergency Prohibition Order
DIMS	Disaster Information Management System
DoH	Department of Health
DRI	Disaster Recovery Institute of Canada
DT	Department of Transport
DTI	Department of Trade and Industry
ECCS	Emergency Core Cooling System
ECRA	European Community Reportable Accidents
EEA	Environmental Emergency Aspect
EEI	Environmental Emergency Impact
EHO	Environmental Health Officer
FAA	Federal Aviation Authority
FEMA	Federal Emergency Management Agency of the USA
FSA	Formal safety assessment
FSO	Food Safety Officer
FSSPSA	Fire Safety and Safety of Places of Sport Act 1987
GPS	Global Positioning Systems
HA	Health Authority
HACCP	Hazard Analysis and Critical Control Points
HAZAN	Hazard Analysis
HAZMAT	Hazardous Material
HMCG	Her Majesty's Coast Guard
HSC	Health and Safety Commission
HSE	Health & Safety Executive
HTM	Health Technical Memoranda
IC	internal control
ICAO	International Civil Aviation Organisation
ICE	Institution of Civil Engineers
IDNDR	International Decade for Natural Disaster Reduction
IEC	International Electro-technical Commission
IED	Improvised Explosive Device
IEE	Institution of Electrical Engineers
EHOs	Environmental Health Officers
IDER	International Disaster and Emergency Readiness

IDS	Incomes Data Service
ILO	International Labour Organisation
IOD	Institute of Directors
IPCS	International Programme on Chemical Safety
IPPC	Integrated Pollution Prevention and Control
IPT	Insurance premium tax
ISO	International Organisation for Standardisation
IStrucE	Institution of Structural Engineers
IT	Information Technology
JRC	Joint Research Centre
LA	Local Authority
LCC	London County Council
LEI	Legal expenses insurance
MAFF	Ministry of Agriculture, Food and Fisheries
MAO	Maximum Acceptable Outage
MAPP	Major accident prevention policy
MARS	Major Accident Reporting System
MCA	Maritime and Coastguard Agency
MIC	Methyl Isocyanate
MIS	Management Information Systems
MoD	Ministry of Defence
MSV	Management System Verification
NASH	National Academy of Safety and Health
NDM	Naturalistic Decision Making
NFPA	National Fire Protection Association
NORSAR	Norwegian seismological body
NSARDA	National Search and Rescue Dog Association
NTSB	National Transportation Safety Board
NYPA	New York Port Authority
OCG	Outbreak Control Group
OCHA	Office for the Co-ordination of Humanitarian Affairs
OCP	Outbreak Control Plan
OHSAS	Occupation Health and Safety Assessment Series
OGC	Office of Government Commerce
OIM	Offshore Installation Managers
OSH	Occupational Safety and Health
PHLS	Public Health Laboratory Service
PID	Project Initiation Document
PIO	Police Incident Officer
PPE	personal protective equipment
PTSD	Post Traumatic Stress Disorder
RC	Reinforced Insitu Concrete
RIBA	Royal Institute of British Architects
RIDDOR	Reporting of Injuries, Diseases and Dangerous Occurrences Regulations 1995
RNLI	Royal National Lifeboat Institution
ROES	Representatives of Employee Safety
RPDM	Recognition Primed Decision-Making
RPG	Risk Planning Group

RRV	Rapid Response Vehicle
SAR	search and rescue
SEPA	Scottish Environment Protection Agency
SOC	Strategic Outline Case
SOLAS	Safety of Life at Sea
TCCD	Dioxin
TQM	Total Quality Management
TOR	Technic of Operations Review
UCC	Union Carbide Corporation
UCIL	Union Carbide India Ltd
UN	United Nations
UNCHS	UN Centre for Human Settlement
UNDP	UN Development Programme
UNEP	United National Environment Programme
UNESCO	UN Educational, Scientific and Cultural Organisation
UNHCR	UN's High Commissioner for Refugees
UNIDO	UN Industrial Development Organisation
UNITAR	UN Institute for Training and Research
UNU	United Nations University
WHO	World Health Organisation
WTC	World Trade Centre

Table of Cases

Table of Statutes

Table of Statutory Instruments

Table of European Legislation

1

Business Continuity Management: Principles and Approaches

In this chapter:

- Historical background.
- Terminology.
- Reasons for Business Continuity Management.
- Approaches.
- Comparative assessment.
- Conclusion.

Historical Background

'The King is dead. Long live the King' 1.1

'Business Continuity' is not a new discipline. Its origin is probably as old as anthropoids battling with natural and socio-economic catastrophes. Humans are not alone in the endeavour to survive as a social entity and to recover from tragedies. Creatures such as bees, hyenas, foxes, elephants and a range of other mammals employ bio-chemical methods of identifying threats, developing collective responses and recovering after loss.

There have been different global traditions to the development of 'Business Continuity Management' (BCM), some of which are summarised below.

Machiavelli 1.2

In his book, *The Prince*, Machiavelli's answer to the loss suffered by a ruler, his wealth and the state was to behave like a 'lion' or 'fox' and to grab resources from others; continuity or survival does not have to be via legitimate means.

Russell Dynes 1.3

Sociologist Russell Dynes writes that the earthquake in Lisbon, Portugal at around 9.30 am on 1 November 1755, was probably the first instance in post-Enlightenment Europe where the nation-state had to take responsibility for the management of the consequence and the re-establishment of infrastructure, communications and civil defence. With the loss of around 70,000 people, the destruction to property and assets of the nobility and royalty, the State responded by creating continuity for Lisbon and its reputation as a centre for the arts and culture. Whilst this was a sort of reactive and ad hoc Business Continuity Strategy, it was one nevertheless that taught valuable lessons to other European states (International Journal of Mass Emergencies & Disasters, pp 97–115, Vol 18, No 1, March 2000).

Karl Marx 1.4

In *Das Kapital* (Volume 3) and *Theories of Surplus Value*, Karl Marx makes two key observations. First, he notes that when the markets have failed, the motive of re-establishing the factory is contingent on the entrepreneur and the market having a personal gain from investing. In the language of pre-Turnbull 1980–1990s, this meant the FTSE companies and the market itself will only invest in 'business continuity' if a direct, tangible and quantifiable gain can be forecasted. Second, Marx also notes that where there is surplus labour and the market value of labour is low, the entrepreneur does not need to worry extensively about 'key persons' being lost or property being destroyed as this can be recovered due to the low cost of re-establishing the enterprise. In both cases 'business continuity' will only happen and only work if it assists the particular interests of a group, company or nation state.

Malinowski 1.5

The Anthropologist Malinowski in his studies of the South Pacific Trobriand Islanders in the 1920s noted how they had developed rituals and rules that guided them when the fish catch was poor or when nature showed no mercy. These codes explained the event, the impact on other functions of their society, the actions to be taken to restore normalcy and the behaviours to be avoided in the future that displeased the gods.

Kibbutzim 1.6

The Israeli Kibbutzim is a model of how social enterprises share risks and if one was to suffer major loss or disaster, a mechanism for the transfer of resources exists from one Kibbutz to another. This reflects the principle of continuity through co-operation.

Continuity strategies **1.7**

The experience of the USA in the 1800s saw major fires, earthquakes and flooding. For example, the Great Chicago Fire of 1871, The Boston Fire in 1872 (which destroyed 800 buildings) and, in 1886, the Charleston Earthquake in South Carolina which destroyed 90% of buildings in City. There was a growing realisation that without adequate 'continuity' strategies there would be dire economic and political implications.

IT recovery strategies **1.8**

It was in the 1970s in the USA, with the growth of Information Technology (IT) and the large mainframe computers that risks of catastrophic failure were being mooted. The banking sector in particular began to consider IT recovery strategies. Not only was this required for internal financial survival but USA banks formed the basis of national and global financial economic stability. Failures in IT systems in one bank could have chaotic effects worldwide.

'Year 2000 Bug' **1.9**

The most recent and important impetus to the formalisation of BCM as a discipline has been the 'Year 2000 Bug' (Y2K). The prognostication that failures to back-up data, make alternative IT arrangements and to ensure diagnostics of computers would lead to billions of dollars of losses led banks, governments, industry and professional bodies to develop various advisory documents. For example, the Health & Safety Executive (HSE) and the Department of Trade and Industry (DTI) produced various diagnostic guides, identifying how to ensure that computers could continue to function, back up systems, informing suppliers and customers etc.

Terminology **1.10**

BCM is not the same as Disaster Recovery, Emergency Management, Disaster/Crisis Management or Contingency Management. All of these different approaches are dimensions of 'Major Incident Management' (managing those catastrophic but rare events that require special arrangements in the mobilisation of human, financial, physical and informational resources) – see Figures 1 and 2 below.

Note: To exemplify the differences, consider an example of a fire in a school.

Figure 1: Different dimensions of Major Incident management

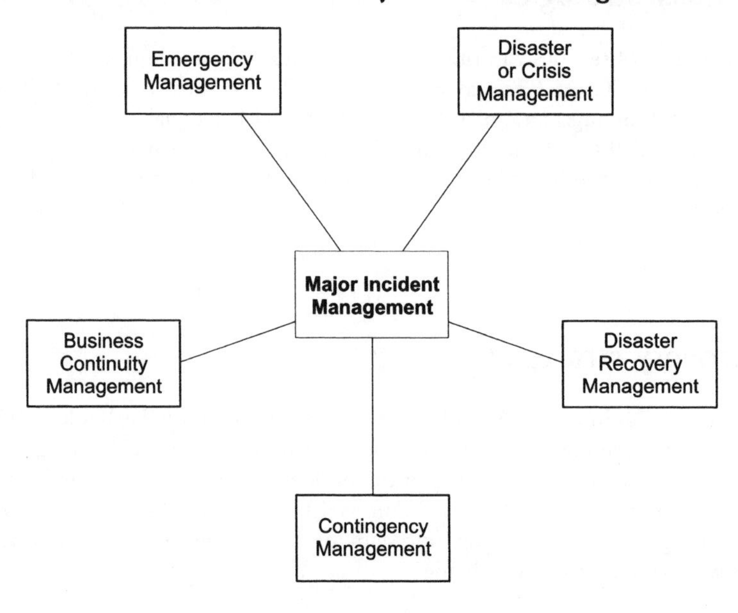

Figure 2: How BCM differs

Management	*Function*	*Summary*	*Example*
Disaster/crisis management	Managing the impact and adverse consequences of an event that is overwhelming and intensely difficult to cope with	'This is how we can *manage* now'	●Communicating with pupils and parents ●Crisis monitoring and reporting ('picture building') ●Strategic impact of fire on school etc
Contingency management	Alternative strategies in the event of the existing systems or facilities not functioning	'This is our back-up strategy or *Plan B*'	●Alternative venue ●Alternative staff ●Resource audit etc
Emergency management	Immediate response strategy and plan that is targeted, bespoke and specific to the major incident	'This is how we will *respond* now'	●Fire evacuation procedure ●Liaison with Fire Service. ●Fire wardens.

Disaster recovery management	The immediate actions needed both during and after the major incident to ensure the establishment will be able to function, mitigating business interruption	'This is how we will start *functioning asap*'	• Facilities safety (electrical, gas, water) • Pens, chairs, desks, board etc • IT recovery • Health and hygiene • Liaison with inspectors etc
Business continuity management	In contrast to the disaster recovery management, the BCP looks into the medium and longer term and focuses on ALL aspects of the establishment. Strategic and systemic impact plan	'This is how we will re-establish key business processes and proactively ensure survival into *the longer term*'	• Audit of school facilities – actual against needed • Key persons, buildings, procedures, finances etc • Insurance • Prevention strategy etc

Meaning of Business Continuity Management (BCM)

Three elements **1.11**

The Manual, Emergency Preparedness, which accompanies the Civil Contingencies Act 2004, defines Business Continuity Management (BCM) as 'a management process that helps manage the risks to the smooth running of an organisation or delivery of a service, ensuring that it can operate to the extent required in the event of a disruption' (Glossary, p. 215).

The Office of Government Commerce (OGC) ('A Guide to Business Continuity Management' (2001) Stationery Office, p 12) tells us that BCM has three main elements:

- prevention;
- planning;
- transfer.

A related document called 'An Introduction to Business Continuity Management' (Central Computer & Telecommunications Agency (CCTA) (1995) HMSO, p 7) says:

> 'BCM is concerned with managing risks to ensure that at all times an organisation can continue operating to, at least, a pre-determined minimal level'.

The CCTA also adds that BCM is about enhancing corporate performance, open ended, flexible, needs to be integrated into and with other core business functions and is an issue that needs to be lead from by senior management.

Multi-disciplinary process **1.12**

Others such as the De Montfort University/IDS Study ('Business Continuity Management: Preparing for the Worst' (1999) IDS, p 5) say that BCM is an open, proactive, preventative, holistic, multi-disciplinary process involving a variety of stakeholders. It's not just about 'response' but about 'adding value'.

Preventative and recovery controls **1.13**

Another definition of BCM is given in BS ISO/IEC 17799:2000 ('Information Technology – Code of practice for information security management', BSI, 2000, p 56), 'Objective: To counteract interruptions to business activities and to protect critical business processes form the effects of major failures or Disasters. A business continuity management process should be implemented to reduce the disruption caused by Disasters and security failures (which may be the result of, for example, natural disasters, accidents, equipment failures, and deliberate actions) to an acceptable level through a combination of preventative and recovery controls'.

A business issue **1.14**

The Business Continuity Institute ('BCM: A Strategy for Business Survival') says that:

> 'BCM is not just about disaster recovery, crisis management, risk management or about IT. It is a business issue. It presents you with an opportunity to review the way your organisation performs its processes, to improve procedures and practices and increase resilience to interruption and loss'.

In addition,

> 'Business Continuity Management is the act of anticipating incidents which will affect critical functions and processes for the organisation and ensuring that it responds to any incident in a planned and rehearsed manner'.

Characteristics of BCM **1.15**

In summary, it can be inferred from these definitions, BCM has the following characteristics:

- **P**roactive – developing continuity solutions beforehand to be prepared when a Disaster or Emergency strikes.
- **R**esourcing and ensuring that money, people, machinery etc can be called upon to ensure business survival.
- **E**fficiency focused – BCM is not a drain on resources. Use cost benefit analysis to show the merits of the approach.
- **V**alue adding – both financial, corporate and social.
- **E**ssential services and personnel need to be deployed to ensure continuity.
- **N**ormalisation being a key goal – to ensure that after the Disaster or Emergency the operation gets back to some element of 'what it was like before'.
- **T**ime focused – BCM is about realistically setting short, medium and long term plans to achieve that normalcy.
- **I**nformation management to key suppliers, contractors and personnel.
- **T**op management led.

Thus, BCM is concerned with **PREVENT IT** – the direct and indirect, insurable and uninsurable losses through better continuity preparedness.

Reasons for Business Continuity Management

Corporate governance reasons **1.16**

Whilst the Turnbull Report (Institute of Chartered Accountants of England and Wales, 1999) does not overtly comment on BCM, the whole thrust of the Report is to ensure that financial crisis that were prevalent in the 1980s and early 1990 on the London Stock Exchange (largely due to 'insider dealing' or lack of corporate governance) can be averted through effective boardroom control. Turnbull noted the financial and human implications of business interruption, employment implications and, indeed, reputational loss. In short, sound business management is not just about finance and marketing, it's also about Boards being able to ensure long-term corporate survival.

The British Standards Institute (BSI) produced BS PD 6668: 2000 ('Managing Risk for Corporate Governance', Robbins & Smith, BSI). This was in response to Turnbull and meant to provide a practical set of actions for managers to ensure effective governance at the 'shop front'. This document does refer more overtly to continuity and factoring this into decision-making processes at strategic, operational and tactical levels of management. This is an important statement in that BCM is not the exclusive domain of senior management but as a concept and approach, needs to flow through all layers of the organisation.

Insurance and loss control reasons 1.17

For those enterprises seeking 'business interruption insurance' or 'key person insurance', most insurers and brokers will want to see evidence of some BCM in place. The premium is correlated to the risk of interruption or the loss of key person(s) and any mitigation of such risk via BCM is rewarded by the insurer through a reduction in premium.

Effective BCM is one way of mitigating future losses (insurable or uninsurable). For instance, the ability of the organisation to start operating after a Disaster or Emergency will impact on the quantum of financial loss – the longer the delay in becoming operational the greater the financial losses (cash flow, balance sheet, bank credit, etc).

Legal reasons 1.18

In the UK, there is now legislation, by virtue of the Civil Contingencies Act 2004 and the accompanying Regulations, which imposes duties on some organisations to develop BCM.

- The Civil Contingencies Act and accompanying Regulations requires Category 1 Responders to maintain plans to ensure that they can continue to exercise their functions in the event of an emergency as far as is reasonably practicable. The Act also requires local authorities to provide advice and assistance to those undertaking commercial activities and voluntary organisations in relation to BCM in the event of an emergency.
- The *Data Protection Act 1998* requires organisations to have 'back-up' strategies in place as well as the capability to retrieve data. This involves making copies, locating copies in different locations, keeping a balance between electronic and hard copy formats.
- The *Control of Major Accident Hazards Regulations 1999* (*SI 1999 No 743*) (COMAH) and accompanying guidance from the HSE/Environment Agency/Scottish Environment Protection Agency (SEPA) does ask COMAH operators to think about the consequence of a Disaster – the clean up process on and off site, the importance of co-ordination centres, informing personnel of the event, developing recovery strategies and safely re-establishing the operation.

There is no European Directive at the time of writing on BCM. There is also a lack of case law in England and Wales on BCM.

Social reasons 1.19

Often ignored in the discussion of BCM is the importance of 'sustainable businesses'. One often hears of 'developing sustainable communities' that are resilient to natural and man-made Disasters. It could be said, resilient and sustainable businesses are the nucleus to ensuring social continuity as businesses become operational, then people get back to work, re-establish their work connections, start interacting with customers and suppliers and in general ensuring that economy and community start to function. Especially where there is the potential for natural Disasters such as flooding and earthquakes, strong business networks, business support from government and banks will yield social dividends. It could be said that in the Turkey or the Gujarat earthquakes there was an imbalance, with a greater focus on 'Disaster/Crisis Management', 'Emergency Management' and very little by local authorities on getting credit, facilities and infrastructure to their businesses, no activation of pre-developed local BCM strategies etc. Therefore, BCM has a value beyond the organisation – it can contribute (whether overtly or covertly) to social survival.

Humanitarian reasons 1.20

Whilst the private sector could argue it does not owe the community a moral duty to possess BCM, the public sector in particular does. As a provider of services to the community, a hospital, school, college, university, power station, road maintenance agency etc will be under pressure to 'get back to normal' and 'start operating as soon as possible'. BCM is not just a private sector preoccupation but also one that public bodies also need to develop separately and fuse collectively. It can be noted the guidance and advice from the CCTA and OGC – both government bodies – that BCM is for organisations irrespective of sector.

Approaches 1.21

There are a variety of approaches to the development of BCM for the organisation. The principal ones that shall be examined include the approaches from:

- BCI/BSI PAS 56 (BS 25999 pending);
- the UK's Department of Trade and Industry (DTI)/Visor Consultants;
- Office of Government Commerce (OGC)/Computer and Telecommunications Agency (CCTA);
- Incomes Data Service (IDS)/De Montfort University;
- Australian National Audit Office (ANAO);
- Disaster Recovery Institute of Canada (DRI);
- BS ISO 17799:2000.

BCI & BSI PAS 56 Approach (BS 25999 pending) 1.22

The Business Continuity Institute (BCI) approach is illustrated in Figure 3 below.

Figure 3: The BCI view of BCM as a five-step process

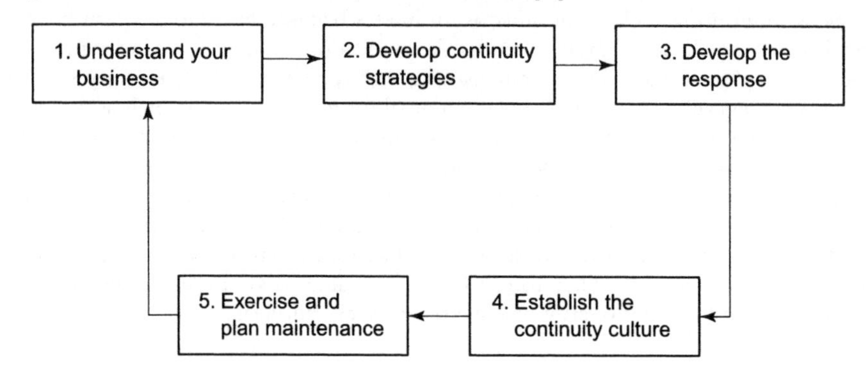

Note: The BCI are suggesting that the organisation needs to appreciate that these steps are guidelines only and need to be made bespoke to one's operation, sector and corporate culture.

Stage I: Understand your business **1.23**

The starting point is to step back and look at the organisation holistically. This is a business-critical step as most organisations will focus immediately on, say, a Risk Assessment or a Business Impact Assessment before embarking on a 'health check' of the organisation. There are four fundamental questions:

- *What does the organisation do – what is it all about?* — Analyse the organisation as a system of inputs (people, material, time, finances), process (mechanical, human, ergonomic), output (the product and/or service), the feedback (stakeholders). Look at the external and internal factors shaping the organisation and the type of corporate culture prevalent. Also consider the philosophy, mission and values the organisation holds. It is a case of defining the 'systems boundary', its goal and objectives. A schematic diagram showing the parts, hierarchies and functions is a helpful aid memoire.
- *When will the organisation achieve its goals?* — This is a case of delineating time scales and determining what plans will be needed in the 'short, medium and long terms'.
- *Next, identify who are the key stakeholders* — both internally and externally to the organisation. That is, the people who make the product or service possible. The range includes employees, contractors, investors, bank, customers, suppliers etc.
- *Identify the techniques, methods and strategies the organisation intends to use to achieve its goal and objectives* — Is the operation highly dependent on technology or people? Is there a Business Plan? Do adequate finances exist, etc?

The organisation can then ask what types and exactly how would 'risks' and 'business impacts' affect the four questions/areas examined? The risks can relate to health and safety related, financial, marketing, reputational, legal, corporate governance as well as others. The Business Impact Assessment could quantity the impact of

Disasters/Emergencies in a scale of 1–10 or more qualitative ('low, medium or high'). Likewise, the BCI suggest Risk Assessments could follow a similar approach. The Risk Assessment focuses on internal and external threats whilst the BIA asks how failures in mission critical functions would affect the organisation and its goal. The latter is more holistic and macro.

Stage 1 requires the support of all stakeholders, in particular the Board and the Directors. There has to be a commitment and acceptance of the BCM path entered into as well as any critical findings. Stage 1 can be summarised as a 'reality check'; know the organisation – its strengths, weaknesses, limitations and delineate/prioritise critical functions without which the organisation cannot survive.

Stage 2: Continuity strategies **1.24**

The next step is to determine the options to ensure the organisation can cope with the risks identified and continue operating. These options are summarised below.

'Do nothing' **1.25**

This is a form of 'risk retention'. The Board may perceive that it can 'take the risk'; that it can cover the risk exposure from either cash flow or liquidating other assets at time of need. This judgement may be due to either the chance of mission critical functions failing as being very low or because retention is more cost effective as against, say, insurance or, indeed, it is custom and practice in that sector to retain risks.

Changing or ending the process **1.26**

The Board may decide to alter the way it operates and its key policies and procedures. Stage 1 may have indicated the organisation is not keeping apace with industry best practice or reflecting latest legislation. The organisation could consider organisational redesign (ie become more flexible), revaluate its use of technology, consider the volume and quality of its operating procedures, decide if it can operate on a smaller basis without a significant marginal loss of income and also focus on the product and service it offers – quality, quantity and deliverability.

Insurance **1.27**

Insurance is the main form of 'risk transfer'. Whilst the organisation cannot transfer the hazard or danger or business threat, it can transfer the associated quantifiable commercial risk. Again, the Board has to consider the cost and benefit of insurance. For example, 'business interruption insurance' is an indemnity against unexpected and predefined perils (strikes, natural, electrical failings, non-payment by customers etc). Whilst this sounds attractive, the Board has to consider the probability of the policy ever being activated, the terms and conditions imposed and the total premium payable as against the payout. In short, the BCI suggest that insurance is one option to consider and not the first and only option.

Loss mitigation **1.28**

Loss mitigation are strategies to cope with the losses when they strike. Such losses can be direct, indirect or uninsurable. This option could consider renegotiating credit with the bank, possible alliance/merger with another, purchasing stock from another, changing markets, assessment of assets and liabilities to dispose of, etc.

Business Continuity Planning (BCP) **1.29**

The Manual, Emergency Preparedness, which accompanies the Civil Contingencies Act 2004, defines Business Continuity Planning as 'a documented set of procedures and information intended to deliver continuity of critical functions in the event of a disruption' (Glossary, p. 215). It should be noted, therefore, that BCP is a stage in the BCM process and should not equated with it. BCP is by definition a 'plan' that specifies a goal (ensuring the organisation remains resilient and recovers as efficiently as possible from the interruptions) and the tactics to achieve this (resources, time-scales, key persons, logistics, networks to name a few). BCP is analogous to a life support machine that keep the patient functioning until normalcy and equilibrium is achieved.

Stage 3: Developing the response **1.30**

It must be noted that the choices outlined in Stage 2 above are not mutually exclusive. Organisations will have to decide which one and how many of these to adopt and adapt to their operation. The response package chosen will need to be flexible, costed, and time scaled and also approved by key stakeholders (Board and shareholders in particular).

Assuming BCP was the preferred option, what are the key sections that the BCP need to contain? The BCI advises the following:

- *Part 1: General introduction and overview–*
    - Objectives – of BCP; terms of reference; where and when applicable? The authority of BCP and its authorisation.
    - Responsibilities – This is a broad outline of the key elements of the BCP and which part of the organisation has responsibility for it, eg the Board, production management, finance management and so on.
    - Exercising the BCP – when, who will be involved, the time and cost involved?
    - Maintenance – the BCP needs to be kept up to date, reflecting corporate and economic-social-political realities.

- *Part 2: Plan invocation* –
  - Disaster declaration – when can/should a 'disaster' be declared?
  - Damage assessment – physical, human, financial, informational.
  - Continuity actions and procedures – how to ring-fence critical functions and/or to activate back-up systems (eg anything from a diesel generator through to moving into a new building).
  - Team organisation and responsibilities – the 'who' aspects of implementation – key personnel and stakeholders both within and outside the organisation.
  - Emergency (crisis) operations centre – where it is, facilities it contains (paper, pens, phone, charts, chairs, etc).
- *Part 3: Communications* –
  - Who should be informed – all key stakeholders (employees, shareholders, suppliers, key customers, media, local authority, enforcers, insurers).
  - Contacts – list of key corporate officers and contact details.
  - Key messages – to be given, the exact wording, the particular use of and who should transmit the message. This section is also important for legal liability reasons.
- *Part 4: Suppliers* –
  - List of recovery Suppliers – up-to-date details of key contractors and support services that can be called to assist.
  - Details of contract provision – ensuring that contractual obligations can be met, alternative ways of meeting obligations identified and also arrangements with alternative sources to enable continuity.

The four Parts above are the core of the BCP with other elements added to suit the organisation's needs. As with all plans, the BCP needs to be flexible and SMART (simple, measurable, achievable, realistic and time-scaled).

Stage 4: Establishing the continuity culture **1.31**

This step is concerned with effectively communicating, training and educating stakeholders about the BCP and the wider BCM process. It is concerned with the theory of the previous steps into action. The key elements include:

- Implementation – ensuring that all stakeholders, internally and externally, understand BCP/BCM, their own roles and responsibilities and will co-operate in its implementation.

- In-house BCP/BCM training – whether specifically or as a part of other programmes, the training needs to be a fusion of principles, case studies and perhaps a simulation exercise. The training also needs to include Board members or a separate programme to meet their specific needs.
- Education and Awareness – to customers, suppliers, insurers and other contractors. They do not need the full BCP or a detailed insight into the BCM, rather a succinct summary of 'what to do', 'what we expect from you', 'how you can assist us' type information.

If the organisation identified the need, then a BCM Group could be established to oversee the process. The group needs to consist of key stakeholders – Board level director, key functional managers, representatives from health and safety, IT, security, finance, communication/media advisor, perhaps external representatives such as from an insurance company and bank. The group may meet quarterly or semi-annually depending on the size, risk exposure and economic setting of the organisation and its operation.

Stage 5: Exercise, plan maintenance and audit 1.32

This step is about reviewing the previous four steps.

Exercising of the BCP could follow a Table Top or live simulation format.

The maintenance of the plan needs to be an on-going activity and not just a year-end task. The BCP is a 'living document'. This role could be allocated to one individual but reviewed by the BCM group. Given that the BCP could be used in legal proceedings, it may also be advisable to deliver copies to insurers and the bank – perhaps in summary form with the option of the full document.

The BCM Audit will enable the organisation to benchmark its performance against the previous steps in the BCM process and identify the gaps and the corresponding solutions necessary. The Audit needs to follow the 'golden rules' of auditing, as is commonly documented in health and safety and management books:

- objectivity;
- recording findings;
- look at the operation and not just the BCP;
- speak to people;
- seek the facts and do not rely on hearsay;
- keep hard and electronic copies;
- keep an audit trail in case of legal or insurance claims etc.

The DTI/Visor consultants BCM approach: 'Business Continuity Management – preventing chaos in a crisis' **1.33**

The Department of Trade and Industry (DTI) propose an eight-stage BCM approach. This is illustrated in Figure 4 below.

Figure 4: Eight stages of BCM

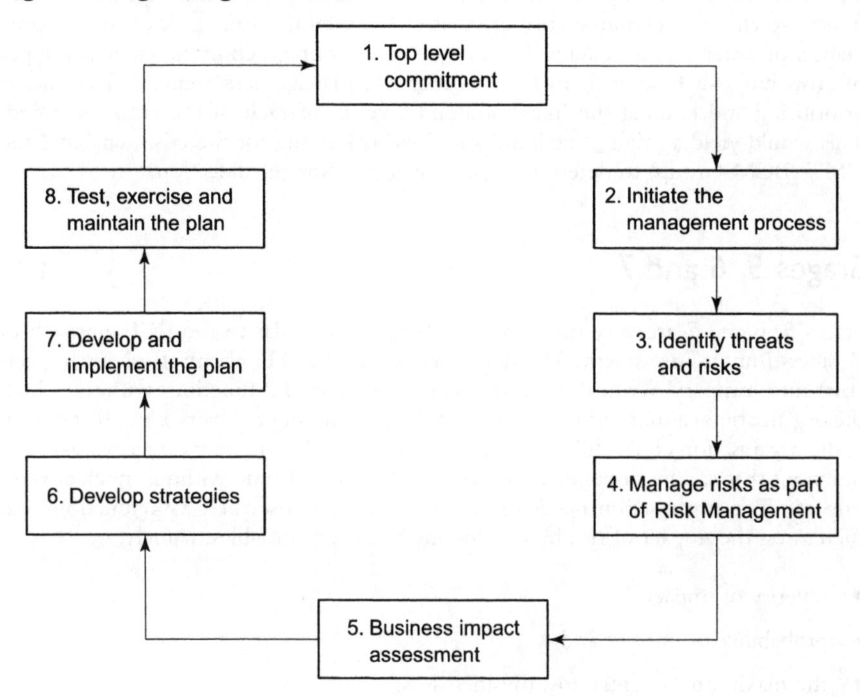

Stages 1 and 2 **1.34**

Stages 1 and 2 are linked and require Board level commitment and a team approach respectively.

The DTI stresses the importance of finding a 'champion' to push the BCM cause at Board level.

There is also an earlier emphasis on ensuring the Crisis Management Team (CMT) is in place with adequate resources, personnel, as well as an understanding of the legal, regulatory and compliance issues. One assumes that the CMT will have a similar function to a BCM Team and the terms are being used interchangeably.

This approach emphasises that BCM is a 'management process' and that BCM needs to be accepted and integrated into business thinking at action.

Stages 3 and 4 **1.35**

Stages 3 and 4 are also linked. Stage 3 is a case of identifying the key 'risks and threats'. The guidelines say that, in fact, Stage 3 is a case of determining the range of 'crisis' that could impact on the organisation. The classification includes Internal, External, Technical/Economic and People/Social. This yields a 2×2 matrix. Examples of an internal-technical/economic crisis are that of IT systems breaking down or contamination of water supply systems. It is not just a case of predicting the range and types of crisis but also linking them to a broader risk management strategy. This means prioritising and ranking the likelihood and severity of each of the crisis identified. This would yield a 'high', 'medium' and 'low' risk rating for the crisis, enabling the CMT (BCM Group) to determine resource and personnel allocation.

Stages 5, 6 and 7 **1.36**

Stages 5, 6 and 7 are very much related. Stage 5 is similar to the BCI approach of Business Impact Assessment. The difference is that the DTI classify Business Impacts (BIs) into 'primary' (critical, revenue-generating tasks and functions without which the organisation cannot survive); secondary (important, but recovery is not that critical as the organisation could still survive); and those that fall in neither category (tasks and functions that can be suspended for a greater duration of time without much adverse impact). The organisation needs to identify and categorise these task/functions and then assess the impact of risk by considering three key variables, namely:

- severity of impact;
- probability of impact; and
- the maximum potential loss (business loss).

The key tasks/functions are then prioritised depending on risk.

Stage 6 is a core element to the DTI approach and stressed throughout the BCM process. There needs be at least three teams/groups, the Senior Management Team (whose focus is long-term sustainability and strategy), the Incident Control Team (focuses on the crisis on site and is more tactical/operational) and finally the Departmental Recovery Teams (department specific with goal of recovering local assets). The three teams need to be working in unison and under a chain of command from the Senior Management Team.

Stage 7 highlights some golden rules concerned with developing a BCP. This includes, keep it simple, know the reader's needs, be objective and state job/function rather than people's names, media management, departmental recovery plans need to be mentioned, up to date etc.

Stage 8 **1.37**

Stage 8 is similar to the BCI. This step overtly mentions 'paper exercises' and Table Top as tools for testing the BCM.

Approach of the CCTA and OGC **1.38**

Being Government departments, the Computer and Telecommunications Agency (CCTA) and the Office of Government Commerce (OGC) have an identical approach, although the approaches are delineated in separate documents (see **1.11** above). Figure 5 below summarises the key stages.

Figure 5: Key Stages in the BCM Process – OGC/CCTA

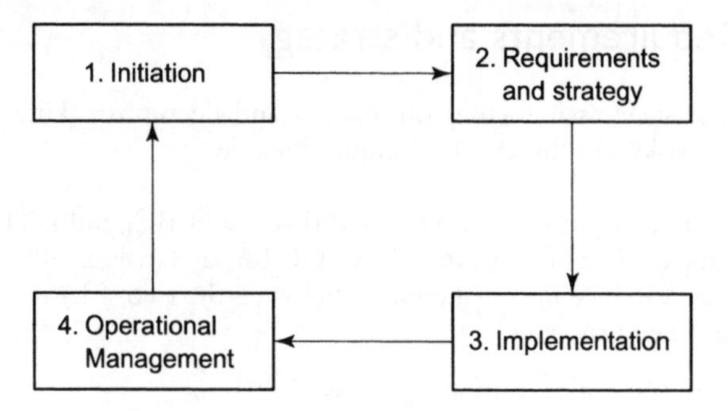

Stage 1: Initiation **1.39**

This is essentially a stage of introducing BCM into the organisation – whether for the first time or as a part of another project. How does one seamlessly and proactively introduce BCM? What are the key variables to focus on? These are the fundamental questions the OCG/CCTA approach addresses. It considers the following factors:

- *Set the Policy* – this requires Board level commitment ('significant and positive' – see **1.34** above). The Policy is more of a mission and vision statement, stipulating what the Board intends to do (similar in logic to Part 1 of a Health and Safety Policy). Thus, 'the when, why, how and where' of the BCM is articulated as well as the commitment to resourcing the process. Note also that the policy will overlap with other policies, in particular, health and safety, security, IT and finance, to name a few.
- *Specify terms of reference and scope* – what is the extent, range, breadth and domain of the BCM process? What does it exclude? The terms of reference and scope needs to consider which key business processes/functions need addressing? Should external stakeholders be included and which ones? Finally, what are the key risks (eg generic, specific, internal, external, insurable, uninsurable) that need to be addressed?

- *Allocate resources* – this would be the necessary financial, human, technical and informational resources.
- *Define project organisation and control structure* – this is a project management issue. It requires consideration of the chain of command, committee and local teams ('who') and issues of documentary control, information management, quality control, review and assessment of progress etc (the 'what' and the 'how'). The control of the project could be via a Project Initiation Document (PID) they suggest, eg a summative document of the key aims, objectives, assumptions, terms of reference, deliverables etc of the BCM project.
- *Agree project and quality plans* – this final part of the initiation process refers to start-review-completion dates, key milestones (drafts, meetings, presentations), performance criteria, key outputs as well as person(s) assigned for quality reviews.

Stage 2: Requirements and strategy **1.40**

This is a critical stage as it focuses on analysis and decision-making techniques to identify the key risks and threats. Techniques include:

- *Business Impact Analysis* – It can be noted that whilst BCI and DTI use the terms 'Business Impact Assessment', the OCG/CCTA say 'analysis'; there seems to be no significant variance in actual content between the two. BIA is concerned with identifying 7 key issues:
  - What are the critical business processes?
  - What is the potential loss that could hit these processes?
  - What form the loss may take (monetary, property, human, reputation etc)?
  - How may the loss vary (increase) with time following a Disaster/Emergency?
  - What is the minimum level of resources required to ensure operation of critical processes at some acceptable level?
  - What is the recovery time for such resources?
  - What is the recovery time for the business process as a whole.
- *Risk Assessment* – this is similar to that of the BCI and DTI approaches, although the CCTA has its own 'Risk Analysis and Management Model' (CRAMM) available from the HMSO.
- *Business Continuity Strategy* – the above two techniques will then enable a bespoke, flexible and up-to-date recovery option (IT, facilities, human, systems etc). For example, the BIA may have concluded that worst case scenario of the building being flooded and the Risk Assessment suggest the risk is 'medium', then the strategy could include recovery options of working from home, relocation, finding a temporary site or using a contractors facility. The relative benefits, costs and risks will need to be assessed. This also raises issues of risk retention as against risk transfer (insurance) and the relative merits of each.

Stage 3: Implementation **1.41**

This is more of a practical and 'doing' stage; its focus is on developing the relevant recovery plans and procedures to meet local departmental or site needs. Key issues include the following:

- What is the chain of command for the implementation of recovery plans? What is the role of internal and external stakeholders? What are the lines and channels of communication between them? How will recovery strategies be implemented and what are the key issues/challenges?
- What 'stand-by arrangement' is in place with regards to IT, accommodation, contractors, key personnel, facilities etc?
- Are there any other risk reduction/mitigation measures that the strategy has identified which need implementation? These would be over and above the stand-by arrangement. They could include activating insurance, use of protective clothing, use of barriers, ergonomic controls to name a few.
- Have the Business Recovery Plans been developed and communicated? Such plans focus on three key dimensions:
  - the Key Content (roles and responsibilities of plan activators/doers, the 'Action List', Key Data such as contact points and inventories);
  - the time dimension – as the Recovery Plans will be used, emphasise and focus on different variables according to the time period after the Disaster/Emergency. These time periods are referred to by the CCTA/OGC as the Alert, Invocation/Recovery and Normalcy Phases;
  - the key Recovery Activities – the Recovery Plans will list the key recovery activities (depending on the phases identified). These can range from the immediate Emergency response to the process of creating stability and getting back to 'normal operations'.
- Apart from the 'Action Lists', does the organisation have detailed procedures (mechanical, systems, emergency, communication etc) to give specific guidance to the organisation when Disaster/Emergency strikes? For example, in the event of a major flooding of a building, the Business Recovery Plan will delineate the actions needed to reactivate particular business functions (IT, accommodation, finance etc) but detailed procedures aid management to ensure how to use the facilities of a new but temporary building, how to establish and use new communications, deal with issues of people management as well, for example, the use of limited financial resources.
- Has Business Recovery Tests been carried out? This will help ensure that the business recovery plans, procedures, stand-by arrangement and the wider strategy is likely to succeed when the real event occurs. The CCTA/OGC identify four types of recovery tests, for instance:
  - walkthroughs (paper exercise);
  - technical component tests (parts of the strategy, eg machinery, IT, accommodation are forensically assessed);

- business component tests (eg finance or production failures); and
- full tests of the strategy.

Stage 4: Operational management 1.42

This stage is concerned with creating an on-going culture of vigilance and alertness. The message from the CCTA/OGC is that producing documents in detail and initial testing is one thing, but keeping systems and strategies live, functioning and taken seriously by organisations requires education, awareness, training, reviews, testing, change control and assurance. This is very similar to the BCI's 'Continuity Culture'.

Incomes Data Service (IDS)/De Montfort University (Elliot, Swartz & Herbane) 1.43

The diagram in Figure 6 below summarises the key stages in the De Montfort approach:

Figure 6: Key stages in the BCM process – IDS/De Montfort

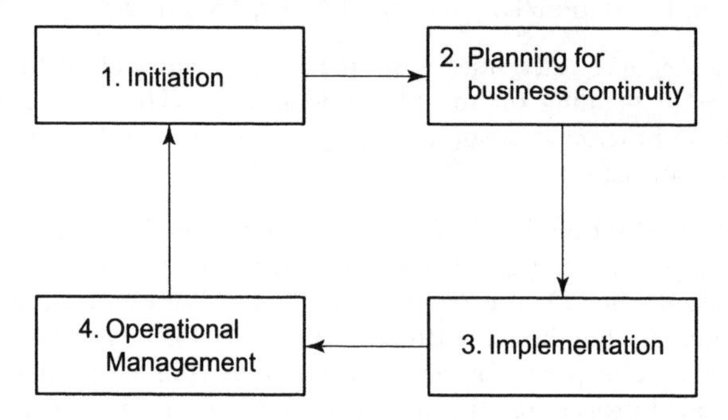

Note: Indeed, this is similar to the CCTA/OGC approach. The differences are in terms of examples, extent of analysis and templates provided.

Stage 1: Initiation 1.44

The key issues to consider include developing a BCM Policy that is lead by the Board, the key human, managerial and technical structures to enable the policy, ensure that adequate resources exist as well as a mechanism for the initiation and dissemination of BCM across the organisation and its sites (committees, teams, groups).

Stage 2: Planning for business continuity **1.45**

This stage focuses on the goal and objectives of a BCM process for the organisation ('what are we trying to protect and continue?'). In addition, issues of the range of crisis that could affect the operation(s) need to be identified. Thirdly, Business Impact Analysis needs to be carried out of the internal and external factors that could affect operations. From this an evaluation is made of:

- key risks;
- what the priority activities are;
- the range of scenarios that can affect operations and key controls.

Finally, after such analysis the BCP can be developed. De Montfort provides a summary (see Figure 7 below) of the key components of a BCP, which are similar to that of the BCI.

Figure 7: Components of a BCP – IDS/De Montfort

Section 1: Overview – the key goal, objectives, Terms of Reference, BCM Policy, key priorities and systems.

Section 2: Recovery procedures – the key arrangements (the 'what', 'how' and 'where from') such as declaration of an event that warrants activating the BCP, logistics, facilities, inventory needed to run operations and recover, key procedures, action lists and systems, communication etc.

Section 3: Operating requirements and reference – the key organisation (the 'who'), such as key initiator and recovery team details, internal and external stakeholders.

Section 4: Plan maintenance and testing – identify changes needed to the BCP, testing and evaluation of the Plan.

Section 5: Forms, logs and general reference –key information about equipment, suppliers, contracts, service agreements, PR etc.

Appendices – specific BCPs for departments and sites, time-lines for recovery etc.

(Elliott D, Swartz E and Herbane B, *Business Continuity Management: Preparing for the Worst* (1999) for IDS/De Montfort University, p 56, with modifications from original.)

Stage 3: Implementation **1.46**

The key focus in this stage is on creating new management structures to enable the BCM process where there are teams, groups and committees working in unison.

The advice to achieving this is similar to that of the CCTA/OGC (see **1.38–1.42** above). In addition, there is a need to create the 'right' preconditions to enable the Implementation process. This includes 'cultural change, communications, performance management and reward, and training should all be used to integrate business continuity into mainstream management processes'.

Stage 4: Operational management **1.47**

This stage is similar to the CCTA/OGC approach – see **1.42** above.

'Business Continuity Management – keeping the wheels in motion' (Australian National Audit Office, 2000) **1.48**

The approach of the Australian National Audit Office (ANAO) is illustrated in Figure 8 below and the key stages of this approach are examined below.

Figure 8: The ANAO approach

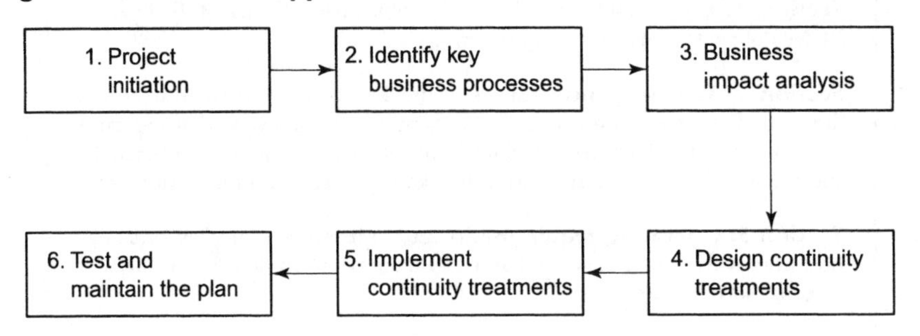

Stage 1: Project initiation **1.49**

This is very similar to stage 1 of the approach of CCTA/OGC (see **1.39** above) and IDS/De Montfort (see **1.44** above). The ANAO approach stresses the importance of Board level 'champions' to push the BCM cause. Also, much earlier in the process than the other approaches, it calls for a Project Plan. This is a 'route map' of the key goal, objectives, targets and milestones to be aimed for. It is similar to the BCM Policy in the other two preceding approaches.

Stage 2: Identify key business processes **1.50**

This is similar to Stage 1 in the BCI approach – understand your business (see **1.23** above). The key output is a 'Resource and Activity Schedule' which highlights the organisation's current physical, financial, human, informational and technical resources, the quantity and quality of resources, how productive/active, life cycles, source and origin etc.

Stage 3: Business impact analysis **1.51**

For ANAO, BIA is about asking:

- What if the following system(s) were missing from the organisation for whatever reason?
- Would the organisation be able to survive and function?
- What impact would the removal of the heart or lungs or a limb have upon a person?

BIA is a form of functional analysis.

BIA is about predicting Outage and Maximum Acceptable Outage (MAO). The former refers to some 'extraordinary event' (not downtime nor a systems failure) that leads to 'significant loss' of primary business processes that results in 'high impact' of the resources, functioning and efficiency of the organisation. MAO refers to the period of time a process/function/operation can survive before it has to activate the BCP and associated procedures.

The ANAO approach provides both qualitative and quantitative approaches to BIA. The former is a tick-box approach that triggers a range of questions about business processes (IT, finance, logistics etc) and enables gaps to be identified. The quantitative approach provides a scale as illustrated in Figure 9 below. This then enables a scoring to be given to each of those business processes, attributes and functions (inputs used in production, IT, marketing, admin, personnel, logistics, reputation, finance, compliance, output, stakeholders). This in turn enables 'Consequence Analysis' whereby scenarios can be constructed as to which of these listed items has the greater probability and frequency of failure and what is the MAO of each item.

Figure 9: Quantitative BIA from ANAO

Level of impact	*Assessment*	*Score*
Extreme	Threatens political and business viability	5
Major	Significant impact on business drivers	4
Moderate	Major impact on short term business operation	3
Minor	Inconvenient but no real ongoing business impact	2
Nil	Reconsider the inclusion of this as a critical resource	1

Stage 4: Design continuity treatment **1.52**

This is similar to Stage 2 of the BCI approach (see **1.24** above). It requires for each of the business processes, attributes and functions a Risk Treatment Plan depending on the MAO identified. For example, it may have been calculated that the MAO is one week of key personnel being inoperable. This has a '4' scoring in terms of

BIA. The Risk Treatment Plan may specify, the key skills missing (ie gaps left), the sources of alternative supply of key persons at short notice (arrangements made), the actions to be taken if such persons or, indeed, in-house persons cannot be found or diverted from other activities and the role of 'key person insurance'. This Treatment Plan can be made bespoke to each business function and site.

Stage 5: Implement **1.53**

This is a practical and proactive stage with a number of key actions:

- Arrangements for managing documentation and other key resources:
  - keys;
  - artefacts;
  - records;
  - reports etc.

  Both in-house and off-site back ups:
  - where?
  - who?
  - why?
  - how delivered and retrieved?
  - when?

  This involves Record Management:
  - hardcopy
  - electronic
  - archive methods
  - data security and information control
  - implement guidelines.
- Arrangements with external parties for information and documentary management is stressed, for example:
  - contractors;
  - suppliers;
  - insurers;
  - bankers;
  - data storage companies; and
  - others who can supply missing information.

  This leads into Data Protection issues.

- Prepare a BCP – whilst the method/steps of developing a BCP differs from other approaches above, the results are broadly similar:
  - develop and define the Recovery Organisation Structure;
  - establish a Recovery Team;
  - produce and integrate specific Service Area Recovery Plans; also
  - have a generic organisation wide Recovery Plan; and
  - ensure there is a list of contacts, contracts, useful sources of information.

  This then enables the production of a BCP.

Figure 10: The structure of a BCP (ANAO)

Part	*Title*	*What are the key issues?*
1	Cover page	Aim, Objectives, Terms of Reference. Senior Corporate Officer needs to sign off.
2	Table of contents	
3	Event log	This is a log of what happened and consequence following an Outage
4	Management recovery plan	Definition of a disaster, when it will be called etc. The organisation ('who') for managing Outage. Steps from disaster, recovery, interim and restoration.
5	Service area recovery plans	As above but for specific services, functions, processes.
6	Referenced procedures	This can range from redirecting messages, through to buying in services.
7	Technical recovery items	IT and computer arrangements for restoring and enabling communication.
8	Contact lists	Contact information of key internal and external stakeholder.
9	Inventory	Existing and additional resources that may be required.
10	Limitations	Parameters of this plan – when and where it applies, issues of reliability and validity, any data protection issues.
11	Testing and maintenance	How often, by whom and how the BCP will be tested and reviewed; key milestones and deliverables.
(*Adapted and modified from ANAO*)		

Stage 6: Test and maintain the plan **1.54**

Again, this follows a very similar approach to the others. The additional point mentioned is the external and internal barriers that can affect the successful implementation of the BCP. These need to be identified for each business process/function or service.

Disaster Recovery Institute (North America) 1.55

The approach of the Disaster Recovery Institute (DRI) of North America is illustrated in Figure 11 below.

Figure 11: DRI's 'Business Continuity Planning Model'

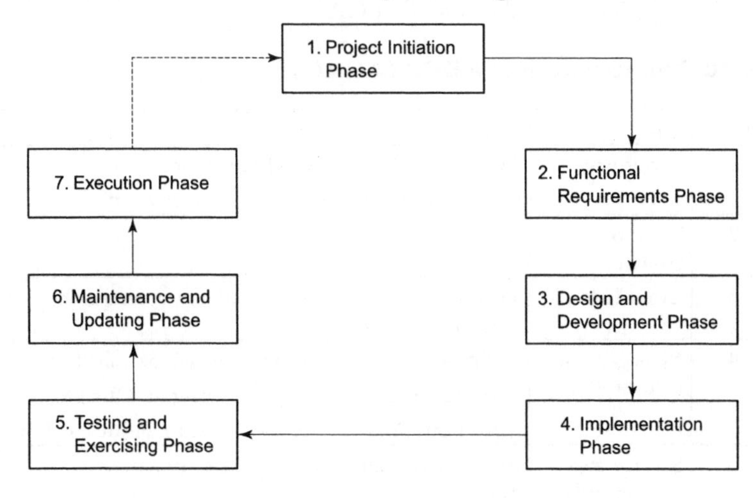

Stage 1: Project initiation phase (objectives and assumptions) 1.56

That is, the aim, objectives, terms of reference, setting a policy and establishing a senior management committee to manage the BCM process. This is very similar to Stage 1 in the BCI approach (see **1.23** above), however, the DRI stage has two other considerations:

- what will the BCM process cost the organisation; and
- how it differs from Disaster Recovery?

This approach stresses the need for careful and realistic budgeting before embarking on the process.

Stage 2: Functional requirements phase 1.57

This stage involves the BIA, risk analysis, developing alternative BC strategies as well as the cost benefit analysis of each of the strategies. Again, there needs to be budget appraisals for each strategy. The DRI focuses on developing a management strategy using conventional business management techniques. Effective BCM has to begin with giving those with the authority, power and resources, the kind of information they need to back BCM.

Stages 3 and 4: Design and development phase **1.58**

The DRI have a two-step approach to the BCP. First, develop the skeleton structure, the key parameters and scope. Secondly, add the arrangements ('how' and 'what') and the organisation ('who'). This approach allows flexibility as departments can follow the design template or make it bespoke. They can then reflect local market and cultural conditions to their department or site.

The BCP needs to have 15 key sections and these are illustrated in Figure 12 below.

Figure 12: Components of BCP (Disaster Recovery Institute)

A. *DESIGN FACTORS – generic*

1. Plan Scope and Objectives – the terms of reference – when and where it will apply.
2. Business Recovery Organization (BRO) and Responsibilities (Recovery Team) – the organisation chart (the 'who').
3. Major Plan Components – for each of the critical business functions identify the risks and recovery strategies.
4. Scenario to Execute Plan (the 'how').
5. Escalation, Notification and Plan Activation – the method, procedure and rules of activation.
6. Vital Records and Off-Site Storage Program –'where' to store.
7. Personnel Control Program – notification, evacuation, management, training, awareness.
8. Data Loss Limitations – off-site and on-site storage, retrieval.
9. Plan Administration (general).

B. *LOCAL FACTORS – create and make bespoke to department/site*

10. Emergency Response Procedures – actions to be taken to evacuate people, key documents, materials or products.
11. Command, Control and Emergency Operations Centre – on and off site.

12. Delegation/Designation of Authority – 'who'?
13. Emergency Response Linkage to Business Recovery – responding to and recovering key assets that can enable business processes to operate.
14. Detailed Resumption, Recovery and Restoration Procedures – the immediate aftermath (Disaster Recovery) and BCM.
15. Vendor Contracts and Purchase of Recovery Resources – source of emergency supplies and contractors in vicinity.

Stage 5: Testing and exercising phase (post implementation plan review) 1.59

This stage is similar to the previous approaches in that awareness, training, education, exercising/scenario analysis, dissemination of information etc are all advocated. There is also scope for external scrutiny of BCP and BCM by competent persons and competent authorities.

Stage 6: Maintenance and updating phase (updating the plan) 1.60

The results from regular testing and exercising will enable gaps to be identified and reviews to be conducted. The review needs to ask:

- Have the chances of the BCP failing been minimised?
- Can software be used to audit and check systems?
- Has the BCM process remained within budget?
- How many copies and who gets copies of the BCP?
- Is there any additional advice we need to seek?

Stage 7: Execution phase (declare disaster and executive recovery operations) 1.61

This is a 'putting the theory into practice' stage. However, it is more than that, as key observations need to be made of how the BCP/BCM operated and what real world factors were overlooked. Also, 'what were the unforeseen factors during this Disaster/Emergency and how can we internalise these into the revision of the BCP/BCM?' This, therefore, requires 'organisational learning' and an openness to improve.

BS ISO/IEC 17799:2000 'Information Technology – Code of Practice for Information Security Management' (2001, BSI) 1.62

This is an ISO Standard that focuses on 'Information Technology' Security. It is based on BS7799 (ceased) and is a Standard of the International Organisation for Standardisation (ISO) and International Electro-technical Commission (IEC).

The Code has twelve key sections relating to how to mitigate the risks and threats associated with IT hacking, data loss and theft as well as data protection issues. Business Continuity Management forms Section 11 of the Code. Although it is aimed at why and how organisations can recover and continue to operate following an IT disaster/emergency, it is sufficiently broadly written to form best practice for organisations in general (for non-IT functions).

The Code is more of a competency checklist than a systemic model. It follows the same logic as ISO 14001; OHSAS 18001; ISO 9000:2000 series etc. It is a part of the BSI/ISO family of Total Quality Management products.

There are five key 'competencies' that the organisation needs to meet in order to conform to 17799, which shall now be discussed:

Business continuity management process 1.63

The Code does not stipulate how the organisation should develop a BCM – rather, the importance of having a 'managed process' is stressed. This managed process needs to bring together (the method of doing so lies with the organisation) the following key elements:

- Understand the risks the organisation could face – assess chance and consequence, 'including the identification and prioritisation of critical business processes'.
- What impact would the interruption have on business? Consider 'small' and 'large' incidents/impacts.
- Consider suitable and sufficient Insurance cover.
- Formulate a Business Continuity Strategy – aim, objectives, terms of reference etc.
- Develop BC Plans in line with the strategy. This is a broad statement of intent rather than advice as to how to construct the BC Plans (which is shown elsewhere in the document).
- The importance of testing and updating of the Plans.
- Integrate BCM into the broader corporate management systems.

This acts as a checklist for the organisation to benchmark itself against.

Business continuity and impact analysis **1.64**

This is a two-stage process, first identify those events that can cause business processes and functions interruption – this can range from health and safety risks, IT, financial crisis, natural events etc. Then carry out a risk assessment that can assess the chance and severity of these perils on the organisation. This then leads to a strategy being devised, which then is supported by senior management and activated.

No methodology is being offered as to BIA or Risk Assessment – the choice of which lies with the organisation.

Writing and implementing continuity plans **1.65**

The Code provides some golden rules of planning:

- Determine, identify, agree key arrangements and organisation relating to Emergency situations.
- The implementation of such Emergency Procedures – these procedures need to be realistic and achievable so that they will enable business recovery in specified time periods. For example, if a fire will destroy documents in one minute, then the plans need to be able to respond in such a time span.
- Key policies, procedures and methods of salvage need to be documented and clearly understood by all.

The importance of Emergency/Disaster training and education to employees is emphasised.

Plans need to reflect the up-to-date internal and external real world factors.

Business continuity planning framework **1.66**

The Code allows for bespoke and flexible BCPs to be constructed according to department, site, function and activity. Each separate BCP needs to be managed by a named person(s). However, there needs to be a single control process – of co-ordination and management. The Code highlights the key elements of a BCP:

- Conditions to activate the Plan(s) – when, why and how?
- Emergency Procedures – on-site and off-site issues.
- Fallback Procedures – how to safely remove key assets to another location or to activate existing processes in a given time period (electricity, water, gas, IT).
- Resumption Procedures – how to establish normalcy.
- Maintenance Schedule for testing the Plan(s) – a schedule and log of testing the BCP(s) and updating and Plan Maintenance.

- Awareness and Education.
- Chain of Command – with 'shadow' nominee also indicated.

Testing, maintaining and re-assessing BCPs **1.67**

The methods stipulated are very similar to the other approaches. Testing includes:

- 'table top' testing;
- simulation;
- testing at an alternative site;
- supplier facilities and services (external); or/and
- complete rehearsal.

A number of factors may necessitate the need to update the BCP:

- changes in personnel;
- relocation;
- changes in legal requirements;
- change of contractors etc.

In short, the BCP needs to be a living document that could be activated at any moment.

Comparative assessment **1.68**

How do these seven approaches compare with each other and what are the key inferences that can be drawn? Figure 13 below summarises the key issues arising.

Figure 13: Summary of the key issues arising

	BCI	DTI/VISOR	CCTA/OGC	IDS/De M	ANAO	DRI	17799	INFERENCES
1. Definition used	'Business Continuity Management is the act of anticipating incidents which will affect critical functions and processes for the organisation and ensuring that it responds to any incident in a planned and rehearsed manner'.	'BCM is a comparatively new approach to looking at your business risks and considering where it is exposed to the effects of disasters. Making judgements about what is critical and planning to maintain the business beyond the event – should a catastrophe happen'.	'BCM is concerned with managing risks to ensure that at all times an organisation can continue operating to, at least, a pre-determined minimal level' (CCTA).	'Business Continuity Management is an organisational business process that seeks to ensure that organisations are able to withstand any disruption to their normal functioning'.	'Business continuity means maintaining the uninter-rupted availability of all key business resources required to support essential business activities'.	A phased approach to identifying critical business functions and processes so as to re-establish the commercial operation.	'Objective: To counteract interruptions to business activities and to protect critical business processes from the effects of major failures or disasters'.	The generic features from these Definitions include: 1. BCM is a process 2. Identifying the mission critical functions/parts/units of an organisation without which operating is difficult or not possible. 3. Concerned with forecasting the array of disruptive events. 3. Diagnosing the impact of these events on mission critical operations. 4. Developing bespoke, 'SMART' BCPs. 5. BCM is a part of a broader Risk Management System.

	BCI	DTI/VISOR	CCTA/OGC	IDS/De M	ANAO	DRI	17799	INFERENCES
2. Summary of key Stages	1. Understand your business. 2. Develop continuity strategies. 3. Develop the response. 4. Establish continuity culture. 5. Exercise and plan maintenance.	1. Top level commitment. 2. Initiate management process. 3. Identify threats and risks. 4. Manage risks as part of risk management process. 5. BIA. 6. Develop strategies. 7. Develop and Implement the plan. 8. Test, exercise and maintain.	1. Initiation. 2. Requirements and strategy. 3. Implementation. 4. Operational Management.	1. Initiation. 2. Planning for Business Continuity. 3. Implementation. 4. Operational Management.	1. Project initiation. 2. Identify key business processes. 3. BIA. 4. Design continuity treatment. 5. Implement continuity treatment. 6. Test and maintain.	1. Project initiation phase. 2. Functional requirement phase. 3. Design and development phase. 4. Implementa-tion phase. 5. Testing and exercising phase. 6. Maintenance and updating phase. 7. Execution phase.	1. BCM process. 2. BC and impact analysis. 3. Writing and implementing continuity plans. 4. BCP framework. 5. Testing, maintaining and re-assessing BCP.	The approaches: 1. Use both forward and reverse systems logic. 2. Are proactive – BCM is performed before the disruptive event and executed during it. 3. Concerned with the 'long-term' or until 'normalcy' is achieved. 4. Require internal and external stakeholder participation. 5. Involve a review process. 6. Involve evaluation, appraisal of alternative options and Cost Benefit computations. 7. Are 'open-ended' – allowing for real world changes to be internalised.

	BCI	DTI/VISOR	CCTA/OGC	IDS/De M	ANAO	DRI	17799	INFERENCES
3.Corresponding stages (benchmarked against BCI)	1	1,2,5 & 6	1 & 2	1 & 2	1,2 & 3	1 & 2	1 & 2	POLICY & ANALYSIS
	2	3 & 4	2	2	4	2	2	STRATEGY
	3	6 & 7	3	3	5	3 & 4	3	PLANNING
	4	6 & 7	4	4	6	5	4	ORGANISING
	5	8	4	4	6	5, 6 & 7	5	AUDIT & REVIEW
4. Sector Focus	All but particularly SMEs.	All but mainly medium and large.	Public and larger private sector organisations.	All but mainly medium to large.	All but particularly SMEs.	All but particularly medium to large.	All.	In practice, mainly medium to large organisations will use approach(es) even though aimed at all types of operations.
5. Integrative value with other standards, codes of guidance eg HSG65, OHSAS 18001, ISO 14001, ISO 9000:2000, BS PD 6668: 2000, Aus Nz 4360:1999 etc	No overt guidance provided other than integration is required and management needs to develop this.	No overt guidance provided other than integration is required and management needs to develop this.	Attempts to link with Total Quality Management in particular.	No overt guidance provided other than integration is required and management needs to develop this.	Advice given on linking with Risk Management Standard 4360	Advice on linking with NFPA Standard 1600 (see below).	Follows the same logic as other BS/ISO standards.	Key issue for industry is how to marry all these diverse OSH, Environmental and other approaches together. BCM approaches seem to have a dual approach – accepting that such integration is important but also suggesting that BCM needs to be kept ring-fenced so it is not dependant on any other system.

	BCI	DTI/VISOR	CCTA/OGC	IDS/De M	ANAO	DRI	17799	INFERENCES
6. Case studies	Available on BCI web site free.	Some available at DTI web site.	Generic, template based example.	Extensive examples from both public and private sector.	Examples on ANAO web site.	Examples in document.	No examples.	Need for examples that show SMEs in particular how to develop BCM and what to focus on in a succinct and bespoke manner.

Conclusion

1.69

It can be noted that there maybe a need to rationalise these approaches into a fewer number. Users may be confused by the purpose and value of having these (and indeed many other) diverse BCM approaches that broadly, have a common approach. Two points can be made. First, Figure 13 indicates that there is a generic method underlying the seven approaches examined, which can be termed 'PASPOAR' (as above in Figure 13). Secondly, there is a move to develop a global standard on BCM that also integrates with Emergency Management, Disaster Management and Contingency Management. This is the USA standard developed by the National Fire Protection Association (NFPA) called 'NFPA: 1600 – new standard on Disaster Recovery, Emergency Management and Business Continuity Planning' (2000). It is commonly called 'NFPA 1600' and was developed after consulting a number of global agencies (including the BCI and DRI). The main sections are summarised in Figure 14 below.

Figure 14 An integrated and universal business continuity approach

Chapter 1 Introduction

- Scope
- Purpose
- Definitions

Chapter 2 Program Management

- Policy
- Program Co-ordinator
- Program Committee
- Program Assessment

Chapter 3 Program Elements

- General
- Laws and Authorities
- Hazard Identification and Risk Assessment
- Hazard Mitigation
- Resource Management
- Planning
- Direction, Control, and Co-ordination

➡

- Communications and Warning
- Operations and Procedures
- Logistics and Facilities
- Training
- Exercises, Evaluations, and Corrective Actions
- Crisis Communications, Public Education, and Information
- Finance and Administration

Future development **1.70**

The key method that binds this NFPA 1600 approach is that of Risk Assessment which also enables a much easier integration with Health and Safety Management. Currently at consultation stage in the USA but there is a move to develop this into an ISO standard.

A key inference from the seven approaches seems to be that the more sites, more extensive the production process and higher the process risks, then the more detailed and involved the BCM process may need to be. Although, there is research that shows (even for larger organisations) that plans and procedures do not guarantee success (improved performance) and in fact may have little direct benefits. The emphasis needs to be on competence and sustainable business rather than complex planning (Kleiner B M, *Macro-ergonomic Analysis of Formalisation in a Dynamic Work System* (August 1998) Applied Ergonomics, Vol 29, No 4, pp 255–259).

2

Case Studies on International Disasters

In this chapter:

- Introduction.
- Case study 1: Nigerian ammunition dump explosion.
- Case study 2: Ammonium nitrate explosion, Toulouse, France.
- Case study 3: The sinking of the Kursk submarine.
- Case study 4: Campos Basin oil platform, Brazil.
- Case study 5: Japan Air Lines Boeing 747 crash, 1985.
- Case study 6: Ellis Park Stadium soccer disaster.
- Conclusion.

Introduction 2.1

Human-made and industrial Disasters are a global phenomena. Such events know no political, economic nor geographic boundaries. Disasters such as Chernobyl did not restrict itself to the Ukranian border. Neither do aviation disasters such as Lockerbie or Japanese Airlines Flight 123, as the victims are usually from many countries, thus requiring a multi-national response.

This chapter looks at six case studies from different parts of the globe and from different politico-economic levels of development. The rationale for the six reflects different sectors, cultures and points in time. Usually, the chosen case study reflects a 'major event' that sits as a milestone in that country's disaster history. The case studies do not represent, by any means, a comprehensive list. In other parts of this book treatment has been given to a variety of other significant human-made and industrial disasters.

The six case studies are:

- Nigerian Ammunition explosion;
- Toulouse Ammonium Nitrate explosion;

- the Kursk nuclear submarine;
- Campos Basin P-36 Oil Platform;
- Japanese Airlines Flight 123;
- Ellis Park Football Stadium.

Case study I: Nigerian Ammunition Dump Explosion

The event 2.2

A series of explosions occurred at the Ikeja Cantonment in the capital city of Lagos at approximately 18.00 hrs on 27 January 2002, which persisted till around 14.00 hrs into the 28 January. Eyewitnesses reported a 'popcorn effect' with explosions resulting every 20 to 30 minutes. The Cantonment was surrounded by a sprawling urban district, itself composed of a number of smaller shanty town-like districts, with a combination of residential and commercial activity. The nearest residency to the Cantonment was literally a few metres. The Cantonment contained 'high calibre bombs', tank rounds, shells and other ammunition. Whilst Nigeria went through a bitter civil war that ended officially in 1970, there has been religious and ethnic conflict between Christian and Islamic fundamentalist groups. Nigeria has a democratically elected Parliament but regions also follow local laws, eg Sharia law in Muslim areas.

The total blast area was approximately 1.5 km^2 with a 'core' of 400 m^2 where the intensity and severity were the greatest. This core also contained unexploded ordnance that was randomly scattered. The tremors from the blast were also felt some 50 km away.

The loss 2.3

This is one of the world's worst human-made disasters with estimates of 1,100 fatalities; the actual figures according to the UN Office for the Co-ordination of Humanitarian Affairs (OCHA) are as high as 2,000.

There were 4,500 personnel (largely military) that had to be relocated from the Cantonment site and were temporarily displaced. In addition, OCHA says that 20,000 people were directly affected or displaced. These were mainly the inhabitants of the immediate blast area.

OCHA also indicates that the majority of the known fatalities were due not to the series of explosions, rather, firstly the chaos and panic that ensued. This led to people being crushed and young children being trampled on. Secondly, the largely misinformed public thought that political insurgency had resulted and began to move away from the blast area towards the Oke Afa Canal around which they thought they would be safe. Fearing the risk of burns and fire hazards, elements of the crowd jumped into the Canal. This was an erroneous move resulting in around 600 deaths.

There was considerable secondary damage to infrastructure from the explosions:

- some 33 ammunition warehouses were destroyed in the Cantonment area;
- there were also residential flats destroyed with around 100 damaged;
- buildings some 18 km away reported being hit by shells;
- three of the nine schools in the Cantonment were destroyed completely. In the blast area as a whole, some 27 schools had to close due to damage;
- approximately 200 children were displaced due to the chaos and panic;
- the impact on the natural environment has not been fully quantified.

OCHA reports that the Lagos site is not unique and there are 'many more' in Nigeria and other developing countries.

Basic cause 2.4

This Major Incident was investigated by both the United Nations and local Nigerian authorities (including their military). The full facts of the causative processes are unclear but it is reported by OCHA that most likely a fire in one of the market stalls that had emerged to serve the local population and Cantonment seems the basic cause. The timber made stands and combustible cloth coverings used by the market stallholders resulted in a domino effect with the fire spreading from stall to stall towards the Cantonment.

Underlying causes 2.5

Accepting the 'fire in the market' as the most plausible hypothesis, OCHA has identified a number of underlying causes that contributed to this Major Incident. These are summarised below.

Hazardous material 2.6

The failure to label, segregate and manage hazardous material. The perception of risk indicates that the military and the population did not regard the ammunition dump as HAZMAT. It was in the Cantonment, therefore 'safe' – and could not affect the local economy. There seems to have been an erroneous understanding of ordnance issues, i.e. the storage, temperature regulation and quantity of explosive materials at any one location.

Major Incident arrangements 2.7

The 'arrangements' for coping with Major Incidents by the military authorities has been questioned. Were such scenarios practiced during exercising and training sessions? Were there On-Site and Off-Site Emergency Plans in place by the military? What were the strategic arrangements between the military, Lagos State Government and federal authorities? However, the UN Disaster Assessment team did credit the military for its Emergency response and evacuation of Cantonment personnel and others.

Urban planning and population dynamics 2.8

A third underlying cause is that of urban planning and population dynamics. Cantonments in many developing countries attract economic migrants and commercial services. This seems to have developed without planning guidelines, consideration for facilities, community safety and fire safety issues. There are some two million people living in the sprawling developments around the Cantonment and shanty towns that have grown in that part of Lagos.

Communication 2.9

Another root cause has been put as a lack of effective communication by the Lagos authorities. There were no immediate public broadcasts, no marshalling of the public to safety, no early warning mechanisms as to community response if disaster strikes and indeed some key public officials were amongst those that were vacating Lagos State buildings. The Nigerian Trade Union Congress has been critical of the lack of information management by State authorities.

Emergency Management 2.10

There was no Integrated Emergency Management between the authorities and emergency services according to OCHA. Even though procedures and systems were existing at federal and State level, they were not put into operation with the result that strategic control of the Major Incident was lacking.

Lack of trained personnel 2.11

OCHA has also noted the reliance by authorities on international expertise of the Red Cross/Red Crescent, UN disaster bodies, ordnance detection agencies and other European and North American assistance. This dependency is an index of a lack of locally trained personnel with the key competencies to respond and manage such a Major Incident.

Organisational learning 2.12

This Major Incident has implications beyond Nigeria and beyond the military. The 'isomorphic' issues (that is, comparative learning) that arise for public and private sector organisations include:

- The importance of strategic co-ordination when combustible and hazardous material is on site. In particular the need to check whether the material falls under regulatory parameters such as the Seveso II Directive (see **6.4** and **6.27** below) in the European Union or equivalents elsewhere in the world. This requires the organisation to liaise with the enforcing authorities and seek competent person advice.
- The Lagos Major Incident raises issues of On and Off-Site Emergency Planning by the organisation. 'Leaving it to others' is not an option and the organisation needs to initiate, develop, lead and control the Emergency Planning process with external stakeholders (schools, local government, emergency services).
- The centrality of communications by the State authorities to notify the public of Disasters/Major Incidents is also critical. This requires the organisation to develop an in-house media policy, liaise with the local media and ensure that arrangements are in place to disseminate information.
- Exercising and testing plans and procedures by involving as many of the stakeholders in this process. The importance of order and structure during a process of chaos and panic is critical in averting fatalities.

Case study 2: Ammonium Nitrate Explosion, Toulouse, France

The event 2.13

On 21 September 2001, at approximately 10.15 am, explosions occurred in an ammonium nitrate plant on the outskirts of the city of Toulouse, south-west France. The factory was located some 3 km from the City Centre. Toulouse is regarded as the 'space and aviation capital' of France, attracting major transnationals and is noted as France's leading research centre for aeronautical sciences.

The ownership structure of the enterprise consisted of:

- TotalFinaElf, who were the parent company. They were at the time of the Major Incident the fourth largest oil company in the world. The company is an amalgamation between Total, PetroFina and Elf Aquitaine.
- TotalFinaElf owned Atofina, which is the chemicals part of the overall group.
- Grand Paroisse is France's largest fertiliser manufacturer and is owned by Atofina. Grand Paroisse trades under the name of Azote de France (AZF). It should be noted that AZF was formerly called ONIA (Nitrogen National Industrial Office).

The factory where the Major Incident occurred was built in 1924. The plant was one of France's 1,250 Seveso rated sites. The factory was clustered in an industrial area with the Tolochimie chemical works and an explosives maker, Societe Nationale des Poudres et Explosifs (SNPE) nearby.

Ammonium Nitrate (AN) formed the basis of the factory's operation. It appears as an odourless and colourless crystal as well as white solids. In its liquid form it is a stable compound. It has a melting point of 170 degrees Celsius.

It is reported that 6,300 tonnes of liquefied ammonia were kept at the site. In addition, 6,000 tonnes of AN and 30,000 tonnes of solid fertiliser were stored. The factory had the capacity to produce 400 tonnes per day of industrial grade AN (for explosives use) and 850 tonnes per day of fertiliser AN.

Further to their investigations, the Ministere de l'Amenagement du Territoire et de l'Environnment reported that the factory/AZF was a Seveso II operation. There existed a Safety Management System, Safety Reports (with some 36 worst case/accident scenarios being considered), On and Off Site Emergency Plans and a regime of inspections by the management (twice per annum) as well as the key risks being communicated to the nearby population (the Off Site Plan would affect some 16,000 people if it had to be operationalised).

The loss 2.14

At the time of the explosions there were some 470 persons on site including contractors. There were 22 workers killed, as well as a further eight civilians. There were around 2,000 persons (some reports say 2,400) that were hospitalised due to cuts, abrasions and burns with at least seven persons deemed to be 'critical'.

The explosions that occurred measured 3.4 on the Richter scale. The destruction also included:

- 11,177 flats/houses that were 'severely damaged';
- glass breakage resulted into the City Centre and the Ministere de l'Amenagement du Territoire et de l'Environnment said that even up to 7 km, breakages were reported.
- some 85 schools/colleges damaged;
- 500 homes approximately were uninhabitable;
- there was indirect implications for Toulouse's economy as flights and travel were affected.
- a crater of 50 m diameter and 10 m deep was created by the explosions.

Basic cause 2.15

The basic cause seems to have been the method of storage of the non-conforming industrial and fertiliser AN stored in building 221; around 300 to 400 tonnes was stored. Commercial fertilisers are required to be tested under a *European Council Directive 80/876/CE*. This enables segregation of acceptable and below quality products to be made. The latter are then stored and either 'destroyed' or reconstituted.

AN is usually resistant to detonation but when there are increases in temperature or contamination, then increases in the likelihood of explosion can occur. In addition, improper stacking or loading can increase the likelihood of bubbles being formed which increases the chance of explosion. Also, shock waves (for example due to sudden or improper movement) can increase instability, form bubbles and generate the potential for explosion.

What is certain is that the temperature must have increased creating the instability that triggered the first explosion. The French investigators have mooted the possibility of a second critical explosion after examining the mode of destruction of the plant, where the tower may have been destroyed before the storage facility.

Given the proximity of this disaster to that of September 11th in the US, the initial media and AZF response was to leave open the role of human intervention, namely involving terrorism. The French investigators seem to have focused more on the chemical management as the arena of causation rather than terrorism.

Underlying causes 2.16

If the storage of the AN is a basic cause then why did this failure arise? Reasons cited include the following:

- The variance between systems/documentary control and actual inspection and assessment. AZF had quality assurance systems and the management seems to have complied with the requirements of the Seveso Directive. The shortfalls may have been with the frequency and accuracy of the inspections and reporting back into the management cycle.
- There are also issues raised of the competence of the immediate site operators. Did they not know the potential for explosion and detonation if the below grade AN was exposed to higher temperatures or if inappropriately stored? Did the acceptable and below grade become confused in the mind of the operators?
- What of the location of the plant? Whilst it has had a history since 1924, what planning issues and risk assessments were carried out, especially since two other plants were so close to AZF?
- The French authorities and the UK safety media had raised concerns about the depth and value of the existing Seveso II Directive. Is there too much reliance on 'systems' and compliance? Should the emphasis be on specific and operational procedures relating to parts of the plant? The whole emphasis on the testing and exercising of On and Off Site Emergency Planning is raised again. Is it possible to 'test' Off Site Emergency Plans in particular given the logistical issues involved and the cost implications? Is a paper-based or a 'snap-shot' exercise really adequate?

- Finally, should the Threshold Quantity for AN be reduced from 500 tonnes to 150 tonnes? AN had been assumed to be 'relatively' safe. Following Toulouse and other incidents in Europe it is being mooted that a lower Threshold means the operator is obligated to report sooner on the storage and management of AN.

Organisational learning 2.17

The key isomorphic issues arising from Toulouse include:

- Reliance on a multivariate sources – not just systems and procedures. The importance of ensuring that quality standards are seen as a 'means' and not the 'end'. A regime of visual inspection and assessment with more specific scientific analysis on a regular basis will enable the verification of what is actually happening in reality.
- Competence and training of all key operators with a greater emphasis on retraining at regular intervals. Especially where technical/scientific competence is required, the organisation needs to identify or recruit appropriate personnel.
- Also critical is the liaison with the enforcers and competent authorities to ensure that the operational management is compliant with the law and best practice.
- There are wider issues for local and central government with regards to planning policy and whether 'industrial estates' that house many Seveso sites in a cluster is an acceptable planning strategy from a community risk perspective.

Case study 3: The Sinking of the Kursk Submarine

The event 2.18

The Kursk Oscar II K–141 nuclear submarine was built in May 1994 and commissioned in January 1995. It was a guided-missile submarine built with the aim of striking at 'enemy carriers'. It was designed to launch a cocktail of up to 24 nuclear tipped missiles and a range of up to 30 torpedoes. It could reach a dived speed of 28 knots and 15 knots when surfaced. In addition, it could reach a depth of 500 metres. A typical Boeing 747 is 70 metres in length, whilst the Kursk was 155 metres. Its crew capacity was 130 personnel.

The critical features of the physical structure of the Kursk consisted of the following:

- A propulsion at the rear.
- There were two nuclear reactors on board that generated power for both the propulsion and for the facilities for the crew. Russian authorities claimed that both reactors 'had been' shut down. It is not clear whether this only prohibited

the use of nuclear energy for the weaponry or for the entire submarine. If the latter, then back-up batteries would have been required. Contradictory reports claimed that the Kursk had left without these batteries and, instead, had taken on board liquid fuel that would power the launch of the torpedoes. The fuel was a cost-saving device, more economical than the battery option. However, the risk of combustion were also increased.

- Moving gradually towards the front end, after the nuclear reactors, came the living quarters, located on several levels. This was followed by the Command and Control Centre. This leads to the missile part of the submarine. The damaged part of the Kursk was at the front towards one side.

On Saturday 12 August 2000, the Kursk was taking part in a series of exercises in the Barents Sea when it lost contact with the Northern Fleet Command of the Russian Navy. It is claimed by the Russian authorities that the submarine sank on the following day. On that Sunday, 13 August, a sonic device detected the Kursk some 108 metres on the seabed and approximately 85 miles from Severomorsk and some 184 miles east of Norway's Artic coast.

The Russian authorities claimed that there were no nuclear weapons on board and the nuclear reactor that powered the Kursk had 'shut down'. Attempts were made to determine if there were any survivors on 14 and 15 August, but heavy storms, poor underwater visibility and the angle at which the Kursk lay, hindered the search and rescue.

The Kursk was finally dredged to the Russian port of Roslyakova on 10 October 2001. Some 26 giant steel cables were used in the 15-hour lifting operation with the Kursk effectively strapped beneath a giant barge.

The loss 2.19

On Monday 21 August, the Russian authorities announced that 118 persons were on board and no survivors were found. The dead consisted of 86 commissioned and warrant officers, 31 non-commissioned officers and sailors and one civilian. Not all the bodies were found.

Whilst dismissed by the Russian authorities, Norwegian environmentalists claimed that there was a risk of nuclear radiation and damage to marine life.

The rescue operation cost in the region of US$65 m.

There were 22 missiles on board when rescuers accessed the Kursk. A news report on 5 November 2001 suggested there were also six cruise missiles on board that needed to be safely removed before further work could proceed. Also, 150 kg of TNT was discovered inside the Kursk's hull. Further, there were pockets of hydrogen sulphide discovered that hampered efforts to remove bodies.

Basic cause 2.20

US authorities had been monitoring the exercises taking place and had detected two explosions in the vicinity where the Kursk sank. Further, NORSAR (a Norwegian seismological body) had registered two explosions as well. The first explosion was registered by NORSAR as equivalent of around 100 kg of TNT or approximately 1.5 on the Richter scale. The second explosion has been reported as more critical than the first and was equivalent to approximately two tonnes of TNT or around 3.5 on the Richter scale. There is some confusion over the time scales between the two explosions; NORSAR suggests it was two minutes 15 seconds. Other reports claim the second explosion occurred approximately 35 seconds after the first.

Underlying causes 2.21

There have been a range of hypotheses as to the 'cause'. Originally there were 12 versions including the following:

- The possibility of one of the torpedoes exploding, triggering off a series of fires and the more lethal and fatal second explosion that ripped apart the twin pressurised hulls. US authorities have claimed that the first explosion possibly involved fuel from a torpedo.
- There is also the view that the Kursk collided with the seabed which led to the tanks containing pressurised air to explode.
- The hypothesis that the Kursk collided with another vessel in the vicinity. The Russians claim there were two US submarines and one belonging to the Royal Navy in the vicinity.
- It has been suggested that maybe the Kursk hit a World War II mine.
- Was the Kursk hit by a radar-guided granite missile fired by the Peter the Great nuclear cruiser?
- Did the Kursk hit a surface vessel?

A Government Commission was set up by the Russian authorities on 14 March 2000 to investigate the causative processes. It was headed by the Deputy Prime Minister, Illya Klebanov, with a range of military specialists as well as politicians and business persons. An independent parliamentary commission was deemed unnecessary.

The Kursk had a 'black box' similar to aeroplanes, which was 'found' in late October (reported on 30 October 2001). The evidence from this has been 'classified'.

There have been a range of claims made in the media that the Russian Navy has been under funded, that personnel were not appropriately trained and that the exercise was poorly planned. This is difficult to verify other than possibly looking at historical evidence. This shows that the former Soviet Union and the emerged Russia had a poor incident rate with submarine disasters. There have been in total nine reported

Major Incidents mostly involving fatalities throughout the 1980s and 1990s.Most have involved collisions or explosions.

Organisational learning 2.22

The Kursk raises a number of lessons for both industry and the public sector:

- The importance of careful pre-planning of exercises (see further **CHAPTER 20: TRAINING AND EXERCISING FOR EFFECTIVE PREPAREDENESS AND RESPONSE**). Critical actions for the exercise planners are:
  - the health and safety of those taking part;
  - issues of dynamic risk assessment;
  - having back up communications technology; and
  - knowing where the key human and physical resources are.

  In addition, the importance of taking insurance and legal advice before embarking upon a detailed exercise can identify potential problems and risks.
- Could the exercise be computer simulated or be carried out as a Table Top (see **20.16** below)? Is it necessary to do a live exercise (see **20.21** below)? The planners need to address this question not only for cost purposes but also for logistical and safety reasons.
- Historical isomorphic learning needs to be considered. The Russian authorities seem not to have reflected back in time and learnt from the previous Major Incidents and planned the Barents Sea exercise accordingly.
- There have been considerable problems of knowing what 'really happened'. The Russian authorities have been concerned about reputation loss, not disclosing confidential information to third parties and the political implications. However, for private industry a distinction needs to be drawn between disclosing the facts as soon as reasonably practicable and commercial confidentiality. The former is critical at the earliest stage so other expertise can be called upon; other specialists evaluate data and advise accordingly. The Disaster and Emergency Policy needs to clarify the timescales, who and how much will be disclosed to employees, the media and others – rather than a blanket 'no information' strategy.

Case study 4: Campos Basin Oil Platform, Brazil

The event 2.23

Three critical explosions occurred on the P–36 oil platform on 15 March 2001 in the Campos Basin's Roncador Field, some 125 km from the Brazilian coast near Macae and some 270 km northeast from Rio de Janeiro. The platform was owned by the Brazilian State owned company, Petrobras. The platform sank on 20 March 2001 and lay 1,368 metres deep on the Atlantic Ocean floor. Campos Basin is the main hub of Brazil's oil industry, producing around 80% of the country's oil needs.

P–36 was the world's largest and tallest oil platform. It had been designed and built mainly for drilling and production operations. It was the same height as a 40-storey building (some 210 metres high) and weighed approximately 31,400 tons. It had only been operational for less than a year prior to the Major Incident. The rig had been built by an Italian engineering firm at a cost of approximately US$400m. It had the capacity of producing 84,000 barrels of crude oil per day and 1.3 million cubic metres of natural gas per day. The rig could accommodate 175 employees at any one time.

It needs to be noted that the rig had undergone modification and conversion to enable the processing of up to 180,000 barrels per day, from the originally planned 84,000, in order to meet the growing oil needs of Brazilian industry.

The loss 2.24

Eleven persons died in total relating to this event, mainly from the second and third explosions.

P–36 had some 1.5 million litres of crude oil and some 1,200 cubic metres of diesel fuel in storage tanks or in the 21 main pipelines attached to it. There were risks of rupture of the storage tanks and its contents leaking out. Petrobras claims that the oil slick reached and remained 90 miles away from the coastline.

Some 81,600 gallons of diesel and crude was said to be on the surface covering an area equivalent to 5 square miles.

The underwater wells that are some 1.3 km beneath the surface had been capped to minimise release but some release had occurred according to Greenpeace.

The loss of P–36 is equivalent to a loss of US$450m in profits in one tax year. On the Sao Paulo Stock Exchange the share price fell by 6–7%.

Basic cause 2.25

P–36 is no different from most other platforms other than for its size. It possessed two large Drain Storage Tanks (DSTs) one for the port side and the other for the starboard. The pumped gas, oil and water found its way to the starboard DST. The port side DST contained mainly water.

There seems to have been several but connected actions leading to the event:

- Maintenance work was required on the starboard DST pump. Accordingly, this pump was removed by a contractor without the safe systems in place to assess the impact of the pump removal.
  - How would the removal impact on the starboard DST intake valve?
  - What impact does it have on pressure, flow, speed, rate of release of gas, oil and water?

A range of theoretical assumptions were made about impact of removal of the pump without validating the actual impact.

- The relationship between the two DSTs is one of homeostatics – retaining a smooth balance and flow and ensuring that excess pressure does not build up in one part that could adversely affect the other DST. This requires, therefore, the emptying of the port side DST. The water in this DST was pumped through a production header. In order to do this safely, this requires the closure of the starboard DST's intake valve. Otherwise the water from the port side DST could flow into the starboard DST and without the starboard DST's pump, this could create an explosion potential. Either 'human error' or a regime of poor maintenance meant that the starboard DST's intake valve had not been completely closed.

The two critical events above led to the first 'mechanical explosion' with the starboard DST rupturing, followed by two further explosions as the oil and gas was released. Pictures of the P–36 rig show the submergence of the starboard side. Eventually, by 20 March, the rig tilted at an angle three times that of the Leaning Tower of Pisa.

Underlying causes 2.26

Platform related Major Incidents are not new in Brazil, nor for Petrobras. In 1984, a fire on a petroleum platform killed 37 employees. In January 2000 there was the Guanabara Bay oil spill when 1.3 million litres of oil was released. The other major event was the 4 million litres of oil in July 2000 (Araucaria oil spill). From a period of 1987 to 2000, there have been 81 fatalities at petroleum sites owned by Petrobras, of which 66 belonged to contractors.

The Brazilian investigators identified a range of deep-rooted causes. They were as follows:

- Operational Health and Safety Management needs improving. There is a lack of a systematic approach to safety and health and work. Whilst systems and procedures exist, they are fragmented. There was no auditing regime in place.
- The safety culture in the oil sector as a whole – the emphasis has been on cost cutting, leading to the reduction of personnel and contracting of self-employed contractors without a common command and control over their safety documentation and implementation.
- Issues of a lack of safety training and competence of operational staff and supervisors on P–36, the lack of use of dynamic risk assessment, safe systems and Permit type systems and the failure to communicate risks between contractors and Petrobras staff have been focused on by Brazilian investigators.
- One of the critical issues raised has been the design and modification of platforms without consideration of the safety implications.
- The investigators have also criticised the fragmented approach between the Brazilian Maritime Authority and the Ministry of Works.

Organisational learning 2.27

The key lessons for industry and the public sector arising from P–36 were in many ways anticipated and forecasted by Lord Cullen in the Inquiry into Piper Alpha in 1988. There has been a lack of isomorphic learning by the Brazilian authorities not only from their historic incident rate but also from the conclusions and recommendations from Cullen.

The key lessons are:

- the vetting of all contractors to assess their competence, training, use of safe systems, risk assessments and historic track record in similar work;
- a regime of regular monitoring and oversight of all contractors;
- regular meetings and assessment of contractors;
- senior management taking a proactive and hands-on role in the safety function;
- the 'separation of powers', so that the assessor, supervisor, enforcer and auditor functions are kept separate to avoid a conflict of interest.

Case study 5: Japan Air Lines Boeing 747 Crash, 1985

The event 2.28

On 12 August 1985, a Boeing 747SR (Short Range), owned by Japanese Airlines (JAL) took off from Tokyo's Haneda International Airport en route for the industrial city of Osaka, approximately 400 km away. Designed for relatively short journeys, the aircraft carried 509 passengers, including 12 small children, and a crew of 15. Approximately, ten minutes after take-off, the captain notified Air Traffic Control that he had an emergency on-board. Some ten minutes later the aircraft crashed into a pine forest near the top of Mount Osutaka, exploding into flames and breaking up. Much of the wreckage spilled down the side of the mountain. Flying over the area a little later, the pilot of a rescue helicopter reported seeing ten small fires within an area about 300 square metres but no survivors. It was raining and visibility was poor.

Darkness and the deteriorating, stormy weather conditions forced the authorities to temporarily suspend the use of helicopters. Inaccessibility of the accident site meant that it took rescue parties 14 hours to reach the scene. Later, when the weather improved, Chinook helicopters, using rappelling gear, were able to drop troops onto the wreckage site. As the morning wore on, rescuers found four survivors, two women and two children. All had been seated at the rear of the aircraft. News of the discovery of the four survivors raised the hopes of the relatives assembled in a school gymnasium in the town of Fujioka close to the foot of the mountain. At the site, medical staff found people who appeared to have survived the impact but, wearing only light clothing, had probably died of shock and exposure in the near-zero overnight temperatures.

JAL arranged for 1,000 employees to go to the crash site to recover the bodies.

The loss **2.29**

In total 520 people died. This represents the worst aviation disaster involving a single aircraft.

There was also the direct and indirect losses and delays for other flights, the human pain and suffering sustained by the next of kin and the multi-million dollar clean up that followed.

Apart from the investigation time, there were losses sustained by Boeing itself in sending personnel and trying to decipher the causation process.

Feelings in the community also ran high. Some JAL staff were assaulted in the immediate aftermath of the accident and there were instances of others being thrown out of shops and restaurants in Tokyo if they entered in uniform. A JAL maintenance manager at Haneda Airport committed suicide to 'apologise' for the accident.

JAL suffered considerably. The airline's President resigned. Domestic flights dropped by more than a third. It took years for JAL to rebuild its shattered morale, win back public confidence and regain its public image.

Basic cause **2.30**

The initial cause of the accident was the rupture of the aircraft's aft pressure bulkhead. This allowed air to flow from the cabin into the unpressurised tail assembly. The wall of the auxiliary power unit was then breached, allowing air to rush into the vertical tail fin. Within seconds, a large section of the rudder had broken off and all four lines of the hydraulic system had been severed. In less than two minutes the aircraft had become uncontrollable.

An examination of the maintenance records of this particular aircraft revealed that its fuselage had scraped the ground during a particularly heavy landing in 1978. Replacement of the damaged portion of the pressure bulkhead had been incorrectly assembled, using a single row of rivets instead of two, thus reducing its resistance to fatigue by some 30%. This had remained undiscovered until the accident, seven years later. There was a general acceptance that there had been nothing the pilot of the aircraft could do; indeed, the crew had exhibited great skill in keeping the aircraft flying for about 30 minutes after the accident using engine thrust only.

Underlying causes 2.31

One of the key underlying causes was that of assessing and reviewing parts of the plane that had undergone previous repair. There was a presumption that Boeing engineers had carried out the repairs and so no further assessments were needed. The frequency and depth of the inspections by the Japanese authorities was brought into question. This raises issues of how and why 'maintenance' was confused with 'inspection'.

Other underlying causes have been stated as economic pressures. The Japanese economy and JAL were flourishing in the era of the 'free market' and suggestions that quality control and risk oversight took second place have been made.

Organisational learning 2.32

Important lessons arise for industry and the public sector from JAL 123. These are summarised below.

Importance of a unified command and control structure 2.33

Looking at the series of events, one identifies a series of fragmentations. Firstly, there was a dispute amongst Japanese agencies as to who was actually in charge. Japanese regulations stated that all wreckage should be seized and retained by the police. Therefore, the local police of the area in which the aircraft crashed insisted they were responsible for carrying out the investigation. As a result, the Air Accident Investigation Committee, set up by the Ministry of Transport, were prevented from taking key pieces of wreckage to a central inspection facility in Tokyo. According to the police they could only 'borrow' wreckage.

Secondly, when a nine-man team of investigators arrived from the US – two each from the National Transportation Safety Board (NTSB) and the Federal Aviation Authority (FAA) and five from Boeing – they had great difficulty in gaining access to the accident site. When they finally did gain access, they found that critical pieces of wreckage had been removed by rescue workers.

Disaster Recovery Plan 2.34

Further, there were delays by the Government in activating the Disaster Recovery Plan. This required the authority of all members of the Cabinet. Such delays were repeated during the Kobe earthquake. This raises issues of whether Disaster and Emergency Management needs to be kept as a specialised branch with its own response and reaction strategies free from political interference. However, the counter argument raised is that the Cabinet or a 'cabinet office' is the key forum that links government departments together so in order to improve crisis communication, such political and administrative oversight is necessary and, in the case of Lockerbie, proved a success.

Culture **2.35**

Many investigation delays took place as Boeing, NSTB and FAA personnel were denied access to the disaster scene. Issues of understanding culture and norms arise, as the Japanese maintained that dealing with persons of authority, with the 'right' designations on business cards and status were of importance – they needed the assurance that they were dealing with decision makers who could be trusted.

Case study 6: Ellis Park Stadium Soccer Disaster

The event **2.36**

During the evening of 11 April 2001, a large crowd of people came to the Ellis Park Stadium in Johannesburg, to watch a Premier Soccer League fixture between Kaizer Chiefs, the home club, and Orlando Pirates. The two teams have the largest support in South Africa. Both are based in Johannesburg and the game was crucial to the outcome of the championship for that season with both clubs still in the running to be crowned champions. The ground's capacity is 60,000 but it is estimated that between 80,000 and 100,000 people turned up. When the official capacity was reached, signs to this effect were posted around the ground.

Immediately following the disaster, the President of South Africa set up a Commission of Inquiry, chaired by a judge. The Commission was to inquire into and report on the facts that led to the disaster, and make recommendations on how similar occurrences could be prevented in the future.

The loss **2.37**

The tragedy started to unfold before the game commenced shortly after 8 pm. Only when victims were being brought out of the crowd and laid on the pitch behind one of the goals was the game stopped. 43 people died and a further 158 were injured.

Finding their way to the stadium obstructed by the large crowds, many people who arrived in vehicles just abandoned them in the surrounding streets. The resulting congestion hampered emergency vehicles that were called to the scene once the problem developed.

Basic cause **2.38**

Thousands of people who had bought tickets were left outside the ground. In frustration people outside the ground broke down the perimeter fence, set light to ticket cubicles and eventually broke down the entry gates into the stadium. As a result, both ticket-holders and non-ticket holders accessed the stadium in large numbers. Although disputed by some of the people who gave evidence to the Inquiry, the Chairman found that a tear-gas canister was discharged, causing a section of the crowd to panic.

Underlying causes 2.39

The inquiry found that many of the arrangements on the day were defective. Principle amongst these were the following:

- All the agencies grossly underestimated the possible attendance.
- There had been a failure to learn the lessons from previous encounters between the two teams.
- There was no clear agreement as to who was in command and what the specific responsibilities of the various agencies were. For instance, all three organisations, PSL Security, Stallion and the Public Order Policing Unit, who had a roll to play at the outer perimeter fence, denied that it was their responsibility to stem the breach.
- FIFA and SAFA guidelines specifically state that a game should not start until the situation inside and outside the stadium is under control. When the game started, there were still thousands of people outside the stadium.
- No one was tasked with or accepted responsibility for monitoring the crowd outside the stadium.
- Although there was a purpose-built Joint Operation Centre manned by representatives of most of the agencies present, no one person was in command. Instead there was a collection of independent heads of security groupings who did not operate in a co-ordinated manner and all of whom denied that they carried the ultimate responsibility for what happened. Indeed, the Inquiry suggested it was a joint operations centre in name only. The Inquiry suggested this as 'a glaring weakness in the security plan'.
- Too few security officers were deployed at some parts of the ground. This enabled some people to enter the ground without tickets. In other cases, some security officers allowed people in without tickets in return for money.
- The Public Order Police were slow to react to the worsening situation.

Organisational learning 2.40

Again, the issue of isomorphic learning arises in that the South African authorities do not seem to have followed up from disasters such as Hillsborough. Perhaps such disasters are seen as remote – from a different continent and culture and 'not relevant'.

Another issue arising is education and awareness in the wider community of safe behaviour and 'crowd norms' expected from attendees when attending sports matches.

There is also the economic motivation of sports and leisure organisers, who have legal duties under most Commonwealth legal jurisdictions to consider and implement 'safety arrangements' and the risk controls, given the numbers attending. The UK Health & Safety Executive have produced a guidance, *Crowd Safety*.

There are also broader issues of design and architecture of sports and leisure stadiums, the materials used, the access and egress paths, the use of space etc. This has implications for architects as well as urban planners.

Conclusion **2.41**

The above Major Incidents highlight that geography and culture need not be barriers to understanding and 'learning from disasters'. They display some generic causative processes as well as common 'solutions', namely, the importance of:

- the need for strategic oversight by the Board and senior management and ensuring that operational and tactical management are 'linked together' in one system of command and control. This is a key message of the Turnbull Report and systems such as BS PD 6668 – effective management oversight;
- the Training and Exercising Plans and Procedures need regular testing, in particular in conjunction and collaboration with contractors;
- a regime for reporting accurately and analysing historical incidents and communicating the findings;
- a proactive and open culture which encourages the analysing and the asking of critical questions; and
- regularity – ensuring that a regime of continuous checking and feedback is prevalent, involving employees and external agencies.

Conclusion

3

Construction Related Disasters

In this chapter:

- Historical overview of UK and abroad.
- Terminology.
- Principles of design in architecture and engineering.
- Case studies.
- Quebec Bridge Disaster – 29 August 1907.
- Tay Bridge Disaster, Dundee – 29 December 1879.
- Ferry Bridge Cooling Towers – 1 November 1965.
- Progressive collapse: Ronan Point – 16 May 1968.
- The Alfred P Murrah Building, Oklahoma City, USA – 19 April 1995.
- Box girder bridges: The collapse of the Milford Haven Bridge – June 1970 and West Gate Bridge over the Yarra River, Victoria, Australia – December 1970.
- World Trade Centre, New York – 11 September 2001.
- Lessons for 0SH/Disaster Management practitioners.
- Conclusion.

Historical Overview of UK and Abroad 3.1

'Nothing was so instructive to the younger members of the profession, as records of accidents in large works, and the means employed in repairing the damage. A faithful account of these accidents, and the means by which the consequences were met were really more valuable than a description of the most successful works. The older engineers derived their most useful stores of experience, from the observations of those casualties which had occurred to their own and other works, and it was most important that they should

> be faithfully recorded in the archives of the Institution.' (Robert Stephenson, Engineer, President of Institution of Civil Engineers, 1856.)

Constructional disasters have been with us since building began. The Bible warns us not to build our house on shifting sands. Noah was instructed to cover his Ark inside and out with pitch as an early form of damp proof course. At one level it could be argued that these sayings showed us the way forward and allowed progress – a constructional form of survival of the fittest. At another it could be argued they demonstrated a form of hubris sometimes rapidly followed by nemesis, if they were not followed.

Certainly in the days of the Babylonian Kings there was not much room for manoeuvre for the unfortunate builder who could never return to learn from his mistakes. The Code of Hammurabi, King of Babylon 2200BC stated:

> 'If a Builder build a House for a Man and do not make its construction firm and the House which he has built collapse and cause the death of the Owner of the house – that Builder shall be put to Death'.

In the reign of Henry IV in the fifteenth century an English law stated:

> 'If a carpenter undertake to build a house and does it ill, an action will lie against him'. (*Construction Failure* by Jacob Feld.)

Mediaeval failures 3.2

We know little about mediaeval failures apart from major ones like the spire of Beauvais Cathedral, which appears to have collapsed because of asymmetrical buttressing. If the nave had been constructed before the work commenced on the spire then it might be standing today just as the choir and crossing still is. But with the architecture, culture and philosophy of the Renaissance rushing across Europe and threatening to engulf everything in its path, how could those master builders argue against the hubris of the Gothic Bishops who wanted the spire built as a beacon signifying all that they stood for?

Perhaps the most famous building failure that has become a landmark in its own right is the Leaning Tower of Pisa caused by foundation failure. The tilt, which recently reached a point where the Tower was closed on safety grounds, was gently eased back in a giant sling until the Tower reached equilibrium, then the foundations were stabilised.

But it was the huge and unprecedented construction booms of the nineteenth and twentieth centuries principally in Europe and the North America that brought building and structural failures into the public eye through newspapers and the courts. They brought Parliamentary changes in the law, case law and changes to codes of practice and building regulations, often following lengthy and costly public inquiries. The implications of major building and engineering failures have worldwide repercussions.

Human factors 3.3

Why do structures fail? The simple answer is gravity. When structural failure occurs gravity brings the unsupported structure crashing down to the ground. But there is another cause; architects, engineers or contractors leave something out either in the design, in the calculations or in the construction.

There is also the human factor. Although the design of structures has become more sophisticated since the collapse of the Tay and Quebec Bridges, people have remained much the same. This tends to get forgotten. Structures still fall down. Those responsible sometimes state the obvious without realising what they have said. Asked on television why a temporary stand had collapsed after a goal was scored at a football match in France in the early 1990s, a spokesman for the firm supplying the stand said, 'People were jumping up and down on it'.

Terminology 3.4

The following relevant terminology is defined:

- **Codes of Practice:** As the Building Regulations cannot cover all eventualities; separate codes are drafted from time to time. These cover such things as: Reinforced Insitu Concrete (RC), Precast Concrete, Steelwork and Means of Escape in Case of Fire etc. Often they have been drafted when a building failure demonstrates a loophole in the existing legislation. An example of this is BSCP 111, part 304 (see case studies on Quebec Bridge at **3.9** below and Ronan Point at **3.27** below).
- **Progressive collapse:** A form of failure of one part of a structure which leads to the collapse of part or the whole. Highlighted by the partial collapse of one corner of Ronan Point. This was due to its form of Large Panel System although this form of collapse can occur in traditional forms of construction as well. See 'Ronan Point 1968' (HMSO Inquiry Report); and the Alfred P Murrah Building, Oklahoma, 1995 (Wearne P, *Collapse: Why buildings fall down* (1999)).
- **LPS:** Large Panel System consists of the placing of one large concrete panel on top of another 'house of cards' fashion.
- **RC:** Reinforced Concrete.
- **Dead load:** The actual weight of a structure upon itself and the ground.
- **Live load:** The actual weight of fittings such as furniture, machinery, cranes, vehicles and people.
- **Imposed load:** The load imposed on the structure from inside or outside such as wind loading, impact from vehicles or planes or external or internal explosions. The expansion caused by a fire can impose a load beyond safe limits.
- **Fire load:** The amount of combustible material that can be consumed in a fire. It consists of furniture, fittings, the fabric of the building and people unfortunate enough to be caught in the fire. This can cause the collapse of a building (see the World Trade Centre, 11 September 2001 at **3.39** below).

- **Latent defect:** A hidden defect, either in design, materials or workmanship which comes to light often some years after the building is complete.
- **Patent defect:** A defect, either in design, materials or workmanship which comes to light which it is held that any reasonable architect, engineer or surveyor would have seen during the construction.
- **Factor of safety:** When designing the Britannia Bridge over the Menai Straits, Robert Stephenson used a factor of safety of 4. Modern structures have a factor of safety of 1.08. This sometimes results in failure during construction when loads exceed those for which the structure can bear.
- **Kentledge:** An applied load on a structure during construction such as concrete blocks or the blocks that counterbalance a crane. Originally the name applied to scrap metal ballast used in ships.
- **RIBA:** The Royal Institute of British Architects.
- **ICE:** The Institution of Civil Engineers.
- **IStrucE:** The Institution of Structural Engineers.
- **ASCE:** The American Society of Civil Engineers.
- **Murphy's Law:** A law which states: 'if anything can go wrong it will go wrong'.
- **The Peter Principle:** Invented by Professor Peter, the law states that: 'in a hierarchy every employee eventually rises to their own level of incompetence'. The aim in avoiding this is to remain creatively incompetent.

Principles of Design in Architecture and Engineering 3.5

No architect or engineer knowingly designs a substandard building or structure which is likely to fail. No contractor will knowingly construct a building or structure which will fail. However, the courts have many examples of apparently well thought out structures which have failed.

The judgments in the following cases demonstrate salutary lessons and should be required reading for anyone, however remotely interested in the art of construction, who is contemplating erecting a structure:

- *Alderman & Burgesses of the London Borough of Newham v Taylor Woodrow – Anglia 1979 (unreported)*: Ronan Point;
- the Report of the Inquiry into the collapse of the Yarra Bridge in Australia 1971; and
- Royal Commission into the collapse of the Quebec Bridge in 1908.

As the Judge in the Australian Yarra Bridge Inquiry in 1971 said, 'engineers should accept that they do not always know what they are doing'. That also applies to architects, builders, surveyors and the whole of the building industry.

Chin-Lung Bridge 3.6

How does one judge a good design? The answer is if it looks right, it usually is right. Design in architecture and engineering is very rarely rocket science. Very little in building is completely original. Most things have been done before. For instance the use of iron chains for suspension bridges was first used, not in the nineteenth century in Europe or North America, but almost a thousand years before in China. High up in the rocky gorges of the Yangste in Yunnan Province is the Chin-Lung Bridge. The 18 chains of this remarkable bridge famed throughout China and celebrated on one of its postage stamps, span 328 feet across the Yangste, which runs at this point at an altitude of 4,600 feet (*Science and Civilization in China* by Joseph Needham).

To construct such a bridge today would be a major feat involving complex contracts and even more complex machinery. It was built before many of the great Gothic Cathedrals of Europe were even started.

Figure 1: The first law of building

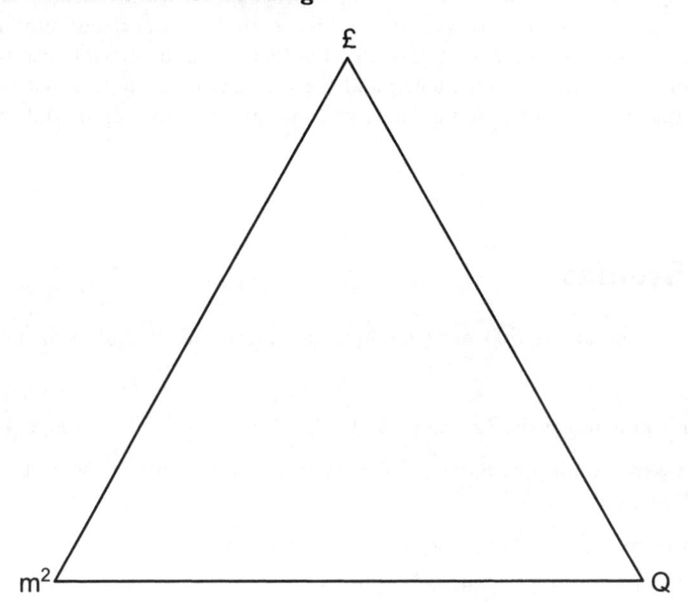

First law of design 3.7

The driving forces on a designer are all too often misunderstood. They are finite. The client wants the maximum area for the minimum cost. The designer wants quality and wants to see his or her work published – that way ensures more jobs. Unfortunately, the first law of design is a law of two fixed variables (£ & m2) and the only variable, the third one of quality (Q), is often driven down. Designers ignore this at their peril. In the past this ignorance has led to many failures, big and small. It will no doubt lead to more unknown failures in the future. Many people and ideas contribute to a design from law maker to code drafter to client's whims. But building regulations and codes of practice are minimum standards not maximum. Often this gets forgotten. Clients are loath to pay for extras if this means thickening steel, increasing safety margins, putting in more fire escapes if the law says they do not legally need to. But these laws are often drawn up after long periods of negotiation on committees where the loudest voice often holds sway. In the end a compromise is reached.

Architects and engineers:

> '. . . should know what not to do rather than give to the untutored and inexperienced mind that blind compliance with minimal code provisions and reports of committees, signed by all members after several years of disagreement but with ultimate consent to the strongest willed minority, is a guarantee of sufficiency' (Feld).

There is always a fine line to be drawn between, on the one hand, the dreams of the client for a building or structure which is built to the maximum area, span or height for the minimum cost and the architect or engineer who wants quality. Often the three cannot be reconciled and the one factor in the trio that is reduced is that of quality. Architects and engineers should understand they should sometimes say 'no'.

Case Studies 3.8

The case studies of the following disasters and causes are discussed in more detail below.

- **Hubris and nemesis:** The Quebec Bridge 1907 and the Tay Bridge 1879.
- **Progressive collapse:** Ronan Point 1968 and the Alfred P Murrah Building, Oklahoma City 1995.
- **Wind damage:** The Ferrybridge Cooling Towers 1965.
- **Box Girder Bridge Failures:** Milford Haven 1970 and the West Gate Bridge, Yarra, Victoria, Australia 1970.
- **Terrorist Attack:** The World Trade Centre, New York 2001.

George and Robert Stephenson, and Marc and Isambard Kingdom Brunel all witnessed failure of their works at one time or another, from the flooding of the Brunel's Thames Tunnel to the failure of a series of Stephenson's small railway bridges. Luckily, there was no major loss of life unlike the disasters that befell Sir Thomas Bouch, the designer of the Tay Bridge, and Theodore Cooper, designer of the Quebec Bridge.

Quebec Bridge Disaster – 29 August 1907 3.9

Although this Disaster occurred at the beginning of the last century, it was, like the Tay Bridge Disaster before it, a classic example of what could go wrong both in the design, calculation, fabrication and management of any structure.

Backgound 3.10

By 1897, Theodore Cooper was one of the most eminent engineers of his age. He had been a Director of the American Society of Civil Engineers. Born in 1839 he served in the Navy during the American Civil War. In May 1872, Capt James Buchanan Eads appointed Cooper Inspector of Steel Manufacturing for the construction of the St Louis Bridge. Impressed with his skill, Eads put Cooper in charge of the erection of the bridge employing a form of cantilever arch construction never before attempted on such a scale. During construction the arch ribs started to buckle. Cooper telegraphed Eads at midnight, received the answer and rectified the problem.

Expertise 3.11

The bridge was completed in 1874 and immediately Cooper's expertise was sought. He became the superintendent of Andrew Carnegie's huge Keystone Bridge Company in Pittsburgh, which had supplied the wrought iron for the St Louis Bridge, before setting up as an independent consulting engineer in New York in 1879. He was then 40 years old. Cooper built many notable bridges in New York, Pittsburgh and Providence. But by 1897 he had never built on the heroic scale of men like Eads or Roebling, designer of the Brooklyn Bridge in New York. Cooper was hungry for something more.

That year the ASCE held its annual convention in Quebec. The Quebec Bridge Company was on the lookout for a consultant. Its own engineer, Edward W Hoare, had never built a bridge with a span more than 300 feet and the company proposed bridging the St Lawrence River. Hoare was unqualified. Cooper introduced himself and saw the bridge as the crowning glory of his life. He expressed an interest. The Quebec Bridge Company saw the bridge as a means of getting out of the dire financial position they were in. Such a situation was hardly a propitious prelude for the design and construction of the largest and longest span bridge in the world.

Finances 3.12

The Dominion Parliament granted the Charter authorising the construction of the bridge over the St Lawrence in 1882. Five years later, an Act of Parliament created the company. For the next 12 years, always with the financial wolf at the door, it achieved little and did even less. Then having met Cooper in New York on a formal basis in 1899, the Quebec Bridge Company asked him to review all the tenders for the new bridge, bearing in mind all the time, of course, the parlous state of the company. A wiser man might have backed off at this point but whether spurred on by his own ambition or buoyed up by claims that the company might one day somehow weather the financial storm once the bridge started, he carried on. What no one had noticed was that once work started on the bridge it would have to bear not only its own load but the financial load of the Quebec Bridge Company as well.

Design approved **3.13**

The Quebec Bridge Company had met up with the Phoenix Bridge Company of Phoenixville, Pennsylvania. Their chief design engineer, Peter Szlapka, had submitted preliminary plans and the Quebec Company were keen to give the firm the work. On 23 June 1899, after having assessed all the bids, Cooper wrote to the Quebec Bridge Company, 'I therefore hereby conclude and report that the cantilever superstructure plan of the Phoenix Bridge Company is the "best and cheapest" plan and proposal'. He thus approved Szlapka's design. On 6 May 1900, Cooper became the consulting engineer for the bridge for the duration of the contract. With the inevitability of a Greek Tragedy, the bridge and Theodore Cooper were moving towards nemesis.

Concerned that the proposed span of 1,600 feet would mean piers far out in the river, subject to damage by ice floes in winter and entailing working at more extreme distances from the shore in deeper water for the divers, he proposed that the span be increased by another 200 feet to 1,800 feet clear span, thus making work in shallower water. This would make the bridge cheaper to build and save money. To keep the weight of steel down because of the increased span he proposed modifying the specifications, which would allow higher working stresses. The Quebec Bridge Company jumped at this offer. After all they thought the risks, if there were any, would all fall on Cooper's shoulders.

Change in design **3.14**

Because the Company was so short of money no checking was made of the stresses of what was, in fact, a completely new bridge design. The company did not want to gamble on development costs of a design they might not use. The directors were reluctant even to enter a contract let alone conduct the necessary tests. They relied on the good name and experience of Cooper to carry them through. Cooper had approved the plans and the bridge and everything to do with it became his legal responsibility. Then with three years lost to inaction, the Canadian Government suddenly came up with a plan to raise $6.7m to build the bridge. The company was saved, work could go ahead and everyone, including Cooper, would be covered in glory.

Increasing the span by placing the supporting piers nearer the shore, while reducing dangers from ice floe damage, was to make it the longest single span bridge in the world carrying the trans-continental railroad, a major road, tramways and footways. The economic benefits to Canada would be enormous. Previous daring examples of cantilever bridges existed over Niagara Falls, built by John Roebling's son, Washington Roebling in 1883 and Benjamin Baker and John Fowler's Forth Bridge 1882–90.

There was no time to lose and even less for development testing or recalculation to see if the new bridge structure could take the higher stresses. Cooper accepted all the figures the Phoenix Bridge Company submitted to him on face value. It was to prove a fatal error. Based in New York Cooper had only visited the site three times. He never went to the site again after 1903. All requests that he visit the bridge under construction fell on deaf ears.

Concerns **3.15**

The man responsible for the safety of the trains crossing the bridge, Collingwood Schreiber, chief engineer of the Canadian Department of Railways and Canals, became worried and expressed extreme concern that no one had checked Cooper's work, or what was happening on the site. He suggested the Department hire its own consulting engineer to check the calculations and construction. Cooper was furious and rushed to Ottawa even though he claimed he was too sick and infirm to look at the bridge. Pride drove Cooper at this point. He wrote to the Quebec Bridge Company claiming he would be subordinate to Schreiber if his drawings were independently checked. He then appointed a young, recently qualified and inexperienced graduate named Norman McLure to oversee the work and report to him. McLure, lacking practical experience, was unable to act on his own accord and Cooper was now in total control of all aspects of the bridge construction.

On 6 February 1906, the Phoenix Bridge Company sent Cooper a report informing him that the weight of the steel built into the bridge far exceeded the original weight on which the calculations were based. By now, all the parts of the south cantilever had been fabricated and much of it built out into space over the river. Cooper decided that the 10% increase in stress that he calculated was acceptable. This was a vast underestimate. Work on the bridge continued. Adding to the stresses on the cantilever arm was the live load caused by a moving crane and its load as it inched its way out on the ever-increasing span.

Overload **3.16**

A year later the folly of Cooper's decision was becoming all too apparent. The lower chord compression members of the cantilever arm, where they sat on the piers rising from the water, were beginning to show signs of distress and overload. Huge jacks weighing 75 tons were employed to force the buckles out of the steel. As the summer progressed and the crane moved further and further out on the span, the ends of both chords buckled as the stress reached intolerable levels. Still work proceeded even though Cooper asked in a telegram, 'how did bend occur in both chords?'.

Within a week in early August 1907, the deflection in one chord increased from ¾ inch to 2¼ inches (18 mm to 30 mm). McLure wrote a letter to Cooper in

New York. The general foreman on the bridge brought all the men off fearful that it could collapse. The next day he ordered work to recommence. Hoare, the chief engineer, was happy with this. Stopping work might mean shutting down for the winter. Hoare lacked the experience to deal with the situation so he sent the young engineer, McLure, to New York to tell Cooper what was happening. McLure arrived at the same time as his letter sent two days earlier. Cooper, reading the letter and thinking work had stopped on site, sent a telegram to the Phoenix Bridge Company saying, 'Add no more load to bridge till after due consideration of facts. Mr McLure will be over at five o'clock'. It was 12.16 pm on the morning of August 29 1907. By then the end of the bridge projected nearly nine hundred feet over the river.

Strategy 3.17

Cooper decided on a strategy. He gave instructions in another telegram to McLure to wire to the site. McLure put it in his pocket and then forgot to send it in his rush back to Quebec. Under the impression that work had ceased two days earlier, Cooper intended to keep it that way until McLure reported back to him. His first telegram arrived at 3 pm. It was read and disregarded by John Sterling Deans, the chief engineer of the Phoenix Bridge Company. The workers remained working high out over the St Lawrence River. The crane continued its work slowly moving out. McLure arrived on site at 5.00 pm and saw Deans and Szlapka who said that the problem could wait until morning. The men were ordered to come down off the bridge at the end of their shift. It was 5.30 pm.

Failure 3.18

A steel erector sat at the very end of the south cantilever arm, underneath the crane and looking out at the view while waiting for the signal whistle to bring him down. As Ingwall Hall gazed out across the river, two sounds like gunshots rang out below and behind him. The two compression chords of the south anchor arm of the bridge had failed catastrophically. 17,000 tons of steel, the crane and 86 workmen on the bridge plunged into the St Lawrence River. An eyewitness was later to tell the Royal Commission that the collapsing columns of the bridge looked like 'ice pillars whose ends were rapidly melting away'. Hall was one of only eleven survivors. 75 workmen were killed.

Inquiry 3.19

In 1908 the Royal Commission of Inquiry published a two-volume report, containing all the relevant documents, letters, telegrams and photographs, plus the calculations for the bridge. It said:

> 'We are satisfied that no one connected with the work was expecting immediate disaster, and we believe that in the case of Mr Cooper, his opinion was justified. He understood that erection was not proceeding; and without additional load the bridge might have held out for days'.

Of John Sterling Deans, who disregarded Cooper's first telegram, the Inquiry had nothing but damnation for his lack of judgement and professional incompetence. The Phoenix Bridge Company was criticised for appointing the unqualified engineer, Edward Hoare. But it was Cooper and Peter Szlapka who bore the brunt of their criticism. Cooper had approved Szlapka's design for the bridge:

> 'The failure cannot be attributed directly to any cause other than the errors in judgement on the part of these two engineers . . . A grave error was made in assuming the dead load for the calculations at too low a value'.

This last value, said the Commissioners, was in itself sufficient to have condemned the bridge even if the lower chords had been strong enough. Collingwood Schreiber had been vindicated, but at the cost of 75 lives.

The bridge was rebuilt. As the central span was being lifted into place on 11 September 1916, watched by thousands of people, part of the lifting gear failed. The central span fell into the St Lawrence with 90 men. Nine of these drowned. The new bridge was finally completed in 1917. In its day it was the longest span bridge in the world. Theodore Cooper never worked again after the Inquiry and never saw the new bridge. He died in 1919 aged 80.

Impact 3.20

The impact of the 1907 collapse was enormous in engineering circles and had an effect all over the world. In the UK the main change was incorporated in BSCP 111, Part 304, which states that any vertical structural member must be able to resist a horizontal load at its base equivalent to 11/2% of the vertical load carried. This principle applies to all construction and includes walls as well as columns throughout the world in one form or another.

The Quebec Bridge Disaster was a classic example of a failure, which embodied virtually every known thing that could go wrong in a design. It was like Murphy's Law and the Peter Principle all rolled into one, as if the lessons of the Tay Bridge Disaster of 1879 were forgotten.

Tay Bridge Disaster, Dundee – 29 December 1879 3.21

The multi spanned railway bridge was designed by Sir Thomas Bouch. On completion, his next project was to have been an even bigger bridge across the Firth of Forth.

Background 3.22

Like the Quebec Bridge the work on site was left to an unscrupulous foreman called Fergus Ferguson who filled blowholes and faults in the cast iron columns with a mixture of lamp black, beeswax and iron filings. This was called 'Beaumont's Egg' and served no structural purpose. Ferguson did this to avoid breaking up the substandard columns. The bridge spanned across the Tay on a number of cast iron piers with the trains running along the tops of the girders until they reached the middle spans where they ran through the inside of the girders. This extra height was to allow shipping to pass underneath. It was a requirement of the Admiralty.

The supervision both during and after construction of the bridge was poor. Bouch seldom visited to see what was going on. As a result many of the bolts holding everything together were never checked for adequacy. Often they were left out or fell out. During inspections by boat after the bridge had opened, loose bolts that had fallen out of the completed structure were picked up from the piers rising from the water. No one connected with these inspections ever drew the obvious conclusions. Added to this, Bouch had little or no idea of the wind loads likely to occur in the middle of the Tay. He approached the Astronomer Royal whose knowledge was on a par with Bouch's. A figure of 10 lbs a square foot was arrived at and this was adopted.

Strong gale 3.23

A year and a half after it opened, when many who crossed it became alarmed at the sway and took the long way round by road and ferry, a train crossed from Edinburgh to Dundee on the last Sunday night of 1869. In the teeth of a force 11 gale, with gusts reputed to have reached over 100 mph, the train with its claret, gold and blue carriages reached the centre spans of the high girders, each one 245 feet long. In the high wind it tipped over as the high girders lifted from their seating. The whole middle section of the bridge progressively collapsed into the raging Tay. The signalman in his box on the south shore saw sparks from the wheels at this point and waited for the signal from the box on the Dundee side to say the train had passed. None came. Over 70 people were drowned.

Inquiry 3.24

An inquiry was held in Westminster Hall. Bouch was held to have failed in his design, failed in his supervision and failed to take account of the sort of wind loading other engineers were employing. Gustave Eiffel and William Le Baron Jenny, the American Architect/Engineer working in Chicago, used 55 lbs and 50 lbs a square foot respectively for bridges and high buildings. Even if the bridge had stood, the defects in its design and construction would sooner or later have brought it down. Professionally, Bouch was destroyed. He died a year after the report was published. He was 58.

Parts of the old bridge were reused in the construction of the new one. The report prompted a more up-to-date approach in wind loading on buildings and structures, although it was not to last. When Baker and Fowler's magnificent Forth Bridge was finished, many engineers felt that it was over designed and the factors of safety far too high.

Ferry Bridge Cooling Towers – 1 November 1965

3.25

In 1962 the contract to design and build eight cooling towers at Ferry Bridge in Yorkshire was given to Film Cooling Towers (Concrete) Ltd. The towers were 375 feet high and had greater shell diameters and surface areas than any other towers then built. They were also built more closely together. In a 75 mph wind on 1 November 1968, three of the towers collapsed within an hour, the first at 10.30 am, the second at 10.40 am and the third at 11.20 am. No one was killed or injured.

Causes

3.26

The causes were:

- the wind loads were too low by 25%;
- basic winds speed was taken over one minute instead of allowing for peak gusts at the higher ten second average;
- there were unknown wind pressure effects from uplift within the towers;
- here was no factor of safety in the design for the unknowns of wind action.

At a 300-foot height the wind pressure for Ferrybridge had been set at 17 lbs a square foot even though similar towers elsewhere had used 35 lbs square foot.

Progressive collapse: Ronan Point – 16 May 1968

3.27

Ronan Point was a 22-storey Large Panel System (LPS) tower block of 110 flats standing on a 2-storey RC podium in Canning Town, east London. It was designed in the architects department of the County Borough of West Ham, just over the London County Council (LCC) border in Essex. The original scheme consisted of four blocks over an underground car park when Thomas North, the Borough Architect, took it to the Ministry of Housing. He was told he would only get permission for a system built scheme. In 1965 the new Labour Government had published Circular 76/65, which proposed building 500,000 industrialised houses in five years.

Background

3.28

At the Ministry's encouragement West Ham approached Taylor Woodrow Anglian (TWA), formed from Taylor Woodrow a major contractor and Anglian Building Products, a subsidiary company of Ready Mix Concrete based in Norfolk. The merger between the two companies had been encouraged by the Ministry and the LCC Architects Department. In a 1962 report to the Council, the Chief Architect of the LCC described the form of construction as, 'large concrete panels piled one on top of the other house of cards fashion'. Like Theodore Cooper's comment that the design by the Phoenix Bridge Company was the 'best and cheapest', it was a description that was eventually to hang round the neck of large panel system building like an albatross.

When West Ham visited TWA they were told they would have to have nine blocks totalling 1,000 flats, otherwise they would have to accept TWA's basic design. Also, they would go to the back of the housing queue. West Ham did not want this. They had the worst housing conditions in London. There were acres of 'jerry-built' nineteenth century slum housing in the southern part of the borough around the Royal Docks. During the war West Ham lost 27% of all its housing stock.

Scheme approval **3.29**

So the scheme went ahead. Phillips Consultants, a wholly owned subsidiary company of Taylor Woodrow was appointed structural engineers for the structure and they submitted their scheme to West Ham's Borough Engineer, Mr Williams, for approval.

Williams was not happy with what he found. The wind-loading figure for the highest LPS flats ever built was 17 lbs per square foot. He asked for an upgrade and the figure was raised to 40 lbs per square foot. So keen was Thomas North, the West Ham Chief Architect, and the Borough Engineer to get on with the scheme that the Borough Engineer granted byelaw approval before Mr Williams had seen all the drawings and calculations.

Gas supply **3.30**

Work started on site in July 1966 and was finished in March 1968. The first tenants moved in and as the buildings were next door to the largest gas works in Europe at Becton, it was inevitable that the North Thames Gas Board would supply the gas for cooking. The central heating for the flats was a relatively untried form of electrical underfloor heating. Ivy Hodge, a tenant on floor 18 in flat 90, asked a former neighbour, Mr Pike, if he would reconnect her second-hand gas stove. He went to the gas showrooms, discussed what he was going to do, told them he was a plumber's mate, bought all the parts and reconnected it.

Miss Hodge was a very careful woman with a keen sense of smell. She did not smoke and had turned off the pilot light in her gas stove in case it blew out and there was a gas leak. On the night of 15 May 1968 she went to bed as usual leaving her bedroom door ajar. In the night, she got up and went to the bathroom disturbed by a noise. She woke at her normal time, went to the bathroom and then into the kitchen to make a cup of tea. She did not smell any gas. There was an explosion which, although it did not kill her, was sufficient to blow out the load bearing flank walls on the outside of the building.

People in the street below saw this happen and watched in horror as the whole side of the block cascaded to the ground in a progressive collapse, as they said, 'like a pack of cards'. People in the block raced to their neighbour's flats to see if they were safe. They looked through the letter boxes and saw only the sky beyond. Thomas North was summoned from his house by the police and taken to the site.

An engineer from Taylor Woodrow said, 'I can see no sign of structural failure' (*The Times*, 17 May 1968). And in front of dumbstruck reporters, North said, 'These flats are safe. They could be lived in if only people would move back into them' (*Daily Express*, 17 May 1968). It was an extraordinary statement considering it was the very first time that Thomas North had set eyes on the building.

Inquiry 3.31

An Inquiry was set up. Two days after the collapse, Dr Chan, the Chief Engineer of the National Building Agency and Cleeve Barr, its Chief Architect, visited Ronan Point. They quickly established that the explosive force required to start progressive collapse was 1.4 lbs a square inch. This was not made public.

The LCC had placed an order for 1,800 flats using the TWA system. On one of these sites the District Surveyor of Wandsworth, had become concerned about the absence of continuity between the load bearing ends of the floor slabs and the flank walls. He asked for a mechanical connection so that the structure complied with BSCP 111, Part 304 which dated back to the Quebec Bridge Disaster.

In mid-August 1968 the results of a structural test at the BRE became public. Within a day the Minister issued an Emergency instruction to all local authorities in the UK banning gas. The flats were to be made all electric. The joint between the floor slab and flank wall had failed at 1.8 lbs a square inch, as Chan and Barr had found two days after the collapse.

The report was published on 8 November 1968, the day Richard Nixon was elected President. It was a very good day to publish bad news. As well as pointing out the obvious dangers of progressive collapse and total lack of structural continuity, the report drew the attention of designers to the need for an increase in wind loading. Sir Alfred Pugsley, the engineer on the tribunal, wanted 45 lbs a square foot for a peak gust wind speed of 105 mph. He also wanted 65 lbs a square foot at the top corners of the block to withstand a peak gust of over 150 mph. In his draft chapter on the structure he had written:

> '. . . it should be remembered that to make the walls and their joints capable of withstanding pressures, for example of 45 lbs per square foot, would be only to make them about as strong as the glass in ordinary windows'.

All this passage in the draft was struck out against his wishes. In order to emphasise what he meant, Pugsley added a marginal note by the side:

> 'However, it would seem to us very unfortunate if in a wind liable to break many glass windows, the inhabitants of Ronan Point should have to start worrying also about its structural stability'.

This too was struck out. In a hand written letter to the Minister, the Chairman of the Tribunal drew attention to the serious wind damage that had occurred in a LPS block of flats in Glasgow. He said: 'we have deliberately not referred to it in the report to avoid undue public anxiety' (*The Times*, 18 February 1985).

Pugsley was very concerned that if high winds broke glass at the top of the building then the wind could play on both sides of the load bearing flank wall. This could move it and trigger progressive collapse.

The report also drew attention to the following:

- the weakness of the joints;
- banned gas;
- warned of dangers of fire which could lead to collapse; and
- required all existing flats to be strengthened.

Rebuilding 3.32

Pressurised by the construction industry, the Government bowed down. Existing buildings could not be strengthened to the recommended 5 psi asked for by the Ministry. A compromise was reached. If the gas was removed and the flats became all electric then the strengthening need only be to 2½ psi.

The shattered corner of Ronan Point was rebuilt in 1971/72 and tenants moved back in. Blocks throughout the UK were strengthened to 2½ psi by means of 4 × 4 inches (100 mm × 100 mm) steel angles bolted under the floor slabs to increase the bearing. The under-floor heating in the end flats against the flank walls in Ronan Point was turned off to prevent the expansion of the slabs causing progressive collapse. And there it rested until new investigations in 1983/84.

Other LPS buildings 3.33

Throughout the UK various things started to go wrong with LPS buildings. The concrete facing of panels on Bison Wall Frame blocks in Birmingham, Liverpool and Glasgow fell off. Although not structural, these weighed several tonnes. Worried about their Bison blocks, the London Borough of Hillingdon carried out a huge exercise under the Borough Architect, to find out what was wrong. They carefully demolished a block as the only way to find out how it had been constructed. What they uncovered was alarming. Many metal fixings connecting the panels had been left out by the builders. Bolts had been burned off. Concrete was missing in joints and the facing panels were found in some cases to be only held in place by the adhesion of the polystyrene insulation behind. All these fixings were missing in some cases and there was no way to find this out until a panel fell off. Hillingdon looked at over 1,400 flats.

Structural tests **3.34**

Amid worries about structural failure the Minister of Housing and Construction issued a letter to all Local Authorities in October 1983, asking for structural checks on all forms of Bison Wall Frame construction from high-rise blocks to two-storey houses. At the same time, the London Borough of Newham decided to upgrade Ronan Point and the other eight blocks on the estate. Following tenant pressure, a series of structural tests were conducted in Ronan Point. Most crucial of these was a full-scale fire test held on 18 July 1984 in flat 26 on the third floor. Scheduled to run for at least 30 minutes it was stopped after ten minutes, on the orders of the engineer in charge because the expansion of the floor slab was moving the load bearing flank wall joint to a point where progressive collapse could reoccur.

Further tests were carried out including opening up the flank wall joints at the bottom of the block. These joints were found to be virtually empty of concrete and filled with the modern day equivalent of Beaumont's Egg (see **3.22** above). Instead, of being full of concrete the joints were found to be full of floor sweepings, cigarette ends, cement bags, newspapers and bottle tops. Instead of resting as a uniformly distributed load on a bed of mortar, the walls were found to be resting on the nuts on the one-inch diameter lifting bolts used to hoist the panels into place and then level them to take the floor slabs. The effect of this was to create a series of extreme point loads, something for which the structure was never designed and one under which it could eventually fail, especially in high winds.

In 1986 work started on the dismantling of Ronan Point and as work progressed it was discovered that the actual conditions found were far worse than had been anticipated. Not one single load-bearing joint complied with the drawings and towards the bottom of the block the lifting bolts were found to be punching up through the bottoms of the wall panels. All nine blocks were taken down. In the later blocks built after Ronan Point conditions were even worse.

Published reports **3.35**

The Minister ordered the Building Research Establishment (BRE) to carry out further research and publish a series of reports:

- 'The structure of Ronan Point and other Taylor Woodrow Anglian Buildings' (1985) ISBN 0 85125 342 3.
- 'Large Panel system dwellings: preliminary information on ownership and condition' (1986) ISBN 0 85125 186 2.
- 'The structural adequacy and durability of large panel system dwellings' (1987) ISBN 0 85125 250 8.

Serious doubts were beginning to be felt about the long-term life of this form of construction. In 1999 surveys were carried out in Bison Wall Frame LPS blocks in Birmingham where it had been found that a number of the high-rise blocks still had a piped gas supply even though they failed to meet the 5 psi requirement of

Circular 71/68. What would have happened if a gas explosion had taken place in a flat followed by progressive collapse of part of the block can only be imagined. Off the record, engineers from the BRE stated that there were other authorities in the UK with a similar situation.

Decisions by the Government following the publication of the Ronan Point Report in November 1968 have left a huge unwanted legacy. The crisis was due to the Government decision to allow work to restart on the other identical buildings to Ronan Point during the Public Inquiry. They were part of the same contract. This was repeated all over the country. By the time the report was published six months later there were twice as many LPS tower blocks. Following the publication of the report the building industry demanded to know what they had to do. In order to cover up one mistake the Government created conditions for another.

The post Ronan Point Ministry of Housing Circulars 62/68 and 71/68 agreed to in a state of panic in November 1968 allowed a dual standard for existing buildings. If the buildings could survive a 5 psi explosion without failure then piped gas could remain. As the vast majority could not, the figure was dropped to 2½ psi if the gas was removed.

Now we know only too clearly what they should have done as many tower blocks are being demolished with 30 years interest still to be paid off on the 60-year loan taken out to pay for the construction.

The Alfred P Murrah Building, Oklahoma City, USA – 19 April 1995 3.36

Just before 9.00 am on the morning of 19 April 1995 Timothy McVeigh drove a Ryder truck hired in Junction City, Kansas and parked it in an unloading bay on NW Fifth Street outside the main entrance of the building. He got out of the cab, locked it and walked away as the federal employees were settling down to work. No one gave it a second glance. Suddenly there was a massive explosion from the 4,800 lbs of explosives packed inside. Where the truck had stood was a crater 7 feet deep (2.1 m) and 28 feet (7 m) across, only 10 feet (3 m) from the front of the building. The blast ripped through the front entrance lobby and half of the building collapsed killing 169 people and injuring 490. Engineers from the ASCE and the Federal Emergency Management Agency (FEMA) were ordered to discover why so much of the building had collapsed.

They had to work while rescue was going on and the FBI was carrying out a criminal investigation. The first question the experts asked was, 'why had so much of this U-shaped, 28-year old, nine-storey, reinforced concrete building collapsed?' After all, the building was solidly built to withstand tornadoes. But it was when they looked in detail at the structural remains of the front façade lying on the ground that the answers began to emerge.

For the first two floor levels there were four large columns spaced 40 feet (12 m) apart across the front of the building. They supported a 5 feet deep (1.5 m) reinforced concrete transfer beam at third-floor level. Above this the north façade was fully glazed. The seven upper storeys had ten equally spaced columns across the façade. The loads from these were then transferred via the beam to four columns. One column carried three of these columns above and in the explosion it was pulverised as the shock waves reduced it to sand and gravel dust in a process known as brisance. In a millisecond the column failed, as the shockwave travelling faster than the speed of sound blew the slabs upwards causing cracking between the tops of the columns and the slabs. Suddenly, what had been a 40 feet span (12 m) at ground level under the 5 feet deep (1.5 m) transfer beam became an 80 feet clear span (24 m) with an enormous force applied to one side from the explosion. The beam fell over into the building with the three remaining columns still attached. With nothing to support the structure above, everything above the transfer beam collapsed.

Because of their structural continuity all reinforced concrete buildings, have an inbuilt structural redundancy. It began to emerge that the Alfred P Murrah Building had no structural redundancy whatsoever and the investigating engineer's thoughts began to go back to the collapse of Ronan Point (see **3.27** above) nearly 30 years before. They were looking at the progressive collapse of a sturdy reinforced concrete building and they could not understand why.

The explanation became clear when they looked for the reinforcing bars between the tops of the columns and the transfer beams. They could not find any. It was not a monolithic building at all with continuous steel reinforcement positively tying all the elements together. In principle, it was like Ronan Point and relied on gravity to keep everything in place transferring all the loads downwards. Until, that is, an explosion sent shockwaves upwards. A traditional reinforced concrete building with continuous reinforcement would not have collapsed in the same way with such huge loss of life. Until the events of 11 September 2001 this was the worst terrorist attack on American soil.

Box girder bridges: The Collapse of the Milford Haven Bridge – June 1970 and West Gate Bridge over the Yarra River, Victoria, Australia – December 1970 3.37

Modern box girder construction was developed as a means of building lighter and stronger bridges with subsequently greater spans using less steel. In 1970 a series of collapses in Germany, Pembrokeshire in Wales and over the Yarra River in Victoria Australia occurred within a few months. Freeman Fox & Partners, a world famous engineering firm responsible for many major bridges, had developed the box girder form of construction.

Background 3.38

Box girder construction is like a series of steel boxes welded together. As each new box is required it is slid out and lifted into place by a crane on the end of the span. The completed bridge structure rest on a series of piers of reinforced slip form concrete, a continuous casting process producing a very strong pier rising either from the ground or from the river bed.

Once all the sections are in place and welded together the bridge has immense strength. Unfortunately, during construction and before full strength is achieved, the bridge structure is subjected to stresses which can be in excess of its ability to carry and transfer them.

The Milford Haven Bridge collapsed at a pier support as a new section was being lifted into place due to the failure of the vertical diaphragm inside the bridge. The Yarra bridge failure was more complex.

The Victorian Government had wanted to build a tunnel but fears over possible fires in the tunnel led to the construction of a bridge consisting of twin box girder beams bolted together with the roadway on top. During construction, one of the girders became higher than the other and in order to remove this difference in level of about 150 mm, a heavy load of kentledge (see **3.4** above) was added to the bridge to bring the two sections together. As the load was applied to one side of the span the connecting bolts were released. Suddenly, without warning the entire span failed with the internal diaphragm buckling at the pier support. All the men on the span including the engineer in charge of the bridge were killed and the workmen's huts under the span were crushed with considerable loss of life.

With so many box girder bridges under construction in the UK due to the 1960s motorway programme the Government set up the Merrison Committee to discover what went wrong and suggest a solution. A new British Standard BS5400 was produced with comprehensive rules for the design of box girders, but many of the existing bridges were found to be less strong than they should have been and restrictions were placed on the amount of traffic they could carry. In some cases four lane bridges were reduced to two single lanes with considerable traffic congestion as a result.

World Trade Centre, New York – 11 September 2001 3.39

A few horror-struck people, too traumatised by what they are seeing, are often the only witnesses to most building collapses. The collapse of the twin towers was very different. It was witnessed by millions of stunned people all over the world. They saw it live on television as TV companies cancelled all other programmes to show it.

Background 3.40

At 08.46 am Eastern Standard Time, a Boeing 767 from Boston with 81 passengers on board flew in a straight line down the avenues of New York from Lower Manhattan and crashed into the North Tower causing a massive fireball, punching a huge hole through the structure, destroying floor plates, putting all the internal firefighting sprinkler systems out of action and destroying most of the central core including the three stair cases and lift shafts. Aviation fuel poured down service shafts, lift shafts and through holes in the structure. Within a couple of minutes the internal temperatures within the building had reached at least 1,200 degrees centigrade and possibly went as high as 1,500 degrees centigrade. The impact of the plane shattered the ceramic fire proofing on the structural steelwork.

At first everyone thought it was an accident. Instructions to the occupants of the second tower to evacuate the building were cancelled. People were called back to their offices. As they climbed the emergency stairs, the second Boeing 757 from Washington with 58 passengers banked round the blazing North Tower and flew headlong into the South Tower, slicing its way through the corner of the structure, causing a second huge fireball and massive damage. Both planes were fully loaded with fuel and enroute to Los Angeles. It was 09.16 am.

At 09.43 am a third plane crashed into the Pentagon in Washington causing widespread damage. At 10.07 am the South Tower of the World Trade Centre collapsed and, at 10.25 am, a car bomb exploded outside the State Department in Washington. Two minutes later the North Tower of the World Trade Centre (WTC) collapsed, so too did part of the Pentagon and then three minutes after that, a Boeing 757 with 45 passengers crashed near Somerset County Airport, Pennsylvania. The United States was under attack.

Specifications **3.41**

Why did the towers fail? At the time of their completion in 1973 the twin towers at over 1,360 feet (415 m) were the tallest buildings in the world. The client was the New York Port Authority (NYPA). They wanted 12 million square feet of floor space on a 16-acre site. Each of the floors of the twin towers measured 209 feet × 209 feet, an acre in area (68 m × 68 m). The NYPA wanted clear floor space with no internal columns. Each façade was a prefabricated steel lattice with window mullions acting as columns at 39 inch centres (1 m). This close spacing was to help people with vertigo. The vertical lattices were bolted together on site.

The external walls acted as wind bracing to resist all overturning forces. The central core took only the dead loads of the building. The floors spanned 60 feet (19.7 m) from the external walls to the central core and were 4 feet deep (1.2 m). They consisted of a 4-inch deep (100 mm) reinforced concrete plate on top of a steel plate fixed to a series of lattice beams with a false ceiling below. These floors acted as horizontal diaphragms to resist wind loads by stiffening the exterior walls. All of the original fireproofing was in sprayed asbestos but during construction this was replaced by ceramic fireproofing. What asbestos remained was being replaced after the 1993 terrorist bombing.

For elevators to serve 110 floors with traditional configuration would mean that the central core would take up half the floor areas on the lower floors. Otis Elevators developed an express and local lift system for the building. Passengers would take lifts to sky lobbies on the 44th and 78th floors. This halved the number of lift shafts. The

WTC was effectively a steel tube inside a much bigger steel tube joined by the floor plates. There was a sprinkler system on each floor and like all high-rise buildings it was designed so that it would not be necessary to evacuate the entire building in a fire but only the floor immediately above and below the fire. The construction was designed to contain the fire to the floor on which it broke out. There were three fire stairs.

Each of the two towers weighed ½ million tons. Unrestrained in winds such tall slender buildings would sway alarmingly. Each tower had a huge water-filled damper, designed to move and counteract the sway. These sat centrally over the central core as did water tanks, lift motors, air conditioners etc.

The towers collapse **3.42**

The south tower was the first to fall, keeling over like a felled tree. It is probable that the plane and fire caused enormous damage to the corner columns and they buckled under the hundreds of thousands of tonnes of the floors above.

The north tower collapse was different. First the communications mast started to slide slowly down inside the central core. Gathering speed it fell faster than the outside structure. The huge damper fell down the core dragging everything with it.

The strength of the exterior columns was based on restraint at each floor level. This reduced the effective height of the columns. But the impact of the planes had destroyed the floor plates and columns were suddenly two, three and even five storeys high. For a time they held the load as it redistributed itself through other redundant parts of the structure but eventually they failed and the weight of the building above sent a shockwave down the building as joint after joint failed. Wall sections slid over one another in a progressive collapse.

Fireproofed steel is rated to withstand 815–871 °C (1,500–1,600 °F). Unprotected steel loses strength at half that. The impact of the plane destroyed the fireproofing on columns and ceilings. In the raging fires, unsupported floor plates fell, shattering floors below. It was only a matter of time before collapse started and once it did, the tops of the towers acted like pile drivers on the floors below.

The engineers **3.43**

When engineer Leslie Robertson designed the building in the early 1960s he was asked to design a building which would withstand the impact of a Boeing 707. It was just over 15 years since a B-25 had crashed into the 79th floor of the Empire State Building in 1945. The ensuing fire on that day was contained and extinguished by the NY Fire Department. Shortly after the first WTC Tower was completed, an electrical fault caused a serious fire at the top of the building. Three floors were burnt out; the fire was brought under control and the building repaired. In 1993 a large car bomb exploded in the basement of the WTC punching a huge hole through the lower floors. The buildings survived this. Robertson, the engineer, and

Minuro Yamasaki, the architect, had designed a similar building in Seattle for IBM but nowhere near as high. In a moving speech Yamasaki said of his building at its opening, 'the World Trade Centre is a living symbol of man's dedication to world peace . . . '.

Inquiry **3.44**

In the aftermath of the disaster some parties looked for someone to blame. But is it possible to design a building to withstand an impact from a plane not yet invented in a scenario no one can imagine, not once, but twice, on a blue sky Tuesday morning?

Lessons for OSH/Disaster Management Practitioners **3.45**

Accounts of past Disasters often lead the reader to think that the people who made the errors, which led to the Disaster, were fools. With hindsight the causes appear so simple. But all too often the people that have foresight, like Colling-wood Schreiber who queried the Quebec Bridge design, find their fears are not heeded but ignored and sometimes ridiculed.

The engineers for the Quebec and Tay Bridges were not stupid men, although their critics before the event may have been treated as such. They were considered among the foremost engineers of their generation. They put more trust in others than was good for them, took second opinion advice and failed to go and see for themselves what was happening on site. Like many powerful people they were arrogant and that proved as fatal for them as it did for the poor workers and train passengers who were killed.

Proactive action **3.46**

Have a plan. Do not be lulled into thinking structural explanations are beyond understanding. People in all walks of life often think, 'if it isn't complex it must be wrong'. Most explanations as to why a structure failed are often quite simple. With structures, failure is always caused by the action of gravity on unsupported loads.

Think – what is the worst that can go wrong? **3.47**

Sometimes Murphy's Law and the Peter Principle do become one. Sometimes the most highly trained individuals fail to react or even understand what is going on even when it is in front of their eyes. An example of this is the inaction of the police at Hillsborough who saw people dying only an arm's length away but did not react in time to save them.

Read the signs 3.48

Why did the crack occur in the structure? What is the quality of the maintenance or workmanship, eg the Tay Bridge 1869 or Ronan Point 1968?

Reactive action 3.49

After the bomb explosion in the World Trade Centre in 1993, one firm practiced evacuation techniques. They stuck to these on 9/11 and ignored messages to return to their place of work in the South Tower. Then the second plane struck. Every member of staff survived even though their office was above the point of impact.

Conclusion 3.50

Although structures and materials have changed since Biblical times, people have remained much the same. The same aspirations still drive them whether money, fame or public acclaim. Sometimes they believe the hype and that is perhaps when they are in greatest danger. That is the point at which nemesis strikes. There are still dishonest men at all levels in the construction industry. What happened to Bouch and Cooper and the way they were taken in could happen to anyone.

Do not rely on hindsight – generate some foresight, however unpopular this may be at the time.

4

Crisis Management

In this chapter:

- Introduction.
- What is a crisis?
- Types of crisis.
- Crisis life cycle.
- Crisis severity.
- Commencement of a crisis.
- Crisis management.
- Planning.
- Response.
- Selection and training.
- Crisis management and public relations.
- Operational reality.
- Conclusion.

Introduction 4.1

Crisis Management is often regarded as a somewhat negative activity. Indeed, in the past it has been regarded as unproductive in terms of value and it can take up an inordinate amount of time. As a result, certainly before the events in the US on September 11th 2001, managers in many organisations were reluctant to think about it. Pointing out that 'crises still only happen very occasionally' involving very few organisations, Lagadec suggested that managers decided that their time was best spent on other important issues (Lagadec P, *Preventing Chaos in a Crisis* (1991) Magraw-Hill, pp 317/318). However, since September 11th, there are signs that senior managers are engaging 'in efforts to prepare their teams for complex, open, uncertain, and unstable situations'. But, says Lagadec, speaking two months after the dreadful events in the US, 'it should be stressed that these efforts are for the most part recent, limited and suffer from poor follow-up and lack of hierachial support . . . ' (Patrick Lagadec, 'Crisis Management in France: Trends, shifts and perspectives', a presentation to the European Crisis Management Academy on 22/23 November 2001 in Stockholm, p 6). And yet, Kash and Darling suggest that undertaking strategic planning in any organisation without the inclusion of crisis management 'is like sustaining life without guaranteeing life' (Kash TJ and Darling JR, 'Crisis Management: Prevention Diagnosis and Intervention' (1998) Leadership & Organization Development Journal, p 180).

Consequences of crises 4.2

Crises involve tremendous risks and uncertainties. At their worst they can bring down governments, send companies into liquidation and result in death and destruction. National and international security is subjected to a wide variety of military and non-military risks that are multi-directional and often difficult to predict. Regional crises exist; others could evolve rapidly. Some countries face serious economic, social and political difficulties. Ethnic and religious rivalries, territorial disputes, inadequate or failed reforms, the abuse of human rights and the dissolution of States can lead to local and even regional instability. The resulting tensions lead to crises and armed conflicts, which could spill over into neighbouring countries, affecting the security of other states.

Enron 4.3

Case study I

In 1984, Kenneth Lay became the chief executive officer of Houston Natural Gas, a company that dealt, as its name implies, in natural gas. Lay quickly doubled the size of the company by acquiring two other companies, Florida Gas and Transwestern Pipeline. Following this, Lay merged his organisation with a similar company, InterNorth, based in Nebraska. The merged companies became known as Enron in 1986.

In 1987, Enron's petroleum marketing company, Enron Oil, reported losses of $85m. However, the true losses of between $142m and $190m were not revealed until 1993, when two Enron Oil executives pleaded guilty to filing false tax returns and conspiracy to defraud. In 1990, Enron hired Andrew Fastow as Account Director; he quickly rose to become Finance Director. Tough and intimidating, it is alleged that Fastow set up a complex network of subsidiaries. His methods, it is argued, were of 'dubious legal status' and no one at Enron appears to have questioned them. ➡

In August 2000, Enron's revenue exceeded $100 billion for the first time, making it the seventh largest company in the Fortune 500. In February 2001, Jeffrey Skilling took up the post of Chief Executive. Kenneth Lay stayed as Chairman. However, six months later, in August, Skilling resigned stating that it was for 'personal reasons'. Lay again took over the role of Chief Executive, at the same time, retaining his role as Chairman. At this time, Sherron Watkins, an accountant in Enron's finance division, sent a letter to Lay warning him of a potential accounting scandal.

Enron's dealings were complicated. Suffice to say that the balance sheets that were available to the public did not appear to show debts of $1 billion. When this became known, Enron collapsed, going from a $70 billion energy giant to the second biggest bankruptcy ever. By the end of November, Enron's stock was trading at less than $1 a share and, on 2 December, the company filed for Chapter 11 bankruptcy protection and laid off 4,000 employees. In January 2002, the US Justice Department confirmed it was conducting a criminal investigation into Enron.

In her testimony to Congress in February 2002, Watkins made two very significant statements:

- first, she suggested that Enron had no Crisis Management Plan;
- second, she suggested that there was a window of opportunity during which, had the senior management at Enron acted, the company might have been saved.

The story of the Enron collapse is likely to run for many years. There are dozens of lawsuits pending against the company and those who ran it. The criminal investigation is still ongoing but, in August 2002, Michael Kopper, described as Fastow's right-hand man, pleaded guilty to money laundering, fraud and conspiracy charges (*Daily Telegraph*, 22 August 2002). In May 2006, Enron's corporate officers, Kenneth Lay and Jeffrey Skilling were found guilty of fraud and conspiracy. Sentencing is due to take place on 11 September 2006.

Enron followed the standard practices that most companies use in ensuring that their finances are presented in the best possible light, however, they then stretched those practices to such an extent that it appeared that the company might be engaged in fraud.

Self-inflicted crises **4.4**

Many organisations suffer from crises that are not of their own making. This includes those resulting from:

- natural disasters such as hurricanes, floods and earthquakes;
- acts of terrorism;

- industrial sabotage;
- accidental damage; and
- market forces.

But some, such as Enron, appear to suffer from crises that are self-inflicted. Consequently, in *Crisis Management for Managers and Executives* (1998) Financial Times Management, p 1, Robert Heath points out that:

> '. . . managers and executives need to acquire the concepts and capabilities of crisis management and then need to incorporate such skills within their everyday responsibilities and activities so that their organisations and their own jobs have a chance to survive when things go wrong'.

All crises are likely to create a huge impact and may have severe repercussions on a community. For instance, most Enron employees had company stock; many were depending on this stock to enable them to lead a relatively comfortable life following their retirement. Now it is worthless. When people are killed the impact and the repercussions are likely to be even greater – it will certainly hit the headlines of the world communication networks – and it can carry very significant immediate and long-term costs.

Meaning of 'organisation' 4.5

The word 'organisation' has been used throughout this chapter. This should be taken to include international organisations, such as the United Nations, regional organisations, such as the North Atlantic Treaty Organisation, governments, companies, industries and the suppliers of services. Similarly, although Chief Executive Officer (CEO) and senior managers has been used, this also includes the officers of international and regional organisations, and senior members of government.

What is a Crisis? 4.6

Many people have attempted to define crisis; some have been more successful than others but it seems no one has come up with a generally accepted definition. However, Lagadec points out that unless decision makers concern themselves with defining crisis 'they will severely limit their capacity to think about crisis and, consequently, to act' (Lagadec (1991), p 24).

In 1972, Hermann described a crisis as 'a situation that threatens high priority goals of the decision making unit, restricts the amount of time available for response before the decision is transformed, and surprises the members of the decision-making unit by its occurrence' (Hermann CF, *Some Issues in the Study of International Crisis* (1972) New York Free Press, p 13).

Writing in 1976, Randolph Starn claims that the term crisis:

> '. . . has a long history, and considering its many varied uses, no one can expect its meaning to be unequivocal. As all journalists and politicians know, crisis is a very useful term in the age of mass media. It suggests drama and the need for decisions, creates emotion without requiring sober reflection, and magnifies the importance of both non-events and events, coup d'états and minor incidents' (quoted in Lagadec (1991), p. 26).

In 1989, Rosenthal and his team defined it as 'a serious threat to the basic structures or the fundamental values and norms of a social system, which – under time pressure and highly uncertain circumstances – necessitates making critical decisions' (Rosenthal, Uriel; Michael Charles and Paul t'Hart (eds) Coping with Crisis: The Management of Disasters, Riots and Terrorism (1989) Charles C Thomas, Springfield, Illinois, p 10). Later, in 1991, Rosenthal and Pijnenburg describe the concept of a crisis as relating to 'situations featuring severe threat, uncertainty, and a sense of urgency' (Rosenthal, Uriel and Bert Pijnenburg (eds) *Simulation Oriented Scenarios: In Crisis Management and Decision Making* (1991) Kluwer Academic Publishing, Norway, p 3).

In 1993, Barton suggested that a crisis is 'a major, unpredictable event that has potentially negative results'. He went on to state that 'the event and its aftermath may significantly damage an organisation and its employees, products, services, financial condition, and reputation' (Barton L *Crisis in Organisations: Managing and Communicating in the Heat of Chaos* (1993) South Western, Ohio, p 2).

In 1995, Clark defined a crisis as 'any unplanned event that can cause death or significant injuries to employees, customers or the public; shut down the business; disrupt operations; cause physical or environmental damage; or threaten the facility's financial standing or public image' (Clarke J, 'Hope for the best, but plan for the worst – the need for disaster planning' (1995) Employment Relations Today, Vol 22, No 4).

A more recent definition appeared in an advisory document issued by the North Atlantic Treaty Organisation to all those countries seeking admission. In it, crisis was defined as 'a national or international situation in which there is a threat to priority values, interests or goals of the parties involved'. As is to be expected, this definition is aimed at the political and military spectrum rather than business and commerce.

Characteristics of a crisis 4.7

It being very difficult to decide on a generally accepted definition of crisis, it may be better to look at the characteristics of such an event. Whilst every crisis is different there are often shared characteristics, namely:

- there is an element of surprise;
- there is a perceived or real loss of control, particularly during the early stages;
- there are no immediate obvious solutions;
- there is a shortage of time;

- the events outpace the response by the organisation, at least during the early stages;
- there is an escalating flow of events that become more intensive;
- there is insufficient information when it is most needed; occasionally there may be too much information;
- important interests are at stake such as a threat to resources or people;
- there is a lack of resources, at least during the early stages;
- there is intense scrutiny from the outside;
- the key players develop a siege mentality;
- there is panic;
- the regular decision-making processes are disrupted;
- there is an urgent need for rapid decision making, particularly in the short-term;
- affected managers focus on short-term planning/decisions/actions.

When does a crisis occur? 4.8

A crisis occurs when the media and/or a National Parliament and/or some other credible or powerful interest group(s) identify it as a crisis. A crisis need not pose a serious threat to human life, but it must somehow challenge the public's sense of appropriateness, tradition, values, safety, security or the integrity of the Government. Often the truth of whether a crisis is actually taking place is irrelevant; it is what the media and/or the people perceive to be is happening that needs to be addressed.

Types of Crisis 4.9

In 2000 and 2001, the UK, or at least a significant part of it, suffered from a series of crises. The period commenced with a fuel crisis in the autumn of 2000, immediately followed by severe flooding in various parts of the country. And then, in March 2001, the country was hit by the Foot and Mouth crisis.

The Foot and Mouth crisis in the UK 4.10

Case study 2

In February 2001, the contingency plan in place within the Ministry of Agriculture, Food and Fisheries (MAFF), anticipated that any outbreak of food and mouth amongst cattle and sheep would involve officials in dealing with no more than ten infected premises. By the time the initial diagnosis was made on 23 February at least 57 premises were infected and, by the time the last diagnosis was made on 30 September 2001, over 2,000 cases had been detected. ➡

It is estimated that the overall cost of the crisis was at least £8 billion. Ten million animals were killed and compensation, amounting to £1.3 billion, was paid to farmers.

In making 80 recommendations for the future handling of such an epidemic, the subsequent report into the crisis suggested, amongst other things, that:

- the Government's Chief Veterinary Officer had been warned in an internal report in 1999 that MAFF would be overwhelmed if the disease broke out in several places at once;
- MAFF, as a whole, had been incompetent. The Minister was eventually removed from office;
- it had taken the Prime Minister 31 days to activate the Cabinet Office crisis management machinery, by which time 471 cases had been confirmed;
- failing to call in the military at the start was a major mistake. One of the main recommendations from a serious outbreak in 1967/68 was to enlist military assistance at once. However, the military were not called in until the 25th day.

In an editorial, *The Times* newspaper suggested that 'with proper and widely circulated contingency plans, speedier decisions and better information, better shared, the emergency need not have turned into a crisis' (23 July 2002). The Inquiry chairman reported that he had detected 'a culture predisposed to decision making by committee' within MAFF 'and an associated fear of personal risk-taking' (reported in *The Times*, 23 July 2002).

Levels of crises **4.11**

The types of crisis are extremely varied and will differ depending on the level at which it occurs and the type of organisation affected. For instance, at regional level, it could include:

- political instability;
- religious and racial conflict between countries;
- the collapse of world financial markets.

At State level it could include:

- a threat to territorial integrity;
- political instability within the State itself;
- political instability or destabilising factors in neighbouring states;

- religious and racial conflict between groups within a State;
- excessive national debt;
- external economic sanctions;
- currency devaluation;
- acts of terrorism;
- serious outbreak of civil disorder;
- organised crime;
- large influx of refugees; and
- large-scale corruption.

Types of incidents leading to crises 4.12

In a 1997 survey undertaken by The Corporate Response Group (quoted in Heath (1998), p 2), a number of companies identified the following as incidents or events that could lead to crises:

- workplace violence (55%);
- kidnap (53%);
- terrorist action (51%);
- fraud (35%);
- product tampering/recall (34%);
- ethics (30%);
- succession of Chief Executive Officer (28%);
- racism-sexism litigation (26%); and
- takeovers (20%).

None of the above lists are exhaustive. What is clear, however, is that unstable conditions or events can lead to many different types of crisis. Very often, a crisis can be a mix of types. For instance, one involving organised crime can have economic implications. Alternatively, an economic crisis, such as a rapid devaluation of the currency, can have social and political implications that might lead to strikes and street protests. Similarly, the discovery of fraud within a company could lead to a change of Chief Executive Officer or, at its worse, send the company into bankruptcy.

Crisis Life Cycle 4.13

The procedures and activities required in order to mount an effective response to such events include information acquisition and assessment, analysis of the situation, establishment of goals to be achieved, development of options for action and comparison, implementation of chosen option(s), and finally, as feed-back information closes the loop, analysis of the reaction of the parties involved. All this goes to make up what is known as the crisis life cycle illustrated in Figure 1 below.

Figure 1: Crisis life cycle

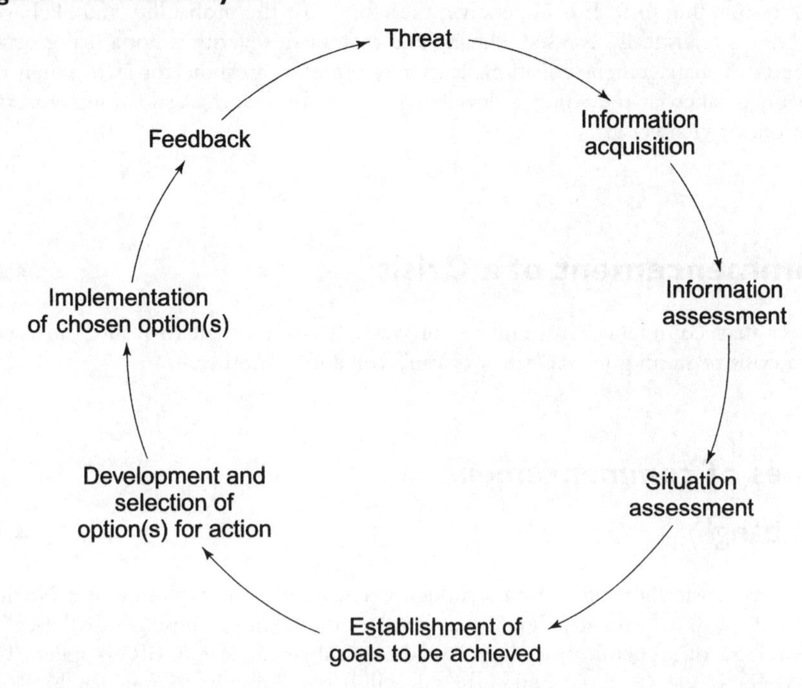

Crisis Severity **4.14**

The severity of a crisis is defined as the size and the immediacy of a threat to priority values, interests or goals. Thus, it relates to the perception of how important the crisis is to an organisation. Qualitatively it can be shown as illustrated in Figure 2 below.

Figure 2: Crisis severity as a function of size and immediacy of threat

Size of threat	**Small**	**Medium**	**Large**
Immediacy of threat	*Severity*	*Severity*	*Severity*
Imminent	Low	Medium	High
Intermediate	Low	Low/medium	Medium
Distant	Low	Low	Low

The severity of a crisis may stay roughly the same throughout. On the other hand, it may change with time, as the crisis progresses, perhaps as a result of actions taken by the parties involved in the crisis. As can be seen from Figure 2 above, a crisis only has high severity when the threat is large and its immediacy is imminent. In many cases, such judgements are likely to be subjective but the concept of crisis severity does matter, in the sense that an organisation must assess the importance of a crisis when responding to it. It is imperative, therefore, that the probability and likely size of a crisis is realistically assessed. Having said that, crisis severity is not a major driver of the crisis management equation. It merely serves as an indicator as to when the decision is taken at the strategic level to move from a normal situation to a crisis situation, or vice versa.

Commencement of a Crisis **4.15**

A crisis may commence in a number of ways. It is important to realise, however, that a crisis beginning in one category can evolve into another.

Types of commencement

'Big bang' **4.16**

A crisis is classically triggered by a sudden event such as an explosion at a Nuclear Power Plant or a terrorist incident that involves the taking of hostages or the deliberate release of a chemical, biological, radiological or nuclear (CBRN) agent. The emergency services and receiving hospitals will become involved immediately. What may not be so immediately obvious, however, are the long-term health implications.

'Tidal wave' **4.17**

The organisation is suddenly hit by an avalanche of problems. Difficulties pile up and combine to make effective management almost impossible – at least during the early stages. There is likely to be insufficient resources, and communications may be ineffective or, at the very least, difficult.

'Rising tide' **4.18**

The problem creeps up gradually, such as occurs in the case of organised crime, corruption, a developing infectious disease epidemic or the receipt of a steady stream of refugees into a country. There is no clear starting point for the crisis and the point at which it becomes a crisis may only be clear in retrospect.

'Cloud on the horizon' **4.19**

A crisis in one place may be in danger of spreading, thus seriously affecting others. Preparatory action is needed in response to an evolving threat elsewhere, such as a major chemical accident in a nearby country, or the collapse of a currency to which other countries are closely aligned, or a breakdown of law and order in a country that an organisation is dependent on for raw materials.

'Headline news' **4.20**

A wave of public or media alarm in reaction to a perceived threat may cause a crisis for a country or an organisation, even if fears proved to be unfounded. The issue itself may be minor in terms of actual risk. It is, therefore, essential to manage the information that is causing the crisis in order to reduce the threat.

'Internal incidents' **4.21**

The Government of a country or the Board of Directors of an organisation may be affected by its own crisis or by an external crisis that impacts on its ability to function normally. A disagreement between coalition partners or an attempt by a dissident group to take control of a government building or an industrial dispute or an attempted take-over, may paralyse the provision of certain services and jeopardise safety arrangements in the short term.

Baring's Bank **4.22**

Case study 3

A merchant bank provides finance and advice for its clients but also trades on its own account as a 'merchant'. In other words, it takes risks by buying or selling land or stock like any other trader. Baring's Bank, founded in the City of London in 1763, was the world's first such bank.

In June 1992, Nick Leeson was appointed General Manager of Baring Futures Singapore. Because of his previous activities in sorting out problems in Jakarta, where, by diligent work, he reduced a £100m hole in the balance sheet to £10m, he was considered to be trustworthy and dynamic. By February 1995, however, he had lost millions of pounds, had left the trading floor in Singapore and vanished. The damage done by this single 'rogue trader' as he became known, was sufficient to cause the total collapse of Baring's Bank. (See also Leeson N *Rogue Trader* (2001) Warner Books.)

➡

As so often happens, attempts were made to lay the blame on the person nearest to where the crisis occurred. Thus, when the bank did collapse in 1995, Leeson was blamed, particularly by the senior managers within the bank. Others thought differently. As the losses mounted in Singapore, money was transferred from London without question. Leeson had no clear reporting lines. In fact, the structure of the bank was such that he reported to four different people and was able to play one off against the others and vice versa as the losses mounted.

In its report into the collapse of the bank, the Bank of England asked two key questions:

- How were such massive losses incurred?
- Why was the true position not noticed earlier?

Their conclusions were that clearly Leeson had incurred losses as a result of unauthorised and concealed trading activities. But more significantly, the Bank of England found that this went undiscovered 'by reason of a serious failure of controls and managerial confusion within Baring's Bank' and that the unauthorised dealings and subsequent losses 'had not been detected prior to the collapse by the external auditors, supervisors or regulators of Baring's'.

Crisis Management 4.23

Darling describes crisis management as 'a series of functions or processes to identify, study and forecast crisis issues, and set forth specific ways that would enable an organisation to prevent or cope with a crisis' (Kash & Darling (1998), p 179). In other words, it involves measures to identify, acquire, and plan the use of resources needed to anticipate, prevent and/or resolve a threatened or actual crisis.

Consequence management, on the other hand, involves measures to alleviate the damage, loss, hardship or suffering caused by crises. This includes measures to protect the company assets and image, restore services if they have been disrupted, and provide support to those subsidiary organisations and individuals that may have been affected.

In practical terms, crisis management can be broadly understood to involve the organisation, arrangements and measures that aim to:

- bring the crisis under the control of the crisis managers; and
- permit the crisis managers, through their actions, to shape the future course of the crisis and thereby bring about an acceptable solution.

Purpose of crisis management 4.24

Clearly, every organisation requires a crisis management system, the purpose of which is to:

- prevent a potential crisis from developing into an actual crisis;
- bring the crisis under the control of the crisis managers;
- permit the crisis managers, through their actions, to shape the future course of the crisis and thereby bring about an acceptable solution.

An acceptable solution is one that returns the situation to normal, with the minimum number of casualties and damage to both property and the environment and, at the same time, retain the organisation's security, integrity and ability to operate. Casualties include people who may be physically hurt, even killed, as well as those who are affected economically or financially. Damage to property includes damage in its widest sense.

Objectives of crisis management 4.25

The objectives of a crisis management system are:

- to co-ordinate information exchange within an organisation and, if appropriate, between organisations, with a view to preventing any crisis from occurring.
- to establish an effective management structure to respond to a threatened or actual crisis; and
- to ensure co-operation between all relevant participants in the field of crisis management.

Management framework 4.26

The management framework for any crisis should always embody the same principles irrespective of the nature of the threatened or actual crisis but it must remain flexible to individual circumstances. To be effective, it must be planned in advance and the subject of focused training and regular exercising.

In the event of a crisis in the UK involving the emergency services and local government emergency planning departments, a three-tier system of command is generally adopted. This is summarised below.

Strategic level 4.27

The strategic level is commonly known as Gold in many, but not all, parts of the UK. In business terms this is likely to be a small group of senior managers, headed by the Chief Executive Officer. Strategy can be defined as the overall plan to combine and direct resources towards managing a potential crisis, and deal effectively with it should it occur. The purpose of the strategic level, therefore, is to establish a framework of policy within which the other two levels of the system will operate.

Tactical level 4.28

The tactical level is commonly known as Silver. In business terms this is likely to be a middle manager. Tactics can be defined as the method of actual deployment and redeployment of resources in order to achieve the overall plan. Therefore, the responsibility of those at tactical level is to determine priority in the allocation of tasks, to plan and co-ordinate when a task will be undertaken and to obtain resources as required.

Operational level 4.29

The operational level is commonly known as Bronze. Those at the operational level will either have responsibility for a function or an area.

Emergency services 4.30

Often the emergency services will be involved in crises that involve commercial organisations. Examples from the UK during the last twenty years include the fire at King's Cross Underground Station, the destruction of the Piper Alpha Oil Rig, a number of transportation accidents involving aircraft and shipping, and numerous bomb explosions arising from terrorist activity. Commercial organisations, indeed any organisation that may be affected by a crisis that could involve the emergency services, are therefore advised to build their crisis management structure along similar lines. This will enable them to link in with the emergency services with the minimum of delay, thereby, hopefully, providing a more integrated and effective response.

Integrated crisis management arrangements 4.31

The response to a crisis may well be beyond the capabilities of any one department within an organisation; in may cases, it may well be beyond a single organisation. So, whilst it is necessary for each organisation, or department within an organisation, to perfect its own role to a high level of competence this, in itself, is insufficient to ensure an effective response. Only through the joint efforts of all organisations, and departments within each of those organisations, will this be achieved. But mounting an integrated response, as it is known, will not occur unless all the responding agencies have liaised during the planning phase and taken part in joint training exercises.

Concepts of integrated crisis management arrangements **4.32**

Integrated crisis management arrangements embrace four concepts, some of which overlap:

- The emphasis in the development of any plan must be in response to the crisis and not the cause of the crisis. Planning arrangements for a range of crises, whether caused internally or externally, must be integrated. The plan has to be flexible and will need to be tested against specific scenarios.
- Crisis management plans should ideally be built on routine arrangements and it is, therefore, essential for those who will respond to any crisis to be involved in the planning process and subsequent exercises. As far as it is possible, the arrangements should be integrated into an organisation's everyday working structure.
- The overall response to a crisis will invariably need input from a number of departments within an organisation. Effective planning must integrate these contributions and establish protocols in order to achieve an efficient and timely response to a threatened or actual crisis. Failure to adopt this principle is likely to lead to a muddled and ineffective response.
- Many crises are likely to involve more than one organisation. If the response is to be truly effective in meeting the needs of everyone caught up in the crisis, then it is essential that co-ordinating arrangements are in place.

Actions required **4.33**

In addition, a crisis requires specific and often rapid actions. These include:

- alerting all relevant personnel to the threatened or actual crisis;
- the immediate formation of the crisis management team;
- taking immediate action in an attempt to prevent the situation from worsening;
- opening all lines of communication to ensure that data is being collated and suitably analysed so that relevant information can be passed to the decision-makers;
- separating the management of the crisis from everyday operations.

Crisis management activities **4.34**

There are six broad crisis management activities:

Activity 1: Situation monitoring.

Activity 2: Crisis detection support.

Activity 3: Containment.

Activity 4: Response.

Activity 5: De-escalation.

Activity 6: Recovery.

These activities are discussed at **4.35** to **4.40** below.

Activity 1: Situation monitoring **4.35**

The main activity is routine situation monitoring; this includes the gathering of routine data. It involves the processing of that data in order to build up a picture of events as they occur. It involves the examination of that picture in order to determine whether anything untoward either has occurred or is occurring. This is an ongoing activity throughout the year and throughout all phases of the crisis

Activity 2: Crisis detection support **4.36**

The activity involves the use of methodologies to examine the routine situational monitoring undertaken during Activity 1. The aim is to identify events and trends that suggest something untoward may be happening. When suspicion is aroused, there needs to be a positive focus on analysing data in order to produce information that will confirm whether or not a threat exists. This activity will include the presentation of that information to senior managers within the organisation and to the crisis management team if it is formed.

Activity 3: Containment **4.37**

This activity occurs when a threat to high priority goals or values has been identified. There should be a focus on the acquisition of information aimed at understanding the causes of the threatened crisis. It involves the explicit setting of organisational objectives to meet the emerging crisis. Alternative strategies/options need to be identified, assessed and, if appropriate, selected. Following the selection of a strategy, with its appropriate options, a plan should be prepared for implementation. The plan needs to be communicated to those who will be required to undertake appropriate action.

Activity 4: Response **4.38**

This activity consists of the actual response. It involves the implementation of the strategy and options in accordance with the plan developed under Activity 3. The main activities are likely to be carried out at the Tactical and Operational levels with the Strategic level keeping a watching brief. There must be a system of assessing and evaluating the effect of the action being taken and, if it is not producing the desired results, it may be necessary to revise the strategy/options originally taken. At the same time, the Strategic level should be planning the actions to be taken under Activities 5 and 6.

Activity 5: De-escalation **4.39**

Hopefully, there will come a time when the threat to high priority goals or values begins to recede and the intensity of the crisis will start to drop. By the time this stage is reached, the Strategic level should already have defined how the post-crisis situation will be handled. It involves setting up and agreeing a timetable for the return to normality. It involves an assessment of risk at every stage in the implementation of this activity, together with plans of the action to be taken if there is a reversal of the de-escalation activity and a resurgence of the original crisis.

Activity 6: Recovery **4.40**

The situation then reduces even more in its intensity to a level that is regarded as stable or normal. This may be higher than it was originally, but it is the best that can be expected, or it may be lower than it was previously which is all to the good. The criteria for this stage is the removal, disappearance or hibernation of the threat to high priority goals or values that caused the crisis in the first place. The activities here involve discussion with all the agencies involved to establish mutually acceptable actions and a timetable to support the declared post-crisis strategy. Whilst, under certain circumstances, unilateral action may be necessary, a detailed assessment of the risk and consequences of such an approach must be taken.

Planning **4.41**

Augustine suggests that there are two forms of crisis, 'those that you manage, and those that manage you' (quoted in Kash & Darling (1998), p 180). In far too many cases, organisations are managed by the crisis. A major reason for this is that organisations fail to acknowledge the possibility that a crisis will occur.

Proactive planning is essential in assisting managers to prevent, control and resolve crises. The growth in the number of crises in today's complex world demands that organisations must meet the events as they unfold according to a well-conceived plan; if the plan is poor it will become increasingly difficult to rectify mistakes. To ensure that everyone operates within the same framework there must be, at the outset, a clearly defined policy with clearly stated objectives. No Chief Executive Officer should allow his or her staff to 'drift' aimlessly into a threatened or actual crisis or become involved piecemeal. The principle underlying any plan must be to bring to bear, in a co-ordinated team effort, as many resources as appear necessary to either prevent the crisis or, if it actually happens, to bring it under control as quickly as possible.

Stage management of events 4.42

The proper stage management of the events is vital. There are numerous things to consider when developing a plan and too many to list them all here. However, some of the more commonly overlooked ones are:

- The crisis should never be under-estimated. There are distinct advantages in attempting to foresee where and in what form the crisis is likely to occur. Whilst such foresight might not be completely accurate, senior managers should use their imagination and consider all possibilities. Pre-vision is essential in formulating the plan.
- There must be a flexible response capability. This means the ability to respond to each level of threat with just as much reaction as is necessary to achieve an objective – no more, no less. Over reaction or under reaction can exacerbate the threatened or actual crisis.
- There must be a proper command structure.
- The tactical deployment of staff must be balanced.
- Different things matter at different levels. Every manager must be quite clear as to what matters at his or her own level.
- The plan should be kept simple and be capable of being communicated right down to the lowest level of the organisation.
- Failure leads to a lack of confidence; success leads to a growth of confidence.

If the principles of crisis planning are carefully followed it should enable managers to respond positively to meet the crisis as it emerges.

Response 4.43

A variety of practices have been employed successfully in crisis management. Indeed, many organisations have not only survived crises but have enhanced their public and professional images in the process. As for any major management challenge, those responsible for undertaking this particular activity must work as a team, operating within the clearly understood guidelines prescribed in the plan.

Managers must be prepared to participate in and focus on the problem at hand, to know what is expected of them. A crisis is exacerbated if there appears, even briefly, to be confusion or nobody in control. In short, a crisis is a time for exceptional management, not panic, or a suspension of good management practices.

Selection and Training 4.44

It is a fallacy to believe that every manager within an organisation will make a good crisis manager. In the same way that there are those who have a talent for administration and procedures, organisational planning (as opposed to implementation), computer or communication systems, sales or the successful everyday management of an organisation or government department, there are those who have a similar talent for managing a crisis. History is littered with cases were people have been thrown into a crisis, sometimes with extremely serious consequences both to the organisation itself, its staff and to the individuals themselves.

A brief outline of the Piper Alpha disaster is given at **11.21** below. Suffice to remind readers that of the 226 men on board that fateful night, only 62 survived. In his report into the disaster, the Inquiry Chairman, Lord Cullen, concluded:

> 'The failure of the OIMs (Offshore Installation Managers) to cope with the problems they faced on the night of the emergency clearly demonstrates that conventional selection and training of OIMs is no guarantee of ability to cope if the man himself is not able in the end to take critical decisions and lead those under his command in a time of extreme stress. Whilst psychological tests may not appeal to some companies, the processes used and proven successful by the armed forces or the Merchant Navy, who have to rely on their officers to lead under stress, should be seriously considered by operating companies' (paragraph 20.29).

Competence **4.45**

In her book, *Sitting in the Hot Seat* (1996), Wiley & Sons, Rhona Flin points out that 'an integral part of operational efficiency' at the time of a crisis 'is the use of formal competence assessment of key individuals or teams' (pp 54–61) prior to the crisis. But, in order for this to take place, organisations need to document the standard of competence they require in those staff who are likely to make up the crisis management team.

The selection and training of the crisis management team is crucial to the successful management of a crisis. It is essential that those who will be required to act at the strategic, tactical and operational levels are identified well before any crisis. The most capable managers involved in running government or an organisation on a day-to-day basis are not necessarily those who will stand up best to either a short-term or long-term crisis To this end, the Chief Executive Officer (CEO) needs to identify those managers, at their various levels, best suited to being part of the crisis management team.

Experience shows that the events for which a plan has been designed never go exactly as expected. Indeed, the military have a saying, 'no plan survives first contact with the enemy'. It is the same with any crisis. No matter how carefully an organisation has planned, something will always occur, invariably during the early stages, that no one has thought of. Sometimes there are a limited number of options and there may be serious time constraints. Therefore, plans need to be modified as the events unfold. As a result, there is likely to be a need for quick decisions. Inaction or inappropriate decision-making can produce undesirable consequences and may have far-reaching implications.

It is essential that all those who will be part of the crisis management team have confidence in the training they undergo if they are to be an effective unit. In order to achieve this, the CEO, if he or she is to be part of the team, and other senior managers must train with them so each becomes aware of the other's capabilities and methods of operation.

Crisis Management and Public Relations 4.46

It is often suggested that every crisis contains within it opportunities for success as well as the roots of failure. Any crisis is likely to be an immediate focus of both public and media attention. Both will tend to react with a predictable sameness. The five key questions that invariably provide the focus of their attention are:

- Who is to blame?
- When did the Government or the organisation discover the problem?
- What did it do before it became a public crisis?
- What is it doing now that it is a crisis?
- How can the interests at stake be protected and/or compensated?

Successful crisis management is about recognising the potential for success. Those likely to be affected by the crisis, and the media, need to be reassured from an early stage that:

- everything was in place in an effort to prevent the crisis in the first place;
- the organisation had plans in place to deal with the situation if they were unable to prevent it;
- the organisation cares about what has occurred and it is doing everything it can to remedy the situation.

The media 4.47

The media will demand information as soon as the crisis becomes public knowledge. A crisis communication plan is therefore essential. This will, hopefully, ensure that the organisation has a positive, pro-active attitude towards the media. This was not the case in the UK Foot and Mouth crisis in 2001 where the Government was continually on the defensive. As the Foot and Mouth crisis was becoming more serious, the Minister for Agriculture, Food and Fisheries was insisting that he was 'absolutely certain' that the disease was under control. Clearly, to most people it was not. As *The Times* reported immediately after the publication of the Inquiry:

> 'The billions-worth of damage wreaked by foot-and-mouth extended far beyond farms. The less tangible damage done by evasions and half-truths compounded misery and betrayed the public trust' (23 July 2002).

The public perception of the Government's ability to handle any crisis was seriously undermined, not only because the public relations side of the response was badly managed but because government representatives continually gave out misleading and unhelpful information.

Operational Reality **4.48**

By learning from the experience of others, managers in all the various grades can draw historical lessons from the past. As a result, they are better able to draw lessons for the future and base any positive or critical analysis on profound professional knowledge.

By simulating a crisis, managers should be able to put some of their theories to the test and make 'real world' decisions in a relatively safe environment, without having to concern themselves unduly with the consequences of those decisions. The advantage of this is that it encourages managers to experiment with different strategic and tactical options. The danger, on the other hand, is that it is never possible to recreate the events that are likely to unfold and the participants in the simulation may get a false impression of operational realities.

Checklist **4.49**

There are numerous activities that need to be carried out before and during a crisis. Some of the principal ones are illustrated below.

Crisis management checklist

- Identify and list potential crisis situations by carrying out a risk analysis.
- Devise strategies for their prevention.
- Formulate strategies and tactics for dealing with each potential crisis.
- Identify who and what are likely to be affected by them.
- Devise effective communication channels to ensure minimum damage to the organisation.
- Select and train the crisis management team.
- Provide a command and control centre.
- Ensure that there are sufficient telephone lines to cope with the floods of additional incoming calls that the crisis will inevitably generate.
- Test everything.

- When faced with a crisis, consider the worst possible scenario – and act accordingly.
- As soon as it appears that a crisis is imminent, establish a command and control room and call the crisis management team together.
- Be prepared to demonstrate human concern for what has happened.
- Be prepared to seize early initiatives by rapidly establishing the organisation as the single authoritative source of information about what has gone wrong and the steps that are being taken to remedy the situation.
- Look for ways of using the media as part of the organisation's armoury for containing the effects of the crisis.
- Know your audience, listen to them, and ensure you have a clear picture of their grievances.
- Get opponents on your side by getting them involved in resolving the crisis.
- In communicating about the crisis, avoid the use of jargon. Use language that shows you care about what has happened and which clearly demonstrates that you are trying to put matters right.
- When the dust has settled, look to see what lessons have been learned, both for the organisation and the rest of the world.

Conclusion

4.50

According to Kash and Darling:

> '. . . it is no longer a question of 'if' a business will face a crisis; it is, rather, a question of "when", "what type" and "how prepared" the company is to deal with it. Whether it is a natural disaster, such as an earthquake, tornado or flood, or a man-made disaster, such as accidents, wildcat strikes or product tampering, a business will eventually face some form of crisis.' (Kash & Darling (1998), p 179).

The key role of a crisis management team must be to ensure that, as far as possible, a crisis does not occur. But can a crisis actually be managed? Coral Bell suggests that the term 'crisis management' is rather worrying, because use of the word 'management' implies 'rational, dispassionate, calculating, well-considered activity, conducted with judgement and perhaps even at a leisurely pace with a view to long term as against short term interests'. However, 'actual decision making is not like that: it is improvised at great pressure of time and events by men working in a fog of ambiguity' (Bell C *Decision Making by Governments in Crisis Situations* (1978), pp 50–52). Nevertheless, Lagadec points out that 'the dynamics of a crisis should be analysed and practical rules devised for its management' (Lagadec (1991) p xxxi).

Despite Bell's reservations, the short answer is 'yes' providing it is accepted that:

- crises are inevitable;
- like any other management challenge, crises should be mitigated against and prepared for;
- crisis management is an integral part of every manager's responsibility.

Essential elements in dealing with a crisis 4.51

There are three essential elements for success in dealing with any crisis:

- proper launching of the response as soon as the potential or actual crisis has been identified;
- the courage, initiative and skill of the managers within the crisis management team;
- the professionalism and morale of the people who are required to take appropriate actions.

If any of these three elements are missing, the situation may well be lost from the outset.

Selection and training 4.52

Proper launching will only occur if the right people have been selected and trained, if the planning has anticipated the threatened or actual crisis, and there is an appropriate and timely build-up of resources. Managers in the crisis management teams at all three levels, strategic, tactical and operational, must:

- have a good knowledge of the techniques of staging the many and varied response activities that a crisis may require;
- understand and be able to stage manage the situation;
- have the ability to concentrate on the tasks in hand;
- be capable of thinking quickly in order to make rapid decisions because, on occasions, depending on the crisis, there may be little or no time for lengthy consultation when the crisis reaches a crucial state; and
- have the ability to see things with clarity and have a flair for improvisation when it is appropriate.

Involvement of senior management **4.53**

The CEO and other senior managers should never blame their staff for deficiencies in their ability to respond to a crisis. Example is the most powerful of all preceptors and the CEO and other senior managers will find that what they do not themselves observe, with regard to planning, training and conduct, will not be acted upon to any successful degree by those who work for them. It is essential, therefore, that they involve themselves in the whole process of crisis management so that their staff have confidence in their ability to lead them successfully in such times.

5

Disaster and Emergency Management Systems (DEMS)

In this chapter:

- Historical development of Disaster and Emergency Management.
- Origin of Disaster and Emergency Management as a modern discipline.
- The need for effective Disaster and Emergency Management.
- Disaster and Emergency Management systems.
- Organise for Disasters and Emergencies.
- Disaster and Emergency planning.
- Monitor the Disaster and Emergency Plan.
- Audit and Review.
- Conclusion.

Historical Development of Disaster and Emergency Management 5.1

The events of 11 September 2001 in New York (see also **3.39** above) has seen a paradigm shift in the thinking, planning and perception of 'man-made disasters'. Throughout the 1990s the Federal Emergency Management Agency (FEMA) of the USA (a competent authority dealing with natural and man-made disasters) was being criticised by Congress and the media for being too resource consuming and inefficient. After '9/11' it was being hailed for its expertise in emergency response management. Suddenly 'corporate America' and the democratic world began to take more seriously the importance of Disaster and Emergency Management, Civil Defence Management, Civil Protection Management and Business Continuity. Since June 2001 we have seen the 'civil contingencies' function being transferred from the Home Office to the Cabinet Office, giving the Prime Minister's Office a greater strategic oversight of national 'major incidents' (similar to many other NATO member countries).

The UK experience broadly indicates that Disaster and Emergency Management (DEM) has evolved in three phases. These are summarised below.

Phase I

5.2

Pre the *Control of Industrial Major Accident Hazards Regulations 1984* (*SI 1984 No 1902*) (*CIMAH Regulations*). DEM was largely confined to Local Government, the emergency services, 'large' organisations and Civil Protection Agencies (environmental, military etc). This phase is marked by large-scale macro plans and detailed, sequential procedures to be followed in the event of a Disaster or Emergency. This is exemplified in the guidance prevalent on 'Incidents' involving radioactivity available from Central Government, the United Nations, the Red Cross and others.

Phase 2

5.3

This is known as the liberalisation phase – with the advent of decentralisation of industry and the growth of privatisation, organisations (large or otherwise) began to focus on 'Procedures' and 'Business Continuity Planning'. British Telecom exemplified this, with the establishment of its Disaster Recovery Unit in addition to offering its expertise to customers at large. Major Incidents such as Chernobyl, Piper Alpha, Kings Cross (see also **20.1** below), The Marchionness, and Challenger etc, which hit the world in the 1980s also focused the minds of planners on effective Disaster and Emergency Management, not just isolated procedures.

Phase 3

5.4

Phase 3 is the holistic phase. This phase has two dimensions, first the role of Europe. The impact of the Framework and Daughter Directives, as transposed into the '6-Pack Regulations' of 1992, introduced for the first time explicit and strict duties on organisations in general to plan for 'serious and imminent dangers'. The European Commission also begins to take the risks of trans-national Disasters more seriously, as highlighted by the greater role given to the Civil Protection Unit, in Director General XI (DGXI) of the European Commission. Second, this phase sees domestic UK legislation becoming more cognisant of Disaster and Emergency issues. For example, the *Environment Act 1995* makes provisions for environmental emergencies; *Control of Major Accident Hazards Regulations 1999* (*SI 1999 No 743*) (COMAH), replaces the *CIMAH Regulations* (*SI 1984 No 1902*) (see **5.2** above) and the need for effective planning when carrying dangerous goods via road legislation is by virtue of the *Carriage of Dangerous Goods by Road Regulations 1996* (*SI 1996 No 2095*).

The most noticeable index of how DEM has changed is the availability of information and templates on Disaster Planning and Emergency Planning to organisations of all sizes. There is also a greater media focus on such issues, with the BBC, for instance, having dedicated web resources on disasters.

Origin of Disaster and Emergency Management as a Modern Discipline 5.5

Disaster and Emergency Management (DEM) is largely a fusion of four branches of knowledge. These branches are discussed below.

Occupational Safety and Health 5.6

Occupational Safety and Health (OSH) – that is 'Internal' or work-based causes of systems failures, with major impact on life and property. This has been exemplified by Heinrich in his *Unsafe Acts and Unsafe Conditions* in the early part of the 1900s, Bird and Loftus with their 'management failing' explanations of Accidents and Incidents and Turner with his 'Incubation' explanation of 'man-made disasters' in the 1970s. OSH Academics have led the forefront in terms of analysing the branch and root causes of major industrial disasters.

Security management 5.7

During the 1970s, security threats in the UK such as bomb explosions, terrorism and electronic surveillance failures have given insights to causation and the motivation behind man-made emergencies. Also, guidance from the Home Office as well as the emergency services has enabled a practical understanding of how to cope with emergency situations.

Business management 5.8

The late 1980s saw a shift in the academic paradigm in economics and business management thought from 'static' or closed business planning – where businesses were told to make the assumption of *cetaris paribus*, that is, assume all things are constant with the business acting as if it was the only one in the market place – to 'dynamic' or open planning. The latter sees uncertainty and risk being factored into decision-making models. This influenced the development of Business Continuity Management – developing strategies when the business faces major corporate uncertainty and crises as well as Contingency Planning.

Insurance 5.9

The fourth significant influence comes from insurance and loss control. The occurrence of Disasters and Accidents has involved loss adjustors and actuarial personnel. The former have developed methods of analysing the basic and underlying causes of an event whilst the latter have been developing statistical methods for calculating the chance of failure and the risk premiums needed to indemnify that failure.

Methods **5.10**

DEM uses both quantitative (statistics, quantified risk assessments, hazard analysis techniques, questionnaires, computer simulation etc) and qualitative methods (inspections, audits, case studies, etc).

Definitions **5.11**

Stan Kaplan (*The Words of Risk Analysis*, 1997, Risk Analysis, Vol. 17, No 4) stated that 50% of the problems with communication are due to individuals using the same words with different meanings. The remaining 50% are due to individuals using different words with the same meaning. In Disaster and Emergency Management this is a basic problem. Some authors have argued that distinguishing definitions has no practical value and can be '. . . highly undesirable to try to control how others use them . . .' ('Managing the Global Consequences of a Disaster', Richard Read, Paper at the 2nd International Disaster and Emergency Readiness (IDER) Conference, The Hague, October 1999, p 130 of IDER Papers). Up until 2004, there was much confusion over the meaning given to core terms. Consequently, it was necessary to go to *The New Oxford Dictionary of English* (1999) to provide the following primary meanings: *The New Oxford Dictionary of English* (1999) Oxford University Press, provide the following primary meanings:

- Catastrophe – 'an event causing great and often sudden damage or suffering: a disaster' (p 287).
- Crisis – 'a time of intense difficulty or danger' (p 435).
- Disaster – 'a sudden event, such as an accident or a natural catastrophe, that causes great damage or loss of life' (p 524).

A catastrophe and a crisis are types of disaster, namely more severe.

- Emergency – 'a serious, unexpected and often dangerous situation requiring immediate action' (p 603).

It should be noted that 'emergency' has now been defined by Section 1 of the Civil Contingencies Act 2004 (see Chapter 14, Annex A for the precise wording of the definition). Whilst both Disasters and Emergencies can be sudden, the former has a macro, large-scale impact whilst the latter requires an immediate response. A Disaster and Emergency is not the same thing, otherwise there would be little reason to have two concepts. Nevertheless, despite these definitions, there does remain some confusion over the precise definitions due to the use of other words in other legislation.

- Accident – '(1) an unfortunate incident that happens unexpectedly and unintentionally, typically resulting in damage or injury. . . , (2) an event that happens by chance or that is without apparent or deliberate cause' (p 10).

It can be noted that the *COMAH Regulations* (*SI 1999 No 743*) had introduced the term 'Major Accident'. This is an event due to:

(a) unexpected, sudden, unplanned developments in the course of the operation;

(b) leads to serious danger to people and the environment both on site at the place of work and off site;

(c) the event involves at least one dangerous substance as defined by the *COMAH Regulations.*

Given the potential for loss, both human and property and given the *COMAH Regulations* requirements for emergency on-site and off-site plans, there seems to be little practical difference between a major accident and an emergency – both are response-based concepts. However, if the legislators wished a 'major accident' to be different from an 'emergency' then they would have either said so or implied so. One can regard a major accident as a type of emergency situation.

- Incident – 'an event or occurrence' (p 923).

The *Reporting of Injuries, Diseases and Dangerous Occurrences Regulations 1995* (*SI 1995 No 3163*) *(RIDDOR)* does not define an accident or incident, rather gives a classification of the types.

- Major Incident – in Responding to Emergencies (Civil Contingencies Secretariat, 2005), a major incident is defined as any emergency that requires the implementation of special arrangements by one or more of the emergency services, the NHS or the local authority for the:

  (a) initial treatment, rescue and transport of a large number of casualties;

  (b) involvement either directly or indirectly of large numbers of people;

  (c) handling of a large number of enquiries likely to be generated both from the public and the news media, usually to the police;

  (d) need for the large scale combined resources of two or more of the emergency services;

  (e) mobilisation and organisation of the emergency services and supporting organisations, eg local authority, to cater for the threat of death, serious injury or homelessness to a large number of people' (p 43 – see above). This definition is accepted by the police, Fire Service, Local Government and, broadly, the NHS (they also have a specific definition of Major Incident).

Thus, the term 'Major Incident' is a broad phrase encompassing an array of events. 'Accident' has not been included as a type of Major Incident given the definition of the latter requiring 'major' mobilisation of human and physical resources, which in most Accidents, is not necessarily the case.

Examples of definitions **5.12**

(a) Two trains missing each other for example would be an Incident (Near Miss).

(b) An employee or member of the public being injured on a train would be an Accident, irrespective of the type of injury or fatality (following *RIDDOR Regulations* (*SI 1995 No 3163*).

(c) An event at a COMAH site where dangerous substances ignite causing damage to the plant, injury to personnel and emissions into the local community, would be a Major Accident under the *COMAH Regulations* (*SI 1999 No 743*).

(d) The immediate event after the collision and the response needed to the chaos – this would be an Emergency. For example, the response by the emergency services to Ladbroke Grove rail crash in 1999.

(e) If two trains collide causing multiple fatalities and immediate property and environmental damage, that would be referred to as a Disaster. For example, the Ladbroke Grove rail crash.

(f) If the collision, with the multiple fatalities and property damage is difficult to access, manage and control, this would be a Crisis. Ladbroke Grove fell short from being a Crisis in contrast to Clapham Junction in 1988 where a triple train crash caused major access and logistical problems.

(g) If the event generated major environmental, public and social harm that has 'longer term' implications, over and above the immediate human and property loss, this would be a Catastrophe. For example, Kings Cross Underground Fire (1987), Chernobyl (1986), Piper Alpha (1988) are some events that had wider consequences over and above the immediate impact.

Events (c) to (g) above would be Major Incidents.

Figure 1 below summarises the essential differences between the above events.

Figure 1: Classification of Incidents, Accidents and Major Incident types

			MAJOR INCIDENTS				
Event/ Characteristic	Incident	Accident	TYPES OF EMERGENCY		TYPES OF DISASTERS		
			Major Accident	Emergency	Disaster Crisis	Crisis	Catastrophe
1. MPL	Low						Very high
2. RISK:							
Severity	Near miss etc						Multiple fatalities
Consequence	Minor						Major
3. NUMBERS	1–5						100 +
4. SOCIO-LEGAL IMPACT	No change likely						New laws or guidance
5. TIME	Short						Longer impact
6. COST:							
Individual	Short term						Irreparable
Social	None usually						Irreparable
Environment	None usually						Irreparable

Key:

MPL = Maximum Potential Loss (economic and property loss).

Risk = Severity x Consequence (Severity refers to the quantum of harm generated by the event whilst Consequence measures the scale of impact).

Numbers = the number of individuals affected.

Socio-Legal Impact = the impact on social attitudes and legislation/guidance as a result of the event.

Time = the length of the event.

Cost = the loss suffered by the Individual, society or nature.

Such classifications are important, first from a philosophical perspective one needs to know how they differ. Second, from a planning perspective, as the resource allocation will accordingly differ. Third, from a response perspective – the response to an Incident differs from an Disaster, with the organisation needing to define and clarify when an event is an Incident and not a Disaster.

The Need for Effective Disaster and Emergency Management

Legal reasons 5.13

There is an array of Emergency-related legislation culminating in the Civil Contingencies Act 2004 and the Civil Contingencies Act 2004 (Contingency Planning) Regulations 2005 (see **CHAPTER 14: LAW RELATING TO EMERGENCIES AND DISASTERS**).

Insurance reasons 5.14

There are also insurance-based reasons for effective Disaster and Emergency Management. Given the positive correlation between risk and premium, the existence of DEM indicates hazard and risk control, consequently the premium ought to be less.

If insurers are dissatisfied with the DEM system in place then they may not insure the operation, in turn increasing corporate risk as well as reducing corporate credibility. It may also prohibit the operation from tendering for contracts.

Corporate reasons 5.15

The corporate experience of companies like P & O European Ferries (Dover) Ltd, indicates that proactive DEM systems would have saved the company considerable money, publicity and reputation. In March 1987, the *Herald of Free Enterprise*, the roll-on roll-off car ferry, left Zeebrugge for Dover, thereafter it sunk resulting in 187 deaths. The case against the company and the five senior corporate officers collapsed for because the court held they were not the 'controlling mind' nor the complete 'embodiment of the company' (see further **CHAPTER 14: LAW RELATING TO EMERGENCIES AND DISASTERS**).

Failure to have sound DEM systems can also have dire financial consequences as Disasters as disparate as Piper Alpha to the Challenger Space Shuttle indicate. In the former case, Occidental Petroleum were generating 10% of all the UK's North Sea output from Piper Alpha. Lord Cullen, who led the Inquiry said 'the safety policy and procedures were in place: the practice was deficient'. Piper Alpha showed that the company had ineffective Emergency response procedures resulting in persons being trapped, dying of smoke inhalation or jumping into the cold North Sea (something the Procedures prohibited but a significant number of those that did jump into the Sea survived).

Such a Disaster had consequences for Occidental's financial reputation, share value, ability to attract investment and growth potential. In contrast, in the case of the Challenger Disaster, when in January 1986, the O-rings that held the two segments of the rocket boosters, which carried the fuel to propel the Challenger into space, fell apart resulting in the Challenger exploding shortly after taking off. This shocked the USA public, resulting in delays and questions as to the viability of NASA's programmes. The failure of the O-rings was well known with the engineers down grading the risk from 'high' to 'acceptable'.

Societal reasons 5.16

Society expects its organisations to plan and prepare for the worst case scenarios and when this doesn't happen, society can seek the closure, forfeiture or expulsion of the organisation from society, for example, the legal and political costs to Union Carbide in India with its mis-management of its plant in Bhopal in India. The company has been banned from operating in India – see **8.3** below for a summary.

Environmental reasons 5.17

Failure to prepare and plan for Disasters and Emergencies will also have environmental costs, no better illustrated then the Chernobyl Disaster in April 1986, with its impact not only on the population of Russia, but as far as Wales in the United Kingdom. The aim was to test if the power reduction of a turbine generator would be sufficient given the use of some suitable voltage generator, to power an Emergency Core Cooling System (ECCS) for a few minutes whilst the stand-by diesel generator becomes operational. The power fell to 7% of full power with a benchmark of 20% being critical for that make of RBMK reactor. Consequently, the reactor exploded with radioactive fall out polluting the physical environment (see also **8.12** below).

Humanitarian reasons 5.18

The human cost for not planning for worst-case outcomes is the most significant. All the major Disasters documented in the media resulted in multiple fatalities – of all demographic and socio-economic groups. After the Ladbroke Grove Disaster in October 1999 when two trains collided resulting in 30 fatalities and 160 being critically injured, the public and the media made a significant outcry for change. This culminated in John Prescott, the Secretary of State for the Environment, Transport and the Regions, establishing a Committee headed by Lord Cullen (of the Piper Alpha Inquiry) to investigate 'safety culture' on the railways.

Disaster and Emergency Management Systems 5.19

A Disaster and Emergency Management System (DEMS) is outlined in Figure 2 below:

Figure 2: Disaster and Emergency Management Systems (DEMS)

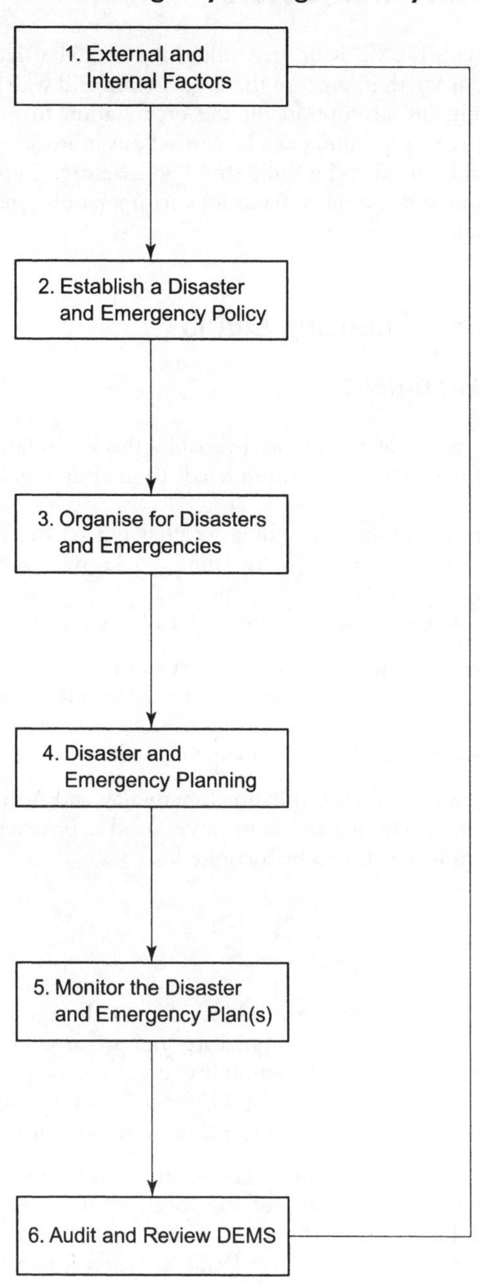

External and internal factors 5.20

The starting point with DEMS is understanding the variables that can influence or affect DEMS. These are both internal to the organisation and wider societal variables. It would be an erroneous assumption for the organisation to make, if it believed that disaster and emergency planning can be carried out in isolation of wider societal variables; these variables need to be understood and factored into any DEMS. The larger the organisation and the more hazardous its operation, the more it needs to provide specific detail.

External factors influencing DEMS

The natural environment 5.21

The organisation needs to identify physical variables that can influence its reaction to a Disaster or Emergency. The organisation needs to identify the following:

- seasonal weather conditions – including weather type(s) and temperature range. Effective rescue can be hampered by failing to know, note and record such variables. A simple chart, identifying the weather conditions, in user-friendly terms on a monthly basis, can assist the reader of any DEM Plans;
- geography around the site – a description of the physical environment such as terrain, rivers or sea, soil type, geology as well as longitude and latitude positioning of the site. These should be brief descriptions unless the site is remote and could have difficulty being accessed in an Emergency; and
- time and distance of the site from major Emergency and Accident Centres, Fire Brigades, police etc. Both long and short routes need to be determined. Longitude and latitude co-ordinates should be identified.

Societal factors 5.22

- Demographics – the organisation needs to identify the age range, gender type, socio-economic structure of the immediate vicinity around the site. A Major Incident with impact on the local community requires the organisation to calculate risks to the community. COMAH (*SI 1999 No 743*) Off-Site Emergency Plans indicates why planning is not just an On-Site 'in these four walls' activity.
- Social attitudes – the organisation needs to investigate briefly and be aware of attitudes (reactions and responses) of the local town or city and the country to Disasters and Emergencies. Union Carbide made the assumption that the population and authorities in Madhya Pradesh, where Bhopal is located, would react like the communities in the USA. The response rates, awareness levels and information available to different communities differ.
- Perception of risk – how does the local community view the site or operation? Is the risk 'tolerable' and 'acceptable' to them for having the operation in their

community? Both Chernobyl and Bhopal highlights that economic necessity can alter the perception of risk in contrast to the statistical level of risk.

- History of community response – in brief the organisation needs to determine how many Major Incidents there have been in the past and how effectively have the community assisted (emergency services and volunteers). Such support will be critical in a Disaster.
- The built environment – a brief description of the urban environment, namely the street layout, urban concentration level, population density, access/egress to railways, motorways etc are useful. These issues could be covered by the inclusion of a map of the area.

Government and political factors **5.23**

- The policy of National Government to Disasters and Emergencies. Are they proactive in advising organisations, what information and guidance do they give? The organisation needs to identify its local or regional office of the HSE, the Environment Agency or a similar body and make contact with officials and seek early input into any planning.
- Local Government – their plans, information and guidance. DEMS need to be aware of any Local Government restrictions on managing Major Incidents. In the case of COMAH Sites, they will need to involve the Local Authority in preparing Off-Site Emergency Plans.
- Committees and agencies – there are many legal and quasi-legal bodies created by legislation which have guidance, information and templates on response (for example, the Flooding Committee under the *Environment Act 1995*).
- Emergency and medical services – making contact, obtaining addresses/contact numbers and knowing the efficiency of the emergency services/Accident and Emergency are critical actions.

Legal factors **5.24**

The organisation needs to know the legal constraints it has to operate within, not only to comply with the law but also to follow the best advice contained in legislation as a means of preventing Disaster and Emergency (see **5.13** above).

Sources of information issues **5.25**

- The local and national media will not only report any incidents but can be critical in relaying messages. Any planning requires the identification of local and national newspapers (their addresses/contact numbers), local/national radio details and any public sector media (through Local Government).
- Local library – they will, in turn, have considerable contact details and networks with other libraries so this can be an efficient medium to relay urgent information.

Making contact with the local librarian and identifying such contact details in any planning will be required.

- Business Associations, ie Chambers of Commerce, Business Links, Training and Enterprise Councils need to be identified for the same reasons cited for local libraries. In the event of a Major Incident they can convey and provide information on a local rescue or occupational health organisation.
- Voluntary Organisations such as the Red Cross and St John Ambulance Association should be identified and noted.

The organisation needs to build information networks both locally and nationally. The experience of major Disasters such as Chernobyl or Kings Cross indicate that letting others know of the incident is not a shameful or embarrassing matter – rather it can warn and halt others from attending the site.

Technological factors 5.26

Be aware of what technology and equipment exists in the local community and what does not, eg lifting equipment, rescue, computers, monitoring equipment measuring equipment etc.

Commercial factors 5.27

- Insurance – liase with insurers from the beginning and ensure that all plans are drawn to their attention and, if possible, approved by them. This can avoid difficulties of any claim that may arise due to a Disaster or Emergency as well as ensuring that best advice from the insurer has been factored into the plans.
- Customer and supplier response – larger organisations need to identify the responses from customers and suppliers, their willingness to work with the organisation in the event of a worst-case scenario and possible alternative suppliers need to be identified. Issues of customer convenience and loyalty also need to be addressed.
- Attitudes of the bank – liquidity issues will arise when a Major Incident strikes. Most organisations do not and cannot afford to make financial provisions for such eventualities. Therefore, if the operation is highly hazardous, establishing financial facilities beforehand with the bank is necessary to ensure availability of liquid cash.
- Strength of the economic sector and economy – larger organisations that are significant players in a sector or the economy need to be cognisant of the 'multiplier effect' that damage to their operation can do to the community and suppliers, employees and others.

Internal factors affecting DEMS 5.28

DEMS also need to be constructed after accounting for various internal or organisational variables. These factors are discussed below.

Resource factors 5.29

Cash flow and budgetary planning whether annual or over a longer period needs to account for resource availability for Major Incidents. This includes physical, human and financial resources. This is wider than banking facilities and encompasses equipment, trained personnel and contingency funds.

Design and architecture 5.30

Issues involving design and architecture include: Is the Workplace physically/structurally capable of withholding a Major Incident? What are the main design and architectural risks? Other issues include layout, access/egress, emergency routes, adequacy of space for vehicles, etc.

Corporate culture and practice 5.31

The organisation needs to identify its own collective behaviour, attitude, strengths and weaknesses to cope with a Major Incident. Being a large organisation does not mean it is a 'coping organisation'; there is also the delay in response that is caused by hierarchical structures. In this case the organisation needs to consider small matrix or project team cell in the hierarchy dedicated to incident response.

Individuals' perceived behaviour 5.32

The perceived behaviour of individuals may be explained by the following:

- The Hale and Hale Model is an attempt to explain how individuals internalise and digest *Perceived Information* of danger, make *Decisions/Choices* according to the *Cost/Benefit* associated with each Decision/Choice and the actual *Actions* they take as well as any *Reactions* that result from their Actions. In short, these five variables need to be understood in the organisation setting and a picture built up of the behavioural response of an individual.
- The Glendon and Hale Model is a macro model of how the organisation (behaving like a *System*), being *Dynamic* and fluid, with *Objectives* and indeed limitations (*Systems Boundary*) can shape behaviour and in turn influence *Human Error.* Following Rasmussen Human Error can be:
  - *Skill based* (failing to perform an action correctly);
  - *Rule based* (have not learned the sequences to avoid harm); or
  - *Knowledge based* (breaching rules or best practice).

  If all three Error Types are committed then the danger level in the System is also greater. The Model is a focus on how wider Systems can contribute to Human Error and how that Error can permeate into the organisation. The inference is the organisation needs to clearly define and communicate its intention and objectives and continuously monitor individual response.

- HSG 48 *Reducing Error and Influencing Behaviour* (1999) Health & Safety Executive, identifies the role of Human Error and Human Factors in Major Incident causation (See further **CHAPTER 11: HUMAN ERROR AND HUMAN FACTORS**). The HSE says 'a *human error* is an action or decision which was *not intended*, which involved a deviation from an accepted standard, and which led to an undesirable outcome' (p 13). Following Rasmussen they classify four types, *Slips* (unintended action) and *Lapses* (short term memory failure) with Slips and Lapses being Skill Based. In addition, *Mistakes* (incorrect decision) are Rule Based and *Violations* (deliberate breach of rules), which are Knowledge Based. Different types of Human Errors contribute in different ways to Major Incidents the HSE say and exemplify. Organisations need to identify from reported incidents the main types of Human Errors and why they are resulting and the negative harm generated.

  The HSE says that understanding Human Factors is a means of reducing Human Error potential. Human Factors is a combination of understanding the *Person's* behaviour, the *Job* they do (ergonomics) and the wider *Organisational* system. Major Incidents can result if the organisation does not analyse and understand these three variables.

Information systems at work 5.33

What types of information systems, how effective, and their accuracy need to be identified. Thus, telephone, fax, e-mail, cellular phone, telex etc need to be assessed for performance and efficacy during a worst-case scenario.

Establish a Disaster and Emergency Policy 5.34

The DEM Policy is a concise document that highlights the corporate intent to cope with and manage a Major Incident. It will have three parts: Statement of Policy; the Arrangements and Command and Control Chart of Arrangements.

Statement of Policy 5.35

The Statement of Policy for managing Major Incidents should be a short (maximum one page) missionary and visionary statement covering the following:

- senior management commitment to be responsible for the co-ordination of Major Incident response
- to comply with the law, namely –
  - the protection of the health, safety and welfare of employees, visitors, the public and contractors;
  - to comply with the duties under the *Management of Health and Safety at Work Regulations 1999* (*SI 1999 No 3242*), *Regs 8* and *9* (see further **CHAPTER 14: LAW RELATING TO EMERGENCIES AND DISASTERS**);
  - to comply with any other legislation that may be applicable to the organisation;

- to make suitable arrangements to cope with a Major Incident and to be proactive and efficient in the implementation process;
- that this Statement applies to all levels of the organisation and all relevant sites in the country of jurisdiction;
- the commitment of human, physical and financial resources to prevent and manage Major Incidents;
- to consult with affected parties (employees, the Local Authority and others if needed).
- to review the Statement;
- to communicate the Statement;
- signed and dated by the most senior corporate officer.

The Arrangements for Major Incident Management 5.36

This refers to *what* the organisation has done, is doing and will do in the event of a Major Incident. Secondly, *how* it will react in those circumstances. The Arrangements are a legal requirement the *Management of Health and Safety at Work Regulations 1999 (as amended 2003)*.

Arrangements should be realistic and achievable. They should focus on major actions to be taken and issues to be addressed rather than being a 'shopping list'. The Arrangements will have to be verified (in particular for COMAH Sites).

Arrangements could be under the following headings with explanations under each. To repeat, the larger the organisation and the more complex the hazard facing it, the more detailed the Arrangements need to be. For example:

- Medical assistance – including first aid availability, first aiders, links with Accident and Emergency at the medical centre, other specialists that could called upon, rules on treating injured persons, specialised medical equipment and its availability etc.
- Facilities management – the location, site plans and accessibility to the main facilities (gas, electricity, water, substances etc), rendering safe such facilities, availability of water supply in-house and within the perimeter of the site etc.
- Equipment to cope – identification of safety equipment available and/or accessible, location of such equipment, types (personal protective equipment, lifting, moving, working at height equipment etc).
- Monitoring equipment – measuring, monitoring and recording devices needed, including basic items such as measuring tapes, paper, pens, tape recorders, intercom and loudspeakers.
- Safe systems – procedures to access site, working safely by employees and contractors under Major Incident conditions (what can and cannot be done), Hazard/risk assessments of dangers being confronted etc, risks to certain groups and procedures needed for rescue (disabled, young persons, children, pregnant women, elderly persons).

- Public safety – ensuring non-access to the Major Incident site by the public (in particular children, trespassers, the media and those with criminal intent), warning systems to the public etc.
- Contractor safety – guidance and information to contractors at the Major Incident on working safely.
- Information arrangements – the supply of information to staff, the media and others (insurers, enforcers) to inform them of the events. Where will the information be supplied from, when will it be done and updates?
- The media – managing the media, confining them to an area, handling pressure from them, what to say and what not to say, etc.
- Insurers/loss adjustors – notifying them and working with them at the earliest opportunity.
- Enforcers – notifying them of the Major Incident, working with them including several types, eg HSE, Environment Agency (or Scottish equivalent) as well as Local Authority (environmental health, planning, building control for instance).
- Evidence and reporting arrangement – to cover strict rules on removal or evidence by employees or others, role and power of enforcers, incident reporting, eg under the *RIDDOR Regulations* (*SI 1995 No 3163*) etc.
- The emergency services – working with the police, Fire, Ambulance/NHS, and other specialists (Red Cross, Search & Rescue), rules of engagement, issues of information supply and communication with these services.
- Specialist arrangements for specific Major Incidents such as bomb explosions – issues of contacting the police, ordnance disposal, access and egress, rescue and search, economic and human impact to name a few issues were most evident during the London Dockland and Manchester bombings in the 1990s.
- Human aspects – removing, storing and naming dead bodies or seriously injured persons during the incident. Informing the next of kin, issues of religious and cultural respect. Issues of counselling support and person-to-person support during the incident.

The list above is not exhaustive. The arrangements should not repeat those in the Safety Policy, rather the latter can be abbreviated and attached as a schedule to the above, so that the reader can have access succinctly to specific OSH Arrangements such as Fire Safety, Occupational Health, Safe Systems at Work, Dangerous Substances, for instance.

Command and Control Chart of Arrangements 5.37

This highlights who is responsible for the effective management of the Major Incident.

- It should be a graphical representation preferably in a hierarchical format, clearly delineating the division of labour between personnel in the organisation and the emergency services/others.

- The chart should display three broad levels of Command and Control, namely strategic, operational and tactical. The first relates to the person(s) in overall charge of the Major Incident. Will this be the person who signed the Statement of Policy for Managing Major Incidents or will it be another (disaster and emergency advisor, safety officer, others)? This person will make major decisions. The second relates to co-ordinators of teams. Operational level personnel need to have the above arrangements assigned to them in clear terms. The third refers to those at the front end of the Major Incident, for instance first aiders.
- It is most important to note that internal command and control of arrangements does not mean *overall* Command and Control of the Major Incident. This can (will be) vested with the appropriate Emergency Service, normally the police or the Fire Authority in the UK. In the event of any conflict of decisions, the external body such as the police will have the final veto. Therefore, the chart and the arrangements must reflect this variable.
- The chart should list on a separate page names/addresses/emergency phone, fax, e-mail, cellular numbers of those identified on the chart. It should also list the numbers for the emergency services as well as 'others' (Red Cross, specialist search and rescue, loss adjustors, enforcing body).
- The chart should also clearly ratify a principle of Command and Control, as to who would be 'In – Charge 1', 'In – Charge 2', if the original person became unavailable.
- The chart and the list of numbers should also be accompanied by a set of 'rules of engagement' in short 'bullet points' to remind personnel of the importance of command and control, eg safety, obedience, communication, accuracy, humanity.

Summary **5.38**

The three parts of the DEM Policy need to be one document. Any detailed procedures can be separately documented ('Disaster and Emergency Procedures') and indeed could be an extensive source of information. However, unlike the Safety Policy and any accompanying safety manual, the same volume of information cannot apply to the DEM Policy. For obvious reasons it must be concise, clearly written, very practical and without complex cross-referencing.

The DEM Policy must 'fit' with the Safety Policy; there can be no conflict so the safety officer and the DEM officer need to cross check and liase. The DEM Policy must also fit with the broader corporate/business policy of the organisation.

The Policy must be proactive and reactive. The former concerning with preventing/mitigating loss and the latter concerned with managing the Major Incident when it does arise in a swift and least harmful manner.

The Policy needs to be reviewed 'regularly'. This could be when there is 'significant change' to the organisation, or as a part of an Audit (semi-annual or annual).

It must be remembered that the DEM Policy is a 'live' document so that it must be accessible and up to date. Although accessibility is important, the Policy should also have controlled circulation to core personnel only (for example those identified in the chart, the legal department). If the Policy was to be accessed by individuals wishing to harm the organisations, this will enable such persons to pre-empt and reduce the efficacy of the Policy.

Finally and most crucially, the core contents of the Policy need to be communicated to all staff and others (contractors, temporary employees and possibly Local Authority).

Organise for Disasters and Emergencies 5.39

Once the establishment has accounted for external and internal factors and has produced a DEM Policy taking account of such factors, next it is necessary to ensure personnel and others are aware of the issues raised in the DEM Policy. The '4 Cs' approach of HSG 65 provides a logical framework to generate this (*Successful Health and Safety Management*, HSE).

Establish effective communication 5.40

Communication is a process of transmission, reception and feedback of information, whether that information is verbal, written, pictorial or intimated.

Effective communication of the DEM Policy, therefore, is not a matter of circulating copies but requires the following.

Transmission 5.41

The whole Policy should not be circulated as it will mean little to employees and others. Rather an abridged version, possibly in booklet format or as an addition to any OSH documentation supplied will make more sense. Such copies must be:

- clear;
- user friendly;
- non-technical as possible;
- aware of the end user's capabilities to digest the information;
- logical/sequential in explanation; and
- use pictorial representation as much as possible.

Being aware of the audience is central to the effective communication of the DEM Policy. The audience is not one entity but will consist of:

- direct employees;
- temporary employees;
- contractors;

- the media;
- the enforcers (COMAH sites);
- the local authority (COMAH sites);
- insurers;
- emergency services (COMAH sites);
- the public (COMAH sites).

This does not necessarily mean separate copies for each of these entities but the abridged copy will need to satisfy the needs of all such groups.

Transmission should start from the Board, through to Departmental Heads, and disseminated downwards and across.

Reception **5.42**

The following question should be addressed:

- What format will the end-user receive the abridged copy in (hard copy, electronic on disc, via e-mail, etc)?
- When will the copy be circulated – upon induction, upon training, ad hoc?

Feedback **5.43**

Will the end user have the opportunity to raise questions, make suggestions or be critical if they spot inconsistencies in the DEM Policy?

There should also be 'tool box talks' and other general awareness programmes to inform individuals of the DEM Policy. This could be combined with general OSH programmes or wider personnel programmes, so that a holistic approach is presented.

The importance of feedback is that the Policy becomes owned by all individuals, which in turn is the single most important factor in successful pre-planning to prevent Major Incidents.

Retaining copies of the Policy **5.44**

In general it may be useful to retain copies of the abridged and the full Policy with other safety documentation in an 'in-house' company library. For smaller organisations, this could comprise of one or two folders on a shelf, through to a dedicated room for the larger organisations. Also, the abridged copy could be pasted onto an intranet site.

Co-operation 5.45

Co-operation is concerned with collaborating, working together to achieve the shared goal and objectives, as follows:

- Between strategic, operational and tactical level management firstly is central. This reflects the chart in the DEM Policy, as discussed at **5.37** above. This could be consolidated as a part of a broader corporate meeting or preferably dedicated time to cover OSH and Fatal Incident issues. This could be a semi-annual event, with a dedicated day allotted for all grades of management to interface. This is not the same as a Board level discussion or a management discussion.
- To give responsibility to either the Safety Committee or the Safety Group to co-ordinate review, debate and assessment of the DEM Policy or to fuse this function within a broader business/corporate review committee. The former has advantages as it is safety dedicated whilst the latter would integrate DEM Policy issues into the wider business debate.
- Involvement of Safety Representatives/Representatives of Employee Safety (ROES) is both a legal requirement as well as inclusive safety management. These persons can be central 'nodes' in linking 'management' and the 'workers' together. In the UK, there has been an increasing realisation that trained safety representatives are a knowledgeable resource with many being trained to IOSH/NEBOSH standard (as with the AEEU (electrical engineers) trade union, with their National Academy of Safety and Health (NASH)).
- Co-operation needs also to extend to contractors. The person responsible for interfacing with contractors needs to update them and make them aware of the DEM Policy and seek their support and suggestions. It may also be valuable to invite Contractors to OSH/Fatal Incident awareness days or the general committee meetings as observers.
- Co-operation is needed between the organisation and external agencies such as the local authority, insurers, enforcers, media etc. This can be via providing abridged copies of minutes or a 'newsletter' (one to two sides of A4) distributed semi-annually informing such bodies of the DEM Policy and any changes as well as other OSH issues. COMAH sites will have to demonstrate as a legal requirement that plans and policies are up to date and effective.

Competence 5.46

Competence is a process of acquiring knowledge, skill and experience to enhance both individual and corporate response. Thus, competence is about enhancing and achieving standards (set by the organisation or others).

Competence is important, so that certain key persons are trained to understand the DEMS process. The above issues, and their link to OSH in particular, require personnel that can assimilate, digest and convey the above issues. Training does not necessarily mean formal or academic training but can be vocational or in-house. Neither does it mean the organisation expending vast sums but can be a part of a wider in-house OSH awareness programme (eg one day per quarter of a year).

For larger organisations, they may be able to recruit 'competent persons' to advise on OSH and Fatal Incident matters.

In short, all employees need to brought up to a minimal standard. Piper Alpha showed that whilst Occidental Petroleum had detailed procedures, the employees generally did not fully understand them nor had they the minimal understanding of Major Incident evacuation. The organisation had failed to impart knowledge, skill and experience sufficient to cope with fires and explosions on off-shore sites.

Control 5.47

Control refers to establishing parameters, constraints and limits on the behaviour and action. This ensures that on the one hand an effective DEM Policy exists and on the other, personnel will act and react in a co-ordinated and responsible manner. Controls can be achieved via, for example:

- contractual means – as a term of a contract of employment that instruction and direction on OSH and Fatal Incident matters must be followed by individuals;
- corporate means – the organisation continuously makes individuals aware of following rules and best practice;
- behavioural means – by establishing clear rules, training, supply of information, leading-through-example, showing top-level management commitment etc;
- supervisory means – ensuring that supervisors monitor employee safety attitudes and risk perceptions.

The cumulative effect of the '4 Cs' should be a positive and proactive culture in which not only OSH issues but the DEM Policy issues are understood. Factors that can mitigate or prevent a positive and proactive culture developing include:

- lack of management commitment;
- lack of awareness of requirements;
- poor attitudes to work and life;
- misperception of the risk facing the organisation;
- lack of resources or the unwillingness to commit resources;
- fatalistic beliefs etc.

Disaster and Emergency Planning 5.48

'Planning' is a process of identifying a clear goal and objectives and pursuing best means to achieve that goal/objectives. Although 'Disaster' and 'Emergency' are two separate but related terms, in the case of planning the two need to be viewed together. This is because in practice one cannot divorce the serious event (the Disaster) from the response to that serious event (Emergency). There is no practical benefit from having a separate Disaster Plan and an Emergency Plan.

Disaster and Emergency Planning can be viewed in three broad stages:

- Stage 1: Before the event.
- Stage 2: Factors to consider during the event.
- Stage 3: After the event.

Stage I: Before the event 5.49

Once the DEM Policy has been established and a culture created where the Policy has been understood and positively received, next one needs to be establish a 'state-of-preparedness', that is addressing issues, speculating on scenarios and developing support services when the Major Incident does strike. This stage can be summarised into three sections, with a special section on COMAH (*SI 1999 No 743*) and the specific legal requirements for COMAH sites:

The risk assessment of Major Incident potential and consequent contingency planning 5.50

What is the probability of the Major Incident resulting? What would be the severity? What type of Major Incident would it be? Figure 3 below depicts a Major Incident Matrix.

Figure 3: Major Incident Matrix

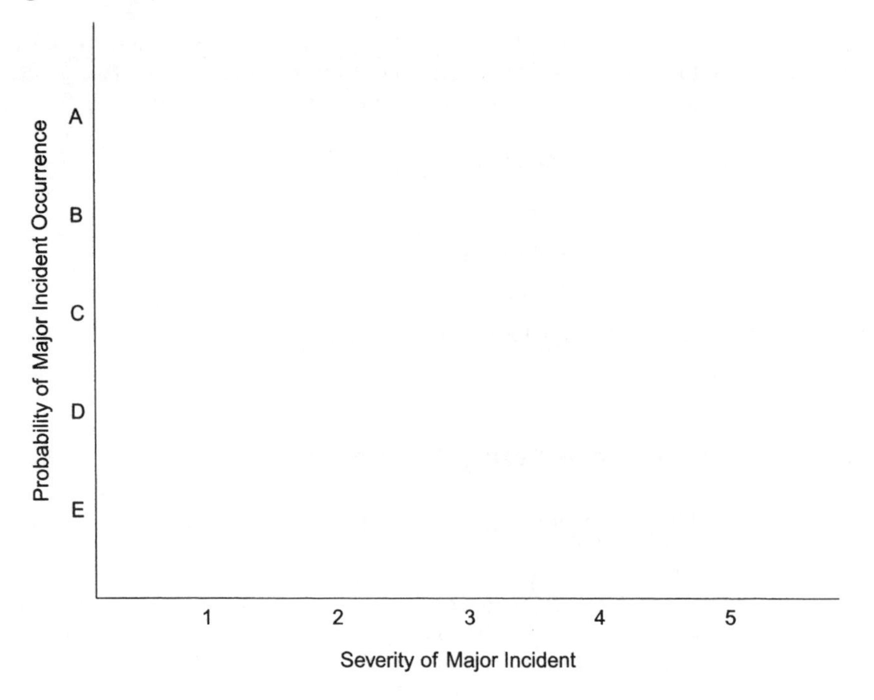

Where:

A = Certainty of occurrence (probability = 1)

B = Highly probable

C = 0.5 probability of occurrence

D = Low probability

E = Most unlikely

and:

1 = Serious injuries

2 = Multiple fatalities

3 = Multiple fatalities and serious economic/property damage

4 = Fatal environmental, bio-sphere and social harm

5 = Socio-physical catastrophe

The planning responses, therefore, will differ according to the cell one finds oneself in. For example, cell 1E requires the least complex state of preparedness in terms of resources, detail of planning and urgency. Whilst 5A is a major societal event that requires macro and holistic responses, in which case the organisation's sole efforts at planning are futile. The organisation needs to assess in which cell the Major Incidents it will confront, will mainly fall in and accordingly develop plans. A conglomerate or an 'exposed operation' such as oil/gas refinery work may require planning at all 25 levels. This is *Contingency Planning* – an analysis of the alternative outcomes and providing adequate responses to those outcomes.

Ten-point strategy **5.51**

During the concerns over the 'Millennium Bug', HSBC Bank provided some general advice on contingency planning ('Tackling the Millennium Bug: A final check on preparations including contingency planning', HSBC, 1999). They suggested a ten-point strategy. 'A contingency plan is a carefully considered set of alternatives to your usual business processes' (p 18). This advice is generic and can be used more broadly for everyday planning:

1. *Identify* – assume things will go wrong – focus on where the probability and severity would be of failure.
2. *Prioritise* – the processes into three categories:
   - Essential for business continuity.
   - Important for business continuity.
   - Non-essential for business continuity.

3. *Analyse* –
   - Identify the effects of risks on operation(s).
   - Assess 'domino effects'.
   - Grade risk using a suitable risk assessment approach.

4. *Develop solutions* –
   - For each operation develop a solution/control measure to the risk.
   - Must be participative – involve suitable and critical stakeholders.
   - Ensure new risks are not created.
5. *Costs* – carry out a 'cost benefit analysis'
6. *Formalise documents* –
   - Write and keep any assessments and inspections.
   - Top Management involvement and 'joined-up business'.
   - Develop an approach to implementation.
   - Regular training needs to be conducted to ensure improved competence.
   - Resources needed – identify and deploy (if available).
   - Communications need to be clear and simple – before, during and after.
   - Contractual issues need to be dealt with so that contractors, customers and insurers are kept informed.
7. *Co-ordinate* – Managers and team to ensure implementation.
8. *Test* –
   - Controlled testing of the contingency plan(s) (see **5.53** below).
9. *Update* –
   - Update the plan.
   - Quality assurance and control of the plan.
   - Accuracy – ensure that the organisation's lawyer and insurer have sight of the plan.
10. *Communicate* –
    - Signed off by top management.
    - Communicate salient features to employees.

Appendices can be included containing any relevant information.

Decisional planning **5.52**

Planning before the event also involves co-ordinating different levels of decision making. Figure 4 depicts a Decisional Matrix:

Figure 4: Decisional Matrix

	Production level	1. Input	2. Process	3. Output
Management level				
A. Strategic				
B. Operational				
C. Tactical				

Where:

A = Strategic or Board/shareholder/controller level (gold level)

B = Operational or departmental level (silver level)

C = Tactical or factory/office level (bronze level)

and:

1. = Inputs or those resources that make production possible, thus raw materials, people, information, machinery and financial resources

2. = Process or the manner in which the inputs are arranged or combined to enable production (method of production)

3. = Output or the product or service that is generated

Decisional planning requires Major Incident Risks to be identified for each cell. Essentially this is assessing where in the production process and at which management level can problems arise that can lead to Major Incidents. This Matrix is not 'closed' but is 'open' and 'dynamic', which means external factors, which were discussed at **5.21** above as well as internal factors need to be accounted for. For example, in a commercial operation at Board level, they need to consider:

- *Inputs*: Purchasing policy of any raw materials used (safety etc), recruitment of stable staff, adequacy of resources to enable safe production, known and foreseeable risks of inputs used, reliability of supply, commitment to environmental safety and protection etc.
- *Process*: External factors and their impact on production, production safety commitment, commitment to researching/investigating in the market for safer production processes etc.
- *Output*: Boardroom commitment to quality assurance, environmental safety of the service and/or product produced, customer care and social responsibility etc.

After consideration at each Level, there needs to be 'fit' and co-ordination. *The end result* should be, in each cell major risks to production and their impact on a Major

Incident needs to be identified. This need not be a major, time-consuming exercise, rather it can be determined through a brain-storming session or each management level fills out the Matrix during their regular meeting and then it is jointly co-ordinated in a short (two to five page) document.

Testing **5.53**

This is considered to be a crucial aspect of pre-planning by all the emergency services in the UK. See further **CHAPTER 20: TRAINING AND EXERCISING FOR EFFECTIVE PREPAREDNESS AND RESPONSE.**

COMAH sites **5.54**

COMAH sites, in addition to the above, also need to ensure that On-Site and Off-Site Emergency Plans are compliant with guidance and are realistic and achievable. See further **CHAPTER 6: EMERGENCY PLANNING FOR COMAH AND NON-COMAH SITES**.

Stage 2: During the event

Activate Disaster and Emergency Policy and pre-planning **5.55**

Major Incident response and management ought to be clear and effective if the previous stages in the DEMS have been followed; external and internal factors and uncertainties have been accounted for; developed a chain of command as well as arrangements to cope; organised the operation to cope with a Major Incident and, finally, pre-planned if the event actually arises.

Co-ordination **5.56**

This is the central action to ensuring efforts are in unison. This includes both liasing with external bodies such as the emergency services, the media and internally at strategic, operational and tactical levels (see **5.36** and **5.37** above). The police will normally have overall command and control, so their guidance must be followed.

Code of conduct **5.57**

A Code of Conduct is a set of 'golden rules' that staff and contractors need to follow if the Major Incident arises. This includes:

- following Instruction from those with Command and Control responsibilities;
- using self initiative;
- facilities management (gas, electricity, water) – making safe, when and when not to switch on or off;

- medical assistance – liaison with health service, first aiders, when and when not to administer;
- rules of evacuation;
- site access and egress;
- site security;
- work and rescue equipment – its safe use and logistics of use;
- record keeping of the major incident (audio-visual, verbal and written);
- resources needed for effective Major Incident management;
- media and public relations management;
- Issues of Care and Compassion of injured persons.

The Code of Conduct is a reflection of the Disaster and Emergency Policy and the pre-planning. It should be reinforced verbally and in writing (one page of A4) and reiterated to the Major Incident Team before and during the Major Incident. Whilst this seems bureaucratic, it must be noted that if the core team (internal and external) fail to follow best practice and safeguard themselves, this increases the risk factor of the Major Incident and could even lead to a double tragedy. A five minute or so reiteration is a minor time cost.

COMAH sites will have to be aware of these issues also as well as activating the On-Site and Off-Site Emergency Plans.

Stage 3: After the event 5.58

Major Incident Planning does not cease as soon as the Major Incident is physically over. There are continuous issues over time that need to be addressed, these issues are summarised below.

Immediate term – immediately after the Major Incident and within a few days

Statutory investigation 5.59

This will involve the HSE, Local Authority, Environment Agency/SEPA for instance. Under statute, these bodies have powers and a duty to investigate a Major Incident.

- The organisation must fully co-operate with these bodies and afford any assistance they require.
- There must be no removal or tampering with anything on the Major incident site (or off-site in COMAH cases). There may be forensic or other data gathering required by these bodies.
- Provision of all documentation and information to these bodies.

Business continuity 5.60

This has to be planned for even whilst the investigation by authorised inspectors is being carried out.

'Business continuity' is a planning exercise to ensure critical facilities, processes and functions are operational and available during and immediately after the Major Incident thereby enabling the organisation to function commercially and socially. A forecasting process of recovery, assessment and ensuring the adequacy of resources for the organisation. See **CHAPTER 1: BUSINESS CONTINUITY MANAGEMENT: PRINCIPLES AND APPROACHES**.

Insurers/loss adjustors 5.61

Assuming that the insurance contract covers direct and consequential damage from a Major Incident (and not all contracts will), the organisation needs to notify the insurer and ensure all paperwork is completed promptly. Most insurance contracts stipulate a time limit by which the paperwork has to be lodged with the insurer; the Major Incident will divert attention to other issues, so ensuring that a person is appointed to activate this insurance task is vital (this should be the organisation's lawyer).

The insurer in turn will notify their loss adjustors to investigate the basic and underlying causes of the event. Again, full disclosure and co-operation are implied 'insurance contractual' requirements. Although, the organisation must check all documents that the loss adjustors completes, to ensure they are accurate and cover all aspects of the event.

The police 5.62

In the suspicion that criminal neglect played a role in the event, then the police will need to interview all core Board members, senior management and others.

Building contractors 5.63

The organisation needs to plan for building contractors to visit the site and make it safe and secure. This may have to be done forthwith after the event, even if insurance issues have not been resolved. Therefore, adequacy of resources, as discussed above becomes vital.

Visitors 5.64

Major Incidents also lead to the public and media making visits. The arrangements made to handle such groups must extend to after the event.

Counselling support **5.65**

Counselling support to affected employees and, possibly, contractors is not just a personnel management requirement which shows 'caring management' but increasing a legal duty of the organisation to provide such support. Issues of 'post-traumatic stress', 'nervous shock' and 'bereavement' means that the organisation has common law obligations to offer medical and psychological support. This should involve a medical practitioner and an occupational nurse. The insurance policy can be extended before the event to cover the cost of such services.

Short term – a week plus after the event

Investigation and inquiry **5.66**

This can be both a statutory inquiry (although most inquiries are called within days) and/or the in-house investigation of the event and lessons to be learnt. Issues to consider include:

- Basic causes – was it a fire, bomb, an explosion or a natural peril?
- Underlying causes – what led up to such a peril occurring? Examine the managerial, personnel, legal, technical, organisational and natural factors that could have caused the event.
- Costs and losses involved.
- Lessons for the future.
- Did the Disaster and Emergency Plan operate as expected? Were there any failings? What improvements are required? There needs to be complete de-briefing and examination of the entire process involving internal personnel and external agencies.

The organisation should weigh up the possibility of external persons carrying out this exercise or whether in-house staff are objective and dispassionate enough to assess what went wrong.

Visit by enforcers **5.67**

The organisation also needs to be prepared for further visits from the enforcers and the possibility of Statutory Enforcement Notices being served to either regulate or prohibit the activity. Multiple Notices are possible, from Health and Safety Officers, Fire Authority, Environment Agency/Scottish Environment Protection Agency (SEPA), Building Control or Planning Officers. This will affect the operation, production process and have economic implications. This ought not to occur if the organisation has taken due care to pre-plan for the Major Incident and had continuous safety monitoring of the operation. Notices will be served if there is a failure to make safe the site.

Coping with speculation 5.68

The public, the media and employees will speculate about causation and there is the risk of adverse publicity. Public relations is a central activity. For legal and moral reasons, it is best practice to disclose all known facts unless the statutory investigation prohibits otherwise.

Medium term – a month plus 5.69

The 'normalisation process' will begin. The organisation needs to carry on the operation, be prepared for further visits from enforcers, loss adjustors and re-assess its corporate/financial health.

Long term – six month onwards

Systems review 5.70

The systems review includes the following:

- A review of the impact of the Major Incident and whether the organisation is recovering from it.
- Impact on reputation.
- Legal threats.
- Any positive outcomes – learning from mistakes, improving technical know-how, wider industrial benefits from knowing the chain reaction of events etc.

Longer term – one year onwards 5.71

The organisation's memory and experience needs to be included in the following:

- in-house training programmes;
- factored into systems and procedures;
- review of entire *espirit de corps* and corporate philosophy.

Disaster and Emergency Planning is an extensive exercise, being dynamic and accounting for a diverse array of phenomena as industrial, man-made, environmental, socio-technical, radiological and natural events.

Monitor the Disaster and Emergency Plan 5.72

Monitoring is a process assessing and evaluating the value, efficiency and robustness of the Disaster and Emergency Plan(s). This involves looking at all three stages as discussed at **5.49** et seq above (before, during and after the event stages) and not just the core document, 'the Plan'. It is comprehensive and holistic in its questioning of the entire Planning Process.

Monitoring can be classified as proactive or reactive. The former attempts to identify problems with the Plan(s) before the advent of an emergency. It is a case of continuously comparing the Plan(s) with even minor incidents and loopholes identified in any training/exercising sessions. Reactive monitoring occurs after a Major Incident or occurrences that could have led up to a Major Incident, thereby reflecting back and assessing if the Plan(s) need improvement. Both types of monitoring are important.

Figure 5 below summarises the different levels of monitoring required with some examples.

Figure 5: Monitoring matrix

Stages in planning/ Monitoring types	1. Before the Major Incident	2. During the Major Incident	3. After the Major Incident
A. PROACTIVE	Risk assessments Testing via training or exercises Major Incident assessment Facilities inspections	Inspections Live Interviews Feedback	Inquiries Systems review Counselling reports
B. REACTIVE	Incident statistics Incident reports Warnings Notices	Critical assessments Incident levels Audio-visual assessment	Brainstorming Loss assessments Enforcement

The cells in Figure 5 are not strictly mutually exclusive; many techniques are both proactive and reactive. A third dimension is added in COMAH cases, that of, On-Site and Off-Site Emergency Plans (types of planning).

Proactive before the Major Incident 5.73

Figure 5, Cell A1 (Proactive before the Major Incident) requires the following monitoring:

- Risk Assessments and Hazard Analysis techniques will identify significant hazards and their risk level. Monitoring such risk is an index of danger, which in turn is a variable in the type and potential of the Major Incident, outlined in Figure 1 at **5.12** above. The HSE's 'Five Steps approach' (see **1.22** above) or their 'Quantified Risk Assessment' methodology provides outlines of assessing risk. Hazard Techniques include Hazard and Operability Studies, HAZard ANalysis (HAZANS), Fault and Event Tree Analysis etc.

- Testing will identify any problems or concerns with the Disaster and Emergency Plan. For example, a live exercise or a synthetic simulation could identify factors the Plan has not considered or which may not be practicable if the Major Incident was to arise. Thus, enabling questioning, critical appraisal and comments should not be perceived as a threat or being awkward with the Plan(s) but can vital information.
- Major Incident assessment, which is a periodical overview of the Plan(s), every quarter or semi-annual by both internal and external persons. This could identify areas of concern. This assessment compares the Plan(s) with the potential threat – can the former cope with the threat? Threats change, as technology and know-how changes, so such assessments become another vital source of information.
- Facilities Inspections, of gas, electricity, water, building structure, equipment available/not available etc. can highlight issues of physical resourcing and adequacy of such resourcing.

Reactive after the Major Incident 5.74

Figure 5, Cell B1 (Reactive after the Major Incident) requires the following monitoring:

- Incident statistics will show the type of incidences, the type of injury, when and where it occurred. This enables the organisation to hypothesise/build a picture of the potential and severity of a bigger incident. One cannot divorce Occupational Health and Safety data from Disaster and Emergency Management.
- Analysing incident reports should enable issues of causation to be assessed. What type of occurrence could trigger a Major Incident? Identifying and developing a pattern of causes will enable assessing whether the plan(s) account for such causes.
- Warnings from employees, contractors, enforcers, public and others of possible and serious problems are to be treated with seriousness. All such warnings are to be analysed and a common pattern and trend spotted.
- Any Notices served by enforcers will identify failings in the operation and the remedial actions required. These can be factored into the Plan(s).

Proactive during the Major Incident 5.75

Figure 5, Cell A2 (Proactive during the Major Incident) requires the following monitoring:

- Inspections will be made even as the event occurs. Inspections can range from the stability of the structure through to how personnel behaved and coped.

As these Inspections are made the Command and Control team needs to evaluate if any aspect of the Plan(s), which is a 'live document' need immediate changes.

- Live Interviews with internal personnel and emergency services' personnel will enable a continuous appraisal of any difficulty with procedures, arrangements and instructions that emanate from the Plan(s). Again, these can lead to immediate changes to the Plan(s).
- Feedback is a proactive technique of requesting regular, interval information on and off-site. This enables a picture to be constructed of what could happen next; trying to anticipate the next sequence and if the Plan(s) can cope with it.

Reactive during the Major Incident **5.76**

Figure 5, Cell B2 (Reactive during the Major Incident) requires the following monitoring:

- Critical assessments are carried out after some unexpected occurrence, which causes uncertainty and may even threaten the efficacy of the Plan(s). The critical assessment is by the Command and Control team as a whole. Why did this happen? Why did we not account for it in the Plan(s)?
- Incident levels – in particular if serious injury or fatalities are increasing, then at a moral or philosophical level the question to ask is whether the Plan(s) have been overwhelmed by reality. All forms of planning, including the statutory COMAH Planning (*COMAH Regulations, SI 1999 No 743*) must not be viewed with rigidity. If the Plan(s) are failing, it is better to re-appraise and re-plan. COMAH does not overtly allow for this, although COMAH stresses flexibility and continuous appraisal of the event. In such a case, there has to be quick and clear decision making, with consequential command and control, as well as immediate communication of this 'alternative plan'. training and exercising sessions need to factor in this dimension and equip people with decisional techniques.
- Audio-visual assessments can be a dramatic means of understanding the actual event. This can be video or photographic footage shot by the incident personnel or from the emergency services. This enables monitoring of the extent, potential and actual threat from the Major Incident.

Proactive after the Major Incident **5.77**

Figure 5, Cell A3 (Proactive after the Major Incident) requires the following monitoring:

- Inquiries are proactive, even though the event has happened. The inquiry (whether internal or external) will identify strengths and weaknesses in the Plan(s), which can lead to future improvements in planning.

- A systems review is an overhaul of the entire reaction and holistic experience of the organisation to the trauma of a Major Incident. This involves developing future coping strategies for personnel and issues:
  - How well did the organisation respond?
  - Were there adequate resources in place to cope?
- Counselling Reports will identify the experiences, perceptual and cognitive issues that affected personnel and others. This can provide probably the most significant information on behavioural response of the Command and Control team and those that were injured. Which in turn can be factored into training and exercise programmes, which itself will lead to personnel skill improvements.

Reactive after the Major Incident 5.78

Figure 5, Cell B3 (Reactive after the Major Incident) requires the following monitoring:

- Brainstorming is an open-ended participative and indeed critical analysis of what went wrong and what was right with the Plan(s). Brainstorming should also be inclusive, involving emergency services and possibly enforcers as well, so that their guidance is factored in.
- The loss adjustors report will be a vital document as to the chain reaction that led to the event and the consequences that followed. For legal reasons, their findings may have to be applied before insurance cover is available.
- Enforcement Notices and enforcers reports will contain recommendations, which need to be viewed as lessons for the future.

Audit and Review 5.79

The final stage of the DEMS process is Audit and Review. An Audit is a comprehensive and holistic examination of the entire DEMS process (see Figure 2 at **5.19** above). An Audit will identify stages in this process that need improvement. A Review is an act of 'zooming in' to that particular stage and carrying out those improvements.

Major Incident Auditing 5.80

Major Incident Auditing can be qualitative or quantitative. The former adopts a 'yes' or 'no' response format to questions. The latter asks the auditor to rate the issue being examined from, say, 0–5.

The Audit must be comprehensive, assessing every aspect of the DEMS process. This means that the Audit will take time to be completed. The Audit is not some 'inspection' that is more random focusing upon a hazard rather than the complete system.

Audits essentially benchmark (compare and contrast) performance. This can be against the DEMS process identified above or against legislation (eg *COMAH 1999* (*as amended 2005*)). The benchmarking could also be against another site or wider industry standards.

Should Audits be carried out in-house or rely on external consultants? There are costs and benefits associated with both, with no definitive answer. The *Management of Health and Safety at Work Regulations 1999* (*as amended 2003*) in the UK emphasises the need to develop and use in-house expertise in relation to general Occupational Safety and Health (OSH) issues, with a reliance on external specialists as a last resort. This may be interpreted as best practice for Disaster and Emergency Management also.

Audits can be annual or semi-annual. The more complex the operation and risk it poses the greater the need for semi-annual Audits.

Audits should be proactive, that is learning from the weaknesses in the DEMS process and reducing or eliminating such weaknesses for the future.

Finally, the results of the Audit need to be fed back into the DEMS process and all affected persons informed of any changes and risk management issues arising.

Audits are holistic (assess anything associated with Major Incidents), systemic (assess the entire DEMS process, ie the 'system') and systematic (that is logical and sequential in analysis).

Review **5.81**

The review of any specific problems needs to be actioned by the organisation. The consultant will identify the areas of concern and make recommendations but the final discussion and implementation lies with the organisation. This needs to be led by senior officials in the organisation.

Reviews are, by definition, 'diagnostic', meaning the organisation needs to look at causation and cure of the failure in any part of the DEMS process.

Budgeting both in time and resource terms is critical in the review, as it will require management and external agency involvement.

A review can be carried out at the same time as an audit. A review can also be a legal requirement, as with *Regulation 11* of the *COMAH 1999* (*as amended 2005*), which requires a review and where necessary a revision of the On-Site and Off-Site Emergency Plans for top-tier establishments.

Conclusion

5.82

All organisations and societies need to prepare for worst-case scenarios. DEMS provides a logical framework to understand the main stages in effective preparation. The larger the organisation, the more detailed and analytical the preparation needs to be. Finally, Disaster and Emergency Management Systems is a live and open system requiring continuous monitoring.

6

Emergency Planning for COMAH and Non-COMAH Sites

In this chapter:

- Introduction.
- Control of Major Accident Hazards Regulations 1999 and the Control of Major Accident Hazards (Amendment) Regulations 2005.
- Non-COMAH sites.
- Lower-tier duties.
- Top-tier duties.
- COMAH Emergency Plans.
- Radiation (Emergency Preparedness and Public Information) Regulations 2001.
- Public Information for Radiation Emergencies Regulations 1992.

Introduction 6.1

The introduction or amendment of directives, rules and regulations to control major accident hazards can almost invariably trace their genesis to disasters resulting in a significant loss of life or injury.

In 1974 an explosion at a chemical plant in Flixborough, England, resulted in 28 deaths.

Two years later at approximately 12:37 pm on Saturday 9 July 1976, a bursting disc on a chemical reactor ruptured. Maintenance staff heard a whistling sound and a cloud of vapour was seen to issue from a vent on the roof. The release lasted for approximately 20 minutes. About an hour after the release the operators were able to admit cooling water to the reactor.

Among the substances of the white cloud released was a small deposit of TCCD (dioxin) a highly toxic material. The nearby town of Seveso, located 15 miles from Milan, Italy, had 17,000 inhabitants.

Over the next few days following the release there was much confusion due to the lack of communication between the company and the authorities in dealing with this type of situation. No human deaths were attributed to dioxin but many individuals fell ill. A number of pregnant women who had been exposed to the release had abortions. In the contaminated area many animals died. (See also **8.9** below.)

Review control of major accident hazards 6.2

These incidents caused the European Commission (EC) to review the control of major accident hazards. The review revealed that there were differing standards of control over industrial activity within Europe and resulted in the EC proposing a directive to control major accident hazards and establish common standards.

The Directive (*Council Directive 8222/501/EEC*) known as the 'Seveso Directive' was adopted in June 1982 and introduced the concept of safety cases. The Directive was implemented in Great Britain by the *Control of Industrial Major Accident Hazard Regulations 1984* (*SI 1984 No 1902*) (*CIMAH*), which required process industries to describe hazards and identify the safeguards necessary to limit the hazards.

The escape of a highly toxic cloud of Methyl Isocyanate in Bhopal, India, in 1984 resulting in 2000 deaths and tens of thousands of injuries (see **8.3** below) and a chemical fire in Basle, Switzerland, in which contaminated fire fighting water caused extensive pollution of the Rhine reminded the world that the consequences of process failure can be disastrous in human and environmental terms. Two amendments to correct technical inaccuracies and omissions and lower the threshold for certain toxic substances followed (*87/216/EEC* and *88/610/EEC*).

In 1988 the Piper Alpha disaster claimed the lives of 167 on an oil platform in the North Sea and resulted in the Secretary of State for Energy directing that a public inquiry, chaired by Lord Cullen, investigate the causes of the accident (see also **20.1** below). The resulting report included 106 recommendations which had major implications for the statutory regime controlling off-shore safety. The recommended approach followed the philosophy of the Seveso Directive, namely one of hazard identification, as well as building upon the concept of risk assessment and its active management through the adoption of best practice procedures found enshrined in the *Health and Safety at Work etc Act 1974*.

Formal safety assessment 6.3

The essential feature of Lord Cullen's recommended off-shore safety regime was the 'formal safety assessment' (FSA) requiring operators to demonstrate safe operation of the installation. An FSA involves the identification and assessment of hazards over the whole life cycle of a project from the initial feasibility study, through the concept and detailed design to construction and commissioning, then the operations phase and finally the commissioning and abandonment.

Lord Cullen's recommendations were implemented in the *Off-Shore Installations (Safety Case) Regulations 1992 (SI 1992 No 2885)* (the *Off-Shore Regulations*) The *Off-Shore Regulations* require that a safety case demonstrate that the hazards have been identified and assessed, appropriate controls provided and that the exposure of personnel to the hazards have been minimised.

A review of the operation of the Seveso Directive by the EC and Member States' enforcement bodies revealed that it had not been strictly enforced, there had been no significant reduction in the accident rate and no change in the average severity of reported accidents. In contrast, major hydro-carbon releases in the North Sea, post the introduction of the *Off-Shore Regulations*, were reduced by half and other less significant releases did not escalate. The off-shore safety regime was therefore able to demonstrate a positive beneficial effect on safety whilst the Seveso Directive could not.

Seveso II Directive 6.4

The 'Seveso II' Directive (*Council Directive 96/82/EEC*) was adopted in December 1996. It seeks to address the limitations of the previous directive by requiring operators to demonstrate through a written description that they have a safety management system that is capable of systematically and continually identifying hazards, assessing them, and eliminating or minimising, in so far as is reasonably practicable, the risk to employees at the facility, those that may be affected by an incident and the environment. The safety management regime must be effective over the life of the facility.

The *Control of Major Accident Hazards Regulations 1999 (SI 1999 No 743)* (COMAH) implement the requirements of the Seveso II Directive in Great Britain and replaces CIMAH. COMAH came into force on 1 April 1999. The provisions of *Article 12* of the Seveso II Directive concerning land use and planning have been implemented in the *Planning (Control of Major Accident Hazards) Regulations 1999 (SI 1999 No 981)*.

The Health & Safety Executive (HSE) and the Environment Agency (England and Wales) or the Scottish Protection Agency are jointly responsible as the Competent Authority (CA) for COMAH.

The Control of Major Accident Hazards (Amendment) Regulations 2005 has resulted in a number of amendments to the 1999 Regulations, which remain in force.

Control of Major Accident Hazards Regulations 1999 and the Control of Major Accident Hazards (Amendment) Regulations 2005 6.5

The *COMAH Regulations (SI 1999 No 743)* gives effect to a safety regime for the prevention and mitigation of major accidents resulting from ultra-hazardous industrial activities. The emphasis is on controlling risks to both people and the environment through demonstrable safety management systems, which are integrated into the routine of business rather than dealt with as an add-on.

An occurrence is regarded as a major accident if it:

- results from uncontrolled developments (i.e. they are sudden, unexpected or unplanned) in the course of the operation of an establishment to which the Regulations apply; and
- leads to serious danger to people or to the environment, on or off-site; and
- involves one or more dangerous substances defined in the Regulations.

Major emissions, fires and explosions are the most typical major accidents.

The duties placed on operators by *COMAH* fall into two categories: 'lower tier' and 'top tier'. Lower-tier duties fall upon all operators. Some operators are subject to additional, top-tier duties.

An 'operator' is a person (including a company or partnership) who is in control of the operation of an establishment or installation. Where the establishment or installation is to be constructed or operated, the 'operator' is the person who proposes to control its operation – where that person is not known, then the operator is the person who has commissioned its design and construction.

As mentioned, the HSE and the Environment Agency (in England and Wales) or the Scottish Environment Protection Agency are jointly responsible as the competent authority (CA) for *COMAH*. A co-ordinated approach will be adopted with the HSE likely to act as the primary contact for operators. A charging regime has been introduced and the fees are payable by the operator to the HSE for work carried out by the HSE and the Environment Agency. Under *COMAH, Reg 19*, the CA must organise an adequate system of inspections of all establishments, and then must prepare a report on the inspection.

By virtue of the amendment to the Seveso II Directive by the European Community in 2003, the UK Government was required to implement this 'Seveso III' Directive through secondary legislation, by way of an amendment to the 1999 COMAH Regulations. The amendments introduced by the 2005 Regulations largely relate to terminology and threshold levels but also reflect the lessons from chemical plant disasters such as Toulouse, France (**see Case Study 2 in Chapter 2**).

The main HSE guidance to the Regulations is contained in 'Guide to the Control of Major Accident Hazards Regulations 1999 (L111)'. Readers can access http:/www.hse.gov/comah/index.htm for a comprehensive update to the 1999 Regulations and can read the full 2005 Regulations at http:/www.opsi.gov.uk/si/si2005/20051088.htm

Application 6.6

The *COMAH Regulations* apply to an *establishment* where:

- dangerous substances are present; or
- their presence is anticipated; or
- it is reasonable to believe that they may be generated during the loss of control of an industrial chemical process.

'Loss of control' excludes expected, planned or permitted discharges.

'Industrial chemical process' means that premises with no such chemical process do not fall within the scope of the Regulations solely because of dangerous substances generated during an accident.

An 'establishment' means the whole area under the control of the same person where dangerous substances are present in one or more installations. Two or more areas under the control of the same person and separated only by a road, railway or inland waterway are to be treated as one whole area.

'Dangerous substances' are those which:

- are named at the appropriate threshold (see Table 1); or
- fall within a generic category at the appropriate threshold (see Table 3).

There are two threshold levels; lower tier (Column 2 of the Tables) and top tier (Column 3) as now defined by Part 2 of Schedule 1 of the 2005 Regulations (note that the thresholds differ when compared with the 1999 Regulations):

Table 1: Dangerous substances

	Quantity in tonnes	
Column 1	*Column 2*	*Column 3*
Ammonium nitrate (as described in Note 1 of this Part; see also Note 8(1) and (2))	5,000	10,000
Ammonium nitrate (as described in Note 2 of this Part; see also Note 8)	1,250	5,000
Ammonium nitrate (as described in Note 3 of this Part; see also Note 8(2) and (3))	350	2,500
Ammonium nitrate (as described in Note 4 of this Part; see also Note 8)	10	50

Potassium nitrate (as described in Note 5 of this Part)	5, 000	10, 000
Potassium nitrate (as described in Note 6 of this Part)	1, 250	5, 000
Arsenic pentoxide, arsenic (V) acid and/or salts	1	2
Arsenic trioxide, arsenious (III) acid and/or salts	0.1	0.1
Bromine	20	100
Chlorine	10	25
Nickel compounds in inhalable powder form (nickel monoxide, nickel dioxide, nickel sulphide, trinickel disulphide, dinickel trioxide)	1	1
Ethyleneimine	10	20
Fluorine	10	20
Formaldehyde (concentration ≥90%)	5	50
Hydrogen	5	50
Hydrogen chloride (liquefied gas)	25	250
Lead alkyls	5	50
Liquefied extremely flammable gases (including LPG) and natural gas (whether liquefied or not)	50	200
Acetylene	5	50
Ethylene oxide	5	50
Propylene oxide	5	50
Methanol	500	5, 000
4, 4-Methylenebis (2-chloraniline) and/or salts, in powder form	0.01	0.01
Methylisocyanate	0.15	0.15
Oxygen	200	2, 000
Toluene diisocyanate	10	100
Carbonyl dichloride (phosgene)	0.3	0.75
Arsenic trihydride (arsine)	0.2	1
Phosphorus trihydride (phosphine)	0.2	1
Sulphur dichloride	1	1
Sulphur trioxide	15	75
Polychlorodibenzofurans and polychlorodibenzodioxins (including TCDD), calculated in TCDD equivalent	0.001	0.001

The following CARCINOGENS at concentrations above 5% by weight: 4-Aminobiphenyl and/or its salts, Benzotrichloride, Benzidine and/or salts, Bis (chloromethyl) ether, Chloromethyl methyl ether, 1,2-Dibromoethane, Diethyl sulphate, Dimethyl sulphate, Dimethylcarbamoyl chloride, 1,2-Dibromo-3-chloropropane, 1,2-Dimethylhydrazine, Dimethylnitrosamine, Hexamethylphosphoric triamide, Hydrazine, 2-Naphthylamine and/or salts, 4-Nitrodiphenyl and 1,3-Propanesultone	0.5	2
Petroleum products: (a) gasolines and naphthas (b) kerosenes (including jet fuels) (c) gas oils (including diesel fuels, home heating oils and gas oil blending streams)	2,500	25,000

[*From COMAH (SI 2005 No 1008), Sch 1, Pt 2*].

Notes to Table 1:

- It will be noted that 'petroleum products' have had their thresholds almost halved resulting in more petroleum product producers and distributors being regulated.
- The Safety Reports and MAPPs required need to be more specific reflecting the impact, severity and scale of effect of the named substances.

Ammonium nitrate

A major amendment to COMAH (SI 1999 No 743) is in the changes and extensions of Ammonium Nitrate (AN) with regards to the Lower Tier (LT) and Top Tier (TT) thresholds as shown in Table 2.

Table 2: Revised levels for ammonium nitrate

AN type	*LT (tonnes)*	*TT (tonnes)*
'Self-sustaining decomposition' fertilisers	5,000	10,000
Fertiliser grade AN	1,250	5000
Technical grade AN	350	500
Those Fertilisers not meeting the 'detonation resistance test' and off-specs	10	50

The implication of these changes is to ensure that AN is now regulated at a much lower threshold. For example, 'reject AN' is now on the 'radar screen' so that the Toulouse type disasters can be prevented or mitigated.

The COMAH (*SI 2005: 1008*) states that the classification of substances shall be based on *Chemicals* (*Hazard Information and Packaging for Supply*) *Regulations 2002*:

Table 3: Categories of dangerous substances

	Quantity in tonnes	
Column 1	*Column 2*	*Column 3*
1. VERY TOXIC	5	20
2. TOXIC	50	200
3. OXIDISING	50	200
4. EXPLOSIVE (see Note 2) where the substance, preparation or article is an explosive within UN/ADR Division 1.4	50	200
5. EXPLOSIVE (see Note 2) where the substance, preparation or article is an explosive within UN/ADR Division 1.1, 1.2, 1.3, 1.5 or 1.6 or risk phrase R2 or R3	10	50
6. FLAMMABLE, where the substance or preparation falls within the definition given in Note 3(a)	5,000	50,000
7a. HIGHLY FLAMMABLE, where the substance or preparation falls within the definition given in Note 3(b)(i)	50	200
7b. HIGHLY FLAMMABLE liquids, where the substance or preparation falls within the definition given in Note 3(b)(ii)	5,000	50,000
8. EXTREMELY FLAMMABLE, where the substance or preparation falls within the definition given in Note 3(c)	10	50
9. DANGEROUS FOR THE ENVIRONMENT risk phrases:		
(a) R50: 'Very toxic to aquatic organisms' (including R50/53)	100	200
(b) R51/53: 'Toxic to aquatic organisms: may cause long term adverse effects in the aquatic environment'	200	500
10. ANY CLASSIFICATION not covered by those given above in combination with risk phrases.		
(a) R14: 'Reacts violently with water' (including R14/15)	100	500
(b) R29: 'in contact with water, liberates toxic gas'	50	200

Where a substance or group of substances named in Table 1 also falls within a category in Table 3, the qualifying quantities set out in Table 1 must be used.

Dangerous substances present at an establishment only in quantities not exceeding 2% of the relevant qualifying quantity are ignored for the purposes of calculating the total quantity present, provided that their location is such that they cannot initiate a major accident elsewhere on site.

Explosives **6.7**

The Enschede fireworks explosion in the Netherlands in 2000 has lead to a more specific definition of explosives.

Note 2 to *COMAH* (*SI 1999 No 743*), *Sch 1, Pt 3*, defines an 'explosive' as:

(a) (i) a substance or preparation which creates the risk of an explosion by shock, friction, fire or other sources of ignition;

(ii) a pyrotechnic substance is a substance (or mixture of substances) designed to produce heat, light, sound, gas or smoke or a combination of such effects through non-detonating self-sustained exothermic chemical reactions; or

(iii) an explosive or pyrotechnic substance or preparation contained in objects;

(b) a substance or preparation which creates extreme risks of explosion by shock, friction, fire or other sources of ignition.

Now with the Amendment Regulations (SI 2005: 1008), the explosive categories are based on their hazard potential (inherent harm) rather than some probability and severity computation. This issue now is whether the firework represents danger to, for instance, a consumer or worker. Also covered are pyrotechnics (see Table 4).

Table 4: The new classification of explosives

Type	*LT (tonnes)*	*TT (tonnes)*
EXPLOSIVE where the substance, preparation or article is an explosive within UN/ADR Division 1.4	50	200
EXPLOSIVE where the substance, preparation or article is an explosive within UN/ADR Division 1.1, 1.2, 1.3, 1.5 or 1.6 or risk phrase R2 or R3	10	50

The UN/ADR reference is to risk classification where 1.1 represents articles and substances with a 'mass explosion hazard' whereas 1.6 represents 'extremely insensitive articles which do not have a mass explosion hazard'. As the thresholds indicate, the majority of explosives are now regulated in the UK as they are across the EU members States that have transposed the Sveso III Directive.

Aggregation

6.8

Changes have been made to the Rule of Aggregation by the Amendment Regulations (SI 2005: 1008). This Rule is used to determine if COMAH applies when there are substances present but, for example, they are lower than the normal TT or LT value ranges stated above.

The Amendment Regulations (SI 2005: 1008), Sch 1, provides the formulae to use in determining application:

> *If the sum –* $q_1/Q_{U1} + q_2/Q_{U2} + q_3/Q_{U3} + q_4/Q_{U4} + q_5/Q_{U5} + \ldots$ *is greater than or equal to 1, where –*
>
> *(a)* qx = *the quantity of dangerous substance x (or category of dangerous substances) falling within Part 2 or 3 of this Schedule; and*
>
> *(b)* QUX = *the relevant qualifying quantity for substance or category x from column 3 of Part 2 or 3,*
>
> *then these Regulations apply.*
>
> *If the sum –* $q_1/Q_{L1} + q_2/Q_{L2} + q_3/Q_{L3} + q_4/Q_{L4} + q_5/Q_{L5} + \ldots$ *is greater than or equal to 1, where –*
>
> *(a)* qx = *the quantity of dangerous substance x (or category of dangerous substances) falling within Part 2 or 3 of this Schedule; and*
>
> *(b)* QLX = *the relevant qualifying quantity for substance or category x from column 2 of Part 2 or 3,*
>
> *then these Regulations, save regulations 7 to 14, apply.*

What does this mean? The formulae are taking the ratio of a dangerous substance that is held by the operator and comparing it to what the COMAH Regulations specifies in the TT and LT Threshold values in Table 1. If the answer is a ratio of 1 or higher this means hazard potential and requires the operator to notify the Competent Authority. This is partly to ensure that those operators who bring dangerous substances in smaller batches or 'temporarily' onto premises are regulated, but even though these qualities are 'small' they represent a serious and imminent danger.

Aggregation needs to be done separately for:

- toxic and very toxic substances;
- oxidising, explosive, flammable, highly flammable and extremely flammable substances;
- substances dangerous for the environment.

This is to ensure that the operator is specifically assessing each article and substance, that they may have manufactured and stored, for their hazard potential.

Non-COMAH Sites **6.9**

The COMAH Regulations 1999 do not apply to the following (subject to some amendments by the 2005 Regulations):

- Ministry of Defence establishments;
- extractive industries exploring for, or exploiting, materials in mines and quaries (but chemical and thermal processing operations involving dangerous substances and their associate storage are brought into scope);
- waste land-fill sites (except for tailing ponds, dams and other operational tailings disposal facilities containing dangerous substances, particularly when they are used in connection with the chemical and thermal processing of minerals);
- transport related activities;
- industrial chemical process establishments not meeting the thresholds identified in Tables 1 and 2;
- industrial chemical process establishments not meeting the thresholds identified in Tables 1 and 2 at **6.6** above.
- to explosives and chemicals at nuclear installations.

Even if establishments are not currently subject to COMAH, operators of such establishments with quantities of dangerous substances which would be brought within the ambit of COMAH if there were modest changes to threshold levels (see also **6.27** below) should nevertheless be aware of COMAH as it:

- is evidence of best practice in the prevention of major accidents;
- provides a mechanism for bench marking their performance and preparedness.

Lower-tier Duties

General duty **6.10**

Under *COMAH* (*SI 1999 No 743*), *Reg 4*, every operator is under a general duty to take *all measures necessary* to prevent major accidents and limit their consequences to persons and the environment.

Although the best practice is to completely eliminate a risk, the Regulations recognise that this is not always possible. Prevention should be considered in a hierarchy based on the principle of reducing risk to a level as low as is reasonably practicable, (ALARP) taking account of what is technically feasible and the balance between the costs and benefits of the measures taken. Operators must be able to demonstrate that they have adopted control measures which are adequate for the risks identified. Where hazards are high, high standards will be expected by the enforcement agencies to ensure that risks are acceptably low.

Major accident prevention policy (MAPP) 6.11

Under *COMAH* (*SI 1999 No 743*), *Reg 5(1)*, every operator must prepare and keep a document (MAPP) setting out his policy with respect to the prevention of major accidents.

The MAPP document must be in writing and must include sufficient particulars to demonstrate that the operator has established an appropriate safety management system. *COMAH* (*SI 1999 No 743*), Sch 2, sets out the principles to be taken into account when preparing a MAPP document and at *para 4* lists the following specific issues to be addressed by the safety management system:

(a) *organisation and personnel* – the roles and responsibilities of personnel involved in the management of major hazards at all levels in the organisation. The identification of training needs of such personnel and the provision of the training so identified. The involvement of employees and, where appropriate, sub-contractors;

(b) *identification and evaluation of major hazards* – adoption and implementation of procedures for systematically identifying major hazards arising from normal and abnormal operation and the assessment of their likelihood and severity;

(c) *operational control* – adoption and implementation of procedures and instructions for safe operation, including maintenance of plant, processes, equipment and temporary stoppages;

(d) *management of change* – adoption and implementation of procedures for planning modifications to, or the design of, new installations, processes or storage facilities;

(e) *planning for emergencies* – adoption and implementation of procedures to identify foreseeable emergencies by systematic analysis and to prepare, test and review Emergency Plans to respond to such emergencies;

(f) *monitoring performance* – adoption and implementation of procedures for the on-going assessment of compliance with the objectives set by the operator's major accident prevention policy and safety management system, and the mechanisms for investigation and taking corrective action in the case of non-compliance. The procedures should cover the operator's system for reporting major accidents or near misses, particularly those involving failure of protective measures, and their investigation and follow-up on the basis of lessons learnt;

(g) *audit and review* – adoption and implementation of procedures for periodic systematic assessment of the major accident prevention policy and the effectiveness and suitability of the safety management system; the documented review of performance of the policy and safety management system and its updating by senior management.

The MAPP is a concise but key document for operators that sets out the framework within which adequate identification, prevention/control and mitigation of major accident hazards is achieved. Its purpose is to compel operators to provide a statement of commitment to achieving high standards of major hazard control, together with an indication that there is a management system covering all the issues set out in (a)–(g) above. The guidance to the Regulations suggests that the essential questions which operators must ask themselves are as follows:

- Does the MAPP meet the requirements of the Regulations?
- Will it deliver a high level of protection for people and the environment?
- Are there management systems in place which achieve the objectives set out in the policy?
- Are the policy, management systems, risk control systems and workplace precautions kept under review to ensure that they are implemented and that they are relevant?

Not only must the MAPP be kept up to date (*COMAH* (*SI 1999 No 743*), *Reg 5*(*4*)), but also the safety management system described in it must be put into operation *COMAH* (*SI 1999 No 743*), *Reg 5*(*5*).

Notifications **6.12**

Notification requirements are set out in *COMAH (SI 1999 No 743*), *Reg 6*. 'Notify' means notify in writing but by virtue of the 2005 Amendments this now includes e-mail.

Within a reasonable period of time before the start of construction of an establishment and before the start of the operation of an establishment, the operator must send to the CA a notification containing the information which is specified in COMAH (SI 1999 No 743), Sch 3. The Amendment Regulations 2005 now specify the timescales for compliance per regulation. For example, notification to the Competent Authority before the construction process begins, of a TT establishment. The HSE has produced an extensive table of the timescales for compliance, which can be seen at http:/www.hse.gov.uk/comah/background/summary.pdf at Appendix 2.

The operator of an existing establishment was required to send a notification containing the specified information by 3 February 2000, unless a report had been sent to the HSE in accordance with the *Control of Industrial Major Accident Hazards Regulations 1984* (*SI 1984 No 1902*), *Reg 7*.

The following information is specified in *Schedule 3* for inclusion in a notification:

(a) the name and address of the operator;

(b) the address of the establishment concerned;

(c) the name or position of the person in charge of the establishment;

(d) information sufficient to identify the dangerous substances or category of dangerous substances present;

(e) the quantity and physical form of the dangerous substances present;

(f) a description of the activity or proposed activity of the installation concerned;

(g) details of the elements of the immediate environment liable to cause a major accident or to aggravate the consequences thereof.

Notification is a continuing duty. *COMAH, Reg 6(4)* provides that an operator must notify the CA forthwith in the event of:

- any significant increase in the quantity of dangerous substances previously notified;
- any significant change in the nature or physical form of the dangerous substances previously notified, the processes employing them or any other information notified to the CA in respect of the establishment;
- *Regulation* 7 ceasing to apply to the establishment as a result of a change in the quantity of dangerous substances present there; or
- permanent closure of an installation in the establishment.

Information that has been included in a safety report does not need to be notified.

Duty to report a major accident 6.13

Where a major accident has occurred at an establishment, the operator must immediately inform the CA of the accident (*COMAH* (*SI 1999 No 743*), *Reg 15*(*3*)).

This duty will be satisfied when the operator notifies a major accident to the HSE in accordance with the requirements of the *Reporting of Injuries, Diseases and Dangerous Occurrences Regulations 1995* (*SI 1995 No 3163*).

The CA must then conduct a thorough investigation into the accident (*COMAH, Reg 19*).

The report 'COMAH Major Accidents Notified to the European Commission, England, Scotland and Wales 1999–2000' provides details of ten major accidents (one of which resulted in the death of an employee) and three near misses notified to the European Commission (EC). The report describes the causes of each accident, their consequences and the enforcement action taken by the CA.

The CA is required to notify certain major accidents to the EC (*COMAH* (*SI 1999 No 743*), *Reg 21*). The criteria include:

- the release of a specified quantity of a dangerous substance;
- specified harm to persons (eg death);
- specified harm to the environment (eg significant damage to more than 10 km of river); or
- in some circumstances, as a 'near miss' of particular technical interest.

[*COMAH* (*SI 1999 No 743*), *Sch 7, Pt 1*].

The notifications are sent to the Major Accident Hazards Bureau of the EC Joint Research Centre (JRC) based at Ispra (Italy). The data is entered onto the Major Accident Reporting System (MARS). The names and addresses of operators are removed and the data is made available to the public on the JRC website at mahbsrv.jrc.it.

The principal conclusion of the UK's first *COMAH* report to the EC are:

- In the period 1999–2000 the UK reported ten EC Reportable Accidents (ECRAs). This is typical of the five to ten accidents reported each year under the previous *CIMAH Regulations* (*SI 1984 No 1902*).
- Whilst recognising that it is difficult to draw conclusions from such a small sample, ECRAs can be used as a crude measure of safety performance. The Accident Frequency Rate (AFR) is 8.3 ECRAs per 1,000 establishments per annum. Alternatively this can be expressed as 1 ECRA per 120 COMAH establishments per annum.
- The AFR rate for ECRAs in the UK appears to be similar to that for Europe as a whole.
- The CA is concerned at both the frequency and magnitude of these accidents and believes that thorough implementation of the *COMAH Regulations* will make a significant contribution towards securing improvements. Such an improvement would mirror the improvement in the AFR following the implementation of the *Off-Shore Installations* (*Safety Case*) *Regulations 1992* (*SI 1992 No 2885*).

Top-tier Duties

Safety report 6.14

Operators of top-tier sites are required to produce a safety report – its key requirement is that operators must show that they have taken all necessary measures for the prevention of major accidents and for limiting the consequences to people and the environment of any that do occur.

Safety reports are required before construction as well as before start-up. The HSE publication *Preparing Safety Reports* (HSG 190), gives practical and comprehensive guidance to site operators in the preparation of a *COMAH* safety report. *COMAH* (*SI 1999 No 743*), *Reg 7*(*1*) requires that within a reasonable period of time before the start of construction of an establishment the operator must send to the CA a report:

- containing information which is sufficient for the purpose specified in *COMAH* (*SI 1999 No 743*) *Sch 4, Part 1, para 3*(*a*) (see **6.15** below); and
- comprising at least such of the information specified in *COMAH* (*SI 1999 No 473*) *Sch 4, Part 2* (see **6.15** below) as is relevant for that purpose.

Within a reasonable period of time before the start of the operation of an establishment, the operator must send to the CA a report containing information which is sufficient for the purposes specified in *Schedule 4, Part 1* and comprising at least the information specified in *Schedule 4, Part 2* (*COMAH* (*SI 1999 No 743*), *Reg* 7(*5*)).

The report sent before start-up is not required to contain information already contained in the report sent before the start of construction.

An operator must ensure that neither construction of the establishment, nor its operation, is started until he has received the CA's conclusions on the report. The CA must communicate its conclusions within a reasonable period of time of receiving a safety report (*COMAH* (*SI 1999 No 743*), *Reg 17*(*1*)(*a*)).

The operator of an existing establishment must send to the CA a report meeting the requirements of *Parts 1* and *2* of *Schedule 4*. Ex-CIMAH top-tier sites that submitted CIMAH safety reports in parts may continue to submit COMAH reports in parts. If a CIMAH report, or any part of it, was up for its three-yearly review before 3 February 2000, a COMAH report was required before that date or such later date (no later than 3 February 2001) as agreed by the CA. If the review would have fallen after that date a COMAH report was required by 3 February 2001. In any other case the safety report must be sent by 3 February 2002.

The CA's Safety Report Assessment Manual is published on the HSE website (www.open.gov.uk/hse/hsehome.htm) and provides details of the criteria used by the CA in assessing safety reports.

Purpose and minimum information 6.15

Figure 1 below illustrates the purpose of a safety report. The minimum information to be included in a safety report is illustrated in Figure 2.

Figure 1: Purpose of safety reports

- Demonstrating that a major accident prevention policy and a safety management system for implementing it have been put into effect in accordance with the information set out in *Schedule 2*.
- Demonstrating that major accident hazards have been identified and that the necessary measures have been taken to prevent such accidents and to limit their consequences to persons and the environment.
- Demonstrating that adequate safety and reliability have been incorporated into the:
  - design and construction; and
  - operation and maintenance;

of any installation and equipment and infrastructure connected with its operation, and that they are linked to major accident hazards within the establishment.

- Demonstrating that On-Site Emergency Plans have been drawn up and supplying information to enable the off-site plan to be drawn up in order to take the necessary measures in the event of a major accident.
- Providing sufficient information to the competent authority to enable decisions to be made in terms of the siting of new activities or developments around establishments.

[*COMAH* (*SI 1999 No 743*), *Sch 4, Pt 1*].

Figure 2: Minimum information to be included in safety report

1. Information on the management system and on the organisation of the establishment with a view to major accident prevention.

   This information must contain the elements set out in *Schedule 2*.

2. Presentation of the environment of the establishment:

   (a) description of the site and its environment including the geographical location, meteorological, geographical, hydrographic conditions and, if necessary, its history;

   (b) identification of installations and other activities of the establishment which could present a major accident hazard;

   (c) a description of areas where a major accident may occur.

3. Description of installation:

   (a) a description of the main activities and products of the parts of the establishment which are important from the point of view of safety, sources of major accident risks and conditions under which such a major accident could happen, together with a description of proposed preventive measures;

   (b) description of processes, in particular the operating methods;

   (c) description of dangerous substances:

   - inventory of dangerous substances including–
     - the identification of dangerous substances: chemical name, the number allocated to the substance by the ➡

Chemicals Abstract Service, name according to International Union of Pure and Applied Chemistry nomenclature;

  - ○ the maximum quantity of dangerous substances present;

- physical, chemical, toxicological characteristics and indication of the hazards, both immediate and delayed, for people and the environment;
- physical and chemical behaviour under normal conditions of use or under foreseeable accidental conditions.

4. Identification and accidental risks analysis and prevention methods:

   (a) detailed description of the possible major accident scenarios and their probability or the conditions under which they occur including a summary of the events which may play a role in triggering each of these scenarios, the causes being internal or external to the installation;

   (b) assessment of the extent and severity of the consequences of identified major accidents;

   (c) description of technical parameters and equipment used for the safety of the installations.

5. Measures of protection and intervention to limit the consequences of an accident:

   (a) description of the equipment installed in the plant to limit the consequences of major accidents;

   (b) organisation of alert and intervention;

   (c) description of mobilisable resources, internal or external;

   (d) summary of elements described in sub-paragraphs (a), (b) and (c) necessary for drawing up the On-Site Emergency Plan.

[*COMAH* (*SI 1999 No 743*), *Sch 4, Pt 2*].

All or part of the information required to be included in a safety report can be so included by reference to information contained in another report or notification furnished by virtue of other statutory requirements. This should be done only where the information in the other document is up to date, and adequate in terms of scope and level of detail.

If an operator can demonstrate that particular dangerous substances are in a state incapable of creating a major accident hazard, the CA can limit the information required to be included in the safety report (*COMAH* (*1999 No 743*), *Reg 7*(*12*)).

An operator must provide the CA with such further information as it may reasonably request in writing following its examination of the safety report (*COMAH* (*SI 1999 No 743*), *Reg 7*(*13*)).

Review and revision of safety report **6.16**

Where a safety report has been sent to the CA, the operator must review it:

- at least every five years;
- whenever a review is necessary because of new facts or to take account of new technical knowledge about safety matters; and
- whenever the operator makes a change to the safety management system, which could have significant repercussions.

Where it is necessary to revise the report as a result of the review, the operator must carry out the revision immediately and inform the CA of the details (*COMAH* (*SI 1999 No 743*), *Reg 8*(*1*)).

An operator must inform the CA when he has reviewed the safety report but not revised it (*COMAH* (*SI 1999 No 743*), *Reg 8*(*2*)).

Current CIMAH top-tier sites that have submitted CIMAH safety reports in parts must review each part within five years from the time when that part was sent to the CA.

When modifications which could have significant repercussions are proposed (to the establishment or an installation in it, to the process carried on there or the nature or quantity of dangerous substances present there), the operator must review the safety report and where necessary revise it, in advance of any such modification. Details of any revision must be notified to the CA.

COMAH Emergency Plans

On-Site Emergency Plan **6.17**

The HSE publication *Emergency planning for major accidents* (HSG 191) is essential reading.

COMAH (*SI 1999 No 743*), *Reg 9*(*1*), requires operators of top-tier establishments to prepare an On-Site Emergency Plan. The plan must be adequate to secure the objectives specified in *Sch 5*, *Part 1*, of the Regulations namely:

- containing and controlling incidents so as to minimise the effects, and to limit damage to persons, the environment and property;

- implementing the measures necessary to protect persons and the environment from the effects of major accidents;
- communicating the necessary information to the public and to the emergency services and authorities concerned in the area;
- providing for the restoration and clean-up of the environment following a major accident.

The plan must contain the following information:

- names or positions of persons authorised to set emergency procedures in motion and the person in charge of and co-ordinating the on-site mitigatory action;
- name or position of the person with responsibility for liaison with the local authority responsible for preparing the Off-Site Emergency Plan (see **6.18** below);
- for foreseeable conditions or events which could be significant in bringing about a major accident, a description of the action which should be taken to control the conditions or events and to limit their consequences, including a description of the safety equipment and the resources available;
- arrangements for limiting the risks to persons on-site including how warnings are to be given and the actions persons are expected to take on receipt of a warning;
- arrangements for providing early warning of the incident to the local authority responsible for setting the Off-Site Emergency Plan in motion, the type of information which should be contained in an initial warning and the arrangements for the provision of more detailed information as it becomes available;
- arrangements for training staff in the duties they will be expected to perform, and where necessary co-ordinating this with the emergency services;
- arrangements for providing assistance with off-site mitigatory action.

[COMAH (SI 1999 No 743), Sch 5, Pt 2].

When preparing an On-Site Emergency Plan, the operator must consult:

- employees at the establishment;
- the Environment Agency (or the Scottish Environment Protection Agency);
- the emergency services;
- the health authority for the area where the establishment is situated; and
- the local authority, unless it has been exempted from the requirement to prepare an Off-Site Emergency Plan.

Off-Site Emergency Plan 6.18

The local authority for the area where a top-tier establishment is located must prepare an adequate Emergency Plan for dealing with off-site consequences of possible major accidents. As with the on-site plan, it should be in writing.

The local authority has six months from:

- being notified by the CA that a plan is needed;
- the date by which the operator must prepare the on-site plan; or
- receipt of the information needed to prepare the plan;

whichever is the latest to prepare the Off-Site Emergency Plan.

The objectives set out in *COMAH* (*SI 1999 No 743*), *Sch 5, Pt 1* (see **6.17** above) also apply to Off-Site Emergency Plans.

The plan must contain the following information:

- names or positions of persons authorised to set emergency procedures in motion and of persons authorised to take charge of and co-ordinate off-site action;
- arrangements for receiving early warning of incidents, and alert and call-out procedures;
- arrangements for co-ordinating resources necessary to implement the Off-Site Emergency Plan;
- arrangements for providing assistance with on-site mitigatory action;
- arrangements for off-site mitigatory action;
- arrangements for providing the public with specific information relating to the accident and the behaviour which it should adopt;
- arrangements for the provision of information to the emergency services of other Member States in the event of a major accident with possible transboundary consequences.

An operator must supply the local authority with the information necessary for the authority's purposes, plus any additional information reasonably requested in writing by the local authority.

In preparing the Off-Site Emergency Plan, the local authority must consult:

- the operator;
- the emergency services;
- the CA;
- each health authority for the area in the vicinity of the establishment; and
- such members of the public as it deems appropriate.

[COMAH (SI 1999 No 743), Reg 10(6)].

In the light of the safety report, the CA may exempt a local authority from the requirement to prepare an Off-Site Emergency Plan in respect of an establishment (*COMAH* (*SI 1999 No 743*), *Reg 10*(7)).

Reviewing Emergency Plans 6.19

COMAH (*SI 1999 No 743*), *Reg 11*, requires that Emergency Plans are reviewed and, where necessary, revised, at least every three years. Reviewing is a key process for addressing the adequacy and effectiveness of the components of the Emergency Plan – it should take into account:

- changes occurring in the establishment to which the plan relates;
- any changes in the emergency services relevant to the operation of the plan;
- advances in technical knowledge;
- knowledge gained as a result of major accidents either on-site or elsewhere; and
- lessons learned during the testing of Emergency Plans.

Testing Emergency Plans 6.20

There is a new requirement to test both On-Site and Off-Site Emergency Plans at least every three years. This is effectively a minimum standard.

It is intended that the testing gives confidence that the plans are accurate, complete and practicable. It should be able to show that people following the Emergency Plan could cope within the range of accidents that could occur. The testing should give an indication of the conditions that may exist on and off the establishment in the event of an emergency. It should also show that the plan would work as proposed in:

- controlling and mitigating the effects of an accident;
- communicating the necessary information; and
- initiating the measures which should lead to the necessary restoration of the environment.

Testing should be based on an accident scenario identified in the safety report as being reasonably foreseeable. Tests should address the response during the initial emergency phase, which is usually the first few hours after the accident occurs. This is the phase of an accident response when key decisions, which will greatly affect the success of any mitigation measures, must be made under considerable pressure and within a short period of time. Therefore, this is where a detailed understanding of the likely sequence of events and appropriate counter measures is of great benefit.

Where there have been any modifications or significant changes to the establishment, operators should not wait for the three-year review before reviewing the adequacy and accuracy of the emergency planning arrangements.

On-site and off-site testing 6.21

Testing On-Site and Off-Site Emergency Plans (or parts of plans) at the same time can produce considerable benefits. These benefits include ensuring that both Emergency Plans work effectively together, and offering potential savings by avoiding duplicate testing. For example, the external agencies' roles in mitigation, both on-site and off-site, are described only in the Off-Site Emergency Plan; there is no such reference in the On-Site Plan. Exercising this part of the Off-Site Emergency Plan with the On-Site Emergency Plan can test the effective co-ordination of all emergency response personnel handling a major accident on the site. Agreement needs to be reached between operators, the local authority and CA on the overall objectives of the testing and the most appropriate route to reach these objectives.

There is no requirement to notify the CA of testing and the current policy is for HSE inspectors to monitor tests by means of informal liaison with operators and local authorities. HSE inspectors have received training in the techniques for observing and monitoring tests and detailed guidance on the conduct of tests has been given in *Emergency Planning for Major Accidents* (HSG 191).

No statistics are currently available to determine whether this policy is effective. No doubt it will be reviewed and, if difficulties are experienced in monitoring the tests and their effectiveness, further powers will be sought by the CA.

If the test reveals any deficiencies, the relevant plan must be revised.

Implementing Emergency Plans 6.22

When a major accident occurs or an uncontrolled event occurs which could reasonably be expected to lead to a major accident, the operator and local authority are under a duty to implement the On-Site and Off-Site Emergency Plans (*COMAH (SI 1999 No 743), Reg 12*).

Local authority charges 6.23

A local authority may charge the operator a fee for performing its functions under *COMAH (SI 1999 No 743), Regs 10 and 11*, ie for preparing, reviewing and testing Off-Site Emergency Plans. This has further been defined and elaborated by the 2005 Amendment Regulations.

The charges can only cover costs that have been reasonably incurred. If a local authority has contracted out some of the work to another organisation, the authority may recover the costs of the contract from the operator, provided that they are reasonable.

In presenting a fee to an operator, the local authority should provide an itemised, detailed statement of work done and costs incurred.

Provision of information to the public 6.24

The operator must supply information on safety measures to people within an area without their having to request it. The area is notified to the operator by the CA as being one in which people are liable to be affected by a major accident occurring at the establishment. The minimum information to be supplied to the public is specified in *COMAH (SI 1999 No 743), Sch 6* and is set out in Figure 3 below.

Figure 3: Information to be supplied to the public

- Name of operator and address of the establishment.
- Identification, by position held, of the person giving the information.
- Confirmation that the establishment is subject to these Regulations (*COMAH*) and that the notification referred to in *Regulation 6* or the safety report has been submitted to the competent authority.
- An explanation in simple terms of the activity or activities undertaken at the establishment.
- The common names or, in the case of dangerous substances covered by *Schedule 1, Part 3,* the generic names or the general danger classification of the substances and preparations involved at the establishment which could give rise to a major accident, with an indication of their principal dangerous characteristics.
- General information relating to the nature of the major accident hazards, including their potential effects on the population and the environment.
- Adequate information on how the population concerned will be warned and kept informed in the event of a major accident.
- Adequate information on the actions the population concerned should take, and on the behaviour they should adopt, in the event of a major accident.
- Confirmation that the operator is required to make adequate arrangements on site, in particular liaison with the emergency services, to deal with major accidents and to minimise their effects.
- A reference to the Off-Site Emergency Plan for the establishment. This should include advice to co-operate with any instructions or requests from the emergency services at the time of an accident.
- Details of where further relevant information can be obtained, unless making that information available would be contrary to the interests of national security or personal confidentiality or would prejudice to an unreasonable degree the commercial interests of any person.

Under *COMAH* (*SI 1999 No 743*), *Sch 8*, the CA must maintain a public register which will include:

- the information included in the notifications submitted by the operators under *COMAH* (*SI 1999 No 743*), *Reg 6*;
- top-tier operators' safety reports;
- the CA's conclusions of its examination of safety reports.

Powers to prohibit use **6.25**

The CA is required to prohibit the operation or bringing into operation of any establishment or installation or any part of it where the measures taken by the operator for the prevention and mitigation of major accidents are seriously deficient (*COMAH* (*SI 1999 No 743*), *Reg 18*(*1*)).

The CA may prohibit the operation or bringing into operation of any establishment or installation or any part of it if the operator has failed to submit any notification, safety report or other information required under the Regulations within the required time. Where the CA proposes to exercise its prohibitory powers, it must serve on the operator a notice giving reasons for the prohibition and specifying the date when it is to take effect. A notice may specify measures to be taken. The CA may, in writing, withdraw any notice.

Enforcement **6.26**

The Regulations are treated as if they are health and safety regulations for the purpose of the *Health and Safety at Work etc Act 1974*. The provisions as to offences of the 1974 Act, *ss 33–42*, apply. A failure by the CA to discharge a duty under the Regulations is not an offence (*COMAH* (*SI 1999 No 743*), *Reg 20*(*2*)), although the remedy of judicial review is available.

Summary **6.27**

The following are the key amendments made to *COMAH* (*SI 1999 No 743*) by virtue of *COMAH* (*SI 2005 No 1008*):

(a) use of e-mail to notify (*Regulation 3*(*b*));

(b) chemical and thermal processing of dangerous substances relating to mines, quarries, boreholes and waste land-fill sites now regulated (*Regulation 4*);

(c) there are now time limits that operators must work to with regards to the preparation of a MAPP, submitting a Safety Report, and preparing On-Site Emergency Plan (*Regulations 5–7 and 9*);

(d) operators are required to notify the Competent Authority when there are modifications to an establishment (*Regulation 6*);

(e) operators are required to notify if the Safety Report is revised, for example (*Regulation 8*);

(f) there is a duty to consult those that work in the establishment when there are changes to the Plan (*Regulation 10*);

(g) when preparing the Off-Site Emergency Plan it is necessary to consult with the Competent Authority which may also include the SEPA and the Environment Agency (*Regulation 11*);

(h) there is a duty on the Local Authority to consult and inform the public when the Off-Site Emergency Plan is reviewed (*Regulation 12*);

(i) hospitals, school, etc., must be provided with safety information (*Regulation 13*);

(j) amendments to the LT and TT values (*Regulation 14* and *Schedule 1*);

(k) duty to ensure that those in the establishment, e.g. staff and contractors are trained to respond to COMAH type emergencies (*Regulation 15*);

(l) relates to petroleum products and the need to inform and notify the Competent Authority of each class of petroleum product held (*Regulation 16*);

(m) use of visual media to represent the major accident and inclusion of the names of those that author the Safety Report (*Regulation 17*);

(n) duty on Competent Authority to notify that a Safety Report has been reviewed but not revised (*Regulation 18*);

(o) operator to provide an amended report to the Competent Authority if information is not to be included in a public register, e.g. out of date, sensitive, in progress, etc. (*Regulation 19*).

Case Study: Buncefield Explosion **6.28**

1. *Time and location*: Approximately 6 am on Sunday 11 December 2005 at Buncefield Oil Storage Depot at Hemel Hempstead, UK.
2. *Source*: The Buncefield Investigation – Progress Report (20 February 2006) available at http:/www.buncefieldinvestigation.gov.uk/report.pdf.
3. *The loss*: Some 20 fuel storage tanks lost due to the explosion and fire; 43 people injured; 2,000 evacuated; traffic chaos as part of M1 Motorway closed and diversions set up onto M25; insurable cost in terms of loss of plant, production and asset loss; 630 businesses affected employing some 16,500 persons; uninsurable losses in terms of pain and suffering of residents, employees and contractors; environmental costs as risks of liquids being released into a nearby acquifer; Prohibition Notices served by the Health & Safety Executive (HSE) on both Hertfordshire Oil Storage Ltd (HOSL) and British Pipeline Agency Ltd (BPA on 16 December 2005); also, the business continuity impacts on fuel supply for London and South East. ➡

4. *The event*: As the occurrence was regulated under COMAH 1999 (as amended), this was a 'major accident' as defined by the legislation.
5. *The basic cause*: The Major Incident Investigation Board (MIIB) appointed on the 12 January 2006 by the Health & Safety Commission (HSC) has said in its initial report that "The explosion(s) that caused the extensive damage on and off site, and the early fires, can probably be ascribed to the ignition of a flammable mixture that is thought to have been associated with the visible mist reported by eyewitnesses and visible on CCTV records." The Report says there is insufficient evidence as yet as to why the fuel stored would have ignited (the exact source).
6. *The underlying causes*: The MIIB will produce a further report where the role of human error, effectiveness of safety management systems, risk assessment, competence and training issues, etc. will be identified and discussed.
7. *Management*: The evidence from other Liquid Petroleum Gas/Highly Flammable Liquids (LPG/HFI) disasters such as Mexico City (1995) could be useful isomorphic comparisons. At Mexico City, a BLEVE effect occurred (Boiling Liquid Expanding Vapour Explosion) when a pipe rupture leads to the release of highly flammable liquid, which almost 'cooks' the surrounding operation leading to an increase in temperature of the storage tanks and in turn leading to an explosion. A large radiating plume was seen over Mexico City. In addition, there was a 'popcorn effect' with a series of explosions when the intense radiating heat, flying particles, etc. hit a nearby LPG bottling plant, which was also destroyed. This was one of the world's worst fuel storage disasters. The question is, did Buncefield learn the lessons from Mexico City? A final report from the MIIB is expected later in 2006.

Radiation (Emergency Preparedness and Public Information) Regulations 2001 6.29

The *Radiation (Emergency Preparedness and Public Information) Regulations 2001* (*SI 2001 No 2975*) revoke the *Public Information for Radiation Emergencies Regulations 1992* (*SI 1992 No 2997*) (subject to savings). The 2001 Regulations came into force on 20 September 2001.

The Regulations implement the emergency planning aspects of *Council Directive 96/29/EC*, which lays down basic safety standards for the protection of health of workers and the general public against the dangers arising from ionising radiation. The Regulations impose requirements on operators of premises where radioactive substances are present (in quantities exceeding specified thresholds). They also impose requirements on carriers transporting radioactive substances (in quantities exceeding specified thresholds) by rail or conveying them through public places, with the exception of carriers conveying radioactive substances by rail, road, inland waterway, sea or air or by means of a pipeline or similar means.

Essential guidance is contained in the HSE publication *Guide to the Radiation (Emergency Preparedness and Public Information) Regulations 2001* (L126).

The Regulations (*SI 2001 No 2975*):

- impose a duty on the operator and the carrier to make an assessment as to hazard identification and risk evaluation and, where the assessment reveals a radiation risk, to take all reasonably practicable steps to prevent a radiation accident of limit the consequences should such an accident occur (*Reg 4*);
- impose a duty on the operator and carrier to send the HSE a report of an assessment containing specified matters at specified times and empower the HSE to require a detailed assessment of such further particulars as may reasonably require (*Reg 6, Schs 5* and *6*);
- impose a duty on the operator and the carrier to make a further assessment following a major change to the work with the ionising radiation or within three years of the date of the last assessment, unless there has been no change of circumstances which would affect the last report of the assessment, and send the HSE a report of that further assessment (*Regs 5* and *6*);
- where an assessment reveals a reasonably foreseeable radiation emergency arising, impose a duty on the operator or the carrier (as the case may be) and, in the case of an operator, the local authority in whose area the premises in question are situated, to prepare Emergency Plans (*Regs* 7, *8* and *9* and *Schs* 7 and *8*);
- require operators, carriers and local authorities to review, revise and test Emergency Plans at suitable intervals not exceeding three years (*Reg 10*);
- make provision as to consultation and co-operation by operators, carriers, employers and local authorities (*Reg 11*);
- make provision as to charging by local authorities for performing their functions under the Regulations in relation to Emergency Plans (*Reg 12*);
- in the event of the occurrence of a radiation emergency or of an event which could reasonably be expected to lead to such an emergency, make provision as to the implementation of Emergency Plans, and in the event of the occurrence of a radiation emergency, the making of both provisional and final assessments as to the circumstances of the emergency (*Reg 13*);
- where an Emergency Plan provides for the possibility of an employee receiving an emergency exposure, impose a duty on the employer to undertake specified arrangements for employees who may be subject to exposures, such as dose assessments, medical surveillance and the determination of appropriate dose levels and impose further duties on employers in the event that an Emergency Plan is implemented (*Reg 14*);
- impose requirements on operators and carriers, where an operator or carrier carries out work with ionising radiation which could give rise to a reasonably foreseeable radiation emergency, and on local authorities, where there has been a radiation emergency in their area, to supply specified information to the public (*Regs 16* and *17* and *Schs 9* and *10*).

Public Information for Radiation Emergencies Regulations 1992 6.30

The *Public Information for Radiation Emergencies Regulations 1992* (*SI 1992 No 2997*) were revoked by the *Radiation* (*Emergency Preparedness and Public Information*) *Regulations 2001* (*SI 2001 No 2975*). However, *Regulation 3* of the 1992 Regulations continues in force to the extent that it applies in relation to the transport of radioactive substances by road, inland waterway, sea or air. Any other provisions of the 1992 Regulations continue in force so far as is necessary to give effect to *Regulation 3*.

An employer (or self-employed person) must supply prior information where a radiation emergency is reasonably foreseeable from his undertaking in relation to the transport of radioactive substances by road, inland waterway, sea or air. He must:

(a) supply information to members of the public likely to be in the area in which they are liable to be affected by a radiation emergency arising from the undertaking with without their having to request it, concerning:

  (i) basic facts about radioactivity and its effects on persons/environment;

  (ii) the various types of radiation emergency covered and their consequences for people/environment;

  (iii) emergency measures to alert, protect and assist people in the event of a radiation emergency;

  (iv) action to be taken by people in the event of an emergency;

  (v) authority/authorities responsible for implementing emergency measures.

(b) make the information publicly available; and

(c) update the information at least every three years.

[*Public Information for Radiation Emergencies Regulations 1992* (*SI 1992 No 2997*)*, Reg 3 and Sch 2*].

7

Historical Context of Environmental Health and Communicable Disease Control within the UK

In this chapter:

- Introduction.
- Concise history of infectious disease in the UK.
- Food Safety.
- Present day management and control of infectious disease.
- Case study: Mary Mallon (Typhoid Mary) – a healthy carrier.
- Legal and practical control of food poisoning.
- Food hazard warnings.
- Legislative control of food borne communicable disease.
- Control of public health pests.
- Legislative control of food safety.
- Severe Acute Respiratory Syndrome (SARS).
- Avain Influenza.

Introduction 7.1

The roots of environmental health are fragmented in nature, whose origin can be traced to the late eighteenth century. The Industrial Revolution resulted in major shifts in populations from rural to industrial environments predominately in the larger towns and cities. This resulted in rapid expansion in the size and population of industrial towns. The need to house the industrial inhabitants was great, placing enormous pressure on the existing industrial housing stock. The problems of housing people and providing them with adequate food, water and facilities to remove sewage allowed several killer diseases to spread with terrifying ease. The need to relocate industrial workers close to their workplace had two major effects in relation to human health. The first was the proliferation of high-density housing. Multi-family occupancy of single rooms were not uncommon, such dwellings often lacked basic amenities such as wholesome water supply and water closets. The second was the general absence of effective planning laws, which allowed the placing of domestic properties in the close proximity of the polluting factories.

Metropolitan life 7.2

Public utilities struggled to maintain adequate water supply and drainage systems for the expanding towns. Excessive demand in relation to the inadequate supply led to generalised water shortages. The transmission of waterborne communicable diseases was facilitated by these conditions. The inadequacy of municipal drainage systems encouraged the depositing of polluting matter from industrial and domestic effluent into watercourses, which were used to transport the effluent to the sea. It has been argued that the development of sewers were neither inevitable nor even the best possible solution to the problem of waste disposal. The first paradigm of sewage being merely inconvenient 'run off' and sewers being open streams had to be redefined as 'dangerous pollution' and 'underground waste disposal pipes' respectively. The growing early community of civil engineers needed to change existing environmental and cultural perceptions with technical ones. Sewage became a hazardous by-product of metropolitan life, which engineers had designed a system of pipes, reproducing and relocating polluted streams that had once constituted the natural drainage systems (see Porter D H *The Thames Embankment Environment, Technology and Society in Victorian London* (1998) University of Arkon Press, pp 149/150).

Proliferation of disease 7.3

Industrial towns provided the environmental conditions that assisted the proliferation of communicable diseases such as smallpox, tuberculosis, typhus, cholera, and typhoid, all of which were endemic. During this period deficiencies in food safety were responsible for the escalation and spread of food borne communicable diseases. The changing situation of the occupants of towns and cities relying on supplies of food from suppliers they did not personally know, who obtained food from producers they had never met.

In an attempt to maximise profits suppliers often adulterated food, safe in knowing that there was little alternative sources of purchase for their customers in this captive market. The overall result was that the nutritional value of and microbiological quality of food consumed was significantly reduced to the detriment of the final consumer.

Concise History of Infectious Disease in the UK 7.4

Outbreaks of infectious disease have affected the human population throughout history. Below are brief outlines of some infectious diseases that have inflicted significant suffering on the population of the UK.

Smallpox 7.5

A highly contagious, often fatal disease spread by droplet infection. It affects only humans and was both endemic and epidemic during the eighteenth century. Smallpox is associated with the first vaccine research pioneered by Edward Jenner. It was also the first disease where medical intervention, as opposed to government intervention resulted in a decline in fatalities (Lane J *Social History of Medicine* (2002) Routledge, p 137). In 1967 the World Health Organisation set up an eradication campaign and in 1979 declared that the disease had been eradicated.

Tuberculosis 7.6

Tuberculosis is a bacterial infectious disease spread by droplet infection or from the consumption of contaminated food. It is a general wasting disease, known in earlier times as consumption. In the UK, the incidences of the disease peaked between the years of 1780 to 1830 when the disease became endemic. The highest death rates were in London and the large industrial towns where, between 1839 to 1843, 60,000 deaths per year were due to tuberculosis. This amounted to 17.6% of all deaths during that period.

Since its peak, the incidence of tuberculosis amongst the population had declined due to improvements in living conditions, vaccination and other related public health control measures. Until 1987, it was believed that the in the UK the disease had almost died out. However, since this time there has been an increase in the number of reported cases. The possible causes being related to social deprivation amongst certain sections of the population. Another reason is thought to be persons entering the country from other parts of the world where tuberculosis is much more prevalent.

Current control measures for the disease include a national vaccination programme and tracing service to locate and treat infectious persons.

Typhus 7.7

Typhus is a disease transmitted by lice and fleas that thrives in poor social conditions. It was a major cause of fatalities in the eighteenth century and was given the name of gaol or factory fever. During this period it was both epidemic and pandemic in the UK. The last epidemic was in 1846/47 and was thought to have arrived with the Irish immigrants fleeing the potato famine, when it was known as famine fever (Lane (2002), p 145). There has been a significant decrease in incidents since the mid nineteenth century, due mostly to improved public health and social conditions.

Cholera 7.8

Cholera originated in Asia and spread to Europe in the mid- nineteenth century. It is an acute diarrhoeal illness with wide ranging symptoms – the most severe resulting in death within a few hours. The illness is transmitted by drinking water or eating food contaminated by the cholera bacterium. The disease can proliferate rapidly in areas where sewage and drinking water are not effectively treated, allowing the cholera bacteria to survive.

During the cholera pandemic of 1831, 21,882 deaths occurred in England and Wales out of a population of 14 million. Following this pandemic, joint action by government and the medical profession resulted in improvements in the sanitary conditions of the population, with a resultant decrease in cholera incidences. An influential report of the time was Edwin Chadwick's *Report of an Inquiry into the Sanitary Conditions of the Labouring Population of Great Britain* (1842). This led to the first *Public Health Act 1848* and the subsequent appointment of Inspectors of Nuisances – the abatement of which are still continued today by local authorities (see **7.14** below). The combined result was the provision of improved sewerage and water treatment supply. Another pioneer in the control of cholera was John Snow MD who proved that cholera was a waterborne disease that could be controlled by the provision of public health control measures.

Typhoid 7.9

Typhoid is a disease transmitted by the consumption of contaminated water or food. It was endemic in England during both the eighteenth and nineteenth centuries, when it was often fatal. Its incidence declined after 1871 due to improvements in water supply, sanitation and hygiene conditions.

Food poisoning 7.10

Food poisoning is an acute illness with varying onset times caused by eating contaminated food. The symptoms of food poisoning are related to the causative organism or agent, but usually include one or more of the following symptoms:

- abdominal pain;
- diarrhoea;
- nausea; or
- vomiting.

Bacterial food poisoning is the most commonly encountered type of food poisoning outbreak. In the majority of infections a large number of bacteria in the food are required to produce the illness. The multiplication of the pathogenic bacteria usually occurs within high-risk foods, which are common sources of outbreaks.

High risk foods 7.11

These are foods that usually have a high protein and moisture content. High-risk foods include:

- all cooked meats and poultry products;
- dairy produce;
- cooked eggs and egg-based products;
- seafood including shellfish;
- cooked rice.

The occurrence of food poisoning 7.12

Food poisoning occurs as a result of a sequence of linked events, which begin with the initial contamination of food and concludes with the presentation of symptoms. The sequence is commonly known as the food poisoning chain. It is a notifiable disease.

Although notified cases of food poisoning are not considered to be a reliable indicator of food-borne disease due to considerable under-reporting, statistics are collated by the Health Protection Agency (HPA). For instance, the HPA has estimated that, in 2000, there may have been as many as 1.3 million cases in England and Wales, of which around 370,000 consulted a doctor. However, in that year, only 86,528 cases were either notified or ascertained. This reduced to 70,449 in 2004 (House of Commons Hansard Written Answers for 24 May 2005: Column 78W (Source: HPA)).

The food poisoning chain 7.13

This is a related sequence of events, which can be considered as links in a chain. A food poisoning outbreak consists of:

- contamination of high-risk food with food poisoning bacteria;
- opportunity for the food poisoning bacteria to multiply.

Such opportunities arise when a contaminated high-risk food is kept out of temperature control at ambient temperatures for a sufficient time period, which allows the multiplication of food poisoning bacteria to occur. Bacterial growth occurs between the ranges of 5 degrees to 63 degrees Celsius. Optimal growth occurs at around 37 degrees Celsius. Bacteria undergo a form of asexual reproduction known as binary fission. This is a simple process where a single bacterium divides and forms two bacteria. The time period for this division is known as a generation time. Under optimum conditions some food poisoning bacteria have a generation time of less than ten minutes. The public health significance of this rapid reproduction rate is that 1,000 food poisoning bacteria can produce 1,000,000 bacteria in one hour and forty minutes. 1,000 bacteria per gram of food is not an uncommon level of contamination of high-risk food.

Food Safety

Introduction 7.14

The quality of food eaten is an essential component in ensuring the well-being of the human race. However, as well as meeting nutritional needs, food can also act as a vehicle for the transmission of food-borne communicable disease.

Within the UK, Food Authorities and their authorised enforcement officers assume a significant role in the legal control of the quality, composition and safety of food. They are also the main players in controlling the spread of food-borne communicable disease.

From its origin in the nineteenth century, food legislation has developed and evolved in response to governmental and public concerns regarding food safety. The latest enabling Act – the Food Safety Act 1990 – was implemented to secure the purity of food and the hygienic control of food establishments. During the last decade, however, the enforcement of food safety legislation in the UK has been repealed and amended by European Regulations.

Food safety legislation in the United Kingdom currently comprises of a series of statutory instruments made under the enabling powers of the Food Safety Act 1990. In the past, EU Directives provided a legislative framework for food legislation in all the Member States of the EU. Each State was required to interpret these Directives and implement them into national legislation, in order to meet the requirements of the EU.

This Directive-based legislative system for food hygiene law has increasingly been replaced by European Regulations. Made under the European Communities Act, 1972, these regulations are directly applicable to Member States. Therefore the subordinate legislation, in the case of England it is the Food Hygiene (England) Regulations 2006, was not made under the provisions of the Food Safety Act 1990. The shift away from Directives to European Regulations reflects the EU's preference for centralised legislation.

The key aim of the new Regulations is to improve food safety and, amongst other things, reduce the number of cases of food poisoning. It sets out more clearly the duty of food businesses to produce food safety and to achieve consistency. The legislation includes most areas of farming for the first time, so it covers the whole food chain, from 'farm to fork'. The food safety legislation is now considerable and what follows is merely an outline of some of the key pieces insofar as they are likely to affect food business operators. The Food Standard Agency has produced a number of guidance documents for different types of businesses within the food chain.

Relevant European Regulations in relation to food safety

EU Regulation (EC) No. 852/2004 **7.15**

Article 1 of the EU Regulation (EC) No. 852/2004 sets out general rules for food business operators on the hygiene of foodstuffs. It states that food safety must be present throughout the food chain, starting with the primary production and that the primary responsibility rests with the food business operator. It also states that where food cannot be stored safely at ambient temperatures, particularly frozen food, it is important to maintain the cold chain (Article 1, paragraph 1(c)).

It requires the general implementation of procedures based on what are generally known as the Hazard Analysis of Critical Control Points (HACCP) principles, together with the application of good food hygiene practice that should reinforce the food business operator's responsibility. HACCP is a recognised and recommended system of food safety management that focuses on identifying 'critical points' in a process where food safety problems (or 'hazards') might arise and putting in steps to prevent them from arising. This is referred to as 'controlling the hazards'. It is by no means a new concept but its inclusion in EU Regulation (EC) No, 852/2004, has resulted in concern being expressed, particularly by small businesses.

Under Article 3, food business operators are required to ensure that all stages of production, processing and distribution of food under their control satisfy the relevant hygiene requirements laid down in the Regulation.

The requirement for the application of a set of the HACCP principles to food business undertakings is found in Article 5, which requires food business operators to put in place, implement and maintain a permanent procedure or procedures based on these principles. Article 5, paragraph 2, states that the principles shall consist of the following:

(a) identifying the hazards that must be prevented, eliminated or reduced to acceptable levels;

(b) identifying critical control points at the step or steps at which control is essential to prevent or eliminate a hazard or reduce it to acceptable levels;

(c) establishing critical limits at critical points which separate acceptability from unacceptability for the prevention, elimination or reduction of identified hazards;

(d) establishing and implementing monitoring procedures at critical points;

(e) establishing corrective actions when monitoring indicates that a critical point is not under control;

(f) establishing procedures, which shall be carried out regularly to verify that the measures outlined in points (a)–(e) are working effectively; and

(g) establishing documents and records commensurate with the nature and size of the food business to demonstrate the effective application of the measures outlined in points (a)–(f).

When any modification is made in the product, or the process, or at any step in the chain, food business operators shall review the procedure and make the necessary changes to it.

There is a requirement for food business operators to provide the enforcing authority with evidence of compliance with HACCP principles in the manner that the competent authority requires, taking into account the nature and size of the food business. Food business operators are required to ensure that any documents describing the procedures are developed in accordance with this Article and are up to date at all times. Any documentation, including records, have to be retained for an appropriate period (Article 5, paragraph 4).

Article 6 requires food business operators to co-operate with the competent authorities in accordance with relevant Community legislation or, if it does not exist, with national law. Under Article 2, 'competent authority' means the central authority of a Member State competent to ensure compliance with the requirements of the Regulation or any other authority to which that central authority has delegated that competence.

Annex I of the Regulations:

- describes the general hygiene provisions for primary production and associated operations, including record-keeping; and
- makes recommendations for guides to good hygiene practice.

Annex II of the Regulations describes:

- the general requirements for food premises;
- specific requirements in rooms where foodstuffs are prepared, treated or processed (excluding dining areas and those mentioned under the next bullet-point);
- requirements for movable and/or temporary premises (such as Marquees, market stalls, mobile sales vehicles) and premises used primarily as private dwelling houses but where foods are regularly prepared for placing on the market and vending machines.

EU Regulation (EC) No. 853/2004 7.16

As a general rule, food premises that produce or process products of animal origin have the potential, if not correctly managed to create a significant risk to public health. These premises are subject to EU Regulation 853/2004. To operate such premises will require approval from the relevant food authority.

In order to obtain an approval, an on-site inspection by an authorised officer is undertaken. An approval is only granted if the relevant requirements of the appropriate hygiene conditions are met and an effective food safety management system are in place. Such premises are subject to audits of good hygiene practice by food authorities.

The enforcing authority can suspend an approval or revoke the approval if the requirements of the regulations have not been achieved. A unique enforcement power is the service of a Remedial Action Notice. This notice is served under the provisions of the Food Hygiene (England) Regulations 2006 (see 7.22 et seq). The conditions that have to be fulfilled are that there is a contravention of the regulations or the authorised officer's inspection has been hampered. The effect of such a notice is the prohibition of the use of equipment, imposition of conditions relating to the way business operates, the regulation of the rate of operation or detention of products of animal origin in approved premises.

EU Regulation (EC) No. 854/2004 **7.17**

EU Regulation 854/2004 lays down specific rules for the organisation of official controls on products of animal origin intended for human consumption.

What the Regulations mean to food business operators **7.18**

The general hygiene requirements for all food business operators are specified in Regulation 852/2004. Regulation 853/2004 supplements Regulation 852/2004 in that it lays down specific requirements for businesses dealing with foods of animal origin whilst Regulation 854/2004 relates to the organisation of official controls on products of animal origin intended for human consumption.

Food business operators are the vital stakeholders in the elimination and control of food-borne disease. It is their actions or inactions that create or prevent the contamination of food by microbiological, chemical or physical agents.

The requirements of HACCP involve food business operators in a risk management approach to food safety, which has several parallels to that of traditional health and safety management systems. Traditional health and safety management has evolved under the ethos of self-regulation, which was a clear intention of the Health and Safety at Work etc. Act 1974. Food safety legislation does not share the same origins, and food business operators need to develop proactive approaches to food safety management.

In order to achieve legal compliance with food hygiene, the legislation requires both the commitment to seek out relevant regulatory obligations and the capacity to comply with evolving food safety legislation. For food business operators that have appropriate resources in terms of competent, trained staff, well-designed premises and procedures, compliance is certainly achievable. For food business operators that are without such resources, compliance with the regulatory obligations may sometimes never be achieved.

Flexibility of the HACCP Requirements 7.19

In the introduction to EU Regulation 852/2004, Recital 15 states that:

> 'The HACCP requirements should take into account the principles contained in the Codex Alimentarius. They should provide *sufficient flexibility in all situations*, including in small businesses. In particular, it is necessary to recognise that, in certain food businesses, it is not possible to identify critical control points and that, in certain cases, good hygienic practices can replace the monitoring of critical points. Similarly, the requirement of establishing 'critical limits' does not imply that it is necessary to fix a numerical limit in every case. In addition, the requirement of retaining documents need to be flexible in order to avoid undue burdens for very small businesses.'

Guidance has been issued on the flexibility of HACCP requirements by SANCO (an acronym for the Directorate General for Health and Consumer Affairs) in SANCO/1955/Rev.3, issued on 16 November 2005. This is particularly applicable to small businesses.

Paragraph 3 of the SANCO Guidance suggests that the HACCP concept allows HACCP principles to be implemented with the required flexibility so as to ensure that it can be applied in all circumstances, pointing out that Annex II gives guidance on a simplified implementation of the HACCP requirements *particularly in small food businesses*.

The SANCO Guidance recommends that, in applying the seven principles, the following activities should be carried out in sequence:

- A team which involves all parts of the food business should be assembled (HACCP team).
- A full description of the product should be drawn up, including relevant safety information.
- The intended users should be identified.
- A detailed flow diagram should be drawn up.
- The HACCP team should confirm the flow diagram during operating hours.
- The HACCP team should conduct a hazard analysis and identify control measures.

Two components that the SANCO Guidelines consider essential to HACCP implementation are the full commitment and involvement of management and the workforce, and the presence of pre-requisite food hygiene requirements. It follows that food business operators are required to ensure that all personnel are aware of the hazards, the critical points relating to food safety, the corrective and preventable measures and the relevant documentation.

The HACCP and Pre-requisite Requirements 7.20

Pointing out that HACCP systems are not a replacement for other food hygiene requirements, the SANCO Guidelines suggest that pre-requisite food hygiene requirements should already be in place under the following headings:

- Infrastructural and equipment requirements
- Requirements for raw materials
- The safe handling of food (including packaging and transport)
- Food waste management
- Pest control procedures
- Cleaning and Disinfection
- Water quality
- Maintenance of the Food Chain
- The health of staff
- Personal hygiene
- Training.

Included in the pre-requisite food hygiene requirements is traceability (see Article 18 of EU Regulation 178/2002). When food is withdrawn, there is a duty to inform the competent authorities under Article 19 of EU Regulation 178/2002.

The SANCO Guidance points out that efficient and accurate record keeping is essential to the application of the HACCP system and reinforces the need to ensure that all personnel are aware of the hazards, the critical points of food safety in the production, storage, transport and/or distribution process, and the corrective and preventative measures.

Enforcement of the Legislation 7.21

In the UK, regulations have been formulated and implemented for the enforcement of the European food regulations. They are:

- The General Food Regulations 2004
- The Food Safety Act 1990 (Amendment) Regulations 2004.

Both pieces of legislation came into force at the end of 2004 or at the beginning of 2005 to facilitate the enforcement provisions contained in EU Regulation 178/2002. They deal principally with the export of food to third countries.

The main requirements concern:

- food safety requirements of such foods;
- the presentation and labelling of such foods;

- the traceability of such foods;
- the requirement to withdraw unsafe food.

Failure to comply with any of these requirements is a criminal offence.

The Food Hygiene (England) Regulations 2006 7.22

Although these relate specifically to England, equivalent regulations have been issued to cover Wales, Scotland and Northern Ireland. The objective of these regulations is the provision of an enforcement framework for the practices and procedures introduced by the three EU Regulation 852/2004, 853/2004 and 854/2005, together with other various European Regulations, a number of which were published in 2005. Whilst relevant to specific trades or products, they are not of a general nature and will only be of concern to a small minority of food business operators. Further details can be acquired from the website of the Food Standards Agency (www.food.gov.uk).

As the Food Hygiene (England) Regulations 2006 have been made under the European Communities Act 1972, the enabling powers of the Food Safety Act 1990 were not used. Therefore the enforcement notices relating to food premises formerly located in the Food Safety Act 1990 are now to be found within the Food Hygiene (England) Regulations 2006.

The regulations identify the relevant enforcing authorities and define the offences and defences. The regulations confer several powers on enforcement officers. These include powers to enter food premises, take samples and serve relevant enforcement notices in the case of non-compliance with statutory provisions.

Hygiene Improvement Notices 7.23

Under Regulation 6 of the Food and Hygiene (England) Regulations 2006, if an authorised officer of an enforcement authority has reasonable grounds for believing that a food business operator is failing to comply with the Hygiene Regulations, he or she may serve a Hygiene Improvement Notice on the operator.

The purpose of the Hygiene Improvement Notice is to require the food business operator to remedy the matters identified in the notice within a specified timescale. The timescale needs to be reasonable. Failure to comply with the requirements is a criminal offence under Regulation 6(2).

Hygiene Improvement Notices may be appropriate in situations relating to:

- the absence of a HACCP-based food safety management system;
- inadequate food hygiene training;

- ineffective pest control measures;
- structural defects within a food premises;

The minimum period for compliance is 14 days from the service of the notice.

Hygiene Prohibition Order 7.24

Under Regulation 7(1) if:

(a) a food business operator is convicted of an offence under these Regulations; and

(b) the court by or before which he is so convicted is satisfied that the health risk condition is fulfilled with respect to the food business concerned,

the court shall, by an order, impose the appropriate prohibition.

Such an order can require the closure of food premises, prohibit the use of equipment or the methods in which food is made or exclude a person from taking an active role in the management of a food business.

A Magistrates Court may decide to prohibit a person, who has just been convicted, from participating in the management of a food business under Regulation 7(4), which states that if:

(a) a food business operator is convicted of an offence under these Regulations; and

(b) the court by or before which he is so convicted thinks it proper to do so in all the circumstances of the case, may, by an order, impose a prohibition on the food business operator participating in the management of any food business, or any food business of a class or description specified in the order.

Hygiene emergency prohibition notices and orders 7.25

Regulation 8 details the use of Hygiene Emergency Prohibition Notices (HEPNs) by authorised officers. The service of such a notice can result in the immediate, but temporary, closure of the food premises or temporary use of food processes or food treatments.

A requirement of the service of a HEPN is that the officer must be satisfied that one or more of the conditions below are in existence:

- non-compliance by the food business operator with the provisions of the Food Hygiene Regulations;
- unacceptable consequences may result if a HEPN was not served.

- an imminent risk to health is present;
- the authorised officer has no confidence in an offer from the food business operator close voluntarily.

Under Regulation 8(6), as soon as practicable after the making of a hygiene emergency prohibition order, the authorised officer of an enforcement authority must:

(a) serve a copy of the order on the relevant food business operator; and

(b) affix a copy of the order in a conspicuous position on such premises used for the purposes of the food business as he considers appropriate.

Any person who knowingly contravenes such an order shall be guilty of an offence.

Once the order is served on the food business operator, the HEPN will take immediate effect.

The authorised officer is required to affix a copy of the notice on the premises or equipment. The officer must then serve a 'Notice of Intention to apply for an Hygiene Emergency Prohibition Order' (Notice of Intent) on the food business operator. This must be at least one day before making the application to the Court for the Hygiene Emergency Prohibition Order (HEPO). The application for the HEPO to the Magistrates court must be made within 3 days from the service of the HEPN.

The authorised officer is required to specify to the food business operator the matters that constitute the imminent risk. The officer should also specify the measures necessary to remove this risk.

Compensation is payable to the food business operator if an application for an HEPO is not made within this time period or if the Magistrates refuse to grant the HEPO.

Cancelling the HEPO 7.26

Regulation 8(8) states:

> 'A hygiene emergency prohibition notice or a hygiene emergency prohibition order shall cease to have effect on the issue by the enforcement authority of a certificate to the effect that they are satisfied that the food business operator has taken sufficient measures to secure that the health risk condition is no longer fulfilled with respect to the food business.'

The food business operator is entitled to request that the HEPO is cancelled at any time. In response to such an application, the enforcing authority is required to issue a certificate within 3 days of being satisfied that health risk no longer exists. The decision by the enforcing authority must be made within 14 days of the application by the food business operator.

Where an Enforcement Authority is not prepared to issue a certificate, then they are required to issue a 'Notice of Continuing risk to health' specifying their reasons.

A temporary prohibition resulting from a HEPN can be made more permanent by a Magistrates Court issuing an HEPO. An application to a Magistrates Court for such an order must be made by the authorised officer within 3 days of issue of the HEPN.

Remedial action notices and detention notices 7.27

Under Regulation 9, where it appears to an authorised officer of an enforcement authority that an establishment subject to approval under Article 4(2) of Regulation 853/2004 is in breach of the Hygiene Regulations or that he or she is being hampered in any way from making an inspection, he or she may serve a Remedial Action Notice on the relevant food business operator or his or her authorised representative which:

(a) prohibits the use of any equipment or part of equipment specified in the notice;

(b) imposes conditions upon or prohibits the carrying on of any process;

(c) requires the rate of operation to be reduced to the extent specified in the notice.

Defence 7.28

Regulation 11 stipulates that, in any proceedings for an offence under the Regulations, it shall be a defence for the accused to prove that he or she took all reasonable precautions and exercised due diligence to avoid the commission of the offence by himself or herself or by a person under his or her control. However, unless the court allows otherwise, notification of the defence must be given to the court at least 7 days before the hearing or, where he or she has already appeared before the court in connection with the alleged offence, within one month of the first appearance.

Powers of entry 7.29

Under Section 14, an authorised officer of a food authority shall, on producing, if so required, some duly authorised document showing his authority, have a right of entry at all reasonable hours for the purpose of discovering whether the Hygiene Regulations are or have been contravened.

Under Section 15, anyone who intentionally obstructs anyone acting in the execution of the Hygiene Regulations, or anyone, without reasonable cause, who fails to assist any person acting in the execution of the Hygiene Regulations, commits an offence.

Rights of appeal 7.30

The recipient of a Hygiene Improvement Notice has a right of appeal to a Magistrates Court under Regulation 20.

The effect of an appeal is that the Hygiene Improvement Notice is effectively suspended until the Appeal is heard by the Magistrates court. The minimum time period for the appeal is 28 days from the service of the notice or the date for compliance specified on the notice, whichever is the shorter.

If the appeal to the Magistrates Court fails there is a further right of appeal to the Crown Court (Regulation 21).

Regulation 22 allows the Court, on appeal, to cancel, affirm or modify the notice.

Following the issue of a Hygiene Prohibition Order, only the Court that issued the Order can lift it. The prohibited person is not allowed to make an application to the court to lift the order until a period of 6 months from the issue of the order.

Voluntary Procedures 7.31

It may be in the public interest to deal with an imminent risk to health by means of voluntary rather than statutory action. In such circumstances, the officer should consider the risk of the premises reopening without consent from the enforcement authority.

The authorised officer may suggest to the food business operator that a voluntary closure is the preferred option. If this course of action is accepted, the officer should explain to the food business operator that his or her rights to compensation would be relinquished.

It is essential to ensure that the terms of closure are written down and that the person entering into the voluntary agreement is authorised to do so.

After a voluntary prohibition has been agreed, the authorised officer should ensure that frequent checks are carried out on the premises in question in order to ensure that the food business does not recommence trading.

Present Day Management and Control of Infectious Disease 7.32

The Department of Health (DoH) issued official guidance in relation to the above in the form of the *Management of Outbreaks of Food Borne Illness* (1994), DoH, Working Group.

Introduction 7.33

The function of the publication is the provision of guidance on the management of food borne illness. It formulates a non- prescriptive framework allowing authorities responsible for control to tailor the guidance to their particular requirements.

The DoH Working Group comprised recognised experts involved in the control of outbreaks of food borne disease, including Senior Medical Officers, the Public Health Laboratory Service (PHLS), Consultants in Communicable Disease Control (CCDC), Microbiologists and Environmental Health Officers (EHO's).

Management arrangements 7.34

An outbreak of food borne illness is defined as either two or more linked cases of the same illness, or a situation when the observed number of cases unaccountably exceeds the expected numbers.

The objectives of controlling an outbreak of food borne disease are as follows:

- to reduce to the minimum the number of primary and secondary cases of illness;
- to prevent further episodes of illness.

Key players in the control of food borne disease 7.35

Joint responsibility for outbreak management and control lies with LAs and Health Authorities (HAs). Both work in a collaborative manner as recommended in the Guidance Document (HSG 93/56) *Public Health: Responsibilities of the NHS and the Roles of others.*

The role of the health authority 7.36

Using the medical expertise of a CCDC, a division of the Health Protection Agency (HPA), the HA undertake disease surveillance, prevention and control of its districts inhabitants. The CCDC provides medical advice and guidance to the LA officers. The CCDC usually co-ordinates and chairs the Outbreak Control Teams.

The role of the local authority 7.37

LA staff – usually EHOs and Food Safety Officers (FSOs), play a key role in outbreak control. They often carry out the fieldwork in the investigation and control of an outbreak, using their powers as authorised officers under UK legislation.

Outbreak Control Plan 7.38

The guidance requires an Outbreak Control Plan (OCP) to be in existence prior to any outbreak. An OCP should include:

- the roles of the key players;
- the procedures for informing and consulting the Central Disease Surveillance Centre (CDSC) of the PHA and other relevant organisations;
- the liaison arrangements between neighbouring LAs and HAs;
- in cases of major outbreaks of disease, the arrangements for setting up an Outbreak Control Group, including defining its duties and the support that will be made available to it;
- if applicable, the physical facilities required, such as an incident room and communication systems.

Outbreak Control Group 7.39

The guidance document does not set out minimum requirements that would instigate the setting up of an Outbreak Control Group (OCG). It does, however, recommend that a joint local agreement be made between the key players, defining the relevant thresholds for the setting up of an OCG. These thresholds should be incorporated into OCPs to avoid ambiguity.

Membership of an OCG 7.40

Membership should reflect the particular circumstances of the outbreak, but will normally include:

- a CCDC;
- a Chief EHO;
- a Consultant Microbiologist;
- administrative and secretarial support.

Other experts, in different but related fields can be included, dependent on the extent and nature of the disease outbreak.

Terms of reference of an OCG 7.41

An OCG must have agreed terms of reference which covers the organisation and arrangements required for the OCG to operate effectively. The terms of reference should be agreed in advance of an outbreak.

Identification of outbreaks 7.42

Effective surveillance is essential for the early identification of outbreaks or potential outbreaks. All doctors with a direct responsibility for the care of patients are under a statutory obligation to inform the proper officer of the local authority – usually the CCDC of all cases of notifiable disease and food poisoning. The Public Health Laboratories who carry out identification of causative organisms in food or fecal samples play a key role in the management and control of foodborne disease.

Complaints from members of the public to LAs or HAs regarding incidences of food poisoning also provide an important source of information. Information obtained from such persons who have suffered an illness due to eating contaminated food from a food premises, can often identify the actual source of an outbreak.

The majority of the fieldwork in the identification and control of food borne disease is undertaken by LA officers. HAs, however, are responsible for collecting surveillance data. Therefore, joint agreements are required between the HA and LA in order to ensure data is collected effectively in a meaningful manner and that duplication of effort is avoided.

How outbreaks can present themselves 7.43

Outbreaks of food borne disease are clusters of cases of identical illnesses connected by common links. Time, place or person associations connect each case. Sometimes there will be an obvious link (eg persons eating at the same location). On other occasions the links are not obvious and detailed interviews and clinical and laboratory investigations are required.

Pseudo outbreaks 7.44

Pseudo outbreaks occur when an apparent outbreak appears and the background levels remain unchanged. Pseudo outbreaks are not uncommon for a variety of reasons, including a simple matter, such as better reporting by medical staff, to the more complex cross contamination of microbiological samples. The latter can arise by sampling officers or laboratory staff using non-aseptic techniques during the taking or analysis of samples.

Investigation and control of an outbreak 7.45

The guidance provides a staged logical approach, set out as phases that may be utilised in the investigation and control of an outbreak.

Preliminary phase 7.46

During this phase a determination is made as to whether there is a real outbreak. A preliminary hypothesis is formulated and a decision is made as to whether a formal OCG is to be set up.

The phase begins by an assessment being made of all available information. The CCDC and the EHOs have to establish that the clusters of disease are the same illness and whether time, place or person associations link the clusters.

A representative number of sufferers would be interviewed in order to obtain details of the illness, its history and possible sources of infection. Fecal samples are obtained from sufferers for laboratory analysis. Depending on the seriousness of the illness and the source of infection, neighbouring HAs and LAs may be alerted in order to react to a cross boundary outbreak. The CDSC and DoH would also be informed.

When the stage is reached when a formal OCG should be convened, it is the usual practice for the district where the outbreak was first identified to lead in the convening of the OCG.

Where a food premises is implicated in an outbreak 7.47

It is essential that the establishment be visited as soon as possible. This allows samples of suspect food and other physical evidence to be obtained. Inspections of food premises are usually undertaken by EHOs or FSOs under the provisions of UK legislation. Representatives of the CCDC or the PHLS, who can provide additional expertise, can also accompany the officers carrying out the inspection.

EHOs/FSOs will focus on obtaining the maximum information in order to identify the cause of the outbreak. As enforcement officers they will also be seeking evidence that may be required in any subsequent enforcement action or prosecution proceedings.

A food hygiene inspection should be carried out in accordance with UK legislation. Information would be obtained from the proprietor, food handlers or any other relevant person implicated in the outbreak. Information is sought which would assist in the identification and control of an outbreak. Such information would normally include:

- details of menus, food suppliers and production schedules;
- the food processes undertaken including storage, preparation, post preparation activities, areas of cross contamination and temperature control of foods;
- the presence of any hazard analysis systems in place;
- staff details including sickness records and procedures employed for excluding staff with food poisoning symptoms. Also included are details of staff who have recently returned from sick leave or abroad;

- food safety policies such as food hygiene training, pest control, cleaning schedules, hygiene audits and the information, instruction and supervision of food handlers.

Where appropriate EHOs/FSOs may take food and/or environmental samples.

Food samples **7.48**

The main objective in obtaining food samples is to ensure that a representative sample of potentially contaminated food is taken and then delivered to the laboratory in a condition that is in the same condition as when it was taken by the investigating officer. This ensures the maximum accuracy is obtained and that meaningful evidence is obtained for any subsequent enforcement action or legal proceedings.

Environmental samples **7.49**

Environmental samples are taken in order to determine the nature and extent of any contamination of the food. Samples are usually taken from areas where contamination is likely, such as food or hand contact with surfaces or food equipment, especially those used for raw and ready-to-eat high-risk foods. The procedures used for obtaining environmental samples require considerable care and expertise, in order to ensure that false results do not arise due to cross contamination of the samples.

Chemical analysis **7.50**

In cases of chemical contamination of foods, physicians may also provide expertise, as would toxicologists where appropriate. Such persons would then form part of the OCG. Further information on such incidents can be found in *Arrangements to deal with health aspects of chemical incidents* (HSG 38 1993).

The formation of a preliminary hypothesis **7.51**

Following the clinical investigations, patient interviews and the inspection of suspect premises, a preliminary hypothesis will be formulated. This will provide an indication as to the cause of the outbreak and its mode of transmission. After considering this information, a decision will be made by the OCG as to whether or not to continue the investigation. Such a decision will be based on the continuing public health risk and be resources that can be allocated to the investigation.

Communications **7.52**

In order that those involved in the investigation can function effectively, it is important for good communications between the key players to be established and maintained. The investigators should set up regular minuted meetings in order to share information and provide mutual support.

Dealing with the media 7.53

Although not detailed in the DoH guidance, good media relations can assist in the control of food borne communicable disease.

In recent years there has been rapid advances in telecommunication technologies. There are now both television and radio stations dedicated entirely to the provision of news. It is, therefore, important in major food poisoning outbreaks, where there is media interest, for the media to be provided with accurate and consistent information regarding the outbreak investigation. It is good practice for a designated person to be appointed to deal with the media.

Nomination of a Media Liaison Officer 7.54

It is essential that a Media Liaison Officer (MLO) be appointed as soon as possible, as the absence of an MLO may well result in the media representatives approaching any person who they perceive as being available. This in turn can give credibility to rumours and related inaccurate sources being quoted as factual information.

Initial statements should focus on what is happening and the measures being taken to fully investigate the outbreak. Where statements are endorsed by a commitment to provide accurate information as soon as possible, the media are more likely to be restrained in their approach.

Outbreaks of food borne disease may result in criminal prosecutions of the persons responsible under the provisions of the *Food Safety Act 1990* or other subordinate legislation. Therefore, care must be taken to ensure any information released to the media does not prejudice any criminal investigation.

It is also important to foster good equitable relationships with the media. Media teams are constantly under competitive pressure to obtain a story. If different media teams perceive they are being treated unequally, they may decide to make their own arrangements in the pursuit of information. Such actions can adversely affect the progress of an investigation and also prejudice related criminal action. An additional reason for good relations is that the media can be used to alert members of the public to specific food borne hazards and finding of targeted groups who may have been infected but have not been traced.

Record keeping 7.55

It is essential that all parties involved in the investigation keep full contemporaneous notes that are carefully documented and controlled. At the conclusion of an investigation, such records may be required in order to justify the actions taken by investigators. Comprehensive records may also be required in subsequent legal proceedings. The compilation and keeping of detailed records are also required for both local and national reporting of outbreaks.

Epidemiology of an outbreak **7.56**

During an investigation those persons at risk need to be identified, with particular attention being given to food handlers, who may pose a risk of spreading the disease to persons consuming food. Of concern in the transmission of food borne disease, are persons known as carriers who have been infected and are excreting pathogens. These persons are identified, subject to biological monitoring, and excluded from food handling until they no longer excrete pathogens. More insidious are healthy carriers – individuals who exhibit no symptoms, yet still excrete the causative organisms, sometimes for considerable periods of time. Whilst undetected and uncontrolled, healthy carriers can cause serious threats to public health. The importance of identifying the health status and controlling the activities of healthy carriers is well illustrated in the case of Mary Mallon (see **7.60–7.65** below). Mallon was a healthy carrier of typhoid organisms and her actions, until her apprehension and detention resulted in a several deaths.

Data analysis and interpretation **7.57**

The results obtained from the investigation have to be collated in a logical manner. The collation phase identifies:

(a) the persons who were exposed to the suspect food;

(b) the dates and times of illness onset;

(c) the main symptoms of the illness;

(d) the time period between ingestion of the food and onset of symptoms.

Both (b) and (c) above can be important, in the absence of clinical data, of ascertaining the causative agent. The common associations of time, place and persons are also defined.

A logical review of this information will produce a tentative hypothesis, which will produce a probable diagnosis of the illness and the source and the route of transmission of the infection.

Control measures **7.58**

Where an investigation has been carried out in a systematic and logical manner, resultant control measures can be implemented in order to ensure the effective control of the outbreak.

A range of control measures can be implemented and are included later in this chapter:

- Seizure of suspect food (see **7.81** below).
- Closure of food premises (see **7.92** below).

- Service of enforcement notices (see **7.84** below).
- Issuing of Food Hazard Warnings (see **7.69** below)
- Exclusion of carriers (see **7.66** below)

The conclusion of the investigation 7.59

Following the decision by the OCG that the outbreak has ended a final debriefing should take place. There are several advantages in carrying out a debriefing, including defining the lessons to be learned that would lead to improvements being implemented for future investigations.

Case study: Mary Mallon (Typhoid Mary) – a Healthy Carrier

The facts 7.60

Mary Mallon was born on 23 September 1869 in Cookstown, Ireland and immigrated to the USA at the age of fifteen. She began her working life as a domestic servant. During her service she became a talented cook, which paid a higher salary than other domestic workers of Irish immigrant decent. She appeared to be in perfect health and was not known to have ever suffered an infection.

Mallon's notoriety as a carrier began in the summer of 1906, when Charles Warren rented a summerhouse in Oyster Bay, Long Island and hired Mallon as a cook. On 27 August 1906, one of Warren's daughters became infected with typhoid fever. Soon after, other members of the family became infected, with a final total of six out of eleven occupants of the house suffering with typhoid fever.

Typhoid was an unusual disease in Oyster Bay and the owners of the property, fearing the financial consequences of this outbreak, hired a civil engineer called George Soper, who had previous experience in the investigation of typhoid outbreaks. Soper began the search for the source of infection by investigating potential contaminated food sources, water supply and environmental discharges such as sewage. Each of these was progressively eliminated.

Soper focused on Mallon's previous employment history and found that outbreaks of typhoid followed Mallon throughout her employment history. Soper noted that 22 persons became infected in the seven jobs where she was employed. The findings convinced Soper that a correlation existed between these incidents and Mallon's activities. In order to prove that Mallon was a healthy carrier, proof needed to be scientifically determined.

In March 1907, Soper traced Mallon to her place of work where she was employed as a cook. Soper attempted to interview and obtain biological samples from Mallon. Her response to these requests was both negative and hostile.

Detention hospital **7.61**

Soper, although an experienced investigator, had no statutory enforcement powers. He therefore decided to present his evidence of Mallon's carrier status to the Medical Officer of Health of the New York City Department of Health. Acting on Soper's evidence, Mallon was forcibly removed to a detention hospital. During her enforced detention samples of her stools and urine were microbiologically examined. The majority of her stool samples were found to contain typhoid bacteria, proving that Mallon was a healthy carrier, although she appeared to be in excellent health.

As with all healthy carriers, her external appearance was misleading. Whilst cleaning herself, following defecation, it was likely that her hands became contaminated with pathogens. Inadequate personal hygiene practices resulted in pathogens being transferred into the food that Mallon handled and the subsequent typhoid infections to persons who consumed the food.

The detention of an apparently healthy person caused an interesting dilemma between the interest of the State and that of the individual. Using statutory powers contained within the New York Charter, Mallon could have been detained in isolation indefinitely.

Mallon was convinced that she was being unfairly victimised. Mallon, as quoted by Leavitt stated: 'I never had typhoid in my life and have always been healthy. Why should I be banished like a leper and compelled to live in solitary confinement with only a dog for companion' (Leavitt J *Typhoid Mary: Captive to the Public's Health* (2002) Beacon Press, Boston, USA).

Deprivation of liberty **7.62**

After two years of confinement, Mallon unsuccessfully sued the Health Department. The essence of her claim was that she was being deprived of her liberty without ever committing a crime and was under a life sentence. Such action by the State was contrary to the US Constitution. In 1910, a new Commissioner of Health was appointed and reviewed Mallon's case. The review resulted in her release on condition that she did not work as a cook on any future occasion and that she would take the necessary hygienic measures so as not to transfer infection. She was also required to submit regular biological samples to the health authority and notify the health department of any change of address.

Fatal outbreak **7.63**

After first complying with her parole conditions, Mallon disappeared and little was known of her subsequent career history. She surfaced again in 1915, when a typhoid outbreak occurred in the Sloane Hospital for Women. This outbreak was the worst incident of her entire career; 25 cases were involved in this outbreak with two fatalities.

An investigation into the outbreak produced evidence that a recently hired cook, a Mrs Brown, was responsible. Mrs Brown was then found to be Mallon using a pseudonym. Before Mallon could be apprehended she again disappeared. Mallon was later traced, apprehended and forcibly removed and detained by the Department of Health.

During her first detention Mallon received a limited amount of public sympathy. This focused around her being an unwitting healthy carrier. Following her second apprehension, public sympathy was lost. This was due to a combination of her then being of known carrier status and her continuance as a food handler using a pseudonym. Mallon remained imprisoned until her death in 1938.

The public health status of Mallon 7.64

Although Mallon was the first carrier found, other healthy carriers of typhoid were also known during this period, none of which were subject to such draconian measures. Various authors have expressed opinions as to why Mallon was detained for life. Leavitt claims that Mallon was discriminated against on the grounds of her sexual, racial and marital status. These factors combined with her bad temper and self-known carrier status resulted in the extreme measures that the state inflicted upon her (Leavitt (2002), pp 96–125).

Lessons learned regarding disease control involving healthy carriers 7.65

Although undoubtedly this was an extreme and unique case, this was due in part to the publicity surrounding Mallon. Her role as a healthy carrier of a food borne communicable disease illustrates the potential dangers to public health that healthy carriers pose. Healthy carriers, when legal restrictions are applied to remove opportunities for their preferred employment type, may perceive that they are being punished for something over which they have no control.

Carriers, who gain employment in the low paid casual 'cash in hand' industries, can easily disappear from health officials, before they can be apprehended. Mallon's case confirms the truth in the adage that the less employees are paid, the less is known about them in the low paid transient industries, such as catering, where such employees are prevalent.

Mallon's case also illustrates a mode of transmission where food borne disease can be spread without the real cause being suspected during the initial stages of an outbreak investigation. It further illustrates the need for investigators to be open minded and objective in arriving at opinions as to the origin of an outbreak.

Minimizing the potential harm to human health from healthy carriers 7.66

It is important to identify such carriers at the earliest possible stage, by the use of biological sampling, possibly as part of pre-employment screening of food handlers. Such carriers who are engaged in high-risk activities, such as food handlers, need to be legally excluded from work whilst they maintain their healthy carrier status and be closely monitored by the relevant authorities.

Carriers must bc told of their condition and must regularly submit biological samples for laboratory analysis. Carriers should receive adequate health education in order to prevent the soiling of hands with excretions. They must also engage in hygienic practices to prevent the transmission of pathogens.

Legal and Practical Control of Food Poisoning

Legal control 7.67

There are two sets of subordinate regulations made under the enabling powers of the *Food Safety Act 1990*:

- *Food Safety (General Food Hygiene) Regulations 1995 (SI 1995 No 1763)* as amended; and
- *Food Safety (Temperature Control) Regulations 1995 (SI 1995 No 2200)* as amended.

These regulations provide the minimum legal requirements for food businesses for the hygienic control of food operations. There are several other sets of regulations, which control these requirements in specific types of food establishments or processes such as dairy products, eggs, fresh meat, fishery products and bivalves.

Practical control 7.68

These general measures, in essence, reflect the requirements of the legislative requirements. Such measures include:

- adequate separation of raw and cooked foods;
- the use of clean food equipment;
- implementing effective cleaning of food premises and cleaning/disinfection of food equipment and food contact surfaces;
- establishing and maintaining high standards of personal hygiene amongst food handlers;
- effective pest control and pest proofing;
- hygienic control of waste food;

- ensuring that food is thoroughly cooked;
- storing high-risk foods outside danger zone temperatures (between 8 degrees and 63 degrees Celcius);
- effective food hygiene training, instruction, information and supervision of food handlers.

Food Hazard Warnings 7.69

In the control of food borne illness and to prevent injurious foods being consumed, it is sometimes vital to inform and alert other food authorities, food companies, food outlets and members of the public to serious problems concerning food which does not meet safety requirements (see **7.82** below). This is the function of the Food Hazard Warning System.

The Food Hazard Warning System operates mostly through the use of voluntary co-operation of the food industry in order to remove such foods from sale or supply. It has proved to be a useful system for the removal of potentially harmful food products.

Food Safety Act 1990 – Code of Practice No 16: Food hazard warnings 7.70

The Code of Practice explains the roles of the key players such as the Food Standards Agency, Government Departments and the relevant enforcing authorities, in the operation of the Food Hazard Warning System. The Code of Practice provides guidance to food authorities as to the appropriate response required when a Food Hazard Warning or Emergency Control Order is issued (see **7.87** below).

The onus is placed on the food authority to make an initial assessment of the scale, extent and severity of the food hazard. Any action taken must be proportionate and public health risk based. Following the classification of the hazard and its public health risk, a food authority should consider whether or not to set up an Outbreak Control Group (see **7.39** above). There is also a requirement where it is necessary for the food authority to liase with the appropriate Government Departments and the Food Standards Agency.

Food authorities need to equip themselves with suitable facilities and communication systems to receive such notifications, including emergencies received outside normal working hours.

Classification of Food Hazards

Category A notifications – for immediate action 7.71

A food authority must give immediate attention to category A notifications. Typical situations for Category A notifications:

- food hazards to consumers;
- potential to affect vulnerable groups;
- where the food poses a cross contamination risk to other food products.

Action required by the Food Authority:

- ensuring voluntary surrender or withdrawal from sale of relevant foods;
- use the *Food Safety Act 1990, s 9* powers to deal with foods not complying with food safety requirements;
- provision of information to the media;
- communication with retailers, wholesalers etc who may be in possession of the food product.

Category B notifications – for action **7.72**

These notifications specify the action required by food authorities.

Typical situations for Category B notifications:

- food hazards to consumers;
- where voluntary surrender needs monitoring and where *Food Safety Act 1990, s 9* powers may be required for non compliance.

Category C notifications – for action deemed necessary **7.73**

These notifications inform food authorities that a problem has been identified.

Typical situations for Category C notifications:

- potential food hazards to consumers;
- a food safety problem relating to quality has been identified;
- food authorities may contact the media or retailers likely to be selling the product or check premises for the presence of the relevant product.

Category D Notifications – For Information Only **7.74**

These are provided for information only.

Typical situations for Category D notifications:

- to provide information following the withdrawal of a specified food;
- food safety matters of general interest to food authorities.

These notifications are often useful when food authorities are dealing with general enquires from members of the public or food businesses.

Malicious contamination of food 7.75

Where a food authority believes that contamination of food may be the result of a malicious action, they should contact the police. Food authorities are required to fully co-operate with any police investigation. Where the police require further specialist information regarding the food, the food authority should contact any relevant experts such as the CCDC, public analyst etc.

Media relations 7.76

The Code of Practice states the importance of good media relations and a consistent approach. The overriding importance of releasing information is the protection of public health. It also stresses the importance of informing relevant food businesses prior to the release of a media announcement in which they may be included, but this must not be to the detriment of protecting public health.

Legislative Control of Food Borne Communicable Disease

Introduction 7.77

As well as meeting nutritional needs, food can also act as a vehicle for the transmission of food borne communicable disease.

Within the UK, food authorities (usually local authorities or Port Health Authorities) and their authorised enforcement officers assume a significant role in the legal control of the quality, composition and safety of food and preventing the spread of food borne disease.

Food legislation from its origin in the in the nineteenth century has developed and evolved in response to governmental and public concerns regarding food safety. The latest enabling Act – *the Food Safety Act 1990 (FSA 1990)* was implemented to secure the purity of food and the hygienic control of food establishments.

Food Safety Act 1990 7.78

Under the Act the definition of food is given a wide meaning and includes drink. The purpose of the Act is to safeguard food during its manufacture, transport, storage and sale.

Enforcement of the Act **7.79**

LAs, including port health authorities are the food enforcing authorities for the purposes of the Act. It is the duty of each food authority, unless a specific legal function stating otherwise, to enforce the provisions of the Act.

In relation to dealing with food borne disease and other food safety matters, authorised officers are given extensive powers as summarised below.

Power of entry **7.80**

FSA 1990, s 32 gives authorised officers power of entry into food establishments at all reasonable times. In the case of refusal or where notice of entry would defeat the object of entry, an authorised may enter a premises with a warrant signed by a Justice of the Peace (JP).

The inspection and seizure of suspect food **7.81**

Under *FSA 1990, s 9(1)*, extensive powers are given to authorised officers to inspect and seize suspect food, which is intended for human consumption. An authorised officer of a food authority may at all reasonable times inspect any food intended for human consumption which:

(a) has been sold or is offered for sale; or

(b) is in the possession of, or has been deposited with or consigned to, any person for the purpose of sale or of preparation for sale.

Procedure used by an authorised officer to deal with suspect food **7.82**

By virtue of *FSA 1990, s 9(3)–(8)* an authorised officer who finds food not complying with food safety requirements may either:

(a) give notice to the person in control of the food that, until the notice is withdrawn, the food:

  (i) is not used for human consumption; and

  (ii) either is not removed or is not to be removed, unless to specified place stated on the notice; or

(b) seize the food and remove it in order that it may be dealt with by a Justice of the Peace (JP).

The effect of such a notice is that any person who knowingly contravenes its stated requirements will be guilty of a criminal offence.

In respect of such a notice, an authorised officer's powers are qualified by the requirement for a decision to made as soon as reasonably practicable. The maximum period for detention is 21 days. An authorised officer is also required to withdraw a notice forthwith as soon as he is satisfied that the food safety requirements have been met.

Where an authorised officer is satisfied that food does not meet safety requirements (see **7.83** below), he is required to seize the food to allow it to be taken before a JP. The person responsible for the food who could be liable for prosecution can appear before the JP and both sides may produce expert witnesses. Where a JP is of the opinion that the seized food does not meet food safety requirements, the Justice will condemn the food and order its destruction or removal from the human food chain.

Food not complying with food safety requirements 7.83

FSA 1990, s 8(2) states that food fails to comply with food safety requirements if:

(a) it has been rendered injurious to health by means of any of the operations stated in *section 7(1)* (see **7.89** below);

(b) it is unfit for human consumption; or

(c) it is so contaminated (whether by extraneous matter or otherwise) that it would not be reasonable to expect it be used for human consumption in that state; and

references to such requirements or to food complying with such requirements shall be construed accordingly.

The use of food safety enforcement notices 7.84

An authorised officer can use several enforcement notices when faced with food safety contraventions which relate to the control of food borne communicable disease.

Improvement Notices 7.85

An Improvement Notice is served when a food business fails to comply with any statutory provisions relating to food safety. The notice requires the proprietor of a food business to take the necessary remedial measures in order to achieve minimum legal compliance with relevant food safety legislation (*FSA 1990, s 10*).

The proprietor has the right of appeal to a Magistrates Court if he is aggrieved. Subject to the right of appeal, it is an offence not to comply with an Improvement Notice.

Prohibition Orders **7.86**

There are two types of Prohibition Order (*FSA 1990, s 11*). The first type is served by the court on a proprietor of a food business who has been convicted of an offence where a health risk condition involves a risk of injury to health. In the control of food borne communicable disease it could be used, for example to prohibit:

- the use of a treatment such as inadequate pasteurisation temperatures being achieved in the treatment of raw milk;
- the use of a food premises contaminated with sewage.

The other type of Prohibition Order prohibits a person from participating in the management of a food business.

Emergency Prohibition Notices and Orders **7.87**

Under *FSA 1990, s 12,* the use of Prohibition Orders are of limited use in the control of food borne communicable disease; this is due to the considerable time period between the actual offence and the trial of the defendant charged with the offence. It is only on conviction that a Prohibition Order can be issued by the court – this intervening period can be up to several months.

The use of Emergency Prohibition Notices allow an authorised officer to take immediate action without the requirement for court intervention. An authorised officer can, therefore, respond immediately to a situation where a food business poses an imminent risk of injury to health. For example where a food business:

- is definitely linked to a serious food poisoning outbreak;
- is infested with rodents.

Under such circumstances an authorised officer can serve an Emergency Prohibition Notice (EPN) on the proprietor of the food premises. As soon as practicable following service, the authorised officer must fix a copy of the notice in a conspicuous position on the food premises. The authorised officer then has to make a formal application to a magistrates court for the court to issue an Emergency Prohibition Order (EPO). This procedure must be undertaken within 3 days of the service on an EPN.

In the case of a successful application the court will issue an EPO. The authorised officer is then required to serve a copy of the order on the proprietor of the food premises and also affix a copy of the order on the relevant food premises. It is a criminal offence to contravene the provisions on an EPO.

Should the Magistrates Court not issue an EPO, then the proprietor is entitled to reasonable compensation from the food authority for the losses incurred.

Emergency Control Orders **7.88**

Although food control is usually effectively controlled at a local level, under certain circumstances it is advantageous for food safety control to be exercised centrally (*FSA 1990, s 13*).

Powers are given to the appropriate minister to issue orders prohibiting operations, which may involve an imminent risk of injury to health. These place conditions on anything prohibited by an Emergency Control Order. For example, if shellfish are harvested from polluted coastal beds, the minister, by Order, can prohibit harvesting from the defined areas, or allow harvesting providing the shellfish are subject to a recognised purification process.

It is an offence to contravene such an order.

Injurious foods **7.89**

FSA 1990, s 7 prohibits practices that have adverse effects on food safety. *Section 7(1)* states that any person who renders any food injurious to health with intent that it shall be sold for human consumption, shall be guilty of an offence.

Any person who renders any food injurious to health by means of any of the following operations namely:

(a) adding any article or substance to food;

(b) using any article or substance as an ingredient in the preparation of food;

(c) abstracting any constituent from food; and

(d) subjecting the food to any other process or treatment,

with intent that it shall be sold for human consumption, shall be guilty of an offence.

Section 8(1) states it is an offence for any person to:

(a) sell for human consumption, offer, expose or advertise for sale for such consumption, or have in his possession for the purpose of such sale or of preparation for sale; or

(b) deposits with, or consigns to, any other person for the purpose of such sale or of preparation for such sale,

any food which fails to comply with food safety requirements.

Although the Act uses the term 'unfit' it does not define its expression. With respect to communicable disease, food could be contaminated, for example, by pathogenic organisms.

Other emergency situations can arise if food is contaminated by chemicals, radio activity or rendered unfit by the presence of moulds or other extraneous matter.

Section 8(3) confers power to an authorised officer to effectively deal with bulk consignments of food. In essence, where the food does not comply, food safety requirements forms part of a batch, and a presumption exists that until proven otherwise, the entire consignment also fails to meet the requirements.

Control of Public Health Pests

Rats and mice **7.90**

Throughout human history rodents have played a significant role in the transmission of epidemic disease. In the UK rats were responsible for the transmission of bubonic plague. Rodents are sometimes responsible for food poisoning incidents such as salmonellosis. Parasitic diseases such as trichinosis which is transmissible to humans is attributable to rats. The presence of infected rat's urine in sewers, ponds and watercourses can transmit a form of leptospirosis in humans known as Weil's disease, a sometimes fatal illness. The physical damage caused by rodents both gnawing and defecating can result in food being rendered unfit for human consumption and the resultant economic losses.

A feature of rodents that poses a threat to human health is their rapid reproduction rate. It has been estimated that even after mortality from various causes are considered, a single pair of rats will produce 130 offspring per annum. During the year life span of a typical female mouse, six litters with an average of six per litter will be produced (Bassett W and Davies F, *Clays Handbook of Environmental Health* (15th edn, 1981) HK Lewis & Co Ltd).

Legislative control of rodents

Prevention of Damage By Pests Act 1949 **7.91**

Under the *Prevention of Damage By Pests Act 1949 (PDPA 1949)*, LAs are given enforcement powers to investigate and eradicate rodents within their administrative area. Under *section 2* of the Act, there exists a legal obligation on a Local Authority (LA) to take the necessary steps to rid its district of rodents. The LA is empowered to inspect land for the presence of rodents and eradicate rodents on its own land. A LA also has enforcement powers to ensure that owners and occupiers of land undertake the same.

Notification of rodent infestations **7.92**

Under *PDPA 1949, s 3*, occupiers of land are under a legal obligation to immediately notify the LA by notice in writing where rodents are living or frequenting such land. Failure to do so can result in the issuance of a Prohibition Notice resulting in immediate closure of the food premises or a designated part of the premises.

Eradication of rodents 7.93

Where it appears to a LA that measures need to be taken to destroy rats and mice on land, or keep such land free from rats and mice, the LA may serve notice on the owner or occupier of land to take reasonable steps to destroy them within a reasonable time period. The notice may specify the eradication treatment to be undertaken. The aggrieved person in receipt of the notice may appeal to a magistrates court (*PDPA 1949, s 4*).

Works in default 7.94

PDPA 1949, s 5 allows a LA to undertake works in default in the eradication of rodents and to claim reasonable expenses for costs incurred.

Block treatments for rats and mice 7.95

An eradication treatment for rats and mice may require rodenticide treatments of groups of premises in different locations. Where rats and mice are found in substantial numbers in different locations, a LA may take such steps as it deems necessary, other than carrying out structural works, to undertake an appropriate eradication treatment (*PDPA 1949, s 6*). Reasonable costs can be recovered by the LA for such works.

Case study: E.coli 0157 outbreak in Wishaw, Lanarkshire, Scotland 7.96

This case study (**7.96–7.106**) is based on the Determination by Graham Cox QC, Sheriff Principal of Sheriffdom of South Strathclyde Dumfries and Galloway into the *E.coli 0157* Fatal Accident Inquiry. The legislation mentioned in the Fatal Accident Inquiry has now been replaced or amended so readers need to be aware that this is an historical assessment of the role played by the Food Safety Act 1990 and associated legislation in the E. coli outbreaks.

The events leading to the Inquiry involved conventional high street butchers, selling both raw and cooked meat products, running what superficially appeared to be a hygienic retail outlet. The consumption of contaminated food produced at the establishment led to the deaths of 21 elderly persons. The Inquiry concluded the deaths were the result of a combination of shortcomings involving:

- poor safety management and hygiene control within the establishment;
- the actions of the Food Safety Officers who carried out food hygiene inspections at the establishment;
- the management of the Food Safety Officers.

The case study will analyse the reasons for the shortcomings which will illustrate the causative factors that led to the loss of both legislative and food safety control that existed at the time and, in the author's opinion, are still in existence.

The nature of E.coli 0157 7.97

E.coli 0157 is a pathogen that produces a toxin, which can result in serious or fatal symptoms in humans. The problem with this pathogen is that only a small inoculus is required to produce the illness. It is highly virulent, requiring only a few organisms to cause ill health in humans. It is often present in raw meat and in the absence of effective hygiene measures, cross contamination between raw meat and ready-to-eat foods can occur.

Legislative Control of Food Safety 7.98

The *Food Safety Act 1990* is the enabling Act, which provides LAs with extensive powers to control foods safety (see **7.78** above). Subordinate regulations made under the *FSA 1990* enhance this enforcement function.

The case study will involve only one set – the *Food Safety (General Food Hygiene) Regulations 1995 (SI 1995 No 1763)* as amended. These regulations place specific requirements on the proprietors of food businesses in relation to food hygiene.

The deficiencies identified by the Inquiry centred around the non-application by the proprietor and poor enforcement of two core components of the regulations training and hazard analysis.

Training 7.99

Schedule 1 (Rules of Hygiene) Chapter X of *SI 1995 No 1763* requires the proprietor of a food business to ensure that food handlers engaged in the food business are supervised, instructed and/or trained in food hygiene matters commensurate with their work activities.

Hazard analysis 7.100

Regulation 4(3) of the *Food Safety (General Food Hygiene) Regulations 1995 (SI 1995 No 1763* as amended), often known as the hazard analysis requirement, states:

> 'A proprietor of a food business shall identify any step in the activities of the food business, which is critical to ensuring food safety and ensure that adequate safety procedures are identified, implemented and reviewed'.

In essence *Regulation 4(3)* can be broadly divided into three simplified components:

- identifying the steps critical to food safety;
- identifying and implementing food safety control procedures;
- monitoring and reviewing the effectiveness of the controls.

Hazard analysis can take several forms from the complex Hazard Analysis and Critical Control Points (HACCP) to the more simplified forms such as Assured Safe Catering (ASC).

Although a hazard analysis is a legal requirement there was, and continues to be, a significant level of non-compliance within the food industry (Mason J, 'Are Numbers Enough' (August 1998) Environmental Health News, p 14).

Cox (1998), p 101 see **7.96** above stated that it is a specific responsibility of the enforcing inspectorate when undertaking food hygiene inspections to identify food safety hazards and to ensure that associated food safety risks are adequately controlled. Cox (1998), p 89 criticised the environmental health officers (EHOs) for failure, prior to the outbreak, to identify hazards relevant to the practices carried out within the food premises.

A combination of lack of compliance by the proprietor and ineffective enforcement by the food authority led to the tragic consequences of this outbreak.

Poor Food Safety Management and Hygiene Control Within the Establishment 7.101

The Inquiry recorded several examples of poor hygienic practices that may have contributed to the *E.coli 0157* outbreak:

- the absence of a safe method of cooking in relation to the correct time/temperature combinations (Cox (1998) p 86);
- the absence of basic food hygiene instructions being given to staff (p 86);
- the assumption by the proprietor that his staff were in possession of adequate food hygiene knowledge (p 86);
- an inability to prevent cross contamination where raw and cooked food processes crossed (p 87);
- the absence of bactericide for use on contaminated food equipment and surfaces (p 89);
- the mistaken belief by the proprietor and his senior staff that biodegradable washing up liquid was a bactericide (p 89);
- senior managers had no idea of the need to use sanitizers/bactericides for the disinfection of food or food equipment and work surfaces (p 90);

- use of contaminated cloths for cleaning purposes which had the effect of applying, rather than removing bacterial contamination (p 91).

If an effective hazard analysis had been undertaken, then these deficiencies would have been identified, remedied or controlled.

Enforcement deficiencies **7.102**

Section 40 of the *Food Safety Act 1990* allows the Minister to issues Codes of Practice (COP) which the enforcing authority must have regard to when carrying out its statutory functions. A COP can, therefore, be considered to be a statutory requirement on an enforcing authority. COP 9 'Inspection Rating – The Priority Classification of Food Premises', requires a food authority to assign a Risk Rating Category to each individual food premises within its administrative area. The COP requires that each food premise be subject to a detailed assessment using defined criteria as summarised below:

- The potential hazard within the food type.
- The current level of legal compliance within the food premises.
- The confidence in management and any control systems relating to food safety.

The principal objective of the inspection-rating scheme is to assign each food premise a risk-rating category, which determines the inspection frequency for each premise. Inspection frequencies range from every six months for Category A premises to every five years for Category F premises. Theoretically, this practice allows food authorities to target their limited resources at the higher risk premises. In order to comply with the COP, a food authority has to carry out their inspection programme of all food premises, in accordance with the minimum inspection frequencies.

There is a statutory requirement for food authorities to report on their performance in relation to the inspection frequencies to both the Audit Commission and Food Standards Agency. These statutory reporting requirements have pressurised food authorities to focus on the number of inspections carried out, rather than focus on the quality and effectiveness of inspections in terms of food safety (Williams B 'A Waste of Resources' (January 1998) Environmental Health News, p 7). In order to meet the target numbers and comply with the minimum inspection frequencies, food inspectors have tended not to carry out in- depth inspections of some high risk food premises and have used 'short-cut' measures (Cox (1998), pp 89 and 91).

In addition, COP 9 requires a food authority to:

- adopt a graduated approach to enforcement, using an educative approach, taking enforcement action as a last resort; and
- have regard to relevant UK and EC Industry Guides on Food Hygiene Practice.

The actions of the Food Safety Officers who carried out food hygiene inspections at the establishment 7.103

Cox stated that 'worthwhile inspections of food premises take time – longer than was devoted to this food premise by inspectors prior to the outbreak (Cox (1998) p 105). He further stated that EHOs, when visiting premises on inspections, must observe what is taking place, discuss food safety matters with staff and spot weaknesses in the operating system (p 87) and also stated that the onus which is placed on food authorities in enormous.

A contributory cause of the outbreak, therefore, was the opposing pressures on the food authority created by striving to comply with COP 9 in relation to meeting the inspection targets and the protocol and quality to be achieved during individual inspections.

Cox raised several specific issues where the results of these conflicting pressures were exhibited in the actions of front line EHOs as follows:

- Concerns noted during food hygiene inspections regarding potential food safety hazards not being recorded on file (p 101) or provided to the proprietor in writing following the inspection (p 105). The lack of such essential information resulted in EHOs who carried out subsequent visits to the establishment being unaware of the long-standing nature of some contraventions. Therefore, when the graduated enforcement approach was undertaken, in the absence of full information contained on file, the resulting situation allowed the practices to continue where *E.coli 0157* contaminated the food premises.

Other recorded deficiencies 7.104

Other deficiencies in the actions of food safety officers were also recorded:

- the failure by Food Safety Officers to assess the competence of food handlers in respect to food hygiene training and safe methods of cooking (Cox (1998) p 97);
- observation of bad food hygiene practices during inspections and failure to question key personnel regarding the same (p 97);
- inadequate attention paid to the method of cooking and cooling of meat products and the contamination of food preparation surfaces by raw meat products (p 97);
- failure to mention or check the need for temperature probes for the core temperatures of meat products (p 98);
- failure to note or inquire as to the time intervals allowed by the food handlers for the disinfectants/sanitizers to carry out their bactericidal effects on food preparation surfaces used for raw or cooked foods (p 98);
- insufficient time taken to inspect a Category A premise. It was noted that the last full food hygiene inspection of the premises took approximately 30 minutes.

A finding of the Inquiry was that a worthwhile inspection takes time – longer than the time that was given to this establishment (p 105);

- lack of the correct aptitude exhibited by a Food Safety Officer when undertaking a food hygiene inspection of the premises (p 100).

The management of the Food Safety Officers **7.105**

Undergraduate EHOs, during the course of their academic study, are required to study the full range of environmental health disciplines – food safety being just one of them. Therefore, during their initial post qualification period, there exists the potential for flawed practices to be missed and, therefore, not recorded by inexperienced EHOs. Good food safety management practice dictates that only experienced qualified officers should be allowed to inspect high-risk food premises or that newly qualified officers should be effectively supervised by experienced officers during such inspections as part of their job specific training.

The Inquiry also recorded specific examples of the lack of management control:

- failure to introduce systems or insist on standards which would result in adequate inspections of premises according to risk (Cox (1998) p 106);
- failure to monitor the performance of EHOs' food hygiene inspections. The Inquiry was of the opinion that this allowed a newly qualified inexperienced EHO to reduce the risk rating from an A to B without consulting management (p 107).

Dr R North (p 103) criticised the Environmental Health Management Team for allowing a newly qualified EHO to carry out a food hygiene inspection of a Category A premises (p 103).

The Inquiry also made an interesting observation of the potential danger of newly qualified EHOs being 'pulled away' from the risk assessment approach required in current food safety inspections by older 'floor, walls and ceilings' EHO's who trained, qualified and practiced pre-1995 (p 106).

Conclusion **7.106**

The consequences of this outbreak were the result of two main areas of deficiency. The first concerned the proprietor's lack of understanding of hazard analysis or his disregard of its relevance to his undertaking.

The second is the actions of the food authority. In response to the pressure to meet two requirements of COP 9 – inspection targets and the adoption of a graduated approach to enforcement – led to conflicting priorities and the resultant shortcuts taken and enforcement deficiencies. Poor management of the field officers exacerbated these shortcomings.

This situation was succinctly stated by B Williams 'A Waste of Resources' *Environmental Health News*, p 7 (1998) who endorsed this view:

> 'Local authorities must stop squandering resources of skilled food inspectors on Mickey Mouse visits that masquerade as food hygiene inspections. I do not want to suffer food poisoning because some EHO is required by his political masters to produce numbers and lists of innumerable visits to assuage the politicians feeling of inadequacy'.

Severe Acute Respiratory Syndrome (SARS)

Introduction 7.107

Severe Acute Respiratory Syndrome (SARS) is a respiratory disease recently reported in Asia, North America, Canada and Europe. Since its appearance in Southern China in November 2000, it has now been reported in 26 countries throughout the world. The causative organism is thought to be a new member of the corona virus family known as SARS corona virus.

Symptoms 7.108

The initial symptoms present a flu like illness with a high fever (>38 degrees Celsius), headache and aching limbs. After two to seven days a dry cough often develops and patients develop respiratory distress. The severity of symptoms range from a mild illness, which clinically does not require hospitalisation to severely ill patients who require respiratory support.

The overall mortality rate is approximately 15%. Most patients, however, especially those under 60 years of age, do recover (see *Health Protection Agency 2003:4 Summary of the UK Public Health Response to SARS. UK SARS Taskforce*, Health Protection Agency (2003) Summary of the UK Public Health Response to SARS).

The World Health Organisation (WHO) regularly publish the numbers of cases of infected persons by country and affected areas. By July 2003 there were approximately 8000 reported cases of SARS, with in excess of 800 deaths.

The emergence of SARS 7.109

The WHO consider SARS to be the first severe and readily transmissible disease of the 21st Century (see *WHO SARS Status of the Outbreak and Lessons to be Learned for the Immediate Future 2003* Health Protection Agency Summary of the UK Public Health Response to SARS. UK SARS Taskforce 24 July 2003 pg 4).

The first known case of SARS arose in November 2002 in the Guangdong Province of China, 305 persons were affected and 5 deaths were recorded. An infected medical physician who had treated patients in the province became infected and brought the virus to the 9th floor of a 4 Star hotel in Hong Kong. Later visitors to this area of the hotel also became infected and were hospitalised in Hong Kong, Vietnam and Singapore. These locations then also became seeded with the SARS Infection.

Concurrent with outbreaks, epidemiological analysis undertaken by the WHO indicated that the disease was being transmitted along international air travel routes due to infected guests travelling to such places as Toronto and other locations. The disease was also spread by medical staff that became infected after treating earlier SARS cases travelling internationally outside the initial outbreak zone.

Rapid increases in SARS cases were particularly prevalent in hospital settings where medical staff who were initially unaware of the significance of SARS fought to save the lives of patients without adopting adequate infection control procedures. As these persons moved outside the initial outbreak environments the transmission of SARS continued.

The transmission of SARS 7.110

The most important routes of transmission are respiratory droplets, direct and indirect contact with bodily fluids from infected to non-infected persons. The majority of cases involve health care workers or family members caring for the patient (see Poutanen S et al *Identification of SARS in Canada*, New England Medical Journal: 384 (20):1995–2003).

Environmental transmission has been postulated which provides an explanation for cases that occur in apartment complexes, where diarrhoea was an obvious clinical feature concomitant with sewage problems (see Health Protection Agency (2003) *Summary of the UK Public Health Response to SARS*).

The global SARS concern 7.111

SARS provides an excellent example of the global mayhem that a new emerging infectious disease can cause.

The emergence of SARS resulted in public panic. Government officials lost their jobs and some areas of the world suffered social instability, which affected the lives of thousands of people. For example, in Singapore the military were deployed to assist in the tracing of contacts and enforcement of SARS quarantine controls.

Global measures against SARS 7.112

There is currently no vaccine or specific treatment for SARS. Under such circumstances the control of the disease centres around the elementary control tools of isolation and quarantine.

In 2000, the WHO formulated a Global Outbreak and Response Network (GOARN) in relation to infectious disease outbreaks. This system links in excess of

100 existing networks which processes the relevant data, expertise and skills to alert the international community for outbreaks of infectious diseases, including SARS. The linking of these networks allows the WHO to closely monitor the evolution of an infectious disease outbreak and to formulate verification and response activities.

A specialised search engine, known as the Global Public Health Intelligence Network (GPHIN), has been used by the WHO since 1997, and its effectiveness was acknowledged for providing the earliest alerts to the SARS outbreaks in China. The advantages of GPHIN in infectious disease control is its timeliness over traditional reporting systems, which begin at a local level and gradually pass to a national level for formal notification to the WHO. The GPHIN has allowed the WHO to rapidly intervene in order to verify or repudiate outbreaks of SARS before they result in social and economic disruption. (see *SARS Status of the Outbreak and Lessons to be Learned for the Immediate Future* (2003) World Health Organisation, Geneva, 20 May 2003).

Control measures for SARS in the UK **7.113**

The control measures implemented in the UK are based on the principles set out by the WHO. The Health Protection Agency (HPA) have acknowledged that preventing infected persons entering the UK is not possible (see Health Protection Agency (2003) *Summary of the UK Public Health Response to SARS*). Optimum containment of the disease transmission is achieved by implementation of the following procedures:

1. Rapid detection of infected cases.
2. Effective infection control measures.
3. Rigorous identification of close contacts of the infected persons.
4. The provision of health education information to raise awareness amongst the general public (especially those linked to SARS infected areas) and health care workers.
5. Effective infection control measures in hospitals where SARS patients are treated.

A combination of these measures can, therefore, be utilised as the appropriate control measures in the prevention of the transmission of SARS within the UK. Their practical implementation can be considered under the following headings:

(I) Surveillance **7.114**

A national SARS surveillance system has been put in place by the Communicable Disease Surveillance Centre (CDSC), a division of the HPA. Medical staff are required to report any potential SARS cases immediately to the CDSC (see *CDSC: Cases Definitions, Case Reporting and Follow up Procedures (Surveillance) of Cases and Controls* [online] HPA 27 May 2003). The information is then assessed by the SARS Team at the CDSC. Where appropriate, local public health teams are alerted of any new cases in order that control measures can be implemented at a local level.

The WHO have recommended that passengers leaving infected areas are screened. This is not considered to be entirely effective as those who have the less severe form of the disease and those who are still incubating the disease may elude identification (see WHO 2003: 97–9 *Weekly Epidemiology Record 78* (14)).

(2) Case management: isolation, infection control, treatment, discharge and follow up **7.115**

An essential component of the control of SARS is early isolation until diagnosis has taken place. The Chief Medical Officer has issued guidance to all chief executives of NHS Trusts and Primary Care Trusts in relation to isolation, infection control for hospitals, ambulance services and primary care establishments. Part of the guidance is the inclusion of several annexes relating to these controls, as discussed in paras **7.116–7.221** below.

Annex C **7.116**

Annex C contains guidance for hospitals, other healthcare settings and ambulances on the prevention of the spread of SARS.

Hospitals **7.117**

In a hospital, the Accident and Emergency Department should transfer infected or suspected infected persons to a single room immediately, without the need to enter the department. Persons that present themselves should be identified at triage and remain in the department for the shortest possible time. A case history should be taken and infection control measures implemented by staff attending the patient.

All healthcare workers must ensure a good standard of hygiene, especially those in contact with bodily fluids.

Patients with suspected SARS being assessed or transported by ambulance **7.118**

Patients with suspected SARS being assessed or transported by ambulance should wear surgical masks to reduce the spread of droplet infection. Healthcare workers, including ambulance staff should also wear surgical masks.

Hospital visiting **7.119**

Close contacts of the patient should be provided with appropriate advice regarding relevant control measures. Such persons should also be medically assessed for fever or respiratory symptoms.

Decontamination procedures for the accommodation and equipment used by patients 7.120

All hospitals are required to adhere to the decontamination procedures as set out in the Hospital Infection Control Manual that all hospitals are required to keep and maintain and also liaise with their Hospital Infection Control Team. A high standard of cleaning is required for all clinical areas. Special attention is required for nubulisers, which have the potential to create an aerosol spread of infection.

Post mortem (PM) examinations of patients that have died from suspected or probable SARS infections 7.121

The range of personal protective equipment that should be used by Post Mortem (PM) staff fall under two main categories:

1. Protective Garments – A surgical scrub suit, surgical cap, impervious gown (with full sleeve coverage), face shield, shoe covers and cut proof gloves are required for PM staff.
2. Respiratory Equipment – PM personnel should wear respirators to the US standard N-95 or N-100. Alternatively, air purifying respirators with high efficiency particulate air filters can also be used.

In addition to PPE and respiratory protection, staff involved in a SARS PM should implement additional safety procedures:

- avoidance of percutaneous injury to the skin;
- removal and correct disposal of outer garments when leaving the immediate autopsy area; and
- effective hand washing following glove removal.

(3) Management of contacts 7.122

Close contacts with probable or suspected SARS cases should be provided with relevant health education information on SARS and informed of the importance of reporting respiratory and SAR like symptoms noted within 10 days of their last contact with the patient.

(4) Information, education and travel advice 7.123

The effective provision of information to the public and health care workers is vital for the early detection and reduction of SARS transmission.

The HPA, through its website (http://www.hpa.org.uk), provides a source of relevant advice, which is updated when appropriate. The Department of Health and

Foreign and Commonwealth Office provide travel alerts and leafleted information for distribution at airports and by port health authorities. The Department of Health has also issued posters for display in all relevant arrival ports in the UK.

(5) Laboratory diagnosis of SARS 7.124

All suspected or probable cases of SARS require laboratory investigation in order to correctly identify the specific SARS infection.

Annex D provides extensive information and advice of the arrangements for the laboratory testing of SARS. The sampling methods used vary according to the severity of the SARS illness in the patient. Testing laboratories have to be advised on the type of sampling required and the procedures required for the sampling of respiratory samples, blood, blood cultures, urine and other relevant samples obtained from clinical diagnosis.

(6) Broader public health measures

Quarantine of people returning from SARS affected areas 7.125

In general, quarantine measures for those returning to the UK have not been considered necessary due to the measures currently in place to ensure the early detection, isolation and treatment of SARS cases.

Closure of schools and other institutions 7.126

The effectiveness of the closure of schools and similar institutions as a control measure is currently unknown. There is no evidence at the present time of SARS transmission in schools or children transmitting SARS infection to adults (see Hon et al 2003: WHO 2003 *Clinical Presentations and Outcomes of SARS in Children Lancet*).

Curtailment of public meetings 7.127

In the UK this measure has not been considered necessary when assessed against the adverse affects of public concern, normal community life and economic activity. Such advice would change in the case of areas affected by recent or current outbreaks of SARS.

Widespread use of masks in the community 7.128

There is no current evidence that the use of masks in the community is effective at reducing the transmission of SARS. The use of facemasks, therefore, is restricted to certain situations, for example patients being managed in the community having to travel to hospital for treatment.

(7) Contingency planning 7.129

As part of their standard emergency planning procedures, health authorities and local authorities should already have in place Infectious Disease Outbreak and Control plans. (see para **7.32**).

The lessons learned from SARS 7.130

It has been acknowledged that the information provided by the WHO led to the rapid detection and isolation of SARS cases [*WHO 2003:6 Severe Acute Respiratory Syndrome-Status of the Outbreak and Lessons for the Immediate Future*]

In addition, the lessons learned from the bio-terrorist attacks in the USA, where anthrax tainted mail was deliberately distributed through the US postal System in October 2001, ensured the control systems were in a state of readiness to control SARS Transmission (see WHO 2003, *Severe Acute Respiratory Syndrome – Status of the Outbreak and Lessons for the Immediate Future*).

The SARS outbreak also showed that the current International Health Regulations, which provide the basis for global surveillance, reporting and control of infectious disease, require urgent update and revision.

Although SARS has now been effectively contained there are still medical concerns regarding its identification, control of transmission and treatment. There is still no reliable and robust diagnostic test, its modes of transmission are not fully understood and there is no specific treatment available.

In the earliest phase of the SARS outbreak, under reporting allowed SARS to become established with the predictable result of its international transmission. What has become apparent, therefore, is that deliberate concealment of SARS cases, through fear of social and economic disadvantages has no long term benefit. The WHO have stated that such action carries a very high price, including loss of international credibility, adverse economic impact, a threat to the health of that geographic area and widespread transmission of the infectious disease (see WHO *SARS Status of the Outbreak and Lessons to be Learned for the Immediate Future* (2003) WHO, Geneva).

SARS, like any other type of novel infectious disease, can result in widespread public panic, in the case of SARS the fear of the virus often spread faster than the virus itself.

However, where good international collaboration exists together with strong but neutral political leadership across the globe, scientists and medical professionals can be motivated to combine their knowledge and skills to face and combat the SARS threat.

Avian Influenza ('Bird Flu')

Introduction 7.131

Avian influenza is a worldwide infectious disease of birds that was first recognised over a century ago. It is a highly contagious viral infection. The susceptibility to infection affects all birds, especially wild waterfowl, and the symptoms of the disease exhibit a wide variation from mild symptoms to a highly infectious disease of epidemic proportions. Such epidemics if not effectively controlled can continue for long periods of time. The manifestation of symptoms is largely dependant upon the strain of virus involved and the species affected with the condition.

Originally known as 'fowl plague', it first appeared in Italy in 1875. However, in April 2006, a dead swan in Fife (Scotland) was confirmed as a carrier of H5N1, the highly pathogenic form of Avian influenza. Also, in April 2006, a Norfolk chicken farm was found to have the H7 strain of Avian influenza. Whilst less virulent towards humans, the H7 strain can spread rapidly amongst birds and it resulted in some 35,000 chickens having to be slaughtered. The UK's Health Protection Agency maintains a daily 'risk watch' of the epidemiology of Avian influenza (http://www.hpa.org.uk).

Avian influenza infections in birds are divided into two groups. Their classification reflecting their ability to cause disease is as follows:

1. Highly Pathogenic Avian Influenza (HPAI) – a serious disease with an extremely high mortality.
2. Low Pathogenic Avian Influenza (LPAI) – a mild form of the disease.

In humans, typical symptoms of Avian influenza include fever, cough, shortness of breath, headache, sore throat, sore eyes and muscular aches. At the present time there is no confirmed evidence of Avian influenza being transferred from person to person.

The causative organism is a type A strain of influenza. Of public health concern is the fact that 15 subtypes of the virus can infect birds. This number of pathogen types means that at any one time there is a large reservoir of subtypes that can be transferred amongst bird populations.

Influenza viruses have evolved properties, which can circumvent the natural defensive mechanisms of the hosts. The genetic make-up of the virus can undergo alteration during replication within the host, producing new antigenic variants through a process known as antigenic drift. Influenza viruses can also exchange and rearrange genetic material and coalesce. This produces new subtypes, which may be unlike the parent viruses. The adverse effects of such new subtypes are that the antigenic drift may both evade the defensive mechanisms of the host and not be susceptible to vaccine control. Such antigenic drifts have often resulted in fatal pandemics where both natural and medical interventions are rendered ineffective.

Avian influenza can spread rapidly between birds, especially in areas such as poultry farms and live bird markets where birds are kept in close proximity. The infected birds transmit the virus through their saliva, nasal fluids and droppings.

Although of public health concern, the financial impact on the poultry industry can be catastrophic. For example, in Europe, a serious epidemic of Avian influenza in the Netherlands in 2003 spread to Belgium and Germany. In total, 250 farms were affected, and in implementing control measures an excess of 28 million poultry were destroyed.

The history of human infection with Avian influenza 7.132

The first recorded incident was in Hong Kong in 1997. Eighteen people were affected. All suffered severe respiratory disease and resulted in 6 deaths. An epidemiological investigation revealed that the human infection ran parallel with an epidemic of the highly pathogenic Avian influenza of an identical subtype that had infected the poultry population located in Hong Kong. The cause of the outbreak was found to be close contact between live infected poultry and members of Hong Kong's population. Genetic studies into the subtypes confirmed that there had been direct transmission of the virus from infected poultry to humans. Since the initial recorded incident there have been further outbreaks in other areas of the world such as Southern China, the Netherlands and Northern Vietnam.

Factors that facilitate the spread of avian influenza from animals to humans 7.133

The close proximity of humans to poultry and pigs produce favourable conditions for an antigenic shift across species. Pigs are especially susceptible to infection with both avian and mammalian viruses. Pigs therefore fulfil the role of a biological mixing vessel, which allows the re-assortment of mammalian and avian viral genetic material and the formation of novel and sometimes highly pathogenic subtypes.

The most pathogenic avian influenza subtype 7.134

Epidemiological studies have identified subtype H5N1 to be of significant public health concern. This subtype has acquired properties of rapid mutation and a predisposition to procure genes from viruses infecting other animal species. It has a high human pathogenicity and has already resulted in severe diseases in humans. Infected birds that survive the infection can excrete the virus, resulting in the transmission to other birds. Therefore poultry farms and markets are particularly vulnerable to the infection.

An increase in the extent of infection in birds will increase the likelihood of increased infection in humans. Humans who become concurrently infected with human and avian influenza strains will form reservoirs for the transfer of genetic material and creation of novel subtypes with the potential of human-to-human transfer. Such a sequence of events may be the start of an influenza pandemic with severe impacts on human health.

The Control of Avian Influenza in the UK

Legislative Control **7.135**

There are a number of legislative controls:

- *EU Council Directive 2005/94/EC on the Control of Avian Influenza*
  - The Directive implements strict requirements for the notification of Avian influenza, the slaughter of birds and the control of movements. It incorporates a flexible risk-based approach to respond to individual situations.
- *Diseases of Poultry (England) Order 2003*
  - The Order introduces Community measures for the control of Avian Influenza. It also gives additional powers to the Department for the Environment, Food and Rural Affairs (Defra) to carry out the necessary checks and surveillance to verify that the disease is not present
- *The Avian Influenza and Newcastle Disease (England and Wales) Order 2003*
  - This order implements the powers of the Animal Health Act 2002 to these diseases. It allows for a preventative cull of poultry and gives Defra officials powers to enter premises and test and sample for these diseases. The slaughter of vaccinated poultry is allowed with the payment of reasonable compensation. Similar legal instruments exist for Scotland and Northern Ireland.
- *The Avian Influenza (Preventative Measures) Regulations 2005*
  - The regulations place a regulatory obligation on any person keeping 50 or more poultry on commercial poultry premises to keep a written record and submit it to Defra.

Other Controls **7.136**

Defra has produced key possible control measures for the prevention and management of Avian influenza. In essence the measures involve the following:

1. *The isolation of poultry*
   This is to ensure that farmed birds and wild birds do not mix.
2. *The surveillance and sampling of poultry*
   This would involve entry into property and sampling to detect the presence of the Avian influenza virus
3. *Notification to Defra by any person keeping 50 or more poultry on commercial poultry premises.*

4. *The surveillance and sampling of wild birds*
   The sampling is for the presence of the Avian influenza virus. Defra believes that this is most likely to detect pockets of LPAI in migratory birds. The detection of H5 or H7 strains would be indicative of the possibility of transmission into domesticated birds and mutation into highly pathogenic strains.

5. *The prohibition of fairs, markets and shows*
   By prohibiting the gatherings of poultry, the transmission route caused by faecal contamination of vehicles, clothing or equipment is eliminated or greatly reduced. Such action would prevent the potential wide transmission from a single infection point.

6. *Controlled movements of poultry*
   The movement of poultry, if effectively controlled, reduces the risk of infected birds spreading infection.

7. *Controlled movements of people, plant and machinery etc.*
   The Avian influenza virus is readily transmitted mechanically through contact with faeces. Restrictions in the movement of people, plant and machinery will control the spread of the disease.

8. *Controlled removal of manure*
   The prohibition of the spreading of poultry litter or manure will prevent the spread of the disease. Therefore effective control needs to be exercised regarding the storage, transport and disposal of manure.

9. *Controlled removal of eggs*
   By restricting the removal of hatching eggs, the spread of the disease will be reduced through breeding.

10. *Disinfection at entrances and exits*
    The spread of the disease is likely to be reduced when the occupier of premises implements effective disinfection at points of access and egress.

The World Health Organisation (WHO) has produced interim measures for the protection of persons involved in the mass slaughtering of animals potentially infected with highly pathogenic Avian influenza viruses.

The WHO deemed this necessary as the exposure to infected poultry and to dust and soil contaminated with poultry faeces can result in human infection. The measures include the following:

1. The provision of suitable personal protective equipment for cullers and transporters.

2. High levels of personal hygiene, including hand washing for cullers and transporters.

3. Effective environmental cleaning procedures in areas of culling.

4. Health surveillance of persons exposed to infected poultry.

5. The exclusion of vulnerable persons, i.e. immunocompromised, persons over 60 and those with chronic heart or lung conditions.

6. Vaccinations with the current WHO influenza vaccine for the avoidance of simultaneous infection by human influenza and Avian influenza and to reduce the likelihood of genetic mixing and the resultant antigenic drift.

7. The thorough and rigorous post mortem examination of specimens in order to detect new viral isolates.

The NHS in their publication 'Bird Flu (Avian Influenza) Public Health Advice for those going to or returning from Bird Flu–Affected Area' outline the precautions that should be taken. Such precautions, for those who are visiting countries with reported outbreaks of H5N1 among poultry, include the following:

1. Avoidance of poultry farms and markets.

2. Avoidance of close contact with live or dead poultry.

3. Not consuming raw or poorly cooked poultry or poultry products.

4. Frequent washing of hands.

8

Environmental Management: The Consequences and Environmental Monitoring of Industrial and Technological Disasters

In this chapter:

- Introduction and general principles.
- Environmental impact.
- Environmental disasters, causes and effects.
- Management systems.
- Important environmental principles.
- Aspects and impacts – environmental risk assessment and defence strategies.

Introduction and General Principles 8.1

On a global scale, even the sceptics have to concede that businesses are heading for serious problems if they fail to consider their environmental impacts. Legislative pressure grows all the time and limits set are getting firmer by the year. Acid rain, melting polar ice caps, holes in the ozone layer, desertification, climate change – all these sayings are no longer in the realms of Governments or so-called 'pressure groups', they are part of the national curriculum taught to our school children.

The first issue that needs to by addressed prior to further reading is the scope of this chapter and the definition of an Environmental Disaster. This raises the question of 'what exactly is a disaster?' Many dictionaries seem to define a disaster as a 'sudden event bringing great damage, loss, or destruction' or similar. This is a broad classification, and the term may mean different things to different people. The second question may therefore be 'what is the environment?' Again, dictionaries define the environment as 'the circumstances, objects or conditions by which one is surrounded'. Environmental Disasters may represent many different types of events. Those related to weather and the earth's geology are the most widely recognised, however several episodes of mass destruction have their links to our own actions. A third question to be addressed at this point is 'what is an accident?' The most commonly accepted definition revolves around 'an unplanned, unwanted event that results in a loss'. In general, most Environmental Disasters also require the need for internal or external specialist assistance. This may be by local, governmental or global assistance, depending on the severity of the disaster.

Taking all of the foregoing information into account, the aim of this chapter is to provide the reader with information on how an Environmental Management System (EMS) may help manage:

> 'an unplanned, unwanted industrial event that may result in a significant impact on the circumstances, objects or conditions by which an organisation is surrounded that may require internal or external specialist assistance'.

Environmental impact **8.2**

It has long been established that a company can gain a competitive advantage and command a higher product/service price by delivering a superior product. Logic may then suggest that companies who adopt a proactive approach to environmental management, rather than mere legislative compliance, should be able to gain the same advantage over competitors. Other key drivers for industry can also be proven through well-managed eco-efficiency. One of the health and safety manager's key skills is to try and 'sell' safety at a senior level. Within environmental matters, key business drivers are a little easier to quantify. A sound EMS will enable companies to manage their downside risks, but also to capitalise on upside opportunities stemming from environment-related innovation and so enhance the business on strategic and operational levels – with these enhancements leading to higher financial returns to the stakeholders. Key business drivers may include stakeholders, society, customers, the workforce and the regulators. In *Cases in Environmental Management and Business Strategy* (1994), Pitman Publishing, London, R Welford wrote:

> 'many companies are exercising their own rights both as purchasers and as vendors and are demanding that all companies within their supply chain seek to minimise their own environmental impacts'.

The costs of cleaning up after a major Environmental Disaster are significant with the 'Polluter Pays Principle' (see **8.29** below) being applied throughout the world. Governments will no longer accept the 'hands-off' approach to Environmental Disasters that may have been apparent in organisations in the past. In other words, the burdens of technology could often have been transferred away from producers and immediate consumers into a universally shared but unprotected natural environment or into specific poor communities (local or overseas) – the $470 million Bhopal settlement (still considered inadequate by some) is a stark warning indeed.

Environmental Disasters do not just incur financial costs however. Costs involve huge loss of human life and suffering (sometimes for generations afterwards). The highly acclaimed BBC television series 'Life on Earth' stated that 46 US States have issued warnings against eating local fish because of dioxin contamination, and in Europe, human breast milk passes on more dioxin to babies than is legally allowed for cows. The effects on animal and plant species cannot be ignored. Industrial pollutants have been responsible for affecting the reproductive organs of fish, alligators and polar bears, preventing them from producing offspring. Up to 50% of species on the planet may be lost if global resources are used continuously at the current rate.

Environmental Disasters, Causes and Effects

Bhopal Disaster – Isocyanate release 8.3

The Bhopal facility was part of India's Green Revolution aimed at increasing the productivity of crops. Considered an essential factor in the effort to achieve self-sufficiency in agricultural production, pesticide production use increased dramatically during the late 1960s and early 1970s. In 1969, the Union Carbide Corporation (UCC) set up a small plant (Union Carbide India Ltd (UCIL)) in Bhopal, the capital city of Madhya Pradesh, to formulate pesticides. A Methyl Isocyanate (MIC) facility was located in the plant to the north of the city, adjacent to an existing residential neighbourhood and approximately two kilometres from the railway station. MIC is one of many 'intermediates' used in pesticide production and is a hazardous chemical. It is a little lighter than water but twice as heavy as air, meaning that when it escapes into the atmosphere it remains close to the ground. Until 1979, the Indian subsidiary of Carbide used to import MIC from the parent company. After 1979, it started to manufacture its own MIC.

Cause 8.4

On the night of 23 December 1984, a dangerous chemical reaction occurred in the Union Carbide factory when a large amount of water (approximately 500 litres) got into an MIC storage tank. The MIC reacted with the water and, at about 11.30 pm, workers reported a leak. A large amount (about 40 tons) of MIC, continued to pour out of the tank for nearly two hours and escaped into the air, spreading over eight kilometres downwind, covering Madhya Pradesh and its 900,000 inhabitants.

Many separate reports point to serious communication problems and management gaps between Union Carbide and its Indian operation. It was reported that the parent companies 'hands-off' approach led to the failure to communicate hazards. These can possibly be related to the hazards that may appear in multi-national operations because of 'the absence of common values, norms, and expectations among managers in different nations'. (Jamie Cassels, *The Uncertain Promise Of Law*, 1993, Bhopal. University of Toronto Press Incorporated. 1993.) Just one example of these problems was that the operating manuals at Bhopal were printed only in English.

Effects 8.5

Thousands of people were killed in their sleep or as they fled (estimates ranging as high as 4,000), and hundreds of thousands remain injured or affected (estimates range as high as 400,000) to this day. The most seriously affected areas were the densely populated shanty-towns immediately surrounding the plant. Victims were almost entirely the poorest members of the population. Exposure to MIC has resulted in damage to the eyes and lungs and has caused respiratory and other ailments.

Not very much is known about the chronic environmental impacts of the gas leak from the Bhopal plant. The Indian Council of Agricultural Research issued a preliminary report on damage to crops, vegetables, animals, and fish from the accidents, but this offered few conclusive findings since they were reported in the early stages after the disaster. This report however, did indicate that the acute impacts from the Carbide plant were highly lethal on exposed animals. Large numbers of cattle, as well as dogs, cats, and birds were killed. Plant life was also severely damaged by exposure to the gas, such as the widespread defoliation of trees, especially in low-lying areas.

On 14 January 1987, the US Second Circuit Court of Appeals in Manhattan upheld a decision by the US District Court to send the legal case against Union Carbide to India and, in February 1989, the Supreme Court of India directed UCC and UCIL to pay a total of $470 million in full settlement of all claims arising from the tragedy. The Government, UCC and UCIL agreed and the two companies paid in full. In November 1994, UCC completed the sale of its 50.9% interest in Union Carbide India Ltd to McLeod Russel (India) Ltd of Calcutta. Indian Government financial institutions owned 26% of the shares and some 24,000 Indian private citizens owned the balance. UCIL is primarily a dry-cell battery manufacturer. UCC sold its battery business, worldwide, in 1986.

The Sea Empress – oil release – offshore 8.6

The use of large sea-borne tankers remains one of the most viable ways of transporting oil around the world. The sad fact is that so long as oil is transported by sea, the risk of the following events (which are not isolated in their nature) may always be present.

Cause 8.7

Shortly after 8.00 pm on 15 February 1996, the oil tanker, *Sea Empress*, ran aground in the entrance to Milford Haven in Pembrokeshire, one of the UK's largest and busiest natural harbours. In the days that followed, while the vessel was brought under control in a salvage operation beset with problems, some 72,000 tonnes of crude oil and 480 tonnes of heavy fuel oil spilled into the sea, polluting around 200 km of coastline recognised internationally for its wildlife and beauty.

Effects **8.8**

The *Sea Empress* could be described as the 'what if' disaster. *If* the accident had happened a few weeks later, *if* the oil had been of a heavier type, *if* the wind had been blowing in a different direction, then the Pembrokeshire wildlife and economy would be suffering much more than they are. It is a fact that the environment locally is recovering remarkably well. The most enduring image of any oil spill is probably the sight of blackened seabirds washed up dead or dying on polluted beaches. As an example, from a population of 15,000 scoter ducks in the area, around 5,000 are estimated to have died and other species such as razor-bills and guillemots were killed in significant numbers. Recent studies indicate that these species are recovering, but the numbers are still down. There have been more subtle effects on the ecology of the area and it will be several more years before the full significance of these is known, and before it will be possible to diagnose a complete 'recovery' from the oil pollution.

Seveso – dioxin release **8.9**

The Vietnam War brought about the widespread knowledge of dioxin when it was identified as a component of the defoliants used on a large scale by the US Government. Before the Seveso release, agricultural workers had mounted campaigns to have certain dioxin chemicals banned because of the alleged toxic effects on humans and several industrial accidents involving dioxin were known to have occurred. Among others, these accidents affected the following firms and countries:

- 1949 Monsanto (USA);
- 1953 BASF (Germany);
- 1960 Dow Chemical (USA);
- 1963 Phillips Duphar (Netherlands);
- 1968 Coalite Chemical Productions (UK).

For some people the name Seveso is tied to the experience of a seriously mismanaged toxic chemical release, but for others it is firmly and positively linked with a set of public policies for managing industrial disasters (see also **12.1** below). One of the most remarkable features of the Seveso experience was reports that neither the residents nor the local and regional authorities suspected that the plant was a source of risk nor did they know much about the type of production processes and chemical substances used there. (See also **6.1** above.)

Cause **8.10**

Around midday on Saturday 10 July 1976, an explosion occurred in a reactor of a chemical plant on the outskirts of Meda, a small town about 20 kilometres north of Milan, Italy. A toxic cloud containing TCDD (2, 3, 7, 8-tetrachlorodibenzo-dioxin) was accidentally released into the atmosphere. This cloud covered a densely populated area about six kilometres long and one kilometre wide lying downwind from the site.

Effects **8.11**

This event became internationally known as the Seveso Disaster, after the name of a neighbouring municipality that was most severely affected. Communities between Milan and Lake Como were directly involved in the toxic release and its aftermath. Particularly traumatic was the chronic nature of the Seveso Disaster. Human health effects of the disaster have been obscure and the process of recovery has been unusual. Though initially slow and conflicted, management of the incident was a process of overcoming initial traumas (eg chloracne, fear of genetic impairments, evacuation, animal deaths) and re-establishing patterns of social and economic lifestyle. To a significant degree, however, local recovery was achieved by exporting parts of the problem whereby seriously contaminated materials were disposed of abroad – their ultimate fate is still unravelling.

In light of this (and similar) disastrous accidents it was clear that something was needed to improve the safety of industrial sites, to plan for off-site emergencies, and to cope with broader aspects of industrial safety. The Seveso Directive is the result of those efforts. A central part of the Directive is a requirement for public information about major industrial hazards and appropriate safety measures in the event of an accident. It is based on recognition that industrial workers and the general public need to know about hazards that threaten them and about safety procedures. This was the first time that the principle of 'need to know' had been enshrined in European Community legislation. See further **6.4** above.

Chernobyl – radiation release **8.12**

The Chernobyl Power Complex consisted of four nuclear reactors, Units 1 and 2 being constructed between 1970 and 1977, while Units 3 and 4 of the same design were completed in 1983. Two more reactors were under construction at the site at the time of the accident. To the southeast of the plant, an artificial lake of some 22 square kilometres was constructed to provide cooling water for the reactors. About 3 km away from the reactor, in Pripyat, there were 49,000 inhabitants and the town of Chernobyl, which had a population of 12,500, is about 15 km to the southeast of the complex. Within a 30-km radius of the power plant, the total population was between 115,000 and 135,000.

Cause **8.13**

The Unit 4 reactor was to be shutdown for routine maintenance on 25 April 1986. It was decided to take advantage of this shutdown to determine whether, in the event of a loss of station power, the slowing turbine could provide enough electrical power to operate the emergency equipment and the core cooling water-circulating pumps, until the diesel emergency power supply became operative. The aim of this test was to determine whether cooling of the core could continue to be ensured in the event of a loss of power. At 01.23 am on 26 April 1986, a failed controlled experiment carried out by Chernobyl Power Plant technicians allowed the power in the fourth reactor to fall to low levels. The reactor overheated causing a meltdown of the core.

Two explosions blew the top off the reactor building and, for over 10 days, clouds of radioactive material were released.

Effects **8.14**

The people of Chernobyl were exposed to radioactivity 100 times greater than the Hiroshima bomb. 70% of the radiation is estimated to have fallen on the Belarus and babies are still being born with no eyes, arms or only stumps for limbs. Estimates put the figure over 15 million people being affected by the disaster in some way and that it will cost over 60 billion dollars. More than 600,000 people were involved with the cleanup – many of whom are now dead or sick.

The Chernobyl accident was attributed to the lack of 'safety culture' and the poor design of the reactors. The operators weren't aware that the particular tests they performed could have brought the reactor into explosive conditions but neither did they comply established operational procedures. The combination of these factors provoked a nuclear accident of maximum severity in which the reactor was totally destroyed within a few seconds.

Baia Mare – cyanide (and other pollutant) release **8.15**

The Somes and Tisza Rivers in Northwest Romania cross the border into Hungary and flow into the Danube River and further into Yugoslavia. The Baia Mare gold mine, an Australian and Romanian Government-owned venture, was situated close to this extremely diverse freshwater ecosystem – the Tisza River itself is home to 19 of Hungary's 29 protected fish species. Otters, other mammals and birds are also heavily dependant on this sensitive local natural environment. Waste waters and sediment from mines sometimes contain high levels of cyanide and other pollutants such as lead, zinc, copper, aluminium. The cyanide is used to separate gold from tailings (residue) left by other mining operations.

Causes and effects **8.16**

An estimated 100,000 cubic meters of polluted mud and wastewater escaped from a break in the dam of the Baia Mare gold mine. By the time the wave of contamination had reached Hungary, the level of cyanide in the Tisza River was still peaking at 300 times the threshold of the Hungarian 'highly polluted' standard. Not long after the spill, 100 tonnes of dead fish were swept from the river's polluted surface. It has been estimated that 300 to 400 otters living in the Somes and Tisza Rivers have completely disappeared, probably the result of eating contaminated fish.

A team of United Nations Environment Programme experts had just finished sampling and assessing the 500-mile stretch of affected waterways, when mine water spilled again from another Romanian dam. This time, more than 20,000 tons of heavily polluted sediment washed from a mine near the city of Borsa, flooding the upper segment of the Tisza River that had been spared by the previous spill. A third, smaller spill occurred from the same mine a few days later.

Guanabara Bay – oil release – pipeline 8.17

The Guanabara Bay is a protected area of the Guapimirim and Jequia swamps in Rio de Janeiro, Brazil. It is the home of the breeding grounds for many fish, birds and crustaceans. Close to the area are the tourist beaches of Copacabana and Ipanema.

Cause and effects 8.18

Brazil's largest oil spill occurred in January 2000 when a corroded Petrobras (the state oil giant) pipeline leaked almost 1.3 million litres of oil near Guanabara Bay, ruining the protected areas and breeding grounds. The nearby beaches remained unharmed but this did not stop the Government fining Petrobas the maximum fine of $29 million and hundreds of local fishermen are demanding compensation for the damage.

Management Systems

Introduction 8.19

An environmental management system (EMS) is designed to allow a company to integrate a planned, co-ordinated and organised approach to managing its environmental aspects and impacts. All industries have processes or products they use in their facility that have the potential to negatively affect the environment. These facilities are a major concern in the fight for improved environmental conditions. By adopting and committing to an EMS, a company may benefit in many ways.

EMSs are a step in the right direction to helping to control industrial Environmental Disasters. They serve to provide a framework that companies can follow to maintain and improve upon their contributions to and impacts on the environment.

Overviews – local management systems

Management System Verification (MSV) 8.20

MSV is a process used by the American Chemistry Council (ACC) and its member companies to evaluate five major management system elements that apply to Codes of Management Practices (CMPs). There are six specific codes:

- Community awareness and emergency response;
- Pollution prevention;
- Process safety;
- Employee health and safety;
- Distribution;
- Product stewardship.

The above codes comprise of 106 specific management practices and questions have been derived for each of the practices. The questions are designed to be open-ended to evoke discussion, demonstration and example. The MSV system looks for specific attributes that describe necessary systems, organisational structures, and policies for each of the management system elements.

The basic technique of using the MSV system is to have a verification team apply the open-ended questions in a panel or group interview environment. The team is typically comprised of a lead verifier, a supporting verifier, a facilitator and a representative of the public. The verifiers are from other companies and are selected on the basis of qualifications, availability and acceptability to the company being verified. The company selects the public participant, often coming from a Community Action Panel (CAP). All the panel members receive one and one-half day training in the process. The scope of the MSV varies according to the needs of the company, but typically will require from two to three days on site for the verification team to complete its work.

Green Zia **8.21**

The Green Zia Environmental Excellence Program is a voluntary system designed to support and assist businesses in New Mexico to achieve environmental excellence through continuous environmental improvement. It is not a conformance system (unlike ISO 14001) but a performance standard. The system encourages integration of environmental excellence into business management practices through the establishment of a prevention-based environmental management system. The Governor of New Mexico makes recognitions and awards annually to organisations that successfully participate in the program.

The New Mexico Environmental Alliance, a partnership of State, local and federal agencies, academia, private industry and environmental advocacy groups administer this program. The Green Zia Environmental Excellence Program is a multi-year program that emphasises continuous environmental improvement. The Green Zia Core Values and Criteria provide a valuable self-assessment framework to help organisations understand environmental excellence and measure their progress toward its achievement.

The system administrators say that an organisation that works through the Green Zia Program from the beginning Commitment Level, through the Achievement Level and then, ultimately, to the Green Zia Environmental Excellence Award, will gain a thorough understanding of environmental issues that will affect its bottom line. The participating organisation, they say, will also establish a system that helps them address environmental issues in cost-effective ways, based on sound business practices.

HSG65 **8.22**

The title of this document is 'Successful Health and Safety Management'. It was first produced as HS(G)65 in 1991, as a practical guide to health and safety management by the UK Health & Safety Executive's (HSE) Accident Prevention Advisory Unit. The second edition was published in 1997 and reprinted with amendments in 2000. The key elements of the document are:

- Policy.
- Organising.
- Planning.
- Measuring Performance.
- Auditing and Reviewing Performance.

The document states that it is aimed at directors and managers with health and safety responsibilities, as well as health and safety professionals. It also states that employees' representatives should find it 'helpful'.

Overviews – International Management Systems

OHSAS 18001 (Note: PHSAS 18001 is to become BS 18001 from 2006 onwards) 8.23

The importance of managing Occupational Health and Safety is recognised by all interested parties:

- employers;
- employees;
- customers;
- suppliers;
- insurers;
- shareholders;
- the community;
- contractors; and
- regulatory agencies.

OHSAS 18001 is an Occupation Health and Safety Assessment Series (OHSAS) for occupational health and safety (OHS) management systems that enables an organisation to control OHS risks and to improve performance. OHSAS 18001 was released in April 1999 and was developed in response to customer demand for a recognisable standard against which their management system may be assessed. OHSAS 18001 is compatible with other systems such as ISO 14001. The standard guides the reader to consult other relevant documents, in particular:

- OHSAS 18002 – Guidelines for the implementation of OHSAS 18001;
- BS8800 – Guide to occupational health and safety management systems.

The specification takes a structured approach to OHS management. The emphasis is placed on practices being pro-active and preventive by the identification of hazards and the evaluation and control of work related risks.

Section 4.4.7 of OHSAS 18001 refers to 'emergency preparedness and response' and states that the organisation shall:

> '... establish and maintain plans and procedures to identify the potential for, and responses to, incidents and emergency situations, and for preventing and mitigating the likely illness and injury that may be associated with them'.

It also requires the plans and procedures to be reviewed, 'in particular after the occurrence of incidents or emergency situations' and tested where 'practicable'.

In section 4.5.2 the document refers again to 'accidents, incidents . . . and corrective and preventative action'. To comply with OHSAS 18001, organisations must define responsibilities and authority for:

- the handling and investigation of accidents and incidents;
- action to be taken following accidents and incidents to mitigate any consequences (which need to be reviewed *'through the risk assessment process prior to implementation'*); and
- the initiation and completion of corrective actions and confirmation that such actions have been effective.

ISO 14001 8.24

ISO 14001 is a standard that provides guidance for a company in managing their impacts on the environment. It works by identifying impacts resulting from products, services, processes, and operations. Unlike laws and regulations, which impose specific requirements that a company must adhere to, ISO attempts to establish a systematic approach for identifying and evaluating a company's environmental aspects and impacts. ISO also provides a framework for setting targets and objectives that are designed to solve or minimize the identified detrimental environmental impacts. ISO 14001 suggests environmental impacts need to be considered for normal, abnormal and emergency situations.

The standard states that an organisation:

- 'shall establish and maintain procedures to identify potential for and respond to accidents and emergency situations, and for preventing and mitigating the environmental impacts that may be associated with them;
- 'shall review and revise, where necessary, its emergency preparedness and response procedure, in particular, after the occurrence of accidents or emergency situations;
- 'shall periodically test procedures where practicable . . .'.

ISO 9001 8.25

This is primarily a quality management system being made up of a number of elements. Of particular relevance is the subdivision entitled 'Corrective Action'. The main focus, as one would imagine, is corrective action relating to quality problems. It can be utilised, however, to assist with the planning and implementation of an EMS Emergency Procedure. The elements of relevance are:

- The series requires responsibility and authority to be defined and assigned.
- Significance of problems needs to be evaluated.
- The relationship between cause and effect needs to be determined.
- The series requires an analysis of root causes and determination prior to the planning of preventative measures.
- Preventative actions need to be implemented in proportion to the magnitude of the potential problems.
- Where preventative action is taken, monitoring should be carried out to ensure goals are met.

International Safety Rating System 8.26

The International Safety Rating System (ISRS) is published by the International Loss Control Institute, Inc. As its name suggests, this is a management rating system whereby trained auditors assess compliance with a very detailed conformance standard. The system analyses an organisation's performance over 20 separate areas and gains a score against each of the set criteria. Of particular note is the section dealing with emergency preparedness that looks in detail at the following areas:

- Administration.
- Emergency Response Analysis.
- Emergency Plan.
- Off-site Emergencies.
- Sources of Energy Controls.
- Protective and Rescue Systems.
- Emergency Teams.
- Lessons Learned System.
- First Aid.
- Organised Outside Help and Mutual Aid.
- Post Event Planning.
- Emergency Communication.
- Communications Within the Community.

The system utilises three main methods of checking the existence and effectiveness of the management programme. They are record checks, on-site interviews and physical conditions sampling.

Important Environmental Principles

Sustainable development 8.27

Destruction or damage in one part of the world will affect us all in the longer term, if not immediately. The key principle for environmental management and policy is sustainable development (SD). In 1987, a report of the World Commission on Environment and Development, entitled 'Our Common Future', defined SD as:

> 'meeting the needs of the current generation without compromising the ability of future generations to meet their own needs'.

This definition is founded on the concept that development cannot take place if the environment is destroyed or irreparably damaged whilst still accepting that economic and social development is an essential part of life. Many countries have adopted this as the starting point for their work by incorporating sustainability into strategic planning policies. As an example, the UK published 'Sustainable Development: the UK Strategy' which highlighted the need for integration and included a number of key indicators that were to be used to measure progress towards achieving the strategy objectives. The strategy looked at the UK economy and environment as a whole and looked beyond national boundaries to consider the effect this country has on other countries. The strategy stressed the importance of setting sustainability targets and identifying the means of achieving them.

Agenda 21 8.28

In June 1992, at the United Nations Earth Summit Conference held at Rio de Janeiro, the idea of sustainable development was taken further. At this conference the UK Government endorsed the Rio Declaration setting out 27 principles of sustainable development and signed, with many other countries, Agenda 21, an action plan specifying what would be needed to achieve sustainable development worldwide in the twenty-first century. Stemming from Agenda 21 are the Local Agenda 21 initiatives. They aim to promote sustainable development at a local level and to work towards implementation of Agenda 21. Local Agenda 21 emphasises the involvement of the entire local community – residents and business, as well as local authorities.

At the time of writing, global leaders are meeting in South Africa to discuss sustainable development issues.

'Polluter pays' principle 8.29

The basic operation of this principal is via purchase of consents for disposal or via fines and clean up costs. The elements as set up by the European Union are:

- Tax charges levied directly at the processes that cause the pollution or the purchase of specific licences which entitle the holder to pollute at certain levels.
- Payments via grants or subsidies that come from the public purse to assist with the development of new processes and clean technology.

UK – BATNEEC 8.30

Best Available Techniques Not Entailing Excessive Cost (BATNEEC) is a requirement for emissions to the environment by certain processes covered by the *Environmental Protection Act 1990*. BATNEEC is a process whereby the industry concerned has to use a technique (which includes such things as competence, methods of work, design, construction, layout, maintenance etc) that is proven to be the most effective in preventing or controlling relevant emissions and is readily accessible at reasonable cost when weighed against the possible environmental damage from the process if the technique was not to be used (the reader should make reference to relevant documents for official definitions). For the health and safety professional, it may be considered similar to 'as far as is reasonably practicable' whereby a risk (the likelihood of a hazard being realised and the severity if it was) is weighed against the cost of mitigation (including time, manpower, finance etc). It is being replaced via Integrated Pollution Prevention and Control (IPPC) with Best Available Techniques (BAT).

BAT 8.31

Best Available Techniques (BAT) (definitions from 'Integrated Pollution Prevention and Control – A Practical Guide', DETR) is the basis for standard determination under Integrated Pollution Prevention and Control (IPPC) required under the *Pollution Prevention and Control Act 1999*. In relation to BAT, 'best' means the most effective in achieving a high general level of protection for the environment as a whole. 'Available techniques' can be defined as those techniques (see BATNEEC at **8.30** above) developed on a scale which allows implementation in the relevant sector, under economically and technically viable conditions, taking into consideration the cost and advantages, whether or not the techniques are used inside the United Kingdom, as long as they are reasonably accessible to the operator.

Integrated pollution prevention and control 8.32

Brought about by the *Pollution Prevention and Control (England and Wales) Regulations 2000 (SI 2000 No 1973)*, themselves made under the *Pollution Prevention and Control Act 1999*, IPPC is a system of regulation of certain industrial activities whereby emissions and waste production are to be considered for all environmental media (air, water and land) plus a range of other environmental effects. This integrated approach (a response to the *European Community Directive 96/61/EC*) is the future for environmental management of all processes and should be applied when developing disaster management strategies. IPPC aims to prevent significant environmental impacts and where this is not practicable, reduce them to acceptable levels. IPPC can be considered as a 'cradle-to-grave' system by not only permitting and regulating industrial emissions but by controlling the restoration of sites when industrial activities cease. The conditions imposed by IPPC are based on the use of BAT and eventually IPPC will replace Part 1 of the *Environmental Protection Act 1990* (*EPA 1990*).

Aspects and Impacts – Environmental Risk Assessment and Defence Strategies

Environmental emergency aspects and impacts 8.33

Environmental Emergency Aspects (EEAs) may be described as any part of the business undertaking or activities that may have an effect on the environment during an emergency. The Environmental Emergency 'Impact' (EEI) is therefore an evaluation of the effect on the environment that is caused by the particular EEA during a foreseeable emergency situation. The EEAs and EEIs of a process may change throughout the process lifecycle and the organisation may have to have different levels of control for different situations. An example of an EEA and EEI of an activity is illustrated in Figure 1.

Figure 1

Activity	*Environmental Emergency Aspect*	*Environmental Emergency Impacts*
On-site storage of isocyanate paints close to controlled water source.	Significant accidental spillage.	Adverse effects on local ecosystems, local air quality, depletion of ozone layer.

The example in Figure 1 only gives one aspect from the activity. In reality, there may be a number of aspects and impacts from the same activity. The example does not evaluate the significance of the impacts. The scope of EEAs and EEIs are very wide and the resultant Emergency Plan need only consider those where the organisation may be deemed to have 'influence'. This does not, however, relieve the organisation from the moral consideration thereby exerting 'influence' over their suppliers, customers etc to have their own emergency procedures to deal with foreseeable EEAs and EEIs.

Environmental emergency impact evaluation

Qualitative 8.34

The criteria illustrated in Figure 2 below may assist when carrying out a qualitative evaluation of the significance of an environmental impact. The business concerns and

Figure 2: Qualitative evaluation

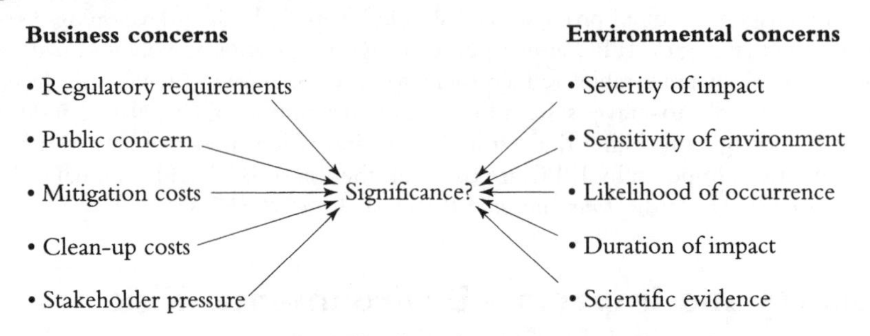

environmental concerns are weighed against the measures required to mitigate any possible emergency and determine the significance criteria.

Quantitative 8.35

More sophisticated assessment techniques can be used to evaluate the environmental impacts of an emergency situation. Any quantitative assessment is of particular use when developing a prioritised mitigation and control system. The following is an example of a quantifiable assessment formula:

S(O + D) + L = ...

Where:	**D**	Detection of occurrence	1. Immediate detection 2. Rapid detection 3. Likely Detection 4. Slow detection 5. Unlikely detection
	L	Legal requirement breached as a consequence of occurrence	Constant 50 if applicable
	O	Occurrence Likelihood	1. Very unlikely to occur 2. Unlikely to occur 3. Possibly occur 4. Likely to occur 5. Very likely to occur
	S	Severity of Occurrence	1. Impact negligible 2. Impact slight 3. Impact severe 4. Impact very severe 5. Impact catastrophic

Environmental risk assessment **8.36**

Prevention of environmental emergencies should always be considered as the best option and includes several components. The most important of these components is the knowledge gained from evaluating risk associated with everyday activities involving the substance(s) of concern. Implementation of a Risk Management programme is key to reducing the frequency and severity of Environmental Disasters. It is also more cost-effective than having to restore any resulting damage to the environment. The most effective Risk Management actions include combining prevention activities, with appropriate levels of preparedness and effective response. The application of environmental risk assessment, similar to health and safety risk assessment, could be applied to assess, eliminate, reduce and control environmental emergency impacts. A procedure needs to be developed to identify any potential situations that could cause significant environmental impact. The HSE's 'five steps to risk assessment' is one system that can be used to great effect.

Five-step approach **8.37**

A development of the 5-step approach is as follows:

- **Step 1 – Identify the environmental hazards**. In this case, the hazards can be defined as any activity or process that, in the event of an emergency, has the potential to lead to significant environmental harm.
- **Step 2 – Identify which environment medium could be harmed and how**. Each foreseeable emergency hazard should be assessed to see how the environment (land, air or water) could be affected and how that effect could happen.
- **Step 3 – Evaluate environmental risk**. The risk can be defined as the likelihood of the specified hazards being realised and, if they were, the significance of the environmental impact in the short, medium and long term. A qualitative or quantitative assessment can be used.
- **Step 4 – Record the significant findings**. An adequate documentation procedure needs to be developed to outline the assessment process and to enable conclusions to be drawn.
- **Step 5 – Review the Assessment**. The emergency system needs to have relevant review periods and the reviews themselves, when carried out, should be documented. Examples being:
  - After a significant environmental incident.
  - Where a process or activity has changed creating further or different risks.
  - The introduction of new technology or techniques.
  - After legislative changes that may affect the assessment.
  - After a change in 'best practice'.
  - Routine review – a programmed look at the procedure to check that it remains relevant.

Environmental Emergency Plans 8.38

From the environmental risk assessment, the site should then go on to develop Emergency Plans that deal with identified emergencies. Effective preparedness for environmental emergencies is built on trust and co-operation between people at all levels in communities, industry and government. Through working together, they must accomplish four things:

- identify potential risks;
- develop environmental Emergency Plans to deal with the risks;
- train personnel to apply the environmental Emergency Plans; and
- review and practice these strategies time and time again.

To enhance the level of preparedness, all key people, including those who may be affected, should be involved with the development and implementation of the environmental Emergency Plan. By completing effective prevention measures (such as Risk Management programs which identify all possible emergency situations), persons preparing and implementing an environmental Emergency Plan can determine the necessary level of preparedness for their situation. The Emergency Plan should be developed with reference to two main areas – the response and the recovery.

Response 8.39

Response to an Environmental Emergency covers many areas and can vary greatly in scope, depending on the nature and magnitude of the emergency. Quick and effective response relies on sound planning and partnerships. Effective emergency response requires teamwork between industry, communities, local organisations and government through partnerships best formed during non-emergency periods. Effective emergency response includes rapid assessment of the emergency and activation of the plan, proper notification of the emergency to the relevant personnel and adequate resource mobilisation. Response is intended to include all aspects of dealing with and managing an emergency, until the emergency portion of the event can be considered completed.

Recovery 8.40

In the context of environmental emergencies, recovery can be seen as 'restoring damage caused to the environment'. The issue of recovery is best managed through discussions between all involved parties. While the ideal situation may be the recovery of an area to its natural state, this may not always be possible. Thus, restoration plans would need to be defined in terms of being acceptable to affected stakeholders and could be specific to each situation.

For ISO 14001, the plans may form part of the management manual. The plans should include detailed call-out lists ensuring that management control can be regained as soon as possible after an event. Specific awareness training is required. This should include all personnel being made aware of their individual responsibilities for emergency preparedness and response. ISO 14004 (Environmental Management Systems – General Guidance on Principles, Systems and Supporting Techniques) gives guidance on what should be included in an Emergency Plan. They are:

- emergency organisation;
- a list of key personnel;
- details of emergency services;
- internal and external communication plans;
- actions to be taken in the event of different types of emergency;
- information on hazardous materials, including each materials potential impact on the environment and measures to be taken in the event of accidental release;
- training plans and testing for effectiveness.

Training of personnel 8.41

For the EMS Emergency Plan to be effective, personnel need to be competent. Training assists with the development of competence, particularly with the skills, knowledge and attitudes required for effective implementation of the EMS. Training should not, however, replace the development of suitable proactive control systems.

Testing of the system 8.42

It is essential that tests of the systems to eliminate or control environmental impacts following an emergency are carried out. Although realism may be required, periodic tests should not be carried out that themselves could cause adverse environmental effects. Reports of tests should be documented and the findings utilised to review the risk assessments and to pro-actively improve the system overall.

The chart in Figure 3 below summarises the environmental risk assessment and Emergency Plan development loop.

Figure 3: Environmental Risk Assessment/Emergency Plan Loop

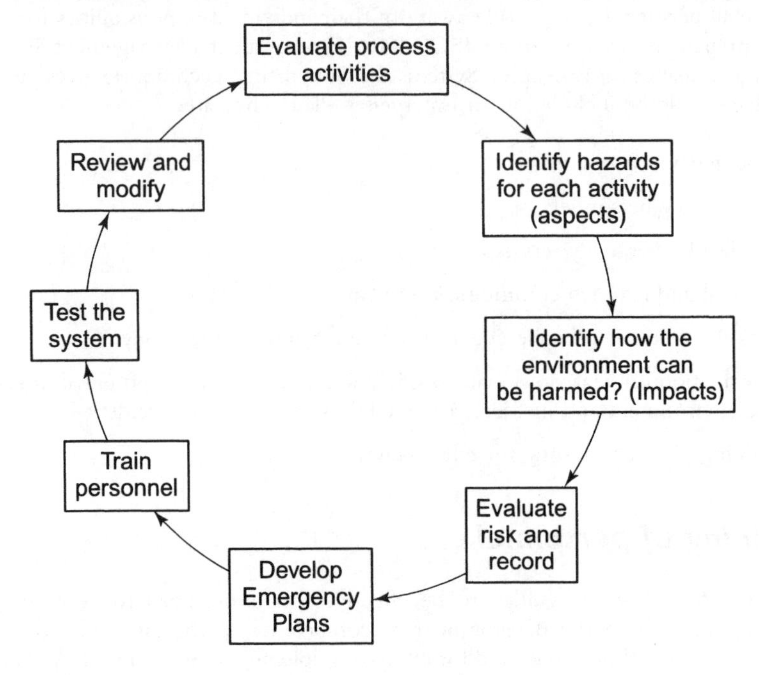

9

Fire Safety Management for 'Public' Buildings

In this chapter:

- Introduction.
- Fire Safety Manual – Section 1: Statutory controls.
- Fire Safety Manual – Section 2: Policy and procedures.
- Fire Safety Manual – Section 3: Description and general layout of estate and buildings.
- Fire Safety Manual – Section 4: Risk assessment.
- Fire Safety Manual – Section 5: Responsibilities.
- Fire Safety Manual – Section 6: Fire prevention.
- Fire Safety Manual – Section 7: Command and control.
- Fire Safety Manual – Section 8: Security.
- Fire Safety Manual – Section 9: Fire Safety Log Book.
- Fire Safety Manual – Section 10: Telephone trees.
- Fire Safety Manual – Appendix 1: Case study: Walsall Hospitals NHS Trust.
- Fire Safety Manual – Appendix 2: Procedures for fire safety in an organisation's premises.
- Fire Safety Manual – Appendix 3: Emergency facilities and equipment.

Introduction 9.1

This chapter is designed to lead the practitioner through the various stages of compiling a Fire Disaster Emergency Management Manual by emphasising the isomorphic relationship that exists in organisations that operate in the built environment. Examples and partial case studies of the Heritage and Healthcare industries are used to encourage the practitioner to look further than the immediate situation when preparing risk management assessments, policies and procedures.

A Fire Safety Manual provides a formal record of the arrangements that have been devised for the premises. It should detail the fire safety strategy and fire safety measures which have been devised for the protection of the building, its occupants and contents. The Manual is, therefore, fundamental to managing fire safety efficiently. The document has no statutory authority in its own right, but can assist all parties to understand exactly how statutory requirements should be met.

Detailed management 9.2

The detailed management of fire safety is the most fundamental aspect of a fire safety strategy for the building. Unless fire equipment and fire warning/suppression systems are correctly installed, checked and maintained in line with statutory requirements and staff are adequately trained in fire safety, including evacuation procedures, residents, patients (in hospital or residential homes), visitors, contractors and staff will be at serious risk in a fire situation. The provision of a Fire Safety Manual will ensure that there is a planned basis on which to ensure that training, checking and maintenance are carried out and that these items are properly recorded.

The Manual is a working, viable, document and not an end in itself. It will require consistent updating and should be formally reviewed every six months, as part of the Fire Precautions Inspection and altered to take account of changes of building use, staff, visitors and the presence and function of contractors.

Aim 9.3

The aim of Fire Safety Management is to minimise the risk to life and property from fire, and ensure that all staff are aware of their duties in the event of fire and are trained accordingly.

Objectives 9.4

The objectives of Fire Safety Management are as follows:

- to develop an integrated risk management and structural fire prevention system for all premises;
- to ensure that in the event of a fire adequate provision is made for the occupants of the building whether; patients, residents, visitors, contractors, or staff to be able to leave the building and be accommodated in a safe location;
- provision is made that in the aftermath of a fire support will be provided for those displaced by it;

- provision is made that in the event of a fire, damage to the fabric of the building and its contents will be controlled;
- to minimise the disruption to the function of the organisation.

The measures to be put in place to achieve those objectives should be appropriate to the risk present in the building and, therefore, the requirement for all measures should be based on a risk assessment approach (see **9.21** below).

The Regulatory Reform (Fire Safety) Order 2005 9.5

Due to amendments and revocations to Fire Safety Legislation, the introduction of The Regulatory Reform (Fire Safety) Order 2005 has resulted in the creation of a Risk Assessment approach to fire safety hazards at work with employers and occupiers being deemed the 'responsible persons' for the management of fire risks. The 2005 Order Part 2 requires a 'POPMAR approach' to fire safety issues by responsible persons covering:

- Policy on Fires and Fire-related Emergencies (Order, Sections 10 and 11)
- Organising to create a fire safety culture (Order, sections 19–23)
- Planning to prevent serious and imminent dangers (Order, Sections 15)
- Monitoring (Order, Sections 9, 12, 13, 14, 16, 17 and 18)
- Audit and Review (Part 3 and 5 related issues).

It is also important to note that the Order revokes completely the Fire Precautions (Workplace) Regulations 1997.

The full content of the Order can be seen at: http:/www.opsi.gov.uk/si/si2005/20051541.htm

Fire Safety Manual – Section 2: Policy and Procedures 9.6

Section 2 of the Manual will list example policies and procedures relating to fire safety and major emergency.

To enable our emergency decision makers to cope with disaster effectively the support systems available must be appropriate, adequate and understood.

> 'The identification of an unstructured event, and the management of that event as it creates itself can only be facilitated if the decision maker has flexible procedures, and an effective communication system, enabling continued gathering of new information, from initial incident conception through to the containment

of that threat.' (Ody K *Facilitating the 'right' decision in crisis: Supporting the decision maker through analysis of their needs* (1995) Safety Science, pp 125–133.)

The *Management of Health and Safety at Work Regulations 1999* (*as amended 2003*) requires employers to establish and implement procedures to be followed in the event of fire or other serious and imminent danger. The focus of those procedures in organisations is generally the evacuation of the building – that is only the starting point in disaster management. As vital as it is to train front line staff in their roles to evacuate and report, of equal importance is the training of those senior managers that will take key roles. The establishment of a Risk Planning Group (RPG) set up to co-ordinate the production, implementation and revision of Emergency Procedures, will ensure consistent liaison with the emergency services. The group would also facilitate interaction between departments in the organisation to ensure that risks are not missed between procedures.

Co-ordinated response 9.7

The management of a disaster requires the decision makers at the incident to focus on and think around the problem. Therefore, the organisation must ensure that procedures are developed and tested to enable staff to have the confidence to innovate and use their own initiative.

> 'If any unforeseen event occurs the operative, or manager, is not helped by algorithmic procedures – a set of procedures that could have been foreseen.' (Reason JT, 'A Systems Approach to Organizational Error' (1995) *Ergonomics* Vol 38 pp 1,708–1,721.)

Most Emergency planning is oriented toward increasing the centralisation of authority and the formulising of procedures. In other words, co-ordination by plan is considered to be normative. The mode of co-ordination is seen as most appropriate, since a military model of organisational functioning in crises is assumed to be most effective in such circumstances. In a major incident, the local emergency services will eventually be assisted by personnel from other areas – none of whom will be familiar with the immediate area. If they do not have the same command procedures and co-ordinated training then it would be impossible to mount an efficient attack on the fire or for the other services to develop a co-ordinated response. It may sound contradictory to advocate a military model whilst emphasising the need to have flexible procedures, but there are certain tasks that have to be performed in a certain order for which systems have been tried and tested, thereby leaving the decision maker with the time to think and liase with emergency service professionals.

The evidence from the Windsor Castle fire (see **9.18** below) is that a military model of organisational functioning need not inhibit individual initiative – in fact, the opposite was the case. As individuals had clear responsibilities and roles, which had been practised, there was a framework that enabled key decision makers to be flexible within their role and delegate tasks clearly to untrained help.

Fire Safety Manual – Section 3: Description and General Layout of Estate and Buildings 9.8

Section 3 of the Manual should start with a description of the premises, concentrating on those aspects which will affect fire safety. Much of the information to be contained in this chapter may be provided in the form of drawings. They provide a succinct and efficient means of conveying the information required, providing that they focus on the fire safety aspects of the property and do not clutter the picture with unnecessary detail. The style of drawings needs to be interpreted consistently by all bodies that will need to access them. Written descriptions and the instructions, which accompany these drawings, need to be clear and unambiguous. A brief description of the layout of the entire estate, accompanied by a simple site plan, is very useful when planning overall firefighting strategy.

Site plan 9.9

The site plan should include:

- safe evacuation routes;
- assembly points for visitors, contractors and staff;
- all site entrances, fire access roads, together with any special hard standing areas, which are provided for fire fighting vehicles;
- details of water supplies;
- non-emergency service vehicle access parking and egress should be clearly indicated;
- media assembly and briefing area;
- incident management control area for Fire Service, Ambulance and Police Command areas; and
- outer and inner security circles (if site permits).

Construction and layout of buildings 9.10

The statement, which details the basic construction methods used for various parts of the building, is a useful reference to include in this chapter of the Manual. It can be kept very short and simple, as its purpose is to aid understanding of fire safety strategies, which have been used, and the level of protection that may be required.

It is most useful to have a set of drawings showing fire resistance, usually referred to as 'compartmentation' drawings. Once they are available any future building or refurbishment work can be planned to improve or maintain compartmentation as necessary. The compartmentation of the building is designed to protect safe exit routes and prevent rapid fire spread. Travel distances to *protected areas* and final exit should also be shown on the plans. The drawings will form an important tool in training staff in Emergency Procedures. They are also a key element in any risk assessment as any puncturing or inappropriate fire stopping of these compartments affects the integrity of the compartment and increases risk.

Layout drawings showing the interior of buildings should also include room numbers for ease of reference. The system should include every space, including stairways and cupboards. The building layouts should show the room doors, with the correct direction of swing indicated. The changes of level within the building also need to be clearly indicated.

The location of special waiting areas, (safe refuge) required to accommodate people with disabilities and any evacuation lifts or manual handling aids also need to be highlighted. When considering the management of disabled persons on site it is necessary to consult with the local fire authority as to their preferred strategy. In some circumstances it may be necessary to plan for vertically assisted escape rather than use a refuge. Part 8 of the BS 5588 provides fuller guidance.

Use of buildings 9.11

Since the uses to which various rooms and areas are put tend to change relatively frequently during the life of a building, it is beneficial to record this information in the form of a schedule, listing the areas by room number. Windows and doors should also follow the same numbering procedure, as should any fire detection and intruder alarm system.

The Property Fire Safety Manager (PFSM) will ensure that the schedule also separately details at least the following types of occupancy:

- residential areas;
- medical areas;
- staff/residential corridors, stairways and lifts;
- catering and service areas;
- service corridors, stairways and lifts;
- public or visitor areas;
- visitor facilities, including sleeping accommodation, catering and toilets;
- areas used for special events and presentations;
- offices;
- shops;
- workshops
- storage areas;
- switch rooms, electrical intakes, and other plant rooms;

- laboratories;
- roof spaces;
- basements and cellars.

Contents of buildings 9.12

Some buildings are of historic importance and/or contain precious artefacts. Where this is the case thought must be given to pre-planning a secure means of removing the items from situ to a safe store in the event of a disaster. It is also most advisable to photograph and properly record all items, including interior decors, so that details are kept to assist with insurance claims, restoration or replication should it become necessary.

Computerised inventory systems 9.13

At one National Trust property a database linked to Computer Aided Design (CAD) drawings has been introduced to integrate Emergency Procedures, Collections Management, Building Services and Steward/Visitor information books. Emergency procedures are not effective unless, like any other aspect of health and safety, they become part of the day-to-day management of the organisation. The development of CAD room plans linked to a computerised inventory system was introduced as part of a Total Quality Management (TQM) approach to heritage management. There is now a common design and layout for room steward information folders, conservation cleaners' manuals and salvage documentation. The rationale was that by common usage of their particular documents, room stewards and cleaners would absorb how the salvage information is compiled and, in an emergency, would not need in-depth training to be able to check inventories or condition reports for insurance purposes. The introduction of diaries and condition reports to the cleaners' manuals gives dated evidence of the condition of the collection should an insurance claim be made, for whatever reason. In addition, the continuous updating of condition reports enables management to gain a more detailed summary of conservation requirements and plan budgets accordingly.

Documentation format 9.14

The format of the documentation gives a layered approach to emergency procedures. The information exists both as software and hard copy. Any of the three sets of information can be used as part of the command and control procedures agreed with Cornwall County Fire Brigade. The Disaster Management Manual is divided by fire compartment and contains four sets of room plans, each encapsulated, plus full inventories of the compartment and the total equipment required to salvage all contents. The design of the system is based on the National Trust Emergency Procedures Manual and is compatible with procedures developed by English Heritage and The Museums and Galleries Commission. Each of the above organisations also has offices, shops, restaurants and workshops on the same site.

Heritage organisations 9.15

Some hospitals are housed in listed buildings, the methods used by heritage organisations to identify and locate artefacts can also be applied to locating people within a building. In America the Federal Emergency Management Agency has endorsed, and Heritage Emergency National Task Force publish, the 'Emergency Response and Salvage Wheel' (http://www.heritageemergency.org/), which gives guidance to immediate priorities in the first 48 hours following a disaster, both for response and the salvage conservation requirements of artefacts. This includes electronic data, books and paper – items that may be vital to the continuity of any business.

Following the Piper Alpha disaster data from the platform was recovered by a conservation organisation.

Back-up 9.16

If salvage is a priority then it should be planned and practiced as part of the Emergency Procedures Training and sleeping contracts arranged to provide the support that will be required. (See further **9.72** below.) Daily back-up of data to a site at least a mile away is required by a number of insurers, it should be the norm for all organisations and that site should also house a back-up copy of the Fire Safety Manual and ideally become the base for the Crises Management Team.

Building works 9.17

Over a ten-year period at one property, £380,000.00 was spent on capital projects for physical fire protection and a further £50,000.00 for security systems. The monitoring and maintenance of the systems accounted for a further £12,000.00 per annum. In the same period, less than £30,000.00 was invested in Risk Management and Emergency Procedures training.

Over time maintenance costs will significantly increase, the efficiency of the systems decrease and replacement costs come earlier than planned. This is a process which is being repeated, to a greater or lesser extent, dependent of the size and type of building, throughout the world. Part of the induction training of any new member of staff must be to manage the disruption to normal function caused by such projects.

Risks during refurbishment 9.18

At the time of 'the fire' in 1992, Windsor Castle was in the midst of a comprehensive works programme, in which the wiring and the heating were being replaced, fire compartmentation and a new fire-detection system were being installed. The project, known as Kingsbury, had commenced in June 1988 as a rolling programme and involved partial dismantling of rooms, removing works of art to a purpose built store in the California Gardens for conservation work to be carried out. This emptying and refilling of rooms was a complex, highly phased programme. The fire-detection was due for completion within ten days of the disaster but at the time there was none.

Similarly Uppark (1989), a significant National Trust house in Hampshire, had been encased in scaffolding for a year as the builders worked their way around the roof, replacing the lead flashings at the foot of the chimneys and relining the gutters. The work was due for completion the day after the fire occurred.

Windsor and Uppark clearly demonstrate that properties are at their highest risk when in the process of refurbishment. It is, therefore, vital that property staff receive specific risk management training and support. The PFSM must ensure that adequate standards of fire safety are enforced and that fire safety arrangements for the building and lightning protection are not compromised when building works or maintenance take place either within or around a building. A copy of the fire precautions procedures to be implemented during building works should be included in the manual.

Construction (Design and Management) 9.19

Under the *Construction (Design and Management) Regulations 1994 (as amended 2003)*, anyone for whom work is being carried out will have to appoint a planning supervisor and a principal contractor to co-ordinate health and safety during design and construction phases respectively. Restoration of historic buildings falls within this legislation, the principal contractor and planning supervisor may be the owner of, or a representative of the owner of, the building, ie The National Trust or English Heritage. They have to notify the Health & Safety Executive (HSE) if the construction phase is to last more than 30 days, or if it will involve more than 500 person days of construction work.

Principal contractors are required to take over and update the Health and Safety plan developed by the planning supervisor and, if necessary, introduce new rules. They have to co-ordinate the activities of the different contractors working on site and ensure that each complies with the relevant Health and Safety legislation and any rules, including fire precautions, set out in the Health and Safety plan. They must also provide contractors with information on risks to Health and Safety and take the necessary steps to control access to the site.

> 'By enforcing strict controls on project contractors, and adopting clearly defined work method statements and the use of hot work permits (if hot work is carried out at all), basic contract administration and management will also go some way toward reducing the risk of accidental fire occurring in the first place.' Maxwell (1997).

These principles need to be applied to refurbishment in all buildings, even when the scope of the project falls outside of the *CDM Regulations*.

Listed Building Consent 9.20

Where Listed Building Consent is required, owners or their advisors must seek advice from the local authorities' conservation or historic buildings section in addition to an application to English Heritage. Access to the site may require disturbance of gardens or approach gates, these aspects would need to be included in the application and may also require the involvement of the local archaeological unit.

Historic buildings do not, by their nature, possess the high standards that one would expect of a building today with regard to structural fire protection. Technological advances made in fire detection, fire warning and firefighting systems, can provide us with unobtrusive and highly efficient solutions, suited to the special features of heritage buildings. New fire resisting materials are available which can be made to successfully blend into existing features.

The conservation principals advanced by the International Council on Monuments and Sites, embraces the principal of 'minimum intervention' with historic fabric and 'reversibility' of new work whenever possible. These procedures can lead to quite protracted debates on the aesthetics of alterations; in some instances, the cost of alterations can be prohibitive without grant aid or specific fund raising. During this period of time, Risk Management strategies will need to be developed to avoid slips and lapses.

Fire Safety Manual – Section 4: Risk Assessment

History 9.21

The fluid use and occupancy of buildings and their change of use over time creates a living risk history of the structures. Differing styles of building have built-in risks and hazards, as do the functions and processes, which are performed within them. When planning to assess the risk and develop an action plan, it is important to take the wider view and not become blinkered by the uniqueness of an individual building, organisation or process within it. Buildings have a commonality by time of construction, location (urban/rural), materials used, original function and present content.

Organisations have an isomorphic risk relationship simply by the fact that they function within a built environment powered by electricity, which in turn is the primary cause of nearly 25% of fires in non-dwelling buildings. HTM 86 provides a set of criteria to assess risk in hospital buildings based on the requirements of HTM 85, it also provides the assessor with tables of compensatory factors to manage identified risk. The thinking behind the strategy can be very useful when applied to other buildings that do not comply with the provisions of *AD B* of *Part B of Schedule 1* to the *Building Regulations 1991 (as amended)*.

Having taken a holistic view of risk, compile a safety history of the individual organisation and process within the building, including a review of policies and procedures. This site history should also reveal any incidents of arson and whether adequate policies and procedures are in place to protect the building and its occupants. In 1991 the Association of British Insurers and the Home Office set up the Arson Prevention Bureau.

Given the high percentage of fires attributed to arson it is worth visiting their website (http://www.arsonpreventionbureau.org.uk/) to gain an overview of the problem and to check the incidence of arson by places of origin and geographic area. If the building is shared occupancy, the *Management of Health and Safety at Work Regulations 1999 (SI 1999 No 3242)* requires the disclosure of risk information between organisations. There is a higher incidence of arson fires in buildings in multiple occupancy than in those owned or leased by a single organisation. The risk of arson is not solely a fire risk – it also involves security and human factors.

Risk assessment process **9.22**

Risk assessment is the appraisal of one or more risk parameters. Risk assessment can be used to implement a safety policy in the most cost effective way – that is, to direct spending in order to address the most significant safety issues. Alternatively, risk assessment can be used to influence policy making by identifying areas where safety policy is needed. Risk assessment is not an alternative to management competency, although it may make the fire safety of the operation more manageable. Similarly, some quantified risk assessments do not take human factors into account such as wedging open fire doors or key staff simply 'switching off' to the constant requirements of fire safety. However, human factors would be included in the statistics to take into account the physical and mental abilities of staff and other users of the building. The safety culture and behaviour of the organisation can then be used as a weighting factor in the risk assessment. Risk assessments and action plans should ensure that tasks and responsibilities given to individuals are manageable within the timescale available. (See the case study at **9.46** below.)

A recent root cause analyses of system failures in seven NHS Trusts identified risk assessment as the most frequent system failure '. . . most commonly associated with poor planning, implementing, measuring and reviewing'.

Risk assessment can be 'simple' or 'complex'; the latter involving detailed fault trees, indices and statistics. For the purpose of this guide an introduction to 'simple' risk assessment is needed.

Four main tasks 9.23

The risk assessment process involves the following four main tasks (see Figure 1):

- Hazard identification – to identify hazards that could give rise to incidents or events of concern.
- Frequency analysis – to estimate how often such events occur.
- Consequence analysis – to estimate the potential severity of such events.
- Risk evaluation – to decide what, if anything, should be done to address the predicted levels of risk.

Figure 1: Schematic of risk assessment

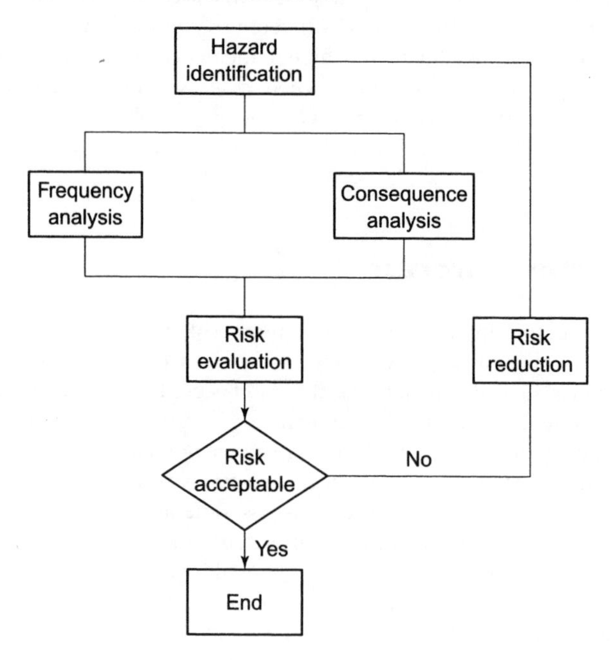

Risk assessment matrix 9.24

A matrix method may be used to evaluate the fire risk of each floor or room within a building. Figure 2 illustrates a simple risk assessment matrix, used in conjunction with a two-stage risk assessment (see Figure 3), and is recommended for evaluating the potential life risk from fire in large and complex buildings. It may also be used to evaluate lesser risks and to act as an aid memoir for future building uses and occupancy.

- Identify flammable and combustible materials.
- Identify any means by which they could be ignited.
- Identify any groups or individuals that may be at special risk.
- See how a fire could spread within a property.
- Assess the likelihood of a fire breaking out.
- Assess the maximum number that can be in the property.
- Identify if there are sufficient staff available to safely evacuate the building.

Figure 2: Fire risk assessment

Risk Number	**Threat to life** (5)	**Very serious** (4)	**Serious** (3)	**Not too serious** (2)	**Not serious** (1)
Very likely to happen (5)	High (25)	High (20)	High (15)	Moderate (10)	Low (5)
Likely (4)	High (20)	High (16)	High (12)	Moderate (8)	Low (4)
Moderately likely (3)	High (15)	High (12)	Moderate (9)	Moderate (6)	Low (3)
Unlikely (2)	Moderate (10)	Moderate (8)	Moderate (6)	Low (4)	Low (2)
Very unlikely (1)	Low (5)	Low (4)	Low (3)	Low (2)	Low (1)

Note:

Multiply vertical axis (likelihood) factor by horizontal axis (seriousness) factor to obtain risk assessment number:

11+ = Needs immediate remedy.

6–10 = Needs attention.

1–5 = No action required.

The data used in the frequency and consequence analysis should be as applicable as possible to the particular circumstances of the study. Where the data is partially applicable, the analysis should contain pessimistic assumptions that ensure that the result errs on the side of safety. The example in Figure 3 assumes that all persons are fit and able to walk at ground level to a place of safety within two minutes.

Figure 3: Fire risk assessment Tables

Table 1: Fire potential			
	Low ignition hazard	**Medium ignition hazard**	**High ignition hazard**
Low fuel hazard	Low	Low	Low
Medium fuel hazard	Low	Medium	High
High fuel hazard	Medium	High	High

Table 2: Lives at risk			
Floor	**Number of people**		
	0–20	21–50	50+
Basement	Medium	High	High
Ground	Low	Medium	Medium
First	Medium	Medium	High
Second and above	Medium	High	High

Table 3: Fire risk category			
	Low life risk rating	**Medium life risk rating**	**High life risk rating**
Low fire potential	Low	Low	Medium
Medium fire potential	Low	Medium	High
High fire potential	Medium	High	High
Note: Use the result from Tables 1 and 2 to evaluate Table 3.			

Dynamic risk assessments 9.25

Risk assessment is also a key element of the management of an emergency site. From the outset and as the incident progresses dynamic risk assessments are required to ensure the safety of personnel close to the incident. Cornwall County Fire Brigade use a mnemonic of the first question that one should ask oneself when faced with danger – Stand or Run – as a basis for the site safety management.

Figure 4: 'Stand or Run'

S	**SITUATION**	What am I faced with? What urgency is there? What information do I need? How can the situation be made safe for personnel?
T	**TACTICS**	What is my plan to deal with the situation? Where do I place my resources? What are the priorities?
A	**ASSISTANCE**	What additional resources are required to complete the task? Estimate requirements to manage safely. DO NOT COMPROMISE ON SAFETY!
N	**NEEDS**	Identify and detail the safety needs for all persons on site. Liase with other organisations on site.
D	**DEPLOY AND SUPERVISE**	Deploy teams, supervise by direct or indirect action … the higher the risk … the closer supervision required!
OR		
R	**REVIEW**	Review your assessment regularly as circumstances change or occur.
U	**UPDATE**	Update as circumstances and the situation changes.
N	**NEVER**	Never commit teams without completing a team protection appraisal

Fire Safety Manual – Section 5: Responsibilities **9.26**

'A little fire is quickly trodden out which, being suffered, rivers cannot quench' (Shakespeare 1586).

All staff has a responsibility to be familiar with the role that they are required to play should an emergency or major incident be declared. As part of the induction training all new staff should demonstrate that fire prevention requirements and emergency evacuation procedures for their role are clearly understood. When staff move location within the building or take on more responsibility they should demonstrate that the procedures for a new area or responsibility are understood and that the person is competent in their execution.

Staff movement appears to be the most consistent inhibiting factor when developing a Fire Action Plan, as staff priorities are to understand their work roles prior to carrying out risk assessments before they embark on such a project. Rarely is the plan tested with external organisations before key staff leave, and the process starts again.

Research of one organisation showed that the majority of managers will complete their procedures preparation between 19 and 36 months of being in post and that after that time there is no significant improvement in preparation. The position of the person in the property with responsibility for Emergency Procedures would appear to affect whether procedures are practised, evaluated and the resources to maintain an effective emergency response are budgeted for.

Changing working practices 9.27

Budget reviews and changes to working practices within the Fire Service have resulted in a much leaner service than ten years ago. Within the County fire brigades there has been a substantial change from whole time to day-manned stations with retained crews providing supporting cover. The opportunity to hold joint training sessions with the brigades at individual properties has been diminished. The requirement by fire brigades to have a systematic approach to emergencies in the built environment has therefore increased.

The vertical and hierarchical flows of information of any kind can be very complicated and difficult. In relationship to fire there are diverse perceptions of risk and consequence by organisations because of the various financial, bureaucratic, political and social interests and cultural values they reflect, this always contributes to the intra and inter-group co-ordination needed at times of crises. This is the current not-too-good situation. Yet such co-ordination is at the very heart of effective and efficient disaster management.

The role of the decision makers will almost certainly fall to people with no operational experience of dealing with emergencies. Are we demanding inappropriate levels of response from our emergency decision makers in the face of crisis, considering the skills and levels of strategic abilities we expect them to display when disaster threatens?

> 'Uncertainty and turbulence in the organisational environment pose a special problem in the co-ordination and decision making of disaster management'. (Therrien M-C 'Interorganizational networks and decision making in technological disasters' (1995) *Safety Science* Vol 20.)

Furthermore, H Dennis (Strategies d'enterprises et incertitudes environmentales (1990) Les Editions Agence D'Arc, Montreal) suggests that 'faced with turbulence, the decision maker will determine the scope of the uncertainty'.

Uncertainty therefore arises from a subjective diagnosis and evaluation of the decision maker. The management style of each decision maker will influence the way in which an incident is managed. 'Decisions made in the first few minutes of an emergency may well determine whether the situation emerges into a disaster' (Craig J, 'The Value of Private Brigades', *Fire*, January 1993, p 7).

Risk Planning Group **9.28**

The Risk Planning Group (RPG) will be able to pool knowledge, spread responsibilities and provide support. (Designating one person to take responsibility is a recipe for disaster as he or she is bound to be absent when disaster strikes and there are too many specialist requirements for one person to perform effectively if they are present.) The RPG should be drawn from senior managers who will form a Crisis Management Team to support individual site Incident Managers who should also be included in the RPG. If the organisation operate at one site the structure and function of the RPG remains valid.

It will take time to develop RPGs and the performance of these groups needs to be supported by a third party and measured to ensure a national standard. Once established training can be rotated by property and organisation within each RPG. This would ensure consistency of procedures for the emergency services, an increased number of people that are able to respond to each property in an emergency and a cost saving to each organisation, as resources and expertise would be integrated.

There are a number of methods that may be employed to measure safety performance including safety audit scores, benchmarking, accident statistics, insurance costs, behavioural measurement and near miss data and 'soft systems'. These methods can be divided into 'proactive (preventing the incident or reducing frequency) and reactive (after the event),' HSG (65) 2000, but there is no one method that can be used to the exclusion of all others given the diversity of the organisations. Financial planning is based on rigorous audit and evaluation. These principles can be applied to safety management to create the environment in which measurement will be seen as an aid to staff, contractors and visitors alike and becomes a positive tool to Risk Planning Groups.

Safety advisors **9.29**

The Health and Safety and Fire Safety advisors have key roles to play. It is they that are charged with:

- identifying the hazards;
- providing legal compliance guidance;
- developing the strategy;
- facilitating the RPG and all inter and intra-organisational communication;
- agreeing the communications media and language to be used;
- training;
- testing and reviewing;
- validating;
- researching new evidence;
- monitoring existing evidence and documentation;
- co-ordinating the lead decision makers;

- monitoring and providing support for the safety of staff during the event;
- ensuring the safety of the site after the event;
- providing evidence at any enquiry;
- providing support to staff after the event.

Fire Safety Manual – Section 6: Fire Prevention 9.30

Technological developments typically lead to the notion that there can be a technological fix for whatever problems to which they are applied. However, Reason has shown that:

> 'the real problem was the flow of information content that was inaccurate, incomplete or misdirected, difficulties not solvable by more communication equipment'. (Reason JT, 'A Systems Approach to Organizational Error' (1995) *Ergonomics* Vol 38 pp 1,708–1,721.)

Practical applications of technological breakthroughs take time to ripen and to become integrated into an organisation. It is important to ensure that developments respond to real rather than perceived or intellectual needs. In this context, there may be a tendency toward applying Disaster/Management technology to those problems that are most easily addressed by the technology and to ignore those problems that cannot be easily handled. The Risk Planning Group:

> '. . . acts as a corporate memory, they have to know who the people are, who knows what, at what point in time . . . no amount of technology can make up for the inadequacies of training, quality, motivation and energised leadership'. (Heymen, quoted in Chartrand RL, *Information Technology Utilization in Emergency Management* (1985) Congressional Research Service, Library of congress, Washington DC.)

As Noble says:

> 'close inspection of technological development reveals that technology leads a double life, one which conforms to the intentions of designers and interests of power and another that which contradicts them – proceeding behind the backs of their architects to yield unintended consequences and unanticipated possibilities'. (Noble D, Forces of Production: A Social History of Industrial Automation (1984) Knopf, New York.)

Inter-organisational relationships 9.31

Emergency response usually calls for the co-operative effort of a broad range of agencies. Typically, police, fire brigade, ambulance, and property staff work together, but other support services may be called upon to assist as required by circumstances. These may include water, gas, telephone and electricity utilities, commercial hire companies and contract labour. Property, regional and national structures are needed if the emergency is of significant dimension. Mental health and social services may be called upon for help. Financial and insurance institutions, salvage and damage mitigation specialists may also be needed.

Most organisations have an inter-dependence of resources and rely on assistance from specialist agencies in addition to their own staff at times of emergency. The emergency services need assistance from outside their normal operating areas to respond to major incidents. It is clear that inter-organisational relationships need to be planned ahead and exercised before actual need occurs. Great attention needs to be directed to maintaining the active involvement of policy making personnel to ensure latent failures do not develop in the relationships as changes to the law, insurance liability and other extraneous factors impact on the participating organisations over time.

Latent and potential active failures will only come to light when organisations develop and review their procedures. A specific Performance Influencing Factors (PIF) taxonomy based on agreed inter-organisational systems and documentation, particularly when properties are in the process of refurbishment, would be a key element in identifying error mechanisms with more certainty.

Fire Safety Manual – Section 7: Command and Control 9.32

At the site of a major incident the senior police officer present will become the Police Incident Officer (PIO) and will take overall control – *except when fire is present, when the Senior Fire Officer is in control until there is no further risk of fire*. Should the incident involve large numbers of casualties, it is the Ambulance Service that instigate and stand-down the NHS Major Emergency Plan.

A forward control point will be established near the scene of the incident and the Incident Manager should be based at this point to liase with the emergency services. As part of the site emergency planning the designated Incident Managers should familiarise themselves with the command structures of the various services and the precise language that is used.

Dependant on the size and nature of the incident the Fire Service will set up several levels of control. The first being a Forward Control sited close to the incident and under the control of the most senior officer present, who will be known as the Incident Commander. This officer will take direct command of the inner cordon and responsibility for all fire fighting and rescue operations. The safety of all personnel involved in these operations will also be under his command.

As more resources arrive the Incident Commander will appoint Sector Officers to take charge of specific functions. An Incident Control Post will be established, usually being run from a mobile control unit. This control will form the principal operational command post for the three emergency services, its function being to control and co-ordinate Fire Service activities. The possible locations for the Incident Command Post should be agreed and tested as part of the planning strategy.

Should the police decide to open a Major Incident Control Room, the Fire Service will be represented by a senior officer at principal level to form part of the police's Overall Incident Control Team. The primary areas of police responsibilities are the protection and preservation of the scene to enable investigation of the incident to be carried out. It is vital that all staff involved in the incident maintain incident logs/diaries and preserve any evidence that may assist in this process. Should there be any fatalities, the police act on behalf of the Coroner who becomes the principal investigator.

Fire Safety Manual – Section 8: Security

Security/fire precautions log books 9.33

All contractors, advisors and staff from other properties should sign in on arrival at reception. The sign-in identity pass will clearly state the date, time, name of the organisation and purpose of visit on the front. The reverse will clearly state fire evacuation procedures, assembly points and the names of those nominated to be responsible for evacuation procedures. The pass folds into a clear plastic badge that all visitors will wear.

A separate, similar, log will be kept for all visitors.

All passes are to be returned and the visitor to sign out before leaving. All identification passes will be valid for one day only.

On activation of the fire alarm these log books will be taken to a designated assembly point and used to account for all visitors.

A signing-in book will be kept at all ancillary buildings and will operate independently from the reception system.

Incident security 9.34

Security at the site is of the utmost importance and must be maintained at all times during any incident. The following security precautions should be considered for implementation as part of the Disaster Management Plan:

- Perimeter security must be established as soon as any evacuation is ordered to prevent unauthorised access to the site or any buildings on it, or the removal of any property from the site or any buildings on it.
- Any areas that can be isolated and can be left unattended, eg peripheral buildings, should be securely locked and any alarms set.
- Any person removed must be transferred to a secure area. Any belongings must also be securely stored.
- Any recovered or salvaged material, or any removed to a safer area, must be protected against theft or vandalism, where practicable, by locking such temporary storage areas; or if this is not possible by ensuring the presence of sufficient security staff patrols to safeguard all such areas and material.

- It may also be necessary to provide security staff escort for any vehicles removing material from the site and for any designated off-site storage areas.

Fire Safety Manual – Section 9: Fire Safety Log Book 9.35

The Introduction to Section 10 of the Manual contains information on the Fire Safety Log Book:

- The Log Book must be kept in a secure area in a fireproof container.
- The designated location should be recorded in the Fire Safety Manual.
- A duplicate of this book should be kept off site and the location of the duplicate recorded in the Manual.
- It is the responsibility of the Property Manager to update and check the accuracy of both Log Books.
- Checking of Log Books shall form part of all fire safety inspections
- When further information is required or a contractor submits an independent report then a copy of the document should be attached to the relevant log sheet.

Contents 9.36

The contents of this Fire Safety Log Book includes record sheets for the following:

- fire evacuation drills;
- fire safety training;
- fire safety inspections;
- fire extinguisher/hose reel inspections;
- fire alarm test;
- testing of fixed fire suppression systems;
- emergency lighting tests;
- visits by Fire Service;
- investigation of fire incidents;
- issuing hot work permits;
- issuing of permits to work;
- testing of isolation valves.

Record keeping for the Fire Safety Log Book is illustrated in Figures 5 to 16 on the following pages.

Figure 5: Fire evacuation drills

Fire Evacuation Drill Record						
DATE	FULL/PART EVACUATION (IDENTIFY AREA)	TIME INITIATED AND BY WHOM	TIME TAKEN TO EVACUATE	NUMBER OF STAFF PRESENT	TOTAL NUMBER OF PEOPLE PRESENT	LEARNING OPPORTUNITY ATTACH COPY OF REPORT

Figure 6: Fire safety training

Staff Instruction and Training Record				
date	nature of instruction or training	number of staff present	duration of instruction& training	NAME OF PERSON GIVING instruction or training

Figure 7: Fire safety inspections

Fire Safety Inspections				
Six-monthly 'in house' and audit				
DATE	TYPE OF INSPECTION AND AREA INSPECTED	BY WHOM (SIGNATURE)	REPORT PASSED TO MANAGER DATE	REMEDIAL ACTION AND SIGNATURE

Figure 8: Fire extinguisher/hose reel inspections

Fire Extinguisher/Hose Reel Inspections				
DATE	NAME OF ORGANISATION TESTING	FAULTS (SPECIFY)	DATE FAULTS CLEARED	SIGNATURE

Figure 9: Fire alarm test

Fire Warning System – Record of Tests						
DATE	FIRE ALARM (PROVIDES EFFECTIVE WARNING)		AUTOMATIC DOOR RELEASES	AUTOMATIC DETECTORS		REMEDIAL ACTION Taken and signature
	Call point used	Satisfactory YES/NO	Satisfactory YES/NO	Detector actuated	Satisfactory	

Figure 10: Smoke detection tests

Testing of Fixed Fire Suppression Systems			
This includes smoke management, smoke control or smoke venting facilities.			
DATE	TYPE OF EQUIPMENT	SATISFACTORY YES/NO	REMEDIAL ACTION TAKEN AND SIGNATURE

Figure 11: Emergency lighting tests

Emergency Lighting – Record of Maintenance and Testing				
DATE	AREA TESTED	SATISFACTORY YES/NO	TYPE OF DEFECT AND LOCATION	REMEDIAL ACTION TAKEN AND SIGNATURE

Figure 12: Visits by Fire Service

Visits by Fire Service and Local Authority Fire Inspectors				
DATE	PURPOSE OF VISIT	NAME OF PERSON AND ORGANISATION	DEFECTS FOUND (ATTACH COPY OF REPORT)	REMEDIAL ACTION REQUIRED BY DATE

Figure 13: Investigation of fire incidents

Investigation of Fire/Near Miss Incidents		
DATE	LOCATION AND DETAIL OF INCIDENT	REMEDIAL ACTION TAKEN AND SIGNATURE

Figure 14: Issuing hot work permits

Issuing of Hot Work Permits				
DATE OF ISSUE OF PERMIT	PURPOSE OF WORK Is there an alternative method of work?	OPERATIVE / ORGANISATION (NAMES) Has read and agreed conditions	ALL CONDITIONS ADHERED TO YES/NO If no – attach detail	DATE WORK COMPLETED AND SIGNATURE OF FIRE WATCHER

Figure 15: Issuing of permits to work

Issuing of Permits to Work				
DATE	PURPOSE OF WORK	OPERATIVE / ORGANISATION (NAMES) Has read and agreed conditions	ALL CONDITIONS ADHERED TO YES/NO If no attach detail	DATE WORK COMPLETED SIGNATURE

Figure 16: Testing of isolation valves

Testing of Isolation Valves					
To include fused link systems					
DATE	TYPE OF VALVE	SATISFACTORY YES/NO	ACCESSIBLE YES/NO	CLEARLY SIGNED YES/NO	REMEDIAL ACTION AND SIGNATURE

Fire Safety Manual – Section 10: Telephone Trees 9.37

'Telephone trees' derive the name from the concept of the spreading routes or branches of a tree. Each person in the system is responsible for calling at least two others and, therefore, the warning and response is supposed to spread rapidly.

The call rate of a manual system is approximately 30 calls per hour per telephone tree. Messages passed can be inconsistent as the cascade system created its own form of'Chinese whispers'. Having started the telephone tree there is no way of knowing who had received the message and who can attend.

The response rate to calls out of working hours has been shown to be less than 20% it is, therefore, important that telephone trees do not rely on A passing information to B in small groups. The tree must be wide enough to support a major incident.

Responsibility for contacting utility and other support organisations should be via a separate tree which can be activated by the Crisis Management Team.

Remote system 9.38

By using a remote system, reports are available at any time to show who has been contacted and who has not. If required, a broadcast can be started remotely by dialling into the system from a touch-tone telephone, a feature which is vital for properties with only one member of staff on call.

The system used by police forces to activate their neighbourhood watch schemes has been adapted to meet the needs of Emergency Planning. A database of the personnel that may need to be contacted in an emergency is created in a simple database program that runs on a standard computer that can be used for other applications when an incident is not in progress. People may be grouped in various ways, such as department, job, role etc. Each person can have up to three contact telephone numbers.

Templates can be set up which define different selections of people for different types of emergency. There could be one template for 'Property Fire' another for 'Estate Fire'. When an incident occurs a message (or messages) is recorded onto the system and the appropriate templates selected. The system would then telephone each of the people selected and relay a consistent message. The system has the capacity to operate up to 24 lines and contact 960 people per hour.

Fire Safety Manual – Appendix I: Case Study: Walsall Hospitals NHS Trust – Risk Management – East Wing Project

Background 9.39

Walsall is an urban district with areas of high deprivation and unemployment. In November 2000, the Walsall Health Partnership presented its Strategic Outline Case (SOC) proposals to modernise the health care system in Walsall to address the health needs of local residents and create an integrated model of health care across the Borough, which is a Health Action Zone. This scheme was forwarded for 5th wave funding. The total capital requirement being circa £45 million.

The NHS Plan sets out challenges for the NHS to modernise its services and develop patient pathways that put patients at the centre of diagnostic, treatment and rehabilitation processes. The SOC will enable the delivery of key aspects of the National Plan in Walsall through the provision of modern, flexible facilities that enable the re-design of clinical services to meet the needs of different patient groups.

The Walsall Health Partnership has a successful record of jointly delivering high quality services, which has been recognised nationally through the HAZ, ERDID, NPAT collaborative programmes in CHD, critical care and primary care.

Service improvements 9.40

In the Community and Hospital sectors there is a considerable requirement for a mix of NHS and privately funded capital. The Community Trust requires a number of schemes to progress mental health rehabilitation, learning disability and child development services. The Hospital Trust wishes to modernise paediatric, gynaecology, rehabilitation and outpatient facilities. Primary care will be able to provide minor injuries, and diagnostic and treatment facilities.

The service improvements detailed in the SOC will deliver modernised, integrated health care to the most vulnerable patient groups and target the health needs of Walsall residents. The document is consistent with the Black Country Strategy, wherewith the majority of tertiary services are provided from Wolver-hampton.

There is a need to improve access to health services, particularly at primary care level, and to target resources to major health needs in order to reduce inequalities in health and improve health outcomes.

The modernisation and improvement of services is restricted by outdated buildings in acute, community and primary care settings. General acute services are provided on two sites, by Walsall Hospitals NHS Trust, in a mixture of old and more modern buildings at the Manor Hospital and Goscote Hospitals. Many are functionally unsuitable or non-compliant with statutory standards eg fire and safety), and they are costly to operate.

The partners' vision is to centralise emergency, major interventional and one stop diagnostic clinics on the Manor Hospital site, and to centralise intermediate and intensive rehabilitation services and GP beds on the Goscote Hospital site.

East Wing 9.41

By April 2001, the surveys from various specialist companies and estate staff engaged in the research for developing the Strategic Outline Case for Walsall identified risks at East Wing of Manor Hospital that could not wait for 5th wave funding.

East Wing is the oldest Victorian building used for patient care within the Walsall Hospitals NHS Trust, dating from 1852. It consists of a traditional 'H' block, which has been modified over time, on three stories (ground, first and second floors) with a mixture of clinical and non-clinical usage. There is a theatre block extension to the north façade and there are services basements and ducting.

The modifications and general wear-and-tear had, over time, weakened the structure. There were concerns regarding the electrical services (including the fire detection system). Fire compartmentation was compromised and the Fire Brigade were concerned that there were two paediatric wards on the first floor with limited emergency egress. A risk assessment was, therefore, commissioned with independent consultants in the light of the above concerns. Their brief was to analyse all surveys and to take a holistic view of the risk.

Risk assessment 9.42

The assessment had to consider the wider risks to the population should the Board decide to close the wing and transfer patients elsewhere. That closure would, in the short-term, compromise the District General Hospital status of Manor Hospital, leading to removal of support services from the site. It would have meant the loss of Gynaecology and Paediatric services to the community and the possible closure of Maternity.

There is a strong link between ill health and deprivation in Walsall. Health problems include high morbidity and mortality for coronary heart disease, with increased risks due to the high levels of poverty. People living in the most deprived areas are far more likely to have poor mental health, and there is a higher incidence of teenage pregnancy.

The risk assessment concluded that the continued use of the building posed *a serious threat to life* that would require both physical and procedural strategies to manage and reduce the risk. Given the high standards of Health and Safety on site and the continuity of staffing, it was concluded that the safest option for the community served by the Manor Hospital was to manage the risk and maintain the service.

The Board unanimously agreed that the existing range of patient services being provided within East Wing should be maintained. However, a number of changes to the use of accommodation within the building needed to be made immediately to reduce the risk whilst a programme of refurbishment and reappraisal was instigated.

A Project Team led by the Director of Estates and Support Services produced detailed plans for future short-term accommodation requirements. Membership of this Team was drawn from key staff working within East Wing. The Director of Operations lead a parallel Project Team to plan and effect changes necessary for accommodation to be released.

Legislation

Duties to others **9.43**

Within health and safety generally there is a requirement to communicate known hazards to anyone that may be at risk from a situation and there is also a requirement to assess the impact of the work of others on staff and patients. This approach to health and safety is in line with the requirements placed upon organisations by the 'European Convention for the Protection of Human Rights and Fundamental Freedoms' (The UK's *Human Rights Act 1998*), as it states that 'It is unlawful for a public authority to act in a manner which is incompatible with the rights and freedoms that are guaranteed by the convention'.

It is also a requirement under the *Management of Health and Safety at Work Regulations 1999* (*SI 1999 No 4242*) that risk is properly assessed and communicated to those who may be affected by the actions or situations.

Duties to staff **9.44**

The *Health and Safety at Work etc Act 1974* (*HSWA 1974*), *ss 2–4* put specific duties on the Trust to provide a safe place for people to work and use, including those who are not employed by the Trust. These sections are backed up by *section 37*, which sets out the terms of the 'responsible officer' who may be prosecuted if failure occurs, especially if they have issued instructions or failed to ensure that faults were corrected if they lead to the problem occurring.

It is a requirement under *HSWA 1974, s 2* that employers provide a safe place of work for their staff. Therefore, there was a clear duty to inform staff immediately and to involve them in the risk management process.

Duties to visitors **9.45**

Visitors are covered by *HSWA 1974, s 3* in so much as it is a requirement that the safety of anyone who has access to the premises must be taken into consideration.

Within standard surgical or medical establishments it is reasonable to assume that people will have a basic comprehension of a required reaction to instructions, if not actually possessing a clearly defined ability to properly understand the roles and the requirements placed upon them. It is not possible to make that assumption for visitors or services users and this renders many of the normal heuristics and working methods, associated with the safety requirements in health service provisions, invalid.

The lack of compartmentation for the prevention of the spread of fire and smoke and the possible panic that may arise in an emergency required the Trust to ensure staffing levels were maintained and those staff were to be trained in the specific risk strategies required to manage East Wing. Anything less would have meant that the Trust was not fulfilling the duties placed upon it by this section of the Act.

Duties to patients **9.46**

Patients should be afforded the same level of safety protection as is applicable to visitors and, in some cases, due to the incapacity of the patient, they will require specifically designed safety regimes. The NHS Estates Firecode was developed to meet this requirement but due to the age and design of the building East Wing did not meet the standards.

General safety requirements had been addressed but were compromised by the deficiency of the structure. There were also concerns that, in the case of an emergency, the requirements of some patients for specialist medical treatment might be compromised. This aspect required a revision of the hospital Major Incident Plan.

Duties to other users **9.47**

It was noted that the security was provided by an outside company (these people were to play a major role in risk management strategy) and this along with the general access to the building, for reasons of work, by persons not directly employed by the Trust, meant that the provisions of *HSWA 1974, s 4*, which again requires the provision of a safe place to work, were applicable.

Duties to the Fire Brigade **9.48**

Fire Brigade officers needed to be informed of the structural condition of East Wing to enable them to produce their operational risk assessment in the knowledge that:

- water poured into the Nightingale Wards in quantity would accelerate internal collapse;
- fire would compromise the integrity of compartments due to the stresses to joists;
- the Fire Brigade would need to review Fire Certificate, as the Fire Certificate Schedule No 1 para (1) Fire Resistance of Floors states 'with regard to 30 minute fire resistance Fire Authority must be satisfied at time of inspection with the standard provided by the floors'.

However, as services pierced ceilings, walls and partitions plus the load bearing of the floors, which had been compromised, fire compartmentation could not be guaranteed to half an hour standard;

- close liaison with the Fire Brigade was required to manage the risk on East Wing, as there is a legal requirement to inform the Fire Brigade of material alterations to the premises.

The Fire Brigade's response was very supportive of the strategy that was being developed, but given the structural information passed to them, they stipulated that all patients should be removed from the first floor by January 2002.

Risk strategy 9.49

The risk strategy had to be a belt, braces and knotted string approach to ensure that risk was managed at a 'reasonably practicable' level.

The overall strategy was defined in a 'Scope of Works' document, which formed part of the tendering documentation:

> 'The purpose of the works is to extend the useful life of the Victorian East Wing at Manor Hospital for a period of five to six years to allow the Design Procurement of its replacement to be tested using Government Private Finance Initiative'.

The scheme proposed the relocation of all clinical activity (other than the Theatre suite) to the ground floor and to reduce the occupancy of the first and second floors to a level, which could be supported by the existing structural and fire constraints. The Scope of Works consisted of three separate streams A, B and C, each with its own critical path. Of key importance to the delivery of this scheme was the need to keep both the Gynaecology and Paediatric departments fully functioning at all times throughout the works, as such a close working relationship with end users and Hospital Managers was essential.

Project Team 9.50

The Project Team included representatives from all end user groups plus a West Midlands Fire Safety Officer, the Hospital Fire Advisor, the Risk Assessor and the Architect responsible for the strategic design. Initially, the team met weekly and in the interim worked in various sub groups to identify the most efficient safe options for the development of the scheme.

Staffs were consulted about all proposals and generally there was a commitment by them to make the scheme work. There were inevitably some detractors who had heard it all before but they soon realised the seriousness of the situation when the Training School was closed and the Supplies department was relocated from the first floor by 20 April 2001, quickly followed by the relocation of the IT department from the second floor.

A strategy was by now in place to manage the risks involved in these moves and the disruption that it caused to staff. Nursing staff/patient ratios had been increased (above Firecode recommendations for the number of patients) and patient numbers being treated on Chester and York wards, at first-floor level was reduced. The role of the security guards had changed so that now they became fire wardens maintaining a continuous patrol of East Wing.

Design Team **9.51**

The Design Team had by now been appointed, as required by the *CDM Regulations* (*SI 1994 No 3140*) and a list of suitable contractors compiled to ask to tender for the refurbishment works. The criteria for selection were based not only on price and quality of work, but on the safety record of the company, their sub-contractors and their ability to deliver on time – the time requirements of the Fire Brigade being a key factor. In fact, the company that was successful was required to provide a named clerk of works because of his past safety record and ability to work with NHS staff.

Staff awareness **9.52**

There was invariably a time gap between appointment of the main contractors and their arrival on site. This time was utilised to complete enabling works, remove office staff, reduce the operational area of Salisbury Ward and to train staff.

Staff awareness training was devised to involve greater numbers to manage the risk. It was vital that everyone understood that although the alterations would provide significant safety improvements, the long-term occupancy of the building would not comply with legislation and that the alteration works presented a significant risk increase.

There had been an arson attempt on East Wing the previous year and, therefore, training took a wider perspective. The focus was on monitoring the working environment and preparing for the disruption that would occur when contractors arrived on site. Human Factors, communication and the Hospital Major Incident Plan were also analysed. Examples of how fire spreads and how that can rapidly affect the structure were also shown to staff, including an example of a successful arson attack on a Victorian Hospital building.

Monitoring **9.53**

The Risk Assessor and Fire Advisor then devised a strategy to manage each area as work commenced, was in progress and returned to its new usage. It was also decided that the Risk Assessor would no longer be part of the Project Team but would now take on the role of monitoring the hospital risk management strategy and audit the work in progress, reporting directly to the Director of Estates. The reports were then fed back to the Project and Design teams for evaluation and implementation.

The work's progress 9.54

When work commenced on site in September the main entrance was closed, as was part of Salisbury Ward (ground floor).

Work started to restructure Imaging (beneath Theatre), the undercroft and construct the lift housing to Theatre. A temporary external emergency staircase was constructed for Personnel staff during this phase of works. Work to this area was delayed due to the discovery of asbestos in the entrance cladding and the requirement to inform the Health & Safety Executive (HSE) prior to disposal.

At this early stage of the contract the risks had significantly increased as the contractors were in a totally unknown environment. During this time there were two unplanned fire alarm activations caused by contractors breaking through into the public area without proper preparation or informing the Project Manager.

In addition to the activation of the fire alarm, dust drifted from the site of one breakthrough to the reception area that is close to Lincoln Ward, where there are elderly patients with respiratory problems. The levels of dust were low but because of the risk of Aspagilus, the staff response and site management strategy were analysed by both the Project and Design Teams plus the Risk Assessor and details of the risk training altered accordingly.

October 9.55

Not long after these incidents there was a strong smell of gas on East Wing. There are no gas appliances on the Wing but the concern was that there might be a main running beneath the building that had been fractured by the works. East Wing was evacuated save one paediatric ward at ground level, which was remote from the smell and separated by four fire compartments from the areas of concern.

The incident turned out to be a false alarm as the problem was external to the Hospital site. However, it did prove the systems now in place on East Wing to be effective, including the stand-by status for the Hospital Major Incident Plan.

By the end of October, Imaging was ready to be handed over, the Gynaecology Day Case Ward had been transferred to the first floor and a risk management strategy devised for the new area.

December 9.56

The new Gynaecology Day Case Ward and the main entrance were opened by 3 December 2001.

The fire action manual was being audited and updated by the Fire Safety Advisor and all amendments altered by him.

Staffs were informed of the alterations and the senior nurse on East Wing signed to acknowledge that they understand their revised emergency roles.

Training data now showed a significant variation in attendance of the ongoing risk training sessions according to role.

By Christmas one paediatric ward was ready to transfer to the refurbished Salisbury Ward (ground floor).

January **9.57**

Space was then created on Salisbury Ward to facilitate a temporary move of Chester Ward whilst the Paediatric Assessment Centre was completed. By 14 January 2002 there were no ward areas at first-floor level and, therefore, the primary objective of the project had been achieved.

Fire detection protection of Personnel (second floor) had been completed, a new external fire escape completed and the temporary scaffold staircase removed.

Work was in progress on the new lift shaft to theatre and the exit ramp to the rear of the building.

Work was now in progress to most areas of the wing. The old wards at first-floor level were being gutted and new office space created, the Paediatric Assessment Unit and Home Care department was being refurbished at ground-floor level and signage and snagging was being completed in the new ward areas.

The concern at this stage of the works was that people would have too much of a 'feel good' factor and not be aware that they were at a high-risk time of the contract. Training was altered to address these concerns.

March **9.58**

The Assessment Centre was completed in March 2002 and patients and staff transferred from Salisbury Ward.

The Home Care Team transferred to the ground floor.

Work was in progress in Chester Ward, York Ward, Lichfield and the first floor offices in the central block.

The new lift and emergency exit ramp were installed.

Work was in progress to install fire compartmentation doors in the ground floor corridor.

April

9.59

By the 12 April 2002, exactly one year to the day since the Risk Assessors' report, a sign-off meeting was held as the major works in all areas was complete.

Chester Ward, York Ward, Lichfield and the first floor offices in the central block are now occupied

Fire compartmentation in all areas has been completed and resurveyed following the completion of various cable and service runs.

Conclusion

9.60

The success to date has been due to a number of factors:

- the focus and professionalism of the Project and Design teams;
- the efficiency of the contractors; and
- the commitment and flexibility of end users to manage change.

Primarily the key to delivering the project has been the observations and communication of all involved.

Now that the physical changes to the building have been completed the challenge is to maintain the risk management strategy for the remainder of the operational life of the building because although the building works have reduced the risk to a 'reasonable' level, the building does not comply with Firecode standard.

The strategy will involve all users to build upon the successes to date and prepare them for the upheaval when East Wing closes and the new SOC comes to fruition.

Note: Following the completion of the refurbishment the number of security guards was reduced – the new computers were stolen!

Fire Safety Manual – Appendix 2: Procedures for fire safety in an organisation's premises

Scope

9.61

The purpose of this document (which would form an appendix to the Manual) is to detail the procedures to be taken in respect of the fire safety of any activity in all premises operated by the Organisation. The procedure contains the following elements:

- General.
- Reporting and recording arrangements.

- Preparation of fire safety risk assessments.
- Fire precaution training programme.
- Evacuation procedure.
- Permit to work – hot work arrangements.
- Fire systems and emergency lighting tests and maintenance.
- Legal notices.
- Inspections and audit.

Procedure detail 9.62

Procedure detail includes:

- production of the detailed emergency plan;
- identification of the 'designated person', and deputies, at each of the organisation's premises;
- detailed procedures for the role of the 'designated person' and their deputy and for the 'chain of command';
- detailed procedures for the informing of senior members of the organisation's staff;
- the carrying out and evaluation of simulation exercises to test the local effectiveness of the plan and detailed procedures;
- the review, evaluation and revision of the emergency plan;
- devising and delivering bespoke training to designated persons and their deputies (including refresher training) to ensure a safe level of competence to perform the duties of a designated person;
- detailed procedures for switchboard staff and duty staff and contractors;
- detailed procedures for the safe evacuation and care of visitors;
- effective dissemination of detailed procedures to all staff including display of emergency instructions and procedures;
- training for all staff in emergency plan procedures and their local application;
- liaison arrangements with the emergency services;
- debriefing procedures;
- periodic review of procedure.

General 9.63

The organisation has a duty under the *Health and Safety at Work etc Act 1974, ss 2–4* to ensure, so far as is reasonably practicable, the health, safety and welfare at work of all persons that use the premises.

In addition the organisation's property may be subject to the *Fire Precautions Act 1971*. The regulating Authority is the fire authority in liaison with the local Health Authority that issues the license. All are subject to the *Regulatory Reform (Fire Safety) Order 2005*, which requires a great emphasis towards a fire risk assessment approach.

The organisation's Fire Prevention Officer will provide a copy of the Fire Safety conditions stipulated by the licensing authority to all managers. After consultation with the relevant departmental managers, the Facility/Property Manager shall suspend an activity or close part or all of the site if, in the opinion of the person in charge of that area, the fire safety is compromised by its continued use.

Reporting and recording **9.64**

Reporting and recording arrangements are as follows:

- The Organisation's Fire Prevention Officer shall compile and maintain a site specific Fire Precautions Manual for each of the organisation's premises as detailed in the annexe to these procedures.
- The Property Manager will maintain and manage the Fire Precautions Manual and ensure, in liaison with the relevant unit/departmental managers, that all fire safety matters are recorded in it.
- Any fire safety incidents or legal notices will be *immediately* reported to the Fire Prevention Officer, verbally and in writing.
- The departmental managers shall ensure that all members of staff are aware of the dangers of the tasks which they are to undertake and their responsibilities for the fire safety of others who may be affected by their actions.
- The departmental manager will appoint a fire warden of the day to record all fire safety inspections and to be first response should the fire alarm sound.

Preparation of fire safety risk assessments **9.65**

The preparation of fire safety risk assessments include the following:

- The organisation's Fire Prevention Officer in liaison with the property managers will ensure that fire risk assessment surveys are undertaken at all of the organisation's premises.
- A matrix method will be used to evaluate the fire risk of each floor or room within a building to:
  - identify flammable and combustible materials;
  - identify any means by which they could be ignited;
  - identify any groups or individuals that may be at special risk;
  - see how a fire could spread within a property;

- assess the likelihood of a fire breaking out;
- assess the maximum number that can be in the property;
- identify if there are sufficient staff available to safely evacuate the building.

- The fire risk assessments will be made available to all staff and will form the basis of the fire risk management of the building.

Fire precautions training programme 9.66

The Fire Prevention Officer will arrange for bespoke training for each of the organisation's premises, based on the requirements of the licensing authority and the findings of the fire risk assessments.

The property managers/department managers will ensure that all staff participate in the training and that all training sessions are recorded in the Fire Safety Manual.

All Department Managers will ensure that all new members of staff are aware of the Organisation's Fire Safety policy as part of their recruitment induction training.

Evacuation procedures 9.67

The evacuation procedures for each building will be site specific Property Managers will ensure that:

- fire instructions including evacuation procedures, are displayed or readily available to all staff/occupants and kept up to date. Staff and occupants must be made aware of their content;
- staff and other personnel responsible for implementing emergency evacuation procedures must be physically capable of this role.

Senior Managers will make reciprocal arrangements with other sites within the organisation or in agreement with another organisation to provide temporary emergency location for staff and equipment in the event of a disaster. This arrangement will have budgetary, insurance, staffing and occupancy implications for the organisations involved.

Permit to work/hot work systems 9.68

All contractors, advisors and staff from other properties will sign in on arrival at reception. The sign-in identity pass/permit to work will clearly state the date, time, name of the organisation and purpose of visit on the front. There will also be a statement that the contractor has read and agreed to comply with the organisation's health and safety policy. The reverse will clearly state fire evacuation procedures, assembly points and the names of those nominated to be responsible for evacuation procedures. The pass folds into a clear plastic badge which all visitors will wear.

Hot work is a major cause of fire. Therefore, no naked flame heat-producing apparatus, including hot air, should be used during building or maintenance work, except after the issuing of an authority to carry out hot work by the organisation's Fire Prevention Officer.

Where 'hot work permits' are in operation, the Property Manager will ensure that all conditions laid down are stringently observed and all department heads in areas affected are informed of these conditions in writing.

Fire systems and emergency lighting tests and maintenance 9.69

The organisation's Fire Prevention Officer in liaison with the property managers will ensure that permanently installed electrical and mechanical systems, together with fixed appliances and portable tools/appliances will be regularly inspected, tested and maintained in accordance with procedures to be set out in planned maintenance documents. At large sites a Maintenance Log Book will be required in addition to the Fire Safety Log Book.

All work to be carried out by qualified engineers to British Standard and within statutory time requirements. The relevant British Standards are the minimum standards that will be applied.

Legal notices 9.70

Legal notices, which may be headed as 'Prohibition Notices', 'Improvement Notices' or 'Notice of steps to be taken' can be issued where authorised officers consider fire precautions inadequate.

Notices would normally be received from the Fire Authority, but in some circumstances may be issued by the Local Health Authority, Environmental Health Officer, Building Control Officer or HSE Inspector.

Immediately a notice is received a copy should be sent to the organisation's Fire Prevention Officer to study in detail the works required and ascertain if there are grounds for objection.

Inspections and audit 9.71

The organisation's Fire Prevention Officer in liaison with property managers will ensure that fire safety inspections are carried out regularly to audit fire safety arrangements at all premises. These inspections will cover all aspects of fire safety arrangements and are to be recorded in the Manual.

- **Good housekeeping safety inspections** – to be carried out daily by senior members of staff designated by the property managers.
- **In-house fire safety inspection** – to be carried out initially after three months and then – every six months by the Property Manager and the organisation's Fire Prevention Officer. The inspections will check:
  - fire precautions and housekeeping (risk reduction) standards are being maintained;
  - fire drills are being carried out;
  - fire safety systems are being tested maintained and recorded;
  - fire safety procedures for building works and Hot Work permits are being correctly enforced;
  - any incidents of fire or near miss incidents have been properly investigated, appropriate action taken and all findings recorded;
  - any planned changes to the property identify the requirements of fire safety;
  - a fire risk audit checklist is complete.
- **Fire safety audit** – to be carried out annually with the Property Manager and the Fire Prevention Officer plus an external consultant:
  - the team will carry out the same inspection routine as per the six-monthly inspections;
  - the purpose of the audit is to enable good practice to be disseminated to all properties (reduce the tendency to re-invent the wheel), and to establish benchmarking and safety measurement at all properties;
  - these audits will be timed to coincide with a fire exercise at the property.

Fire Safety Manual – Appendix 3: Emergency facilities and equipment 9.72

The following information will also form an Appendix to the Manual.

The Quartermaster is a member of staff whose role is to co-ordinate materials, equipment and facilities that will be needed in the event of an emergency.

The Quartermaster is closely linked with establishing working procedures and the maintenance of emergency materials and equipment throughout the organisation and, to this extent, should be seen as an ongoing responsibility outside the event of an emergency.

The following pages give examples of the type of information that might be required; they are not definitive.

Vital equipment is cameras and tape cassettes to record the incident.

A. Emergency numbers	**Firm/Contact person(s)**	**Phone (day/night)**
Ambulance		
Fire Brigade		
Police		
Hospitals		
Chemist/Pharmacy		
Health Authority		
Social Services		
Security Company		
Other participating organisations (SAFE HAVENS) … … … … … … … … … … … … … … … … … …		
Alarm Company/Fire Security		
Insurance Company		
Telephone Company (Extra Lines)		
Solicitor		
Surveyor		
Emergency Planning Officer		
Utility Companies Gas Calor gas Electricity Water		

B. Resources	**Firm/Contact person(s)** **Retained – R** **Sleeping contract – SC**	**Phone (day/night)**
Accommodation, temporary offices		
Architect		
Beds/Furniture		
Bedding/Furnishings		
Builders Merchant		
Carpenter		
Computer Engineer Hardware Software		
Cleaning Service		
Data Recovery Service		
Dehumidification Service		
Electrician		
Freeze drying (books)		
Fire restoration		
Fumigation Service		
Locksmith		
Mason/Builder		
Plant Hire Company		
Plumber		
Portable Buildings Portaloos		
Salvage Company		
Space, Storage Company		
Tent/Marquee hire		
Transport: Residents Staff Security Furniture		
Tool hire		
Waste disposal		
Other:		

C. Operational supplies	**Quantity**	**Location/Supplier** **In-house supply – IH** **Sleeping contract – SC**
Absorbent barriers: Water Fuel		
Batteries: Torch Tools		
Blankets – protect furniture		
Boots, rubber Brooms: Hand Sweeping Yard		
Bubble wrap: Large Small		
Carrying straps (webbing type) Cleaning materials		
Drying supplies: Clothes line (30 lb monofilament or nylon) Clothes pegs, plastic Dehumidifiers Hygrometers Paper for interleaving (paper towels or unprinted newsprint)		
Dustbins — plastic (to transport objects and to be used for removal of debris for sifting)		
Dustsheets		
Extension cables (3-core, grd, 50 ft) First aid kits Fungicide Generators, portable Lights: External Portable		
Light sticks, chemical (create safe routes in dark and light hazards)		
Loudhailer		➡

Keys – spare set		
Klaxon – safety alarms		
Manual handling: Equipment/hoists		
Media packs: Statement History/background information Agreed procedure		
Mop buckets		
Mops		
Packing supplies: Boxes, cardboard, plastic, wooden bread trays, plastic buckets (for photos) Freezer or waxed paper Labels, adhesive luggage tags String Milk crates, plastic Plastic sheeting Scissors Suitcases, trunks Masking tape, sellotape Wrapping paper		
Plastic sheeting		
Personal Protective Equipment: Hard hats Head lamps Gloves, Masks, Overalls		
Radios		
Registration: Books Identification badges Notebooks Pens/pencils Permit to work forms Plastic bags Clipboards Scissors Time sheets Tabards		➡

Sack barrows Tarpaulins		
Tools/Maintenance Equipment: Axes Bolt Cutters Chain saw Crowbar Duct tape Drills: mains battery Hammers Handsaw Hatchets Knives – Stanley type Ladders Nails, miscellaneous sizes Plywood, assorted sizes Pliers Scaffold: Tower Poles Boards Screwdrivers: Hand Electric Shovels Tool belts Wire cutters Trolleys: 4 wheel 6 wheel		
Vehicles: Ambulance Cars Minibus Tractors Trailers Vans Other		
Waste sacks		
Wet/dry vacuum(s)		
Wooden blocks		

10

Forensic Fire and Explosion Investigation

In this chapter:

- Introduction.
- Objectives of fire and explosion investigation.
- Roles and responsibilities.
- Adopting a structured approach.
- Scene management.
- Using the systematic methodology.
- Locating the origin.
- Interpreting post-fire and post-explosion indicators.
- Excavation and reconstruction.
- Identifying the fuel and ignition source.
- Related evidence.
- Recording and documentation.
- Interpretation and hypothesis testing.
- Laboratory and other support.
- Conclusion.

Introduction 10.1

This chapter will look at the ways in which investigations are carried out into fire and explosion incidents and the importance of doing so with due consideration to any potential legal action (whether criminal or civil) which may follow. For the purposes of the following sections, the term 'explosion' will refer to dispersed phase combustion explosions, ie those in which the fuel is in the form of a gas, vapour or finely-divided solid or liquid and the oxidant is air.

Condensed phase explosions, such as those involving high explosives, will not be considered. Although this is a very specialised area of investigation, the principles of handling such incidents are broadly similar.

Forensic science in general 10.2

Fire and explosion investigation is only one discipline within the overall area of forensic science. It may be useful to establish exactly what is meant by these terms. The word 'forensic' is often misused to indicate the process of scientific investigation in the legal system. Indeed, it has occasionally been turned into a general descriptor ('forensics') and even a verb ('to forensicate').

In fact, 'forensic' simply means of, or pertaining to, the courts and the adjective can be applied to any profession or activity whose end product is intended for, or used in, potential litigation. Hence there can be forensic medicine, accountancy, psychiatry etc. The accepted modern use of 'forensic' goes a stage further than purely litigation, to include the investigative process which may or may not lead to it.

Forensic science can, therefore, be described as 'the use of scientific methods to identify and evaluate evidence to assist the judicial process'. Note that this implies that the process is not necessarily carried out by a 'scientist' within the normal meaning of that word.

The 'scientific method' involves going through a number of steps in which the problem is defined (what do we need to know about this incident?), data collected and analysed using inductive reasoning and a hypothesis developed. Further data is then sought and gathered to enable the hypothesis to be tested by deductive reasoning. If the hypothesis fails such testing, it is modified or discarded and a new one formed which takes into account the additional data. This procedure continues in a loop process until a hypothesis capable of withstanding any reasonable challenge is developed and all other feasible hypotheses have been tested and rejected.

Many other occupational groups with a wide range of training and knowledge carry out investigations using scientific methods. In the area covered by this chapter, examples include fire officers, loss adjusters and safety specialists; not all of these will necessarily be trained and qualified as 'scientists' in the narrow sense. However, if their findings are obtained in this way and used in a legal framework, they may be considered in the same way as other forensic specialists.

Objectives of Fire and Explosion Investigation

Determination of origin and cause 10.3

The most common reason for carrying out a fire or explosion investigation is to determine the cause. This is generally to prevent a recurrence, eg by identifying some defective equipment or dangerous practice, or identify criminal activity. Examples of the former include many industrial accidents (eg the Flixborough and Piper Alpha investigations) which led to changes in processes and procedures, while the supermarket fire in Bristol which caused the death of Firefighter Fleur Lombard is a good example of the latter.

It is rarely possible to identify the cause with any accuracy unless the area or point of ignition can first be determined. The methods used and types of evidence considered in identifying the origin of a fire or explosion will be described later in this chapter (see **10.35** et seq below). Once the origin has been located, identifying the cause will involve consideration of both physical and witness evidence and a process of elimination, if more than one potential ignition source and/or fuel was present at the origin. Again, this process will be discussed in more detail (see **10.44** et seq below).

Assessment of development and spread **10.4**

The way in which a fire has developed and spread from the initial area of origin to affect other materials or parts of a structure may be of great importance. This involves evaluating the interaction between initial and subsequent fuel packages, the influence of building construction and/or failure, firefighting activities and numerous other factors. This aspect of an investigation may take on considerable forensic significance in terms of litigation involving, for example, architects, builders, sub-contractors and materials suppliers.

Whether the origin and cause is very clear and non-contentious or is unable to be determined precisely, identifying reasons for the development of the fire may be essential to limit damage in future incidents and in financial loss recovery.

Evaluation of response **10.5**

The investigator may also focus on the way in which people, the structure itself or the contents responded to the fire. Again, there are often important lessons to be learned and fed back into general community safety concerning:

- real-fire performance of design features;
- previously unanticipated interactions between materials; and
- the actual behaviour of people when they find themselves in a fire situation.

Human behaviour **10.6**

This is one of the most active areas of fire investigation research at present. It has been recognised for some time that 'people do not do what we expect them to do' (Townsend, N. in 'Blaze', a Darlow Smithson production for Channel 4) when confronted by a fire. Investigators will, therefore, try to determine what people actually did, how they reacted, what dynamic effect they may have had on the initiation and/or progress of the fire. Understanding how people respond to fire alarms, how they use means of escape etc in real incidents gives essential information to designers of buildings and systems.

Combustion behaviour of the contents 10.7

The investigator will consider how the various fuels within a compartment or structure interacted to allow the fire to grow and spread beyond the first item ignited. The contribution of materials such as fittings and furnishings, household items, commercial stock etc is highly dependent upon their combinations with each other.

An example of this is the 'Stardust' disco fire which occurred in Dublin in 1981. Detailed investigation found that, although none of the available fuels within the building was capable of burning on its own in such a way as to cause the catastrophic fire spread which took place, the particular combination of materials and the way in which they were arranged behaved in a hitherto unexpected and unpredictable way. As a result of this and similar investigations, and a better overall understanding of fire science, both investigators and designers can now take such effects into account.

Behaviour of the structure 10.8

Building in the UK has been closely regulated for many years so as to afford protection to occupants and others in the event of fire. Much of this protection has depended upon the passive fire resistance of structures and materials. It has been essential for regulations to be continuously reformed as changes in materials and building styles have evolved. The feedback from fire investigations has long been recognised as an essential part of this process.

In recent years, there has been a trend away from prescriptive regulation into 'engineering based solutions'. Using these, building designers are permitted to ignore many of the former regulations if they can demonstrate that their design will achieve the same effect through a different route. This has been made possible by the major advances in fire science and engineering, especially predictive modelling, over the last twenty years or so. The investigator has a major role to play in validating (or otherwise) the predictive techniques, by supplying data from real incidents and informing the design process.

The investigation of the building collapse which followed the terrorist attack on the World Trade Centre on 11 September 2001 has revealed a great deal of information about the behaviour of certain construction elements when exposed to prolonged high heat flux. It is already known that this will impact greatly on future building design and construction throughout the world.

The forensic setting 10.9

The potential exists for any fire investigation finding or report to form part of the evidence in a subsequent legal process. These can vary in type from criminal prosecutions for arson or health and safety offences, through civil litigation (usually but not always involving insurers) to settings such as employment tribunals, internal disciplinary hearings etc. Major incidents may also result in public enquiries.

It is, therefore, essential that any investigation which is carried out into an incident is properly conducted. Potential evidence must be identified, adequately recorded, preserved if possible and documented in the correct way. The investigator should have in mind at all times that he or she could be required to explain and support the findings and conclusions under robust questioning and that heavy financial burdens, or even personal liberty, may be at stake.

For these reasons, investigations may often seem to those outside the system to be unnecessarily complex, detailed or time consuming when the 'victim' simply wishes to clear up and carry on. Investigators generally try to be as flexible and co-operative as possible in resumption of normal activities but often this will be incompatible with their needs. Their activities may well have a statutory footing or be based on insurance contract arrangements.

Roles and Responsibilities

Public sector agencies

Fire Service **10.10**

Under the *Fire and Rescue Services Act 2004*, Fire and Rescue Services throughout England and Wales have a range of statutory duties that include:

- the promotion of fire safety;
- preparation for:
  - fighting fires and protecting people from fires;
  - rescuing people from road traffic accidents;
  - dealing with other specific emergencies, such as flooding or terrorist attack, which will be set out by Statutory Order and can be amended in line with changes required of the Service in the future.

The Act does not apply to Scotland (in which the *Fire Services Act 1947* remains in force) or Northern Ireland.

Under the Act, an authorised employee of a fire and rescue authority can enter premises to carry out the above duties. It also allows an authorised employee to gain entry in order to investigate the cause and progression of any fire that has occurred. However, such an entry cannot be by force and 24 hours notice must be given to the occupier of a private dwelling, unless authorised by a justice of the peace (*Section 45*).

The Act also sets out the powers and obligations of an employee of a fire and rescue authority to gain information or to investigate the cause and progression of a fire (*Section 46*). The powers and obligations are similar to those applicable to investigations under health and safety legislation.

Police **10.11**

Again the primary objectives and duties of the police are clearly laid down and governed by statute. They are responsible for the investigation of allegations of crime and are also the lead agency in investigations into sudden or suspicious death, reporting to the Coroner or Procurator Fiscal (as appropriate). Successive Home Office Circulars, most recently 44/2000, make it clear that this responsibility extends to leading investigations into incidents resulting in serious injury, since these should be treated as potential fatalities.

Although few police officers are trained fire investigators, many of the police scientific support personnel (often known as SOCOs) have a reasonable level of training and knowledge. Some have undergone enhanced training alongside fire officers and may be specialists in their own right.

Others **10.12**

Agencies such as the Health & Safety Executive (HSE) or the Civil Aviation Authority may become involved in specific circumstances and also have statutory powers. In the case of workplace incidents where death or serious injury is involved, the investigation should be police led but they will work closely with the HSE. If no death results and it is established that no crime is involved, police will normally hand over the investigation to the HSE.

These investigators may not be specialists in fire and explosion, in which case their role is to co-ordinate and collate the investigations carried out by others.

Private sector agencies **10.13**

The major agencies are the insurers and their representatives, such as loss adjusters. They do not usually have scene examination skills available in-house but commission them from external sources. They may also make use of fire service and police findings if proper information sharing protocols are in place. Insurers' powers to carry out a detailed investigation are normally a contractual condition and non-co-operation may well result in refusal of a claim arising from the incident.

Other persons having an interest in the investigation, and who may carry out some part of it, are legal representatives and/or inquiry agents instructed by them and loss assessors acting on behalf of insurance claimants. Again, the latter may commission specialist assistance from a number of sources.

Forensic science and engineering support **10.14**

Both public and private sector agencies may commission the services of forensic scientists and engineers or other specialists as circumstances dictate. It is essential that their examinations are not compromised by the activities of other investigators who may be present at an earlier stage. Any initial investigations should, therefore, be carried out with a minimum of disturbance until sufficient information is available to enable identification of those incidents which require the involvement of a forensic scientist or other specialist.

Although the police tend to use the Forensic Science Service as their main support agency in this respect, this is by no means universal and they may instruct one of the private companies who provide such services. Insurers normally instruct a private consultancy firm to carry out the scientific part of the investigation on their behalf.

Other specialists 10.15

Other bodies or individuals with a legitimate interest include fire researchers, who look to use the findings from real-life incidents to underpin theoretical work or to influence legislation, and fire protection engineers studying the effectiveness of various prevention, detection and suppression techniques. Subject to the requirements of criminal investigation and civil litigation, it is important to be sympathetic to these aims and to accommodate their requirements.

The team approach 10.16

Fire investigations have been likened to jigsaw puzzles, in which many people hold one or two pieces but no one has them all. The task of the investigator is to assemble as many of the pieces as possible into a coherent whole. In this respect, the inter-agency team approach is essential. Important information may come from firefighters, witnesses, specialists in various disciplines and background research as well as from the scene examination itself.

Among investigators from different agencies there will be a spectrum of disciplines, expertise and access to information sources. It is, therefore, essential that all the investigators co-operate and share findings to arrive at the closest approximation to the truth about the incident. Of course, the exigencies of a particular type of investigation (especially if criminal proceedings are being considered) may render some information exchange difficult but proper team working can overcome many of the obstacles.

Adopting a Structured Approach 10.17

Modern fire investigators are agreed that it is essential to base investigations upon a firm and stable methodology. This gives structure both to the scene investigation and the surrounding activities. Many developed countries use as a basis for this approach the USA formulated *NFPA 921: Guide for Fire and Explosion Investigation* (2nd edition published in 2004 by NFPA, Quincy MA) or a local derivative.

In the UK, the *FM Global Pocket Guide to Fire and Arson Investigation* (UK Edition) is a useful summary for practitioners, supervisors and facilitators encompassing both the methodology to be used and the basic technical information required. The methodology is based upon a systematic approach using the scientific method to identify and evaluate evidence of all types.

Planning, protocols and relationships **10.18**

The importance of the team approach involving all agencies has already been stated. National Fire Protection Association (NFPA) 921, *24.1* says:

> 'Thorough investigations do not just happen, but instead are the result of careful planning, organisation and the ability to anticipate problems before they arise'.

Many police forces, fire services and insurers have well-developed protocols or Memoranda of Understanding which set out clearly their levels of response, lines of communication and areas of responsibility within any given investigation. This may extend to financial and other resourcing.

It is rare, however, for the 'victims' of such an incident to have a planned response. In industrial and commercial terms, the requirements of a post-incident investigation should be anticipated and the appropriate systems put into place long before any such incident occurs. This may well involve a nominated manager meeting with the emergency services and insurers, not just to discuss levels of protection and prevention as is normally the case, but also to determine what assistance would be required in the event of an investigation being necessary.

Safety considerations **10.19**

Fire and explosion scenes can be among the most dangerous in which to carry out detailed investigations. The possible hazards are too numerous to list and it is almost inevitable that any attempt to do so would result in omitting something of great importance. Each hazard may result in a more or less serious risk to the investigator and to others affected by his or her activities. It is necessary for these potential risks to be identified and reduced to acceptable levels. Note that this refers to post-incident investigation; safety considerations during active fire-fighting and/or search and rescue operations are a separate issue, well addressed by the emergency services.

As in many situations where the workplace is outside the direct control of the employer, the commonly accepted approach among employers of fire investigators (and self-employed consultants) uses the concept of the 'safe person'. Employers carry out generic risk assessments for the range of duties and responsibilities involved but also train and equip their employees suitably to carry out specific risk assessments for the individual circumstances. The overall risk assessment in any situation will, therefore, be a combination of the generic and specific assessments. For example, the generic risk assessment for fire investigation would include the likelihood of structural instability after a fire; the investigator would then assess the extent and seriousness of that risk in the case of the individual scene in question.

For this reason, it is essential that the investigator is given as much information as possible about materials, products, processes, building factors etc that may affect his or her risk assessment. It is also important to remember that the nature of the hazards and the degree of exposure may change during an investigation. Specific risk assessment is a dynamic process and more than one period of mitigation may be necessary.

Applying crime scene management principles 10.20

The starting point for the recommended methodology is John DeHaan's maxim that 'every fire scene is a *potential* crime scene', from which it follows that every fatal or serious injury fire is a potential murder. However, even when a crime has been ruled out there may be a wide range of organisations and individuals involved in the investigation, each with a perspective and an agenda.

An example of this was the fire at Kings Cross Underground Station in 1987. The investigation was initially police led but at a later stage the HSE took on a wider responsibility; at the same time, London Underground and various contractors and materials suppliers had a legitimate interest while Fire Research Station investigators were also heavily involved.

If the scene is managed in accordance with crime scene principles as described in the next section, the possibility of loss or contamination of evidence or conflict between investigators with different interests can be minimised. Furthermore, as any detective will confirm, it is much easier and more productive to begin an investigation at the highest level as soon as.possible and scale it down subsequently if necessary than to try to make up lost ground some time later. If there has been a presumption of accidental causation in the early stages, and no crime scene management procedures have been employed, later discovery of suspicious factors may be unable to be followed through because the evidence has been compromised.

Using the scientific method 10.21

NFPA 921, 2.2–2.3 outlines the use of the 'scientific method' as the basis for the systematic approach. As already outlined, it is important to stress that this does not mean that only 'scientists' (as narrowly defined by qualification or training) can carry out fire investigations, simply that investigations must be conducted in a scientific manner.

Scene Management 10.22

Crime scene management principles can be applied to all types of incident, including fires and explosions. These refer to the concept of taking a holistic approach to the incident which encompasses protection and preservation of physical evidence, appropriate handling and recording of witness accounts, appreciation of the interests of all parties concerned and the importance of the team approach.

In an analogous way to risk assessment it can be viewed as breaking down into two areas, generic and specific in scope. The generic approach involves forward planning and the development of suitable systems and protocols to address the problems which may occur at any fire or explosion investigation. The specifics of an individual scene can then be addressed within this general framework by the practitioners involved.

Team activities 10.23

It is important to establish at the beginning of a specific incident the priorities of the investigation, the primacy of the various agencies and the methods of co-ordination which are to be used. This may be through a hierarchic system or a more fluid network of communications. In either case, it is essential to ensure that all parties are kept fully informed of both findings and decision making.

Defining the scene 10.24

It is not possible to manage a scene adequately until one establishes what it comprises. It may encompass any or all of the following:

- structure(s);
- environs;
- area(s) of open land;
- vehicle(s);
- people (dead and alive);
- linked locations;
- displaced materials.

Key initial questions include how big is the incident and where does 'the scene' start? This will lead to consideration of approach routes, debris zone, preservation/contamination issues, casualties, consequential damage and pre-fire activity. The recommended guidance for explosion incidents is that the area of interest extends to the farthest piece of debris plus 50%; fire debris is usually less widely dispersed and the perimeter will normally have to be set much wider than the zone of burning.

In forensic terms, the scene or linked locations may extend to include the location of casualties (hospital, mortuary) and the transportation used to move them. It will also include any other items or materials removed from the location, for whatever reason, and their transport mechanisms.

The scene may include a complex mixture of structures, surrounding areas and access routes. All of these need be taken into account when making decisions on how best to preserve the evidence while managing the initial stages of the investigation. It is essential to realise that the investigation process should already have begun even at the search and rescue and casualty handling stage, in the sense of gathering and recording transient information.

Protection and preservation of physical evidence 10.25

Various methods including barriers, series of outer and inner cordons and access restrictions can be applied according to the circumstances. Wherever possible, a single access route should be identified and used by all personnel who need to approach the scene. The numbers of these should be strictly limited to those carrying out essential activities and a log maintained of all personnel entering the protected areas.

Where items which may need to be examined at a later stage are in danger of being lost or destroyed as a result of building collapse, firefighting activities etc, they should be retrieved and taken to a place of safety, but only if this can be achieved without unacceptable risks to personnel. Otherwise, every effort should be made to record their location and condition before loss or destruction occurs.

Beyond this, however, handling and movement of any physical evidence should be avoided as far as possible and certainly minimised. Many types of evidence have very specific requirements for correct handling and packaging, and are best left to the appropriate experts to deal with.

Initial recording and information gathering 10.26

One of the most important ways in which early control and management of the scene can assist a subsequent investigation is initial information recording. Immediate recording activities should be carried out, preferably involving both photography and note taking (including sketches if appropriate); video recording is also very useful if available. If there is equipment which may have made a record of the incident, eg CCTV surveillance or plant monitoring, the recordings should be secured as soon as possible.

The essential immediate observations and actions include noting physical evidence, persons in the area, pre-incident activities (if apparent) and vehicle or personnel movements. Names and contact details of witnesses should be recorded and it may be appropriate to take brief details of their observations and recollections at an early stage. Detailed interviews, however, should not be carried out at this stage. These are best undertaken by specialists from the appropriate agencies, who can assess whether cautions should be administered or other restraints apply to the interview.

Although it may be helpful to attempt further information gathering, it is essential to be aware of the problems of recording, transfer, witness contamination and the potential for introducing investigator bias. The term 'witness contamination' refers to the inadvertent feeding of information to a witness of which they would otherwise be unaware. This may come from the unwary scene manager or investigator or by allowing witnesses to share and discuss their experiences before formal interviews are carried out.

Information handling and communication 10.27

Once the immediate recording has been carried out, the information needs to be collated and passed in an appropriate format to the relevant investigative agency or co-ordinator. The general protocol developed before the incident should, therefore, advise on issues such as inter-agency relationships, legal constraints, the limits of confidentiality and preferred methods and lines of communication.

In the specific circumstances of an incident, it is essential that the information gathered in the initial and later stages is properly co-ordinated and shared. One method commonly used is for the various agencies to agree a 'lead investigator', through whom formal communications take place. This will often be the nominated officer from a statutory body (such as the police or HSE) but, in the case of an insurance investigation, it may be an agreed loss adjuster or claims manager.

Handover issues 10.28

In many cases, the scene is handed to or preserved for another agency for a detailed investigation. For example, the Fire Service frequently hand a matter to the police for further enquiries to be made. Under these circumstances it is necessary to consider what needs to be done about scene security taking into account legal duties, time and resource factors.

If a crime is suspected or apparent, the scene and its security becomes a police responsibility. However, there may be a delay before they can take control. In this case arrangements may need to be made to secure the scene and any other evidence pending handover. Similar considerations may apply to HSE and other statutory investigations. Insurers may apply their own requirements under contractual arrangements.

When investigators have completed their examination, the scene is often handed back to the owners or other responsible persons. There may be substantial health and safety issues outstanding and the general protocol should contain guidance on this process. For example, lock-outs on utilities, warning notices or physical security may need to be in place before the investigator can return the scene to the owner's control.

Making deductions from a preliminary assessment 10.29

There is often enough information from initial witness accounts, inspection of the surroundings and exterior when combined with a knowledge of fire development to form a first working hypothesis. It is essential for the initial investigator to consider how strongly will this inform subsequent decisions about appropriate investigative levels and agencies.

For example, it may appear from all the accounts that the incident is of accidental cause. In the absence of fatalities or other reason for involving a statutory body, there is often a temptation to abandon crime scene management principles on cost or logistical grounds. This can sometimes be justified if the incident is minor and the cause genuinely obvious. However, in many instances a more detailed insurance investigation carried out subsequently into the supposed accident (eg to determine liability) uncovers evidence of wrongdoing that was not apparent at the preliminary stage.

Lessons for health and safety practitioners 10.30

Probably the most important contributions that health and safety practitioners in commercial, industrial or public service environments can make to the investigative process are as follows.

First, ensuring that suitable protocols and contingency arrangements exist prior to any incident occurring. This includes nominating appropriate managers to liaise with emergency services and others, and having in place arrangements to provide the necessary information and resources.

Second, immediate scene preservation is essential and the health and safety practitioner is often best placed to ensure that access controls are introduced and disturbance of debris and other items prevented. If practicable, potential evidential materials can be covered with plastic sheeting or upturned metal bins. When coupled with immediate photography, this can be a crucial factor in a successful investigation.

Third, the health and safety practitioner often has a good relationship with the workforce which facilitates the gathering and recording of information at an early stage. This person is also likely to have detailed knowledge of the pre-fire risks and, therefore, be of great value to the investigator.

Last but not least, the health and safety practitioner is the person to whom the investigator will initially turn for information which may affect his or her risk assessment, eg the location and nature of stored materials, status of electrical systems etc. The practitioner should, therefore, ensure that such information is gathered and made available during or as soon as possible after the incident.

Summary 10.31

Scene management can be a complex and difficult matter and often the investigation will stand or fall upon its successful application. Yet it is by no means a matter only for the senior fire and police officers, nor does it stop when the public services leave the incident. Scene management starts with the first response, from whichever agency, and carries on throughout subsequent interactions until all the interested parties have had an opportunity to inspect the relevant evidence and draw their own conclusions.

Using the Systematic Methodology

Preliminary physical examination 10.32

The investigator will first carry out a preliminary visual examination of the exterior and environs of the incident. This has the dual function of informing the dynamic risk assessment and gathering initial data about the nature of the incident, fire spread and development, potential area(s) of origin, structural response and associated physical evidence.

Once an approach path has been identified and protected, a similar visual assessment is carried out of the interior. Again, this has a safety function as well as gathering data about the fire. The types of information gained at this stage include directional fire spread and gross damage patterns, material and structure responses, occupancy and use factors, processes in use, etc.

Initial hypothesis formation 10.33

The observations derived from the preliminary physical assessment are combined with the initial witness accounts and other available information to form a preliminary or working hypothesis concerning the incident, for example, where the fire began and how it spread. This point of the investigation is critical because the planning and organisation of the work to be carried out will depend upon the nature of the preliminary hypothesis.

However, the investigator must keep an open mind and not concentrate on looking only for evidence to support the initial hypothesis. The next step is to subject this initial scenario to testing, in order to establish which parts of it are not supported by evidence from a more detailed examination or other witness accounts.

Detailed physical examination 10.34

Again this will commence with the exterior and surroundings. Unburnt parts of the structure can reveal information about pre-fire conditions such as security, state of repair, utilities and services, ventilation effects, external fire spread and many other factors. It is also possible, of course, that the area of origin is external to the building.

The detailed interior examination moves from areas of least to greatest damage. In parts of the building which have not been severely damaged by fire, the investigator looks for information about the pre-fire conditions. Examples include the level of 'housekeeping', nature and condition of equipment and fittings, stock levels, the nature and location of fuel, decorative condition etc. The investigator thus gains information about pre-fire conditions which may have had an impact on the fire's origin, cause and development.

The fire patterns will be observed and documented, and followed back towards the area from which the flames and hot gases originate. The investigator must have a thorough understanding of fire dynamics in order to interpret these patterns correctly.

During the whole of this detailed examination, close attention is paid to all types of physical evidence which may be found; measurements are made and plans or sketches prepared to augment the photographic and/or video recording which should also be carried out. Items of evidential significance will be recovered and suitably preserved for further examination, eg in a laboratory.

In the case of an explosion, displacement of structures and the direction and distance travelled of projectiles is noted. Plotting out of this information normally indicates the source of the overpressure and allows the investigator to focus in on the fuels and ignition sources in that area.

Locating the Origin

Common basic techniques **10.35**

There are several basic techniques used to determine the area of origin of a fire which can be applied in most situations. Among the most common is the interpretation of the effects of heat exposure on structures and materials. Put simply, in many cases where the fire has been burning longest is where it started (since it spread out from there). The investigator will, therefore, look for evidence of the longest exposure to the greatest amount of heat. This may be reflected in a structural failure, destruction of contents and/or different degrees of damage to items of similar construction eg furnishings.

Of course, it is essential for the investigator to take into account differences in type and amount of fuel and ventilation which could make the fire develop more or burn more severely in locations other than the area of origin.

In general, flaming fires burn upwards and outwards at a much greater rate than they burn down. The investigator will, therefore, also be looking for the lowest area of burning within the structure. Again, however, this can be complicated by falling materials, structural collapse etc. The investigator must also be aware of the effects of compartment geometry on fire development. For example, fires develop much more rapidly and to a greater degree in the corner of a room than in the centre.

Local electrical failures, such as arcing on trailing leads and flexes, combined with the operation of protection devices including plug fuses can enable the investigator to determine the directional progress of the fire through a compartment. On a larger scale, this can be applied to building circuits where the various points of arcing and the operation of circuit protection as the cable insulation fails can be used to help locate the first area of major flaming. This process, known as arc mapping, is very time consuming but may be the only method of use in a major structure fire.

Witness observations can clearly be of great assistance in many cases, when the fire has been seen in its early stages. Care is necessary in interpreting these, however, since the vantage point, illumination, obstructions and even imagination may play a major part in the account. In addition, where the witness first saw the fire break out may be some way from the actual origin and, of course, some witnesses are far from accurate in their descriptions (for various reasons). Accurate witness accounts were the key to the successful determination of the cause of the fire at Windsor Castle.

More specialised techniques 10.36

There are numerous other methods of locating the area of origin, some of which rely strongly upon a detailed knowledge of fire science and engineering. These take into account such variables as assisted ventilation, fire load discrepancies, plume movements, unusual thermal behaviour of structures and materials, and the effects of firefighting activities or equipment. It may also be possible to interrogate fire and/or intruder alarm systems to determine sequential operations, or interpret operation of fixed fire extinguishing equipment.

Investigators may construct mathematical computer models to test scenarios. However, it should be stressed that the typical engineering models designed to predict fire development from a given origin and growth rate cannot be 'run backwards' to arrive at an origin by inputting the final damage details.

Complicating factors 10.37

The location of the area of origin is relatively straightforward in some cases. However, it is often complicated by a variety of factors. The most common among these are uneven and variable fire loads within compartments or buildings, the presence of certain fuel types which have very high heat release rates and, therefore, distort the overall fire patterns, and the changes in ventilation which take place as the fire progresses and the firefighting takes place. It can be seen that the task of locating the origin of a fire is a good deal more complex than might be supposed.

Interpreting Post-fire and Post-explosion Indicators 10.38

Some of the so-called 'fire patterns' which the investigator considers to help determine the point of origin can more correctly be described as post-fire indicators, that is physical effects brought about by exposure to heat, flame, smoke or any combination of these on materials such as parts of a structure or its contents.

Duration and rate 10.39

Many of these indicators reflect exposure to a level of heat or flame over a period of time and it is possible to assess both the overall degree of damage and the likely duration of the exposure. Some of the older texts contained material which is now known to be misleading, or even completely incorrect, which suggested that the rate of exposure could also be determined from these indicators. In other words, it was considered possible to state from specific pieces of evidence whether the fire had built up slowly or rapidly and on this basis decisions were often made as to whether it had been artificially accelerated, eg by the use of flammable liquids.

As a result of much more intensive study of fire behaviour over the last twenty years or so, it is now known that many of the indicators do not indicate rates of exposure reliably. It is also now clearly understood that no single indicator should be used in isolation to affirm a specific hypothesis.

Types of indicators

Post-fire indicators **10.40**

Post-fire indicators include the degree and distribution of charring to timber surfaces, spalling of plaster and masonry surfaces, fracture and failure of glass (glazing, containers etc), oxidation and melting of metals and the shapes of fire plumes as demonstrated by smoke deposition or heat damage to walls and ceilings. In the past, each of these has been misinterpreted as indicating something specific about the fire.

For example, it was said that wood charred at a specific rate and thus the depth of charring could be used to determine the duration of the fire, whilst the shape and appearance of the char 'blisters' indicated whether it had a fast or slow build up. It is now understood that this is not the case and that there are very many variables involved including wood type, moisture content, grain direction and radiant heat flux levels.

Similar incorrect ideas were held concerning glass fractures, the size and shape of which were also thought to indicate rate of exposure, spalling to plaster which was thought to show rapid heating, and annealing of metals which was believed to indicate prolonged exposure to relatively low levels of heat flux.

The presence of hard-edged, pool-shaped burns at low level was commonly held to be an indicator that flammable liquid had been used to assist the fire. It is now understood by most fire investigators that similar patterns can be produced in a wide variety of ways, including the burning of normal contents.

Post-explosion indicators **10.41**

Post-explosion indicators include the size, shape and extent of the debris field. The size and mass of typical debris fragments and their displacement can indicate the overpressure achieved in the incident, which in turn can enable the investigator to deduce features of the fuel-air mix such as the nature of the fuel and its probable concentration.

Displacement and distortion of structural elements also contributes a great deal of information, both in localising the origin of the explosive event and in determining the forces involved. From these, fuel volumes, concentrations and venting areas can be estimated.

The investigator will also look for evidence of heat or fire damage on displaced or projected items. This will help to determine whether the explosion preceded the fire or vice versa and hence indicate the mechanism responsible.

The holistic approach 10.42

As outlined above, the older approach of relying upon one or two key indicators to localise the origin and determine the nature and development rate of the fire is inherently unsafe. Modern methodologies stress the need for a holistic approach in which all the physical evidence features are considered together. This provides a much more robust and dependable basis for conclusions.

Excavation and Reconstruction 10.43

Once the area of origin has been established from the fire patterns and other considerations, the investigator will normally excavate the locality. This involves removing the superficial layer of debris that has accumulated as the fire progressed and decorative or structural elements failed, falling onto the area of interest. The process is analogous to that of archaeology, in which the accumulated detritus of many years is removed to expose the strata of interest.

In a fire, the overlying debris accumulates in seconds or minutes rather than years but the process is similar. The extraneous material covers, locks in place and protects items and patterns of evidential value. The investigator will, therefore, remove this material before carefully excavating the stratified debris beneath. In the same way that an archaeological dig is a 'slice through time', so the stratification of the fire debris reveals the sequence of events which occurred during the fire. For example, beneath a window, unburnt carpet with clean glass on top and charred curtain above indicates that the window was broken before the fire started. Contrast this with finding charred curtain lying on heavily burned carpet with smoke-blackened glass on top of that, showing that the fire was well developed at floor level before the curtain fell and then the window broke.

Careful excavation may also reveal a good deal of information about the disposition of objects and items of furniture or goods in the compartment prior to the fire. The investigator must then ask if these were as they should have been, what effect they may have had on the fire development, what they demonstrate about pre-fire activity and so on. Excavation may uncover items or materials alien to the premises or demonstrate that items normally present were removed prior to the incident.

During the fire and subsequent activities, items are often disturbed or broken. Replacing them in their original positions and orientations enables the investigator to study the fire progression patterns, eg from one item of furniture to another, compare the relative damage and narrow down the area of origin. It also clarifies pre-fire activities, disturbance etc and may verify or refute witness accounts of the conditions prior to the incident.

Excavation and reconstruction can be physically demanding and time consuming activities at many scenes, especially extensive fires. In large fires it may be necessary to use heavy plant for the initial phase of extraneous debris removal. However, the quantity and value of the information revealed is often the key to successfully resolving an investigation.

Identifying the Fuel and Ignition Source 10.44

Having located and excavated the area of origin of the fire, the investigator must determine what fuel first became ignited there and what was the heat source responsible. Only once this information is available can the next stage in determining the cause be undertaken.

The initial fuel could be part of a malfunctioning appliance, furnishings or building components near a heat source, materials involved in a production process, waste (rubbish) or some imported fuel such as a flammable liquid. The physical characteristics of the fuel play a major part in whether it will ignite and sustain combustion, so the investigator must consider the quantity, nature, physical state/condition and orientation of the fuel as well as its proximity to any heat source.

Ignition sources may be anything that can produce heat in sufficient quantities and transfer it at a high enough rate to ignite the fuel in question. There may be more than one potential ignition source within the area of origin and the investigator then has to determine if one or more can be eliminated.

Direct evidence 10.45

Sometimes there will be direct physical evidence of the first fuel package, the ignition source or both within the area of origin. The presence of timber frame components and flat zigzag springs might indicate to the investigator that the first item involved was a foam-upholstered chair. This is valuable information because there are known limitations on the ways in which such items can be ignited and the rates of development are well understood.

Similarly, there may be an obvious ignition source such as an overturned radiant heater or the hot surface of some industrial process. In some cases, however, the ignition source is destroyed in the fire, eg cigarette ends and matches which are unlikely to survive significant amounts of burning.

In some deliberately set fires, both the fuel package and ignition source may become apparent after excavation. For example, the remains of a broken bottle with a burnt rag in the neck and surrounded by petrol residues provides good evidence of fuel, heat source and mechanism.

If there is direct evidence of only one potential fuel and ignition source combination in the area of origin, the investigator's task is made much simpler. More commonly, however, either the fuel package, the ignition source or both have been wholly or largely consumed by the fire or there is a number of potential heat sources in the area.

Indirect evidence 10.46

When the fuel has been consumed, it may be possible to identify what it was by other means. These may be straightforward, eg asking persons familiar with the area what was stored or kept there prior to the fire, or may involve theoretical considerations. For example, from the burn patterns and other information it might be possible to estimate the initial heat release rate of the fire and this could help to differentiate between, say, papers in a waste bin or a carton of synthetic padded coats as the first fuel.

Because fuels respond in different ways to specific heat sources, this can in itself limit the number of potential causes. Where more than one possible heat source is present, the nature of the interaction between it and the fuel must be considered.

Ignition is not simply a matter of temperature but of energy transfer. The heat source must be capable of providing sufficient energy for long enough to heat the fuel, while the thermal characteristics of the fuel must permit adequate transfer of that energy if ignition is to result.

Interpretation of the physical findings and witness evidence may involve a substantial amount of deliberation, possibly involving mathematical calculations, and may require testing to be carried out using similar materials and heat sources to verify or exclude a proposed mechanism.

Related Evidence 10.47

During the scene examination, the investigator will seek and record other aspects of physical evidence which may or may not relate directly to the origin, cause and spread of the fire. Often it will not be clear at this stage what bearing these observations have on the final outcome, but it is essential that they are noted.

Security 10.48

The building security is of major importance. Sometimes it will be clear that a door or window has been forced, or that some other illegal entry has occurred, but often such evidence is partially masked by firefighting activities. This evidence is, of course, not confined to the building entrances. There may be evidence of forcing to internal doors, drawers, cupboards, safes etc that will indicate pre-fire activities.

Even if the building security does not appear to have been compromised, key holder activities must be considered. Interrogation of modern alarm systems can provide useful information concerning when and by whom they were set or disarmed, which zones were enabled etc. In occupied premises where the external security is not an issue, eg a public building, internal security may be of considerable importance. Access control systems, CCTV monitoring and physical locks will need to be checked in detail.

Building systems **10.49**

The investigator will need to consider aspects of the building systems which may have contributed to the cause or spread of the fire. This will include fire protection systems, utilities, HVAC systems, production processes etc. In many cases, a fire investigator will need to call on experts in other disciplines to complete this part of the work.

This is especially so where complex electrical, fire precautions or alarm systems are installed. In any case, it is essential that unskilled personnel do not make any attempt to dismantle or repair fire-affected installations as they may severely compromise the investigation, in addition to being potentially dangerous. Alarm system back-up power supplies should be maintained wherever possible so as to preserve volatile memory.

Stock or furnishings levels **10.50**

The investigator will be concerned with the levels of stock and/or other contents at the time of the fire, in comparison with what would normally be expected. Both the quantity and quality of contents are an issue, as is the absence of specific items of value or their substitution by lower value alternatives.

In extreme cases, detailed inventories of the fire-damaged contents may be taken and compared with company records (usually by the loss adjuster or insurer). In one quoted case, a claim for several thousand pairs of top-price designer jeans, which allegedly had been destroyed by fire, was refused because the loss adjuster counted all the zips and found only enough for a few hundred pairs, while the rivets did not carry the appropriate manufacturer's logo for the brand claimed. Similar techniques have been applied to equipment such as office furniture, computers and plant.

Location and nature of the fire **10.51**

In the case of deliberately started fires (which currently represent well over 50% of major fires in occupied buildings), factors such as the area of origin, method of ignition, access routes and so on are noted by the investigator. These may indicate the offender's motivation and psychology, as well as linking incidents through modus operandi (MO).

In the modern era of 'intelligence-led policing', linking MO information with other evidence of local trends and criminal activity is often the key to identifying the perpetrator of a commercial fire if it is not someone directly connected with the business.

Other information **10.52**

The detailed examination of the building and its surroundings may have yielded information concerning the state of repair and maintenance, lifestyle or house-keeping issues, associated criminality (vandalism, graffiti, theft etc). All of this needs to be taken into account by the investigator in drawing together the evidence which points towards the cause of the fire.

Reconstructing the Event Sequence 10.53

It is important to realise that establishing the cause of a fire or explosion is not restricted simply to identifying the fuel and ignition source. Arguably the most important part of the investigator's task is to establish how they came together, which usually involves determination of the probable sequence of events. This is, in some ways, the reverse of fault tree analysis, working from a known outcome back through a series of alternative branches to a probable start point.

The investigator uses a combination of physical evidence, witness accounts, documentation and known background information (eg research findings) to establish a sequence of events that agrees with all the known facts and to produce a reliable chronology.

Further witness and other information are then combined with these known facts. In cases where witness accounts conflict with physical evidence, the reason should be sought; a genuine mistake could be responsible but some witnesses have a motive to mislead the investigator. This does not necessarily imply that the fire was deliberately started, but the individual may feel at risk of blame. In addition, witness perceptions of time are often very distorted during an incident but it may be possible to fix them with reference to each other or to accurately known points, such as the arrival of the Fire Service.

Recording and Documentation 10.54

It is essential that all aspects of the investigation are recorded in sufficient detail and in a legally sustainable manner. There are many methods available, ranging from written notes and diagrams through photography and video to modern digital techniques. It is rare for any one technique to be sufficient on its own; most investigation findings comprise at least two or three types of record.

Scene examination and witness interview records must be contemporaneous. They cannot safely and justifiably be prepared long after the event. Other records, such as equipment monitoring logs, CCTV tapes etc should indicate clearly when they were made and/or copied.

All records must be safely preserved and protected in such a way that they can be readily produced when needed. Legal requirements for original records ('best evidence') mean that, although such formats as CD-ROM can be used for ease of reference, archive facilities for hard originals will still be required. All digital images and photographic negatives must be preserved in their original condition.

Interpretation and Hypothesis Testing **10.55**

Having gathered all the available information concerning the incident, both physical and other evidence, the investigator will collate and interpret the findings. This involves a 'continuous loop' process of inductive and deductive reasoning.

As a hypothesis is developed using inductive reasoning, it is tested against all the known information using a deductive process. Any incongruencies are identified and the hypothesis either discarded or refined to take them into account. Elimination processes are important in this chain of reasoning but simply because something cannot be eliminated does not mean it is the final answer unless all other possibilities have been ruled out. In some cases, however, this method does produce the final result.

> 'When you have eliminated the impossible what remains, however improbable, must be the truth.' (*Sherlock Holmes* by Sir Arthur Conan Doyle.)

In practice, a series of interim alternative hypotheses will have been formed and tested during the scene examination. The final interpretation will usually be a matter of reviewing and fine tuning the last of the interim hypotheses. This version will often be a refinement of the original preliminary scenario, especially when experienced investigators are involved who are adept at assessing the initial scene. However, it is important to make clear that other possibilities have been considered and on what basis they were ruled out.

It is essential then to consider ways in which the final hypothesis from the scene and witness information could be further tested and refined.

First, are there any potential fresh data sources which have not yet been explored? If so, what is their likely reliability? How does this compare in accuracy with the data already considered?

Second, are there any test methods which could be used to validate the information gathered or deduced? These may include laboratory examinations and analysis, computer simulations or physical re-enactments of the event. Whichever method is chosen, the investigator must be sure that the results are meaningful. This involves setting appropriate windows for test variables, being aware of the effects of analysis sensitivity and variations in test parameters and, above all, being open minded and ready to consider alternative scenarios which are amenable to testing.

An example illustrating the way in which this system operates involves the fire at the British Antarctic Survey research station during 2001, which the author investigated. Early photographic records and witness accounts helped localise the area of origin of the fire; physical damage was present there which supported one hypothesis as to cause and tended to eliminate others; this evidence was susceptible to laboratory testing; alternative hypotheses could be excluded; and ultimately, a reliable finding was possible despite the degree of destruction of the premises.

Laboratory and other Support 10.56

In addition to the range of specialists who can assist the investigator at the scene, support is available from laboratory-based and other experts. A detailed exploration of the types and nature of such examination is outside the scope of this chapter but a brief description of some of the relevant areas follows.

General evidence 10.57

Many of the physical evidence types available at fire scenes are common to other crime and incident scenes. These include:

- identifiable marks left by body parts, footwear, tyres, tools etc;
- contact traces arising from textiles, structural materials, body fluids etc;
- engineering and metallurgy examinations of failures, operating mechanisms etc;
- documentary and electronic records;
- image capture devices and enhancement.

Fire-specific evidence 10.58

In addition, there are other evidence types associated specifically (or almost so) with fire and explosion incidents. These include:

- presence of flammable liquid residues;
- heat/flame damage and distribution (to clothing, skin, hair etc);
- appliance and equipment failures;
- ignition and combustion behaviour testing;
- pyrotechnic and incendiary devices.

Medical and toxicology evidence 10.59

In the case of fatal and serious injury incidents, substantial information is likely to be available from pathologists and/or other medical specialists. This includes the nature and distribution of burns and other thermal injuries, levels of toxic gases and alcohol and drug concentrations. Such information helps the fire investigator locate and position the victim relative to the fire, understand the state of mind or physical condition which may have prevented escape, etc.

Other specialists 10.60

Forensic evidence may be provided by a range of other professionals including accountants, engineers, fire protection specialists, psychologists/psychiatrists, and even veterinary surgeons. Their contribution may be made long after the scene investigator has completed his or her phase of the overall fire investigation but it is essential that all parties are kept fully informed, since such findings may modify the scenario.

Conclusion **10.61**

This chapter has attempted to show the complexity of the task, the need for inter-agency team work and the importance of the initial and subsequent scene management. It is not intended to turn the reader into an expert fire investigator but to give an overview of the subject and indicate how everyone involved in an incident has a part to play in the investigation team.

> 'A fire or explosion investigation is a complex endeavour involving skill, technology, knowledge and science. The compilation of factual data, as well as an analysis of those facts, should be accomplished objectively and truthfully' (NFPA 921 *Guide to Fire & Explosion Investigation* (2004)).

> 'The subject of investigating causes of fire may involve professional staff from a wide spectrum, each concerned with their own discipline and with a different viewpoint, but sharing a common objective of accurate definition' (Sir Peter Darby, former HMCIFS, in *Principles of Fire Investigation* (1985) Cooke & Ide).

Conclusion 10.6

11

Human Error and Human Factors

In this chapter:

- Introduction.
- Hindsight.
- Human error.
- The human body.
- The involvement of humans in technological systems.
- Organisational failures.
- Management's responsibility.
- Kletz's four groups of human error accidents.
- Decision making.
- Reactions to failure.
- Conclusion.

Introduction 11.1

At the outset of any human-made Disaster, the most pressing question is invariably what caused it. Was the Disaster due to human error or a systems or technological failure?

It is very easy to blame the master of a ship, or the driver of a train, or the pilot of an aircraft, or the engineer of a chemical plant. This is particularly so if the person has died in the Disaster. Accusations of human error are convenient even though many cases in which such claims are made are complex operations. But, as Crainer points out in *Zeebrugge: Learning from Disaster: Lessons in Corporate Responsibility* (1993) Herald Charitable Trust, although 'the act or omission of an individual' may precipitate a disaster (p 1) the cause of the event is likely 'to involve a number of people over a period of time' (pp 46/47) and is 'likely to lie deep within the systems, outlook and personnel of an organisation' (p 1). Indeed, history shows that human-made

Disasters have consistently been caused, not primarily by individuals but by errors, misjudgment, failures or even negligence by senior management because they have failed to instill a culture of safety within an organisation.

Aberfan

11.2

Case study 1

> On 21 October 1966, a huge tip of coal dust that had been constructed over a number of years outside a mining colliery at Aberfan slid down over part of the village, killing 116 children and 28 adults. In declaring that the Disaster should not have happened, the Inquiry Report found it had occurred because of ignorance, ineptitude and a failure to communicate. Elaborating on that statement, the report stated it had occurred as a result of:
>
> > 'Ignorance on the part of those charged at all levels with the siting, control and daily management of tips; bungling ineptitude on the part of those who had the duty of supervising and directing them; failure on the part of those having knowledge of the factors which affect tip safety to communicate that knowledge and to see that it was applied' (para 18).
>
> However, these people were not villains, the report said, but 'decent men, led astray by foolishness or by ignorance or by both in combination' (para 47).

Cause of error

11.3

Crainer's view is supported by David MacIntosh, a lawyer who was involved in the aftermath of a number of disasters, including the fire at the Bradford Football Stadium in 1985 and the Piper Alpha Oil Rig explosion in 1987. He claims that:

> '. . . it is too tempting, even for corporations who ought to know better, to quickly blame human error. It is easier to blame the error of an individual on the front line than to analyse the shortcomings of a corporation or its safety procedures. It is [says MacIntosh] very rare indeed that one simple piece of negligence causes a disaster. [It involves] 'a series of mistakes, some minor in their individual impact, which in the aggregate lead to the disaster'. (Quoted in Crainer (1993), p 48.)

Kletz claims that 'many accidents . . . are blamed on human error, usually an error by someone at the bottom of the pile who cannot blame someone below him.' Managers or designers, he suggests 'are either not human or do not make errors' (Kletz T, *Learning from Accidents* (1994) Butterworths Heinemann, p 240).

Attributing blame **11.4**

Because of the high profile given to any aircraft accident, pilots have probably been the subject of the closest examination when it comes to human error. The list of aircraft accidents attributed to pilot error is extensive. But why? Recording pilot error as a cause of accidents arguably arose for two reasons. First, there was and, indeed, still is, a need 'to attribute an accident to something, to find an outlet for our grief and dismay, to blame somebody – in other words to find as mankind has done down the ages, a scapegoat'. Second, there was an 'implicit belief that flying as a skill was very difficult' (Beaty D, *The Naked Pilot: The Human Factor in Aircraft Accidents* (1995), Airlife Publishing, p 27). However, when a pilot makes an error it is unlikely to be in isolation. As with most human-made Disasters, it is more likely to be one of a series of errors, many of them made by other people long before the pilot's error culminates in the Disaster.

Often there is no principle cause. Sometimes it is a combination of equipment failure, management failure and human error, in that there has been a failure to heed advice and warnings over an extended period, compounded by lack of judgement at the material time.

Nevertheless, there are numerous examples of incidents when people have performed less than effectively when faced with unusual or abnormal events. Kletz points out that:

> 'human error is often quoted as the cause of an accident [but] there is little we can do to prevent people making errors, especially those due to a moment's forgetfulness' (Kletz (1994)).

Conversely, there are examples of incidents where the blame has been attributed to individuals but, on closer examination, it clearly lies elsewhere.

Hindsight **11.5**

It is often said that hindsight is a specific science; that people should not be judged on facts that become known after an event. But what precisely is hindsight? Sydney Dekker identifies three key characteristics (*The Field Guide to Human Error Investigations* (2002) Ashgate, p 16):

- First – people are in a position to look back on a sequence of events that led to an outcome they already know about.
- Second – they are generally aware of the true nature of the situation in which those involved found themselves, as opposed to the situation they thought they were in.
- Third – because they are now in possession of most, if not all of the information, they are able to identify what those involved missed, what they should not have missed, what they did and what they did not do that perhaps they should have done.

This view is supported by Kletz who states that, 'the best anyone can be expected to do is to make "reasonable" or "rational" decisions based on the information available to them' (Kletz (1994), p 68).

In the aftermath of disasters, the danger of judging people in hindsight has often been the subject of comment. Turner suggests that:

> '. . . if we are looking back upon a decision which has been taken, as most decisions, in the absence of complete information, it is important that we should not assess the actions of decision-makers too harshly in the light of the knowledge which hindsight gives us.' (Quoted in Toft B and Reynolds S, *Learning from Disaster: A Management Approach* (2nd edn, 1997) Perpetuity Press, p 69.)

In his report into a riot in Red Lion Square (Scarman, The Rt Hon Lord, 'The Red Lion Square Disorders of 15 June 1974' (1975) Cmd 5919, HMSO) in which a student from the University of Warwick, Kevin Gately, died, Lord Scarman had this to say about police officers and hindsight:

> 'Wisdom after the event is an occupational hazard for judges and armchair critics. Police officers are people of action; their duty requires them to assess a situation quickly, to make up their minds and then to act before it slips out of control. They are not to be criticised if hindsight invalidates a decision properly arrived at on the information available at the time'.

The following year, the Chairman of an Inquiry into a fire at the Fairfield old people's home commented that he had conducted the Inquiry 'with the benefit of hindsight' and it should not be presumed 'that what is clear now would or should necessarily have occurred to anyone considering the situation before the fire' ('Report of the Committee of Inquiry into the Fire at Fairfield Home, Edwarton, Nottinghamshire, 15 December 1974' (1975) HMSO).

In his Report, of the investigation into the Clapham train crash in 1988, in which 35 people died, Sir Anthony Hidden commented:

> 'In my review I have attempted at all times to remind myself of the dangers of using the powerful beam of hindsight to illuminate the situations revealed in the evidence. The power of that beam has its disadvantages. Hindsight also possesses a lens which can distort and can therefore present a misleading picture: it has to be avoided if fairness and accuracy of judgement are to be sought' (p 3).

So, it is with the above comments in mind that the remainder of the chapter should be read.

Human Error 11.6

There are many reasons why human error occurs. Fatigue, information overload, ignorance, carelessness, negligence, stupidity are just some that come readily to mind.

Dekker puts forward two views of human error, the old and the new. The old view suggests that human error is a cause of accidents; that in order to explain failure you must seek failure and find people's inaccurate assessments, wrong decisions and bad judgements. However, history has shown that, despite all the attempts to attribute blame and ensure that people who have been at the centre of a Disaster attributable to human error, are never placed in such a position again, such accidents continue to occur. The new view suggests that human error is a symptom of deeper trouble inside a system; that in order to explain failure it is insufficient to find out where people went wrong. Instead, any investigation should seek to discover how the assessment and actions of people made sense at the time, given the circumstances that surrounded them (Dekker (2002), p vii). This view suggests that because human error is a symptom of deeper trouble, it is necessary to connect the behaviour of people with the circumstances that surround them in order to identify the source of the trouble and explain that behaviour.

James Reason (*Human Error* (1999) Cambridge University Press) distinguishes between active error, ie effects that are felt almost immediately, and latent error, the adverse consequences of which may lie dormant within the system for a long time. This can clearly be seen by people at the sharp end, the pilot, the train driver, the ship's captain, the engineer of a chemical or nuclear plant. All would be classed as making active errors.

Latent error, however, generally lies with management or the system that supports management. Beaty claims that:

> 'there is a growing awareness within the human reliability community that attempts to discover and neutralise those latent failures will have a greater beneficial effect upon system safety than will localised efforts to minimise active errors' (Beaty (1995), p 221).

Dekker refers to the two different errors as 'sharp end and blunt end' (p 20). But, as he points out, 'the blunt end gives the sharp end resources to accomplish what the organisation needs to accomplish'. But, at the same time it puts constraints and pressures on those at the sharp end. Thus, he claims, 'the blunt end shapes, creates, and can even encourage opportunities for errors at the sharp end' (pp 20/21). However, looking for evidence ofblame at the blunt end can be incredibly difficult, time consuming and costly. Closer examination of most of the case studies mentioned in this chapter reveal that they have occurred mainly as a result of latent management errors, although much of the blame has been attributed to those at the sharp end of the operation.

Ladbroke Grove train crash **11.7**

Case study 2

On 5 October 1999, at Ladbroke Grove Junction, two trains collided as a result of which 31 people, including the two train drivers, died and over ➡

400 were injured; some of these were critical. The collision occurred when a Great Western Train coming from the west of England collided at a combined speed of 130 mph with a Thames Train coming from Paddington, which had failed to stop at a red light. The early blame rested with Michael Holder who was driving the Thames Train. Indeed, Railtrack's submission to the Inquiry was that Holder had failed to see the red signal through inattention (Cullen, The Rt Hon Lord, 'The Ladbroke Grove Rail Inquiry: Part 1 Report' (2000) HSE, para 5.107, p 78).

However, whilst the Inquiry Chairman, Lord Cullen, agreed that Holder clearly went through the red light, he took the view that a principle cause of Holder failing to stop at that signal was due to its poor sighting (para 5.111, p 80).

Cullen found that there had been a history of passing signals at red, particularly those on gantries such as the one Holder passed. Indeed, it did not conform to the regulations then in force as to signal sighting time and with regard to its configuration (para 7.44, p 114). And yet Railtrack had done nothing to rectify the matter. He then went on to say that Railtrack had 'lamentably' failed to respond to recommendations of inquiries into two earlier incidents involving trains going through red lights at that location. There was a recognition that a problem existed and a number of groups were formed to consider what should be done. However, the inquiry report suggests, 'this activity was so disjointed and ineffective that little was achieved' and, therefore, 'the problem was not dealt with in a prompt, proactive and effective manner' (para 1.12, pp 3/4).

Within Thames Trains, Cullen found gaps in the driver training programme. He concluded that 'the safety culture in regard to training was slack and less than adequate; and that there were significant failures in communication within the organisation' (para 1.15, p 4). Hearing that the visual attention of drivers was distributed amongst a number of tasks which they had to perform at the same time, eg looking out for signals, monitoring their speed, observing the track, looking for speed restrictions and looking at their documentation, Cullen said it was important to give drivers' instructions on how to assess priorities in complex situations. In considering the many human factors involved, he also suggested that the driver should be given the technical assistance of automatic braking devices to ensure that he/she always stopped at a red signal (para 9.65, pp 167/168).

Cullen suggested that 'human factors can be thought of as the interplay between the operator, the machinery and the working environment' (Cullen, para 9.62, p 167). In order to lessen the chances of drivers making human errors, he suggested that they should be tested against 'specific, relevant and validated criteria' which should include a recommended number of occasions when key steps should be repeated prior to submitting themselves for testing.

The Human Body **11.8**

Martin Moore-Ede suggests that many 'errors and accidents' do not occur by chance; rather they occur because they are the predictable outcome of a failure 'to allow for human limitations' and 'of investing much more in the technology of machines than in the performance of people' (Moore-Ede M, *The 24 Hour Society: The risks, costs and challenges of a world that never stops* (1993) Piatkus, p 4). Across the world, pilots, lorry drivers, plant operators and doctors are working when they are tired; sometimes this is the result of excessive hours – at other times it is because of inadequate sleep for whatever reason. Machines and equipment are designed to run 24 hours a day but the human body is not. It needs rest and sleep. It is for this reason that humans tend to limit their hours of work and, in many cases, rotate in shifts. However, even then the human body may be unable to cope. It is not infinitely adaptable such as a machine might be. As Moore-Ede points out:

> 'our patterns of sleep and wake, of digestion and metabolism, are governed by internal biological clocks, elegantly attuned to the patterns – of sleep and dusk, night and day – of a simpler era. They make us sleepy at night and alert during the day, and they regulate the activity of our digestive tract to be ready for the next predicted mealtime. The twenty-four hour society forces us to operate the human body outside the design specs . . .' (p 8).

Operating a complex piece of machinery outside the specifications for which it has been designed creates unnecessary risks of breakdown and failure. If a train driver, or a pilot, or the engineer of a chemical plant were to deliberately operate it in this way, he or she would be accused of being reckless and irresponsible. And yet, suggests Moore-Ede, the most sophisticated pieces of machinery on a train or an aircraft or in a chemical plant are not the complex computerised or electronic systems, but 'the bodies and brains of the human operators' (p 20).

History has shown that even the most expensive and complex machinery does sometimes break down, or at least malfunction in other ways. It must, therefore, be supervised by humans in order to ensure that the breakdown or malfunction does not lead to more serious consequences. But, as Moore-Ede points out, 'it is the human operator, more than the machine itself that needs watching'. People are hired to watch the equipment in factories, nuclear power plants, and aircraft, but equipment is rarely installed to 'watch the person to ensure that he or she is awake and alert'. In fact, it can be argued that the weak link in the chain is watching the strongest link (p 20).

For example, an aircraft's equipment is wired to something known as the flight data recorder (one of the two so-called black boxes). This enables an accident involving an aircraft to be reconstructed in great detail in an attempt to discover the cause. Pilots, on the other hand, are not wired into any monitoring system, although there is a recording of their voices on the cockpit voice recorder, the second of the so-called black boxes. So, as Moore-Ede points out, there is no recording of the state of the human brain in order to determine whether the cockpit crew were functioning effectively (p. 84). Consequently, when undertaking an aircraft accident reconstruction it is very difficult to discover the precise nature of human failure.

An examination of human-made disasters reveals that quite often the greatest danger lies at times of change. For instance, when there is a shift change in a chemical plant or the chemical process is being altered or when an aircraft is taking off or landing, or when a train is approaching a busy junction. The human brain has shortcomings. It is only capable of assimilating so much. Because of these shortcomings, operators of complex machinery usually follow elaborate checklists when carrying out change. But if the operator chooses to skip over part of the checklist or, indeed, loses their place because of some distraction, then errors are likely to occur.

The Southall train crash 11.9

Case study 3

On 19 September 1997, two trains, one a High Speed Train (HST) and the other a freight train, collided at Southall East Junction, killing seven people and injuring many others. The junction was protected by three signals, one at red, one at yellow and one at double yellow, all of which should have indicated to the driver of the HST, Larry Harrison, that the line ahead was not clear. Harrison apparently ignored the yellow and double yellow. Unfortunately, the Automatic Warning System (AWS) in his train was switched off because of an earlier fault. Harrison braked on seeing the red light but it was too late. According to the subsequent inquiry, 'the inevitable conclusion, bearing in mind the required degree of probability to be established, is that Driver Harrison was inattentive at the critical moments of passing the two warning signals (Uff J, 'The Southall Rail Accident Inquiry Report' (2000) HSE, para 7.11, p 84).

During its investigation the Inquiry heard evidence from a number of experts on the likely behaviour of a driver in this situation and was told that, 'given the large number of visual tasks which drivers had, the absence of AWS inevitably increased the probability that the driver would, at sometime, fail to see a signal.' The inquiry also heard about the dangers of 'microsleeps'. These are periods of inattention of 5 to 15 seconds that could occur as a result of fatigue but could also occur under stimulating circumstances' (para 1.24, p 15).

Consequence of fatigue 11.10

Fatigue is a word that is often mentioned in relation to transportation accidents. Competition dictates that everything must be done more rapidly. But, at the same time, in order to save costs, and, in many cases, maximise profit, the number of humans employed to operate various transportation systems are being reduced. Therefore, there are increasing time pressures. Such pressures mean that people have less time to relax and sleep and one of the obvious consequences is fatigue. Moore-Ede argues that the 'limits of human tolerance in the aviation industry are being stretched to the hilt, and the prospects for reducing human error do not look good unless appropriate attention is paid to the needs of the human machine' (Moore-Ede (1993), p. 82).

British Airways 11.11

Case study 4

In June 1990, the front windscreen of a British Airways BAC–111 suddenly loosened and flew up over the nose whilst the aircraft was flying at 17,000 feet over Oxfordshire. The force of air pulled the captain from his seat and launched him through the window frame. Fortunately, one of his legs caught on the control column, delaying his exit just long enough for another member of the crew to grab him. Meanwhile, the co-pilot continued to fly the plane, reducing altitude and speed in order that the captain could be pulled back into the cockpit. The plane made an emergency landing eighteen minutes after the loss of the windscreen.

One of the causes of the accident was fatigue. During the night, the windscreen had been replaced by a maintenance manager, working under considerable time pressure, who had failed to notice that of the 90 bolts he used to replace the windscreen, only six were of the correct size. It was his first night on the night shift; the work was carried out between 3 am and 5 am and his body had not fully adapted to the changes this brought.

Financial implications of fatigue 11.12

People suffering from fatigue work more slowly, less effectively and are prone to make numerous errors. Many of these errors are likely to be minor in their effect but just occasionally they may have huge implications. Moore-Ede quotes the example of a bleary-eyed operator who was watching two vast paper machines when the paper on machine No 2 suddenly snagged and tore. In order to fix the tear, the operator reached over to switch off the machine. Suddenly, the plant was quiet and he realised that, in his fatigued state, he had turned off machine No 1 instead of No 2. The whole mill had been shut down, therefore losing valuable minutes in its production process (p 68).

Moore-Ede suggests that the world's worst nuclear accident at Chernobyl in the Ukraine in 1986 was due to fatigue (pp 108/109). Less than a year later, when nuclear inspectors from the US Nuclear Regulatory Commission found control room operators sleeping during the night shift at the Peach Bottom nuclear power plant in western Pennsylvania, they immediately ordered the closure of the plant. It remained closed for two years. The sleeping operators were each fined between $500 and $1,000, the managers responsible for the plant's operations were replaced and the company was fined a record $1.25m. In addition, two insurance companies eventually paid out $34m to disgruntled shareholders.

Lack of concentration 11.13

Doctors in many parts of the world consistently work long hours – much longer than pilots, truck drivers or train drivers are allowed to work. And yet towards the end of one of these long stints, a doctor may be required to make a life or death decision. Intensive care patients are attached to all sorts of monitors that bleep at the slightest change in condition but those who are required to make vital decisions are not.

Some of the errors that have led to disaster are of the type people make in their everyday lives. For example, hitting an incorrect button in a lift so it stops at the wrong floor, or taking a shortcut that doesn't quite work out when doing something in the office, or neglecting to do something around the house as a result of which something else happens. Sometimes these are done deliberately, maybe as an experiment, but on other occasions they are done because of a lack of concentration.

The importance of rest during crises 11.14

During crises, there is a dependence on senior management and their advisors to make clear decisions. And yet, as the crisis develops, there is a strong temptation for them to reduce their sleep in order to be present to make those vital decisions. With the increased capability of communication systems, eg mobile telephones, e-mails and the internet, information flows in rapidly in considerable quantity and varying quality from so many different sources that there is no let up when faced with a deteriorating situation. In prolonged crises it is important, perhaps more important than people realise, for the senior managers to ensure that they get adequate rest. The longer they remain involved over a certain length of time, the more likely it is that their ability to make decision will become less effective.

The Involvement of Humans in Technological Systems 11.15

James Reason points out that 'humans are retained in systems that are primarily' automated principally 'to handle "non-design" emergencies' (Reason (1999), p 182). But current attitudes towards human involvement creates a major problem in this respect. For a variety of reasons, not the least of which is to cut costs, attempts are increasingly being made to take humans out of technological systems. But unless the people undergo regular skills training, including exercises that simulate system failures, 'one of the consequences of automation . . . is that operators become de-skilled in precisely those activities that justify their . . . existence'. Even where such training is undertaken, the very nature of Disasters means that they throw up unexpected things. In fact, it is often said that each Disaster is unique.

According to Reason there are two reasons why this use of human resources may not necessarily be the best way forward. First, where:

> 'operators are required to monitor that the automatic system is functioning properly . . . it is well known that even highly motivated operators cannot maintain effective vigilance for anything more than quite short periods; thus they are demonstrably ill-suited to carry out this residual task of monitoring for rare, abnormal events'.

Second, the manual control of automated systems 'is a highly skilled activity'. So, where it is the task of the operator to take charge when the automatic system fails or cannot be controlled, the skills required 'need to be practised continuously in order to maintain them'. However, 'an automatic system that fails only rarely denies operators the opportunity for practising these basic control skills' (p 180).

Following a pattern of response **11.16**

Humans, Reason suggests, are not particularly good at the kind of inspiration that is needed sometimes in emergencies. One reason for this is that when operating in stressful situations, they tend to follow a pattern of pre-programmed responses. But, these are 'shaped by personal history' and tend 'to reflect patterns of past experience' (Reason (1999), p 182). Learning from one's mistakes can be a beneficial process but in the control room of a nuclear or chemical plant, or at the controls of an aircraft carrying 200 passengers, such experiences can be extremely costly both to the individual, the organisation and to the people involved.

Organisational Failures **11.17**

In the past, the Health & Safety Executive has claimed that 'human error contributes to about 90% of accidents but that 70% are preventable'. However, in many cases, although it is the actions of someone low down in the organisation, the primary fault lies at a more senior level. Employee behaviour depends to a large extent on the way a company is organised. Crainer suggests that 'the critical requirement is visible commitment to safety from the most senior members of the organisation so that a management culture is developed which promotes a climate for safety' (Crainer (1993), p 47).

There is, however, a continuous conflict between profit and safety; a point emphasised by Admiral John Lang, one-time senior investigator of the Marine Accident Investigation Branch in the UK. He said, 'There is implacable pressure on the crews of ships to cut corners, to 'sail too close to the wind' – or too close to other ships or to notorious danger spots for commercial reasons, to deliver their cargo, or their passengers on time' (quoted in Faith – *The Perils of the Waves* (1998) Channel Four Books, p 7).

Herald of Free Enterprise 11.18

Case study 5

Shortly before 7 pm on Friday, 6 March 1987, the *Herald of Free Enterprise*, a roll-on, roll-off car/passenger ferry, capsized when it was about two miles from the Belgium seaport of Zeebrugge. The response of management to the disaster was predictable. 'There is human error factor in everything and that is what happened in this case,' claimed Peter Ford, chairman of P & O Ferries (quoted in Crainer (1993), p 46). And, some five weeks later, just prior to the opening session of the inquiry, the Chairman of P & O, Sir Jeffrey Stirling, said, 'I would be very surprised if (the disaster) proves to be other than avoidable human error.' (*The Times*, 24 April 1987) The Inquiry did in fact declare that the immediate cause was the failure of the assistant bosun, Mark Stanley, to close the bow doors before the ferry sailed. But others on the ferry at the time were also to blame, including the loading officer, who was responsible for checking that the bow doors were closed. The inquiry found that he had failed to do this and was seriously negligent as a result.

However, in May 1985, nearly two years prior to the disaster, the marine department of P & O Ferries received a suggestion from the captain of a sister ship, the *Pride of Enterprise*, that lights ought to be fitted on the bridge to indicate whether the bow doors were closed. This followed an incident when the *Pride of Enterprise* had sailed with its bow doors open after the assistant bosun had fallen asleep; precisely what occurred in the case of the Herald of Free Enterprise. Three other ferry masters made similar suggestions in the next 15 months. These requests were ignored by the Board of Directors of the P & O Company and other senior managers; indeed in some cases, the requests were met with a flippancy that was indicative of the general complacency that existed at senior level at the time.

In his report of the disaster, the Inquiry Chairman, Mr. Justice Sheen, said that the underlying or cardinal faults lay with the Board of Directors who did not have any proper understanding of their duties and did not appreciate that they had a responsibility for the safety of their ships'. Sheen went on to suggest that if indicator lights been installed 'the disaster might well have been prevented' (quoted in Crainer (1993), p 78).

Management's Responsibility 11.19

Effective safety lies in the design of systems that are inherently safe. This includes the selection and training of personnel both to operate the systems efficiently and deal effectively with any emergency should it arise, in the equipment they are given and, most importantly, in the creation of an environment that ensures human error is less likely. Ensuring that both employees and members of the public are safe is a vital part of leadership.

Therefore, all these management responsibilities rest ultimately at boardroom level. They are far too important to be delegated to people lower down in an organisation.

Elaborating on this point, top management has three key responsibilities in preventing disasters:

- The first, and possibly the most important, is to create a safety culture within the organisation. Again, this is primarily a question of leadership.
- The second is to ensure that properly designed safety systems and procedures are in place and kept under constant review.
- The third is to ensure that all employees are aware of the safety systems and procedures and that those required to implement them have the appropriate skills. This means proper selection and training.

The last two requirements relate to skills and competency; the first is arguably an attitude of mind and one that needs to be instilled into an organisation by way of example. If any one of the three are missing it is unlikely that an organisation will be able to deal effectively with an emergency.

ValuJet **11.20**

Case study 6

ValuJet Flight 592 took off from Miami International Airport on 11 May 1996. Ten minutes later it had crashed in the Everglades, killing all 110 people on board. The subsequent investigation found that five boxes of oxygen canisters had been in the hold. The shipping label listed the canisters as empty. In fact, they were full and it is generally thought that they overheated causing a fire in the cargo hold of the plane. The oxygen canisters had been incorrectly labelled by an employee of one of ValuJet's heavy-maintenance contractors, SabreTech Corporation. The trigger for this disaster was the incorrect labelling of canisters by a lowly employee. But the investigation found that ValueJet did not give a particularly high priority to maintenance work and there was an absence of a safety culture within the organisation.

The King's Cross Underground Fire **11.21**

Case study 7

In 1987, a serious fire at King's Cross Underground Station led to the deaths of 31 people. Many hundreds were injured. The Inquiry found that there had ➡

been numerous fires on the London Underground system in the preceding years, leading the chairman, Desmond Fennell, to suggest:

> 'It is clear from what I heard that London Underground was struggling to shake off the rather blinkered approach which had characterised its early history and was in the middle of what the Chairman and Managing Director described as a change of culture and style. But in spite of that change the management remained of the view that fires were inevitable in the oldest most extensive underground system in the world. In my view they were fundamentally in error in their approach'.

Fennell added that he had been told by the Chairman of London Regional Transport that 'whereas financial matters were strictly monitored, safety was not strictly monitored by London Regional Transport' and took the view that he was 'mistaken as to his responsibility' ('Investigation into the King's Cross Underground Fire' (1988) Cm 499, HMSO, p 17).

Unsafe procedures **11.22**

In a report entitled *Reducing Error and Influencing Behaviour* (HSG 48), 2nd Edition, the Health & Safety Executive suggest 'many accidents are blamed on the actions or omissions of an individual who was directly involved in operational or maintenance work'. But, suggests the report, 'this typical but short-sighted response ignores the fundamental failures which led to the accident' before going on to point out that the fault invariably lies, deeply rooted, in the organisation's management system and its decision-making functions. (Quoted in 'The Ladbroke Grove Rail Inquiry: Part 1 Report' (2000), Lord Cullen, para 11.28, p 184.)

The list of practices by management that have reduced an organisation's ability to operate safely is formidable. Cost-cutting in fuel and safety measures, practices that prolong the life of a piece of equipment, cutting corners in maintenance and introducing inherently unsafe procedures in order to maximise profit have all been in the background of disasters that have been attributable to human error.

Clapham train crash **11.23**

Case study 8

On 12 December 1988, at 8.10 am, a train from Poole, carrying 468 people, ran into the back of another train, from Basingstoke, which was carrying 906 passengers. The crash occurred just before Clapham Junction. Almost immediately, an empty train coming in the opposite direction collided with ➡

wreckage that was strewn across the tracks. Thirty-five people died and sixty-nine were seriously injured. The Basingstoke train had stopped at a red light. But the light immediately before this one was green when the Poole train passed it. Clearly there had been some kind of signal failure. But why? The signalling system on that particular stretch of line was being renewed and, two weeks previously, it had been the job of a maintenance engineer, Mr Hemmingway, to connect the wires of the new system and disconnect the old wire in two places. In fact, he only disconnected it in one place and even then he failed to cut back the old wire but merely pushed it to one side. On the day of the accident the old wire had moved back into its old position causing the signal to remain green, even though the Basingstoke train had just passed it.

The Inquiry chairman, Sir Anthony Hidden, pointed out that 'the direct cause of the Clapham Junction accident was undoubtedly the wiring errors which were made by Hemmingway in his work in the Clapham Junction 'A' relay room' ('Investigation into the Clapham Junction Railway Accident (1989) Cm 820, HMSO, para 16.5, p 147). However, more importantly, Hidden also found that British Rail had allowed a culture of bad practice to become commonplace (para 16.8, p 148) because they had failed to monitor the work being carried out.

Although it was stated that British Rail's commitment to safety was unequivocal it was not a reality. The accident and its causes had shown 'that bad workmanship, poor supervision and poor management combined to undermine the commitment'. Working practices, the supervision of staff and the testing of new work that had been completed, failed to live up to the commitment to safety. Indeed, suggested Hidden, it was the precise opposite (para 17.18, p 166).

Additionally, an acute shortage of staff, brought about mainly to reduce costs, meant that people were required to work overtime. Indeed, the situation was so acute that Hemmingway had had only one day off in 13 weeks (para 16.11, p 148).

Kletz's Four Groups of Human Error Accidents **11.24**

Kletz divides so-called human error accidents into four groups each of which require different actions (Kletz (1994), pp 240/241). These groups are summarised below.

Mistakes **11.25**

A mistake is where someone did not know what to do, or thought they knew but were wrong. As Crainer points out, responsibility for safety lies not only 'in the selection, training and motivation of personnel' but also in the design of systems and in the equipment they are given. A working environment must be created to such a standard that human errors become less likely (Crainer (1993), p 48). So, in such cases Kletz believes that efforts should be made to either improve the training or instructions, or simplify the job.

Violations 11.26

A violation is where someone knew what to do but decided not to do it. An example of this is the Clapham Junction case. If allowed to go unchecked, the gap between right and wrong is likely to grow. In these cases, the reason why the job must be done in a particular way must be explained to people. Additionally, periodic checks must be carried out to ensure that the job is being done correctly. Alternatively, it may again be possible to simplify the job.

Mismatches 11.27

Mismatch is where the ability of the person asked to do the job does not match with the requirements of the job itself. In such cases, people know what to do and intend to do it but it is beyond their mental or physical capabilities. But, as Rosenthal points out, such 'human failures can be minimised by implementing safety oriented recruitment, training, and personnel policies. Hiring well-qualified workers, giving them necessary safety training, fostering a culture of vigilance and safety, and motivating them to perform well, are all part of human resource management' (Rosenthal, Charles and T Hart *Coping with Crises: The Management of Disasters, Riots and Terrorism* (1989 Charles C Thomas, Springfield, Illinois, p 114). Alternatively, it may be necessary to change the function either by introducing new procedures or by changing the design of the machinery itself.

Slips and lapses of attention 11.28

Slips and lapses of attention occur when someone knew what to do, intended to do it, had the capability to do it but either did it incorrectly or failed to do it at all. Given the make-up of the human brain and body, there is little that can be done to eradicate them totally. But, measures can be taken to reduce them or to ensure that they have less serious consequences. For instance, once again the way the function is carried out can be changed or efforts must be made to reduce the reasons why the slips or lapses of attention occurred in the first place.

The right attitude 11.29

In addition to the categories above, perhaps there is another category – people whose attitude to their job is not necessarily what it should be. Whilst this category might be similar to Kletz's violations, there is a slight difference. For instance, the inquiry into the Southall Train Crash found that there was evidence that the driver of the HST, Larry Harrison, was a man 'capable of a somewhat cavalier disregard of convention' which taken with other events in his past suggested that he was 'capable of irregular behaviour which might lead to the disregarding of safety rules' ('Southall Rail Accident Inquiry Report' (2000) HSE, para 7.13, pp 85/86). The inquiry chairman queried whether the then system of training and assessment was 'capable of identifying drivers who are 'at risk' before suggesting that the answer might lie in 'the emerging study of human behaviour' (para 7.14, p 86).

Decision Making **11.30**

Rhona Flin suggests that 'the decision-making skill of the on-scene commander appears to be one of the most essential components of effective command and control' (Flin R, *Sitting in the Hot Seat; Leaders and Teams for Critical Incident Management* (1996) Wiley, p 140). In suggesting this, she is not just referring to those who are normally associated with 'command and control' such as the military and the emergency services but also to the manager of a chemical or a nuclear plant or an off shore oil rig.

The literature on traditional decision making (often referred to as classical decision making) during familiar, everyday operations is extensive but it is of little relevance to those who are required to make decisions during unfamiliar or abnormal sudden events. In its simplest form traditional decision making generally involves a series of stages:

- Defining the problem.
- Gathering information.
- Listing the options.
- Evaluating the options.
- Making the decision.
- Communicating the decision.
- Evaluating the effect caused by the decision.

However, researchers – most notably Klein and his team in the US – studying decision making by pilots, surgeons, fire and military commanders etc, increasingly discovered that such people did not follow this pattern (see Zsambok CE, and Klein G, (eds) *Naturalistic Decision Making* (1997) Lawrence Erlbaum Associates, New Jersey). Quite often, the events they were required to respond to were complex, ongoing events such as a serious fire. These events had three key features:

- the conditions in which they are required to operate were dynamic and continually changing;
- there was a need for real-time reactions to those changes;
- the goals were often ill-defined, at least during the initial stages and, therefore, quite often, the tasks would be ill-structured.

Lack of Emergency Management skills 11.31

Klein found that people who successfully managed these types of events were knowledgeable and experienced. Unfortunately, such people are not always in command when the wheel comes off. For instance, following the first explosion on the Piper Alpha, Lord Cullen found that the person who had overall responsibility for the operational management of the Rig, the Offshore Installation Manager (OIM), 'took no initiative in an attempt to save life'. He went on to suggest that 'the death toll of those who died in the accommodation block was substantially greater than it would have been if such an initiative had been taken' (para 8.35). At that time, Offshore Installation Managers were selected solely on their ability to get oil and gas from beneath the seabed; little consideration was given as to whether they had the ability to manage an Emergency on board the rig.

Klein describes a number of factors that characterise decision making under such circumstances:

- there is uncertainty, ambiguity and missing data;
- there are shifting and competing goals;
- it is necessary to make real-time reactions to conditions that are often changing;
- time is short – sometimes very short;
- the stakes are high; people might die;
- there are multiple players from a wide variety of organisations;
- each of these organisations is likely to have different expertise and operating procedures.

Recognition primed decision making 11.32

Decision making taking place during a natural sequence of events in the real world, particularly when it is of a complex nature has become known as Naturalistic Decision Making (NDM).

Research has shown that under such circumstances, the decision-maker:

- focuses on making a rapid assessment of the situation;
- based on his experience, rapidly selects an option which will hopefully deal with the situation but at the same time, it may not be the best solution;
- in his/her mind, checks that the option will work; for experienced decision-makers the first option is generally workable;
- focuses on improving the option as he/she gathers more information about the situation.

This process is known as Recognition Primed Decision Making (RPDM) and is extremely relevant to the emergency services, the military and those working in casualty units in hospitals. But it could apply equally to those who might be faced with an impending disaster or crisis whilst working in chemical and nuclear plants, in the whole area of transportation, in the leisure industry, in huge shopping centres, in commerce and industry generally.

People who will be placed in positions where they may be required to make decisions during rapidly moving, serious events within limited time scales should be carefully selected to make sure that they are temperamentally suited and have the necessary skills to respond to such events. Klein suggests that such people should be trained in a way that will improve the speed and accuracy, and the ability to make situation assessments. Such training is likely to involve the 'rapid presentation of situational data' that requires them to make 'judgements about feasible goals . . . and reasonable courses of action'.

Reactions to Failure **11.33**

Dekker points out that the reactions to errors that ultimately result in Disaster have the following features.

- *Retrospective* – This consists of looking back at a sequence of events with the knowledge of the end result (p 16). This is always done after any Disaster either through the eyes of a public inquiry, or by some investigative body such as the Health & Safety Executive, the Civil Aviation Authority, the Maritime Investigation Brach, the Nuclear Inspectorate, or the police if it is likely to end in a criminal prosecution. The list is not exhaustive.
- *Proximal* – It is easy to focus on people who were closest in time and space to the events leading up to the Disaster (p 21). Invariably the investigation starts with the pilot of the crashed aircraft or the driver of the crashed train or the engineer who was in charge at the precise moment the chemical reaction went out of control.
- *Counterfactual* – The investigation lays out in detail what these people should have done to prevent the Disaster (p 25). Invariably the starting point of any investigation is the end result. But Dekker believes 'this puts the investigator at a disadvantage when it comes to understanding why people took certain action when they did'. Why did they take a particular course of action and not another course of action? As has already been suggested in the comments on hindsight at **11.5** above, those involved in the disaster will not have had the same information that the investigator has.
- *Judgmental* – The report of the investigation will suggest what people should have done, or failed to do, to prevent the Disaster. As Dekker points out, 'in order to explain why a failure occurred, we look for errors, for incorrect actions, flawed analyses, inaccurate perceptions.' But he points out that 'these decisions, judgements, perceptions, are bad or wrong or inaccurate only from hindsight' (p 27). Invariably the Inquiry will find that there was a failure to carry out some course of action. But 'the word failure implies an alternative pathway, one which the people in question did not take.' But, by suggesting 'that people

failed to take this pathway – in hindsight the right one' people's behaviour is judged according to a standard imposed by the investigator only with his or her 'broader knowledge of the mishap, its outcome and the circumstances surrounding it' (p 28).

Conclusion 11.34

There is a belief in many quarters that as technology becomes more sophisticated, so the various processes that it is designed to carry out become safer. This may be true in part. For instance, the number of aircraft accidents per miles flown has greatly reduced over the last twenty years. But, aircraft still crash. And human error is still referred to as the principle cause. But, Dekker points out that human error as 'an explanation for disasters has become increasingly unsatisfying' (Dekker (2002), p 30). Kletz goes further in describing human error as being unhelpful because 'it does not lead to effective action' (Kletz (1994), p 241).

The Flixborough explosion 11.35

Case study 9

On 1 June 1974, there was an explosion at the Nypro chemical plant at Flixborough. The plant, owned by the Dutch State Mines and the UK National Coal Board, was engaged in the manufacture of nylon. The process involved oxidising cyclohexane with a mixture of air. Six reactors, each holding about 20 tonnes were involved in the process but one of the reactors, No. 5, developed a crack and was removed for repair. In order to continue with production, a temporary pipe was manufactured to link reactor 4 with reactor 6. The temporary pipe held for two months but, at 5.00 pm on a Saturday afternoon, it gave way and there was a massive explosion. 28 people in the plant were killed. Property for miles around was damaged. Had it exploded on a normal weekday the death toll would have been much higher.

In the subsequent report into the disaster ('The Flixborough Disaster: Report of the Court of Inquiry' (1975) HMSO), the Inquiry Chairman, R J Parker, suggested:

> 'No plant can be made absolutely safe any more than a car, aeroplane, or home can be made absolutely safe. It is important that this is recognised for if not, plant which complies with whatever may be the requirements of the day tends to be regarded as absolutely safe and the measure of alertness to risk is thereby reduced' (para 196).

➡

The report then went on to suggest that:

> 'even when resources have been concentrated on an unacceptable risk, that risk will not be completely eliminated. It can never be reduced beyond a certain point because not only may there be human error in the operation of equipment, there may also be human error in the construction of a safety device built into the equipment' (para 200).

Human factors in management **11.36**

So why do investigations frequently go for the person at the coalface? According to Dekker it is an easy option. It is simple to understand, suggesting that 'failure is an aberration, a temporary hiccup in an otherwise smoothly performing, safe operation'. It does not require a fundamental change in procedures and therefore any additional expense is kept to a minimum. The overall image of the organisation is saved.

Beaty points out that it took a long time for investigations to focus on human factors in management for three reasons (Beaty (1995), p 238):

- management is part of an Establishment that tends to stick together;
- management controls the operation by hiring, firing and promoting people and paying their salaries;
- management controls most of the relevant papers, which sometimes have a tendency to go missing!

Pattern of failures **11.37**

Consecutive disasters have consistently revealed a pattern of failures, including lack of foresight, irresponsibility and plain incompetence. There has been a failure to update procedures and safety systems to cover the changes in technology and operating conditions. Communications between senior management and frontline employees have been found to be poor and those employees who are required to implement the safety systems and procedures have not been adequately trained.

Failures represent opportunities for learning but it seems they are seldom taken. In his report on the Red Lion Square riot referred to at **11.5** above, Lord Scarman warned of the dangers of judging people in hindsight. But he also suggests that there was a need for police officers 'to develop a continuing capacity for analysing, assessing and learning from, their own operations'. The statement applies equally to anyone engaged in operations that may lead to a Disaster; not just to those at the sharp end, but throughout an organisation including those at Board level with whom the ultimate responsibility lies.

12

International Aspects of Disasters

In this chapter:

- Introduction.
- Code of Practice: Prevention of major industrial accidents.

Introduction **12.1**

The Seveso disaster of 1976 in the outskirts of Meda, a small town in Italy, led to a paradigm shift within Europe in the perception, analysis and management of technological disasters. The disaster occurred around midday, Saturday 10 July 1976 when an explosion occurred in a TCP reactor owned by the Industrie Chimiche Meda Societa Azionaria (see also **8.9** above). TCP was used to manufacture a herbicide which was used in agriculture across Italy. An exothermic reaction in the batch production process led to the chemicals (ethylene glycol and sodium hydroxide) reaching an abnormal and unexpected temperature creating in turn a dioxin. The explosion that ensued resulted in the release of one of the most deadly known dioxins according to specialists – containing TCDD (2,3,7,8-tetrachlorodibenzo-*p*-dioxin). There were vast economic implications:

- the process plant was completely destroyed;
- 250 people were affected by the dioxin resulting in chloracne;
- 450 people were burned by caustic soda;
- around 37,000 people were exposed in some form;
- Nearly 17 km^2 of land was contaminated and 4 km^2 made uninhabitable;
- 4% of farm animals died and 80,000 had to be slaughtered. Farmland was significantly affected;

- eleven communities in total were affected in the countryside between Milan and Lake Como – some of Italy's areas of outstanding beauty.

The hypothesis that disasters only affect the socio-economically deprived communities was falsified.

The need for guidance and standards 12.2

Later came Bhopal in 1984 (see **8.3** above) and Chernobyl in 1986 (see **8.12** above) and the international community realised that a new regime was required that provided operators of 'hazardous installations' with the requisite guidance and standards to develop proactive emergency plans, liaise with the enforcers, disclose key risks to the local population and ensure lawful conduct in accordance with national laws. Whilst Europe created the Directives Seveso I in 1982 (*Council Directive 8222/501/EEC*) and Seveso II in 1996 (*Council Directive 96/82/EEC*) following the disaster, what about the rest of the world? In particular, the developing and emerging societies of the world where disaster management was not a major priority compared to attracting inward investment. (For more information on the Seveso Directives, see further **6.2** to **6.4**, above.)

The International Labour Organisation 12.3

In November 1989, the International Labour Organisation (ILO) passed a resolution that a 'Code of Practice' was required that would provide guidelines and best practice and allow world governments either the flexibility of internalising the Code into domestic legislation or following the Code *as if* it was law. This led to the Code on the 'Prevention of Major Industrial Accidents' 1991. A close look at the Code indicates it is similar to the Seveso Directives in logic, format and content; some would argue it is the Seveso Directive II applied globally.

As a note, the ILO are a specialist agency of the United Nations (UN), created in 1919, in fact, before the formation of the UN (24 October 1945). Its primary objectives are:

- the right to work;
- opportunities for women and men – decent employment/income;
- social protection for all;
- to improve social dialogue and involvement.

It is the 'social protection' objective that encompasses the prevention or mitigation of technological and industrial disasters. The ILO has a robust Secretariat in Geneva, a training campus in Turin, Italy and offices in most European countries. Their web site (www.ilo.org) contains many useful sources of information and guidance documents.

United Nations organisations 12.4

The ILO is one of many UN bodies engaged in 'technological and industrial disaster and emergency management'. Some of the other main bodies include:

- OCHA (Office for the Co-ordination of Humanitarian Affairs);
- UNDP (UN Development Programme);
- UNEP (United National Environment Programme);
- UNHCR (UN's High Commissioner for Refugees);
- UNCHS (UN Centre for Human Settlement);
- UNU (United Nations University);
- UNITAR (UN Institute for Training and Research);
- UNIDO (UN Industrial Development Organisation);
- WHO (World Health Organisation);
- World Bank Organisations;
- UNESCO (UN Educational, Scientific and Cultural Organisation).

In fact most UN agencies and departments retain some interest in Disaster and Emergency Management.

Code of Practice: Prevention of Major Industrial Accidents 12.5

The following document is reproduced with permission from the ILO and is the full Code of Practice.

The original edition of this work was published by the International Labour Office, Geneva, under the title, *Prevention of major industrial accidents: An ILO code of practice.*

1. General provisions

1.1. Objective

The objective of this code of practice is to provide guidance in the setting up of an administrative, legal and technical system for the control of major hazard installations. It seeks to protect workers, the public and the environment by:

(a) preventing major accidents from occurring at these installations;

(b) minimising the consequences of a major accident on-site and off-site, for example by:

 (i) arranging appropriate separation between major hazard installations and housing and other centres of population nearby such as hospitals, schools and shops; and

 (ii) appropriate emergency planning.

1.2. Application and uses

This code applies to major hazard installations which are usually identified by means of a list of hazardous substances, each with an associated threshold quantity, in such a way that the industrial installations brought within the scope of the definition are recognised as those requiring priority attention, ie they have the potential for causing a very serious incident which is likely to affect people, both on-site and off-site, and the environment. The list and threshold quantities of hazardous substances should reflect national priorities.

In order to facilitate step-wise implementation of the provisions of this code of practice, the competent authorities may for a transitional period establish increased threshold quantities for the implementation of particular components of the code.

Excluded from the scope of this code of practice are nuclear hazards and those of a strictly military nature, both of which are likely to have existing comprehensive controls of their own. In addition, the code excludes the transportation of hazardous chemicals since its control and management are different from those at static sites.

This code addresses the activities necessary for competent authorities to establish a major hazard control system, and calls for attention to be paid to them by:

(a) competent authorities such as governmental safety authorities and government inspectorates;

(b) local authorities;

(c) works managements;

(d) workers and workers' representatives;

(e) police;

(f) fire authorities;

(g) health authorities;

(h) suppliers of technologies involving major hazards;

(i) other local organisations depending on particular national arrangements.

Depending on the type and quantity of hazardous substance present, the kinds of major hazard installation covered by this code may include:

(a) chemical and petrochemical works;

(b) oil refineries;

(c) sites storing liquefied petroleum gas (LPG);

(d) major storages of gas and flammable liquids;

(e) chemical warehouses;

(f) fertiliser works;

(g) water treatment works using chlorine.

1.3. Definitions

In this code, the following terms have the meanings hereby assigned to them:

Accident consequence analysis: An analysis of the expected effects of an accident, independent of frequency and probability.

Checklist analysis: A method for identifying hazards by comparison with experience in the form of a list of failure modes and hazardous situations.

Code of practice: A document offering practical guidance on the policy, standard setting and practice in occupational and general public safety and health for use by governments, employers and workers in order to promote safety and health at the national level and at the level of the installation. A code of practice is not necessarily a substitute for existing national legislation, regulations and safety standards.

Competent authority: A Minister, government department or other public authority with the power to issue regulations, orders or other instructions having the force of law.

Emergency plan: A formal written plan which, on the basis of identified potential accidents at the installation together with their consequences, describes how such accidents and their consequences should be handled either on-site or off-site.

Emergency services: External bodies which are available to handle major accidents and their consequences both on-site and off-site, eg fire authorities, police, health services.

Event tree analysis: A method for illustrating the intermediate and final outcomes which may arise after the occurrence of a selected initial event.

Failure mode and effects analysis: A process of hazard identification where all known failure modes of components or features of a system are considered in turn and undesired outcomes are noted.

Fault tree analysis: A method for representing the logical combinations of various system states which lead to a particular outcome (top event).

Hazard: A physical situation with a potential for human injury, damage to property, damage to the environment or some combination of these.

Hazard analysis: The identification of undesired events that lead to the materialisation of the hazard, the analysis of the mechanisms by which those undesired events could occur and usually the estimation of the extent, magnitude and relative likelihood of any harmful effects.

Hazard assessment: An evaluation of the results of a hazard analysis including judgements as to their acceptability and, as a guide, comparison with relevant codes, standards, laws and policies.

Hazard and operability study (HAZOP): A study carried out by application of guide words to identify all deviations from design intent having undesirable effects on safety or operability, with the aim of identifying potential hazards.

Hazardous substance: A substance which by virtue of its chemical, physical or toxicological properties constitutes a hazard.

Hot work: An activity involving a source of ignition such as welding, brazing or spark-producing operations.

Major accident: An unexpected, sudden occurrence including, in particular, a major emission, fire or explosion, resulting from abnormal developments in the course of an industrial activity, leading to a serious danger to workers, the public or the environment, whether immediate or delayed, inside or outside the installation and involving one or more hazardous substances.

Major hazard installation: An industrial installation which stores, processes or produces hazardous substances in such a form and such a quantity that they possess the potential to cause a major accident. The term is also used for an installation which has on its premises, either permanently or temporarily, a quantity of hazardous substance which exceeds the amount prescribed in national or state major hazard legislation.

Operational safety concept: Strategy for process control, incorporating a hierarchy of monitoring and controlling process parameters and of protective action to be taken.

Preliminary hazard analysis (PHA): A procedure for identifying hazards early in the design phase of a project before the final design has been established. Its purpose is to identify opportunities for design modifications which would reduce or eliminate hazards, mitigate the consequences of accidents, or both.

Rapid ranking method: A means of classifying the hazards of separate elements of plant within an industrial complex, to enable areas for priority attention to be quickly established.

Risk: The likelihood of an undesired event with specified consequences occurring within a specified period or in specified circumstances. It may be expressed either as a frequency (the number of specified events in unit time) or as a probability (the probability of a specified event following a prior event), depending on the circumstances.

Risk management: The whole of actions taken to achieve, maintain or improve the safety of an installation and its operation.

Safety audit: A methodical in-depth examination of all or part of a total operating system with relevance to safety.

Safety report: The written presentation of the technical, management and operational information covering the hazards of a major hazard installation and their control in support of a justification for the safety of the installation.

Safety team: A group which may be established by the works management for specific safety purposes, eg inspections or emergency planning. The team should include workers, their representatives where appropriate, and other persons with expertise relevant to the tasks.

Threshold quantity: That quantity of a listed hazardous substance present or liable to be present in an installation which, if exceeded, results in the classification of the installation as a major hazard installation.

Workers: All employed persons.

Works management: Employers and persons at works level having the responsibility and the authority delegated by the employer for taking decisions relevant to the safety of major hazard installations. When appropriate, the definition also includes persons at corporate level having such authority.

1.4. Basic principles

Major hazard installations possess the potential, by virtue of the nature and quantity of hazardous substances present, to cause a major accident in one of the following general categories:

(a) the release of toxic substances in tonnage quantities which are lethal or harmful even at considerable distances from the point of release;

(b) the release of extremely toxic substances in kilogram quantities which are lethal or harmful even at considerable distances from the point of release;

(c) the release of flammable liquids or gases in tonnage quantities which may either burn to produce high levels of thermal radiation or form an explosive vapour cloud;

(d) the explosion of unstable or reactive materials.

Apart from routine safety and health provisions, special attention should be paid by competent authorities to major hazard installations by establishing a major hazard control system.

For each country having major hazard installations, competent authorities should establish such a major hazard control system. This should be implemented at a speed and to an extent dependent on the national financial and technical resources available.

The works managements of each major hazard installation should strive to eliminate all major accidents by developing and implementing an integrated plan of safety management.

Works management should develop and practise plans to mitigate the consequences of accidents which could occur.

For a major hazard control system to be effective, there should be full co-operation and consultation, based on all relevant information, among competent authorities, works managements, and workers and their representatives.

2. Components of a major hazard control system

2.1. Definition and identification of major hazard installations

Competent authorities should make arrangements for both existing and proposed new major hazard installations to be clearly defined and identified by a list of hazardous substances or categories of substances and associated threshold quantities, which should include:

(a) very toxic chemicals such as:

- methyl isocyanate;
- phosgene;

(b) toxic chemicals such as:

- acrylonitrile;
- ammonia;
- chlorine;
- sulphur dioxide;
- hydrogen sulphide;
- hydrogen cyanide;
- carbon disulphide;
- hydrogen fluoride;

– hydrogen chloride;

– sulphur trioxide;

(c) flammable gases and liquids;

(d) explosive substances such as:

– ammonium nitrate;

– nitroglycerine;

– trinitrotoluene.

The definition and identification of major hazard installations by the competent authorities should be arranged in such a way that they allow priorities to be set for those installations requiring particular attention.

2.2. Information about the installations

The works managements of all major hazard installations should notify details of their activities to the competent authorities.

For major hazard installations within the scope of the definition, a safety report should be prepared by the works management. This should include:

(a) technical information about the design and operation of the installation;

(b) details on the management of its safety;

(c) information about the hazards of the installation, systematically identified and documented by means of safety studies;

(d) information about the safety precautions taken to prevent major accidents and the emergency provisions that should reduce the effects of such accidents.

This information should be made available by the works management to all parties concerned in major hazard control systems, including workers, workers' representatives, competent authorities and local authorities where appropriate. These parties should respect the confidentiality of information obtained in the conduct of their duties, in accordance with national law and practice.

For works management the information should:

(a) lead to an appropriate level of safety which should be maintained or updated on the basis of new data;

(b) be used for communication with, and training of, workers;

(c) be used as part of the licence or permit application if such is required;

(d) be used for the preparation of an on-site and off-site (where appropriate) emergency plan.

This information should create awareness in workers at all levels to enable them to take appropriate safety precautions on-site.

For the competent authorities the information should:

(a) give insight into the plant and its hazards;

(b) allow the evaluation of these hazards;

(c) allow licence or permit conditions to be determined where appropriate;

(d) allow priorities to be set for the inspection of major hazard installations in their country or state;

(e) allow the preparation of off-site emergency plans (where appropriate).

The information should be systematically arranged in such a way that parts of the installation which are critical to its safety are clearly identified, possibly by the use of rapid ranking systems.

The information should represent the current activity within the installation. Works management should ensure that this information is updated regularly, and in the case of significant modification.

Relevant information in suitable form should be made available to the public nearby.

2.3. Assessment of major hazards

Major hazard installations should be assessed by works management and, depending on local arrangements, by the competent authorities.

This assessment should identify uncontrolled events which could lead to a fire, an explosion or release of a toxic substance. This should be achieved in a systematic way, for example by means of a hazard and operability study or by checklists, and should include normal operation, start-up and shut-down.

The consequences of a potential explosion, fire or toxic release should be assessed using appropriate techniques and data. These will include:

(a) estimation of blast waves, overpressure and missile effects in the case of an explosion;

(b) estimation of thermal radiation in the case of a fire;

(c) estimation of concentration profiles and toxic doses in the case of a toxic release.

Particular attention should be paid to the potential for domino effects from one installation to another.

The assessment should consider the suitability of the safety measures taken for the hazards identified in order to ensure that they are sufficient.

The assessment of major hazards should take account of the likelihood of a major accident taking place, although not necessarily in the form of a full quantified risk analysis.

2.4. Control of the causes of major industrial accidents

Works management should control major hazard installations by sound engineering and management practices, for example by:

(a) good plant design, fabrication and installation, including the use of high-standard components;

(b) regular plant maintenance;

(c) good plant operation;

(d) good management of safety on-site;

(e) regular inspection of the installation, with repair and replacement of components where necessary.

Works management should consider the possible causes of major accidents, including:

(a) component failure;

(b) deviations from normal operation;

(c) human and organisational errors;

(d) accidents from neighbouring plant or activities;

(e) natural occurrences and catastrophes, and acts of mischief.

Works management should regularly evaluate these causes taking into account any changes in plant design and operation. In addition, further available information arising from accidents worldwide and technological developments should be included in this evaluation.

Works management should arrange for safety equipment and process control instrumentation to be installed and maintained to a high standard consistent with their importance to the safety of the major hazard installation.

2.5. Safe operation of major hazard installations

The primary responsibility for operating and maintaining the installation safely should lie with works management.

Good operational instructions and sound procedures should be provided and enforced by works management.

Works management should ensure that workers operating the installation have been adequately trained in their duties.

Accidents and near misses should be investigated by works management.

2.6. Emergency planning

Emergency planning should be regarded by works management and the competent authorities as an essential feature of a major hazard control system.

The responsibility for on-site emergency planning should lie with works management. Depending on local arrangements, the responsibility for off-site emergency planning should lie with local authorities and works management.

The objectives of emergency planning should be:

(a) to localise any emergencies that may arise and if possible contain them;

(b) to minimise the harmful effects of an emergency on people, property and the environment.

Separate plans should be established for possible emergencies on-site and off-site. These should give details of appropriate technical and organisational procedures to reduce the effects and damage:

(a) to people, property and the environment;

(b) both inside and outside the installation.

The emergency plans should be clear and well defined, and available for use quickly and effectively in the event of a major accident. On-site and off-site plans should be co-ordinated for maximum efficacy.

In industrial areas where available emergency equipment and manpower are limited, works management should attempt to make provisions for mutual assistance between the neighbouring industrial activities in the event of a major accident.

2.7. Siting and land-use planning

Competent authorities should make reasonable attempts to ensure that there is appropriate separation between major hazard installations and:

(a) facilities such as airports and reservoirs;

(b) neighbouring major hazard installations;

(c) housing and other centres of population nearby.

2.8. Inspection of major hazard installations

Major hazard installations should be regularly inspected in order to ensure that the installations are operated according to the appropriate level of safety. This inspection should be carried out both by a safety team which includes workers and workers' representatives and separately by inspectors from competent authorities. Both types of inspection may be carried out in other ways where appropriate.

Safety personnel from the installation within this safety team should be independent of production line management and should have direct access to works management.

Inspectors from competent authorities should have the legal right to free access to all information available within the installation that is necessary in pursuit of their duties, and to consultation with workers' representatives.

3. General duties

3.1. Duties of competent authorities

3.1.1. General

Competent authorities should define appropriate safety objectives, together with a major hazard control system for their implementation.

Although the control of major hazards is primarily the responsibility of the works management operating a major hazard installation, this major hazard control system should be set up by the competent authorities in consultation with all interested parties. Such a system should include:

(a) the establishment of an infrastructure;

(b) the identification and inventory of major hazard installations;

(c) receipt and evaluation of safety reports;

(d) emergency planning and information to the public;

(e) siting and land-use planning;

(f) inspection of installations;

(g) reporting of major accidents;

(h) investigation of major accidents and their short- and long-term effects.

3.1.2. Establishment of infrastructure for a major hazard control system

Competent authorities should establish contacts with the industry at various levels. Such contacts should allow discussion and co-ordination of the various administrative and technical issues concerning major hazard installations and their control.

Competent authorities should make available sufficient expertise to carry out their responsibilities within the major hazard control system.

Where the expertise for a particular aspect of major hazard control is not available within the competent authorities, they should arrange for that expertise to be made available from outside, for example from industry or from external consultants.

Those who provided expertise at the request of the competent authorities should not disclose the information which they have learned in connection with their service to any outside body other than the competent authorities.

3.1.3. Establishment of an inventory of major hazard installations

The implementation of a major hazard control system should start with the identification of major hazard installations. The competent authorities should draw up a definition of major hazard installations using criteria selected for their country or state.

These criteria should be established to take account of national priorities and available resources.

Legislation should be established by competent authorities requiring works managements to notify them where their works fall within the scope of the definition of a major hazard installation.

The notification should include a list of hazardous substances and quantities present which qualify the installation to be classified as a major hazard installation.

3.1.4. Receipt and evaluation of safety reports

A deadline should be set by the competent authorities for a safety report to be submitted or made available to them by the works management, and for its subsequent updating.

The competent authorities should make arrangements so that they may adequately evaluate these safety reports. This evaluation should include:

(a) examination of the information, to check for completeness of the report;

(b) appraisal of the safety of the installation;

(c) on-site inspection to verify some of the information given, preferably on selected safety-relevant items.

The evaluation should preferably be carried out by a team of specialists, covering the various disciplines involved, where necessary with the help of external independent consultants.

3.1.5. Emergency planning and information to the public

Competent authorities should establish arrangements for an on-site emergency plan to be drawn up by the works managements of each major hazard installation.

Competent authorities should establish arrangements for an off-site emergency plan to be drawn up by local authorities and works management, depending on local arrangements. Such a plan should be prepared in consultation with the various bodies involved: fire authorities, police, ambulance services, hospitals, water authorities, public transport, workers and workers' representatives, and so on.

These arrangements should ensure that the off-site plan is consistent with the on-site emergency plan.

These arrangements should cover the need for regular rehearsals to be carried out in order to keep the off-site emergency plan in a state of readiness.

Competent authorities should make arrangements to provide safety information to the public nearby.

3.1.6. Siting and land-use planning

Competent authorities should establish a land-use policy to separate, where appropriate, major hazard installations from people living or working nearby.

Consistent with this policy, competent authorities should make arrangements to prevent encroachment of population nearer to existing major hazard installations.

For situations where existing major hazard installations are not adequately separated from populated areas, a plan for gradual improvement should be established.

3.1.7. Inspection of installations

Competent authorities should make arrangements to have major hazard installations inspected regularly.

Competent authorities should provide adequate guidance and training to enable their inspectors to carry out appropriate inspection of major hazard installations.

Inspection by competent authorities should be consistent with the risks from the major hazard installation. Based on the evaluation of the safety report of a major hazard installation, a specific inspection programme should be drawn up. The aim should be to establish a list of specific safety-relevant items in the installation, with the necessary frequency of inspection.

3.1.8. Reporting of major accidents

Competent authorities should establish a system for the reporting of major accidents by works managements.

3.1.9. Investigation of major accidents

Competent authorities should make adequate arrangements to investigate major accidents and their short- and long-term effects.

Such investigations should make use of relevant accident reports and other information available.

Competent authorities should study and evaluate major accidents occurring world-wide in order that lessons can be learnt in relation to similar installations in their countries.

3.2. Responsibilities of works management

3.2.1. General

Works management operating a major hazard installation should:

(a) provide for a very high standard of safety;

(b) organise and implement the on-site component of the major hazard control system;

(c) contribute to the drawing up and implementation of an off-site emergency plan.

3.2.2. Analysis of hazards and risks

Works management should carry out a hazard analysis of the major hazard installation.

This hazard analysis should be sufficient to enable:

(a) the safety system to be analysed for potential weaknesses;

(b) the residual risk to be identified with the safety system in place;

(c) optimum measures to be developed for technical and organisational protection in the event of abnormal plant operation.

To carry out a hazard analysis, a suitable method should be applied, such as:

- preliminary hazard analysis (PHA);
- hazard and operability study (HAZOP);
- event tree analysis;
- fault tree analysis;
- accident consequences analysis;
- failure modes and effects analysis;
- checklist analysis.

This method should be chosen according to the nature and the complexity of the major hazard installation, and should take account of the protection of workers, the public and the environment.

3.2.3. Determination of causes of major industrial accidents

An analysis of hazards should:

(a) lead to the identification of potential hardware and software failures, process and design deficiencies and human error;

(b) determine what action is necessary to counteract these failures.

In determining potential causes, the failure of hardware components should be considered.

The analysis should show whether these components can withstand all operational loads in order to contain any hazardous substance.

The component examination should indicate where additional safeguards are required and where the design should be altered or improved.

Component failures should be avoided by an in-depth examination of the operational procedures and of the behaviour of the entire installation in the case of any abnormal operation, and start-up and shut-down.

An analysis of potential accidents should include outside accidental interferences, both human and natural.

Human ability to run a major hazard installation safely should be studied in detail, not only for normal operation but also for abnormal conditions, and start-up and shut-down.

Workers operating major hazard installations should be adequately trained by works management.

3.2.4. Safe design and operation of major hazard installations

Works management should seek to ensure in the design of their installation that the quantities of hazardous substances stored and used on-site are the minimum consistent with their operational needs.

Works management should ensure that all operating conditions are considered in the design of components for the major hazard installation.

Particular attention should be paid to all aspects of components containing large amounts of hazardous substances.

For the manufacture of these components, works management should pay special attention to quality assurance. This should include the selection of an experienced manufacturer, inspection and control of all stages of manufacturing, and quality control.

When assembling the installation on-site, works management should pay special attention to assuring the quality of on-site work such as welding, third-party inspection and functional tests before start-up of the installation.

After careful design, manufacture and assembly of a major hazard installation, works management should secure safe operation through:

(a) good operation and control procedures;

(b) sound procedures for the management of changes in technology, operations and equipment;

(c) provision of clear operating and safety instructions;

(d) routine availability of safety systems;

(e) adequate maintenance and monitoring;

(f) adequate inspection and repair;

(g) proper training of workers.

3.2.5. Measures to minimise the consequences of major accidents

Works management should plan and provide measures suitable to mitigate the consequences of potential accidents.

Mitigation should be effected by safety systems, alarm systems, emergency services, and so on.

For every major hazard installation an on-site emergency plan should be drawn up in consultation with the safety team.

Depending on local arrangements, in cooperation with the relevant local authorities, an off-site emergency plan should be developed and implemented.

3.2.6. Reporting to competent authorities

The works management of a major hazard installation should provide the competent authorities with:

(a) the notification of a major hazard installation which will identify its nature and location;

(b) a safety report containing the results of the hazard assessment;

(c) an accident report immediately after a major accident occurs.

Works management should provide these reports, and update them, as specified in local arrangements.

A safety report should document the results of a hazard analysis and inform the authorities about the standard of safety and the potential hazards of the installation.

A brief accident report containing relevant information on the nature and consequences of an accident should be delivered to the competent authorities by works management immediately after an accident occurs.

A full accident report containing information on the causes, the course and the scope of the accident, as well as lessons learnt from it, should be given to the competent authorities by works management within the specified time.

3.2.7. Information to, and training of, workers

In view of the crucial role of workers in the prevention of major accidents, works management should make sure that:

(a) workers have a broad understanding of the process used;

(b) workers are informed of the hazards of substances used;

(c) workers are adequately trained.

This information and training should be provided in an appropriate language and manner.

3.3. Duties and rights of workers

3.3.1. Duties of workers

Workers should carry out their work safely and not compromise their ability, or the ability of others, to do so. Workers and their representatives should cooperate with works management in promoting safety awareness and two-way communication on safety issues, as well as in the investigation of major accidents or near misses which could have led to a major accident.

Workers should be required to report forthwith to the works management any situation which they believe could present a deviation from normal operating conditions, in particular a situation which could develop into a major accident.

If workers in a major hazard installation have reasonable justification to believe that there is a serious and imminent danger to workers, the public or the environment, they should, within the scope of their job, interrupt the activity in as safe a manner as possible. As soon as possible thereafter, workers should notify works management or raise the alarm, as appropriate.

Workers should not be placed at any disadvantage because of the actions referred to above.

3.3.2. Rights of workers

Workers and their representatives should have the right to receive comprehensive information of relevance to the hazards and risks connected with their workplace. In particular, they should be informed of:

(a) the chemical names and composition of the hazardous substances;

(b) the hazardous properties of such substances;

(c) the hazards of the installation and precautions to be taken;

(d) full details of the emergency plan for handling a major accident on-site;

(e) full details of their emergency duties in the event of a major accident.

Workers and their representatives should be consulted before decisions are taken on issues relevant to major hazards. In particular, this includes hazard and risk assessment, failure assessment and examination of major deviations from normal operating conditions.

3.4. Duties of the international supplier of technology involving major hazards

The supplier of technology and equipment should indicate to the competent authorities and works managements in the technology-receiving country whether the technology or equipment involves an installation which would be classified as a major hazard installation in the supplier's country, or elsewhere, if known.

Where technology or equipment would create a major hazard, the supplier should provide, in addition, information on the following aspects:

(a) an identification of the hazardous substances, their properties, the quantities involved and the manner in which they are stored, processed or produced;

(b) a thorough review of the technology and equipment in order to show:

- how control and containment of the hazardous substances could fail;
- how accidents could occur;
- the consequences of accidents;
- the vulnerability of the installation to abnormal external events such as power dips and failures, floods, earthquakes, unusual climatic conditions and sabotage, and their effects;
- the measures that can be taken to counteract these potential accidents;

(c) the management of the systems to prevent accidents from occurring, including:

- the use of design standards;
- the provision of protective devices;
- maintenance requirements;
- inspection and testing schedules;
- plant modification controls;
- operating procedures;
- training requirements;
- safeguards against deviations from the process;

(d) emergency planning based on the consequences of possible accidents assessed under (b) above, including:
   - procedure for raising the alarm;
   - requirements and responsibilities for workers dealing with emergencies;
   - necessary fire-fighting requirements and procedures;
   - procedures for limiting an accident and mitigating its consequences;
   - emergency medical services, procedures and supplies;
   - plant shut-down procedures;
   - procedures for re-entering a plant where a major accident has occurred;

(e) safety performance and accident history of similar plants elsewhere, as available.

According to contractual obligations, the supplier should provide updated safety information as it becomes available, and assistance as necessary.

3.5. Use of consultancy services

Works management and competent authorities should make use of consultancy services if their expertise is not adequate to cover all tasks to be fulfilled in a major hazard control system (see Annex I). On the other hand, consultancy services should not be relied upon to the exclusion of local management expertise.

Consultancy services may provide different fields of expertise, such as:

(a) hazard assessment;

(b) safe design and operation;

(c) analysis of potential accidents;

(d) establishment of on-site and off-site emergency plans;

(e) preparation of reports;

(f) training on major hazard control;

(g) assistance in the event of an emergency involving major hazards;

(h) quality assurance.

Consultants should be experienced in the relevant technology of the major hazard installation to enable them to give independent advice to organisations requiring assistance.

4. Prerequisites for a major hazard control system

4.1. General

The prerequisites for the operation of a major hazard control system are:

(a) manpower, within industry as well as within the competent authorities, including external expertise if necessary;

(b) equipment;

(c) information sources.

4.2. Manpower requirements

4.2.1. General

Works management should ensure that it has an adequate number of workers available with sufficient expertise before operating a major hazard installation. The design of jobs and systems of working hours should be arranged so as not to increase the risk of accidents.

For a fully operational major hazard control system, competent authorities should ensure the availability of the following specialised manpower:

(a) government inspectors with specialist support;

(b) specialists on hazard and risk assessment;

(c) specialists on examination and testing of pressure vessels;

(d) emergency planners;

(e) experts on land-use planning;

(f) emergency services, police, fire authorities and medical services.

Competent authorities should not wait for the availability of specialised manpower in all fields before starting a major hazard control system.

They should set realistic priorities based on available manpower.

4.2.2. Government inspectorate

Competent authorities should make available suitable staff, including specialist support for inspection of major hazard installations, and provide them with suitable training in their duties.

4.2.3. Group of Experts

Competent authorities should make resources available to establish a Group of Experts in the country, particularly when there is a shortage of technical expertise within the existing factory inspectorate. This Group should include experienced engineers and scientists.

If appropriate, this Group should be seconded from outside the competent authorities, such as from industry, trade unions or specialised consultancies.

4.2.4. Advisory committee

Competent authorities should consider the establishment of an advisory committee on major hazards. This committee should include representatives from all organisations involved or experienced in major hazard control, including:

(a) competent authorities;

(b) works managements and employers' organisations;

(c) trade unions or workers' representatives;

(d) local authorities;

(e) scientific institutions.

The objectives of this committee should include:

(a) discussion of priorities for the major hazard control system in the country in accordance with any national requirements;

(b) discussion of technical matters with respect to the implementation of the major hazard control system;

(c) making recommendations on all aspects of the safety of major hazard installations.

4.3. Equipment

Competent authorities should consider whether elements of the major hazard control system require the use of computer systems, particularly in establishing data banks and national or state inventories of major hazard installations.

Depending on local arrangements, works management or local authorities should make available technical equipment, for use in an emergency situation, in accordance with the needs of the emergency plans. Such equipment should include:

(a) first-aid and rescue material;

(b) fire-fighting equipment;

(c) spill containment and control equipment;

(d) personal protective equipment for rescue personnel;

(e) measuring instruments for various toxic materials;

(f) antidotes for the treatment of people affected by toxic substances.

4.4. Sources of information

Competent authorities should determine their information needs for establishing a major hazard control system. These may include:

(a) technological developments in the process industries;

(b) developments in major hazard control;

(c) codes of practice of safety-related technical issues;

(d) accident reports, evaluation studies and lessons learnt;

(e) inventory of experts and specialists on major hazard control.

Competent authorities should consider appropriate sources for this information, which may include:

(a) industry experts and researchers;

(b) industry and trade organisations;

(c) national and international standard-setting organisations;

(d) trade union organisations;

(e) consultants;

(f) universities, colleges and research institutes;

(g) professional institutions;

(h) international codes of practice and guiding principles;

(i) national codes and regulations of highly industrialised countries;

(j) reports of accidents;

(k) published reports about major hazard assessments;

(l) proceedings of seminars and conferences;

(m) specific textbooks;

(n) publications and articles in journals dealing with major hazards.

5. Analysis of hazards and risks

5.1. General

Hazard analyses should be carried out primarily by works management, but the same technique may also be applied to the evaluation of safety systems by the competent authorities.

To analyse the safety of a major hazard installation as well as its potential hazards, a hazard analysis should be carried out covering the following areas:

(a) which toxic, reactive, explosive or flammable substances in the installation constitute a major hazard;

(b) which failures or errors could cause abnormal conditions leading to a major accident;

(c) the consequences of a major accident for the workers, people living or working outside the installation, or the environment;

(d) prevention measures for accidents;

(e) mitigation of the consequences of an accident.

The hazard analysis should follow a formalised method to ensure reasonable completeness and comparability.

5.2. Preliminary hazard analysis (PHA)

As a first step in hazard analysis, a PHA should be carried out.

A PHA should be used to identify types of potential accident in the installation, such as toxic release, fire, explosion or release of flammable material, and to check the fundamental elements of the safety system.

The PHA should be summarised in documentation covering, for each accident considered, the relevant component (storage vessel, reaction vessel, etc), the events initiating the accident and the corresponding safety devices (safety valves, pressure gauges, temperature gauges, etc).

The results of a PHA should indicate which units or procedures within the installation require further and more detailed examination and which are of less significance from a major hazard point of view.

5.3. Hazard and operability study (HAZOP)

A HAZOP study or its equivalent should be carried out to determine deviations from normal operation in the installation, and operational malfunctions which could lead to uncontrolled events.

A HAZOP study should be carried out for new plant at the design stage and for existing plant before significant modifications are implemented or for other operational or legal reasons.

A HAZOP study should be based on the principles described in the relevant literature.

The examination should systematically question every critical part of the design, its intention, deviations from this intention and possible hazardous conditions.

A HAZOP study should be performed by a multidisciplinary expert group, always including workers familiar with the installation.

The HAZOP study group should be headed by an experienced specialist from works management or by a specially trained consultant.

5.4. Accident consequence analysis

As the final step of a hazard analysis, an accident consequence analysis should be carried out to determine the consequences of a potential major accident on the installation, the workers, the neighbourhood and the environment.

An accident consequence analysis should contain:

(a) a description of the potential accident (tank rupture, rupture of a pipe, failure of a safety valve, fire);

(b) an estimation of the quantity of material released (toxic, flammable, explosive);

(c) where appropriate, a calculation of the dispersion of the material released (gas or evaporating liquid);

(d) an assessment of the harmful effects (toxic, heat radiation, blast wave).

The techniques for accident consequence analysis should include physical models for dispersion of pollutants in the atmosphere, propagation of blast waves, thermal radiation and so on, depending on the type of hazardous substances present in the major hazard installation.

The results of the analysis should be used to determine which protective measures, such as fire-fighting systems, alarm systems or pressure-relief systems, are necessary.

5.5. Other methods of analysis

Where necessary, a more sophisticated method should be applied to individual parts of an installation, such as the control system or other components that are very sensitive.

To analyse accidents in more detail and according to the frequency of their occurrence, methods should be considered which, for example, allow the graphic description of failure sequences and the mathematical calculation of probabilities.

The following methods should be applied where necessary:

- event tree analysis;
- fault tree analysis.

The aim of these methods should be the optimisation of the reliability and availability of safety systems.

Application of these quantitative methods should be restricted to sensitive components of a major hazard installation.

The interpretation of the results of quantitative methods should take account of the reliability of data used.

6. Control of the causes of major industrial accidents

6.1. General

The primary responsibility for the control of the causes of major industrial accidents should lie with works management.

A hazard analysis should lead to the identification of a number of potential hardware and software failures and human errors in and around the installation, which need to be controlled by works management.

In determining which failure may be of importance for an individual installation, the following list of possible causes should be included:

- component failure;
- deviations from normal operating conditions;
- human and organisational errors;
- outside accidental interferences;
- natural forces;
- acts of mischief and sabotage.

6.2. Component failure

As a fundamental condition for safe operation, components should withstand all specified operating conditions in order to contain any hazardous substances in use.

As examples, the following causes of failure should be included in an analysis:

(a) inappropriate design against internal pressure, external forces, corrosion, static electricity and temperature;

(b) mechanical damage to components such as vessels and pipe-work due to corrosion or external impact;

(c) malfunction of components such as pumps, compressors, blowers and stirrers;

(d) malfunction of control devices and systems (pressure and temperature sensors, level controllers, flow meters, control units, process computers);

(e) malfunction of safety devices and systems (safety valves, bursting discs, pressure-relief systems, neutralisation systems, flare towers).

Depending on the outcome of the analysis, works management should decide on the need for additional safeguards or design improvements.

6.3. Deviations from normal operating conditions

An in-depth examination of the operational procedures (manual and automatic) should be carried out by works management to determine the consequences of deviations from normal operating conditions.

As examples, the following failures should be considered in the examination:

(a) failure in the monitoring of crucial process parameters (pressure, temperature, flow, quantity, mixing ratios) and in the processing of these parameters, eg in automatic process control systems;

(b) failure in the manual supply of chemical substances;

(c) failure in utilities, such as:

- (i) insufficient coolant for exothermal reactions;
- (ii) insufficient steam or heating medium;
- (iii) no electricity;
- (iv) no inert gas;
- (v) no compressed air (instrument air);

(d) failures in shut-down or start-up procedures, which could lead to hazardous conditions within the installation;

(e) formation or introduction of by-products, residues, water or impurities, which could cause side-reactions (eg polymerisation).

When failures with potential major consequences are identified, works management should consider countermeasures such as improvements in process control, operating procedures, frequency of inspection and testing programmes.

6.4. Human and organisational errors

As human factors in the running of major hazard installations are of fundamental importance, both for highly automated plants and for plants requiring a great deal of manual operation, human and organisational errors and their influence on safety should be examined in detail by works management in co-operation with workers and their representatives.

The examination should consider such errors as:

(a) operator error (wrong button, wrong valve);

(b) disconnected safety systems because of frequent false alarms;

(c) mix-up of hazardous substances;

(d) communication errors;

(e) incorrect repair or maintenance work;

(f) unauthorised procedures, eg hot work, modifications.

This examination should also consider the reasons for human errors, which may include:

(a) workers being unaware of the hazards;

(b) lack of or inadequate working procedures;

(c) workers being inadequately trained;

(d) inappropriate working conditions;

(e) conflicts between safety and production demands;

(f) excessive use of overtime or shift work;

(g) inappropriate work design or arrangements such as single-manned work-places;

(h) conflicts between production and maintenance work;

(i) drug or alcohol abuse at work.

To reduce human and organisational errors, works management should provide workers with regular training in conjunction with clear operating instructions, as well as adapting work design and arrangements as appropriate.

6.5. Outside accidental interferences

To ensure the safe operation of major hazard installations, potential outside accidental interferences should be carefully examined by works management including, as appropriate, accidents involving:

(a) road, rail and ship transport (especially carrying hazardous substances);

(b) loading stations for hazardous substances;

(c) air traffic;

(d) neighbouring installations, especially those handling flammable or explosive substances;

(e) mechanical impact such as that caused by a falling crane.

Such outside interferences should be taken into account by works management when designing and locating sensitive parts of the installation such as control rooms and large storage vessels.

6.6. Natural forces

Depending on the local situation, the following natural forces should be considered by works management in the installation design:

(a) wind;

(b) flooding;

(c) earthquakes;

(d) settlement as the result of mining activities;

(e) extreme frost;

(f) extreme sun;

(g) lightning.

If such hazards are known to occur in the natural environment of the installation, adequate precautions should be taken against them.

6.7. Acts of mischief and sabotage

Every major hazard installation can be a target for mischief or sabotage. Protection from such actions, including site security, should be considered by works management in the design.

7. Safe operation of major hazard installations

7.1. General

The safe operation of a major hazard installation should be the responsibility of works management.

Works management should ensure that the major hazard installation is always operated within the limits of intended design.

Works management should take account of all hazards identified in the hazard analysis together with possible technical and organisational control measures.

Measures used to control hazards should include:

- component design;
- manufacture of components;
- assembly of the installation;
- process control;
- safety systems;
- monitoring;
- management of change;
- inspection, maintenance and repair;
- training of workers;
- supervision;
- control of contract work.

7.2. Component design

Each component of a major hazard installation, such as reaction vessels, storage tanks, pumps, blowers and so on, should be designed to withstand all specified operating conditions.

Works management should ensure that the following aspects are taken into consideration when designing a safety-relevant component:

(a) static forces;

(b) dynamic forces;

(c) internal and external pressure;

(d) corrosion;

(e) stresses due to large differences in temperature;

(f) loads due to external impacts (wind, snow, earthquakes, settlement);

(g) human factors.

When designing a safety-relevant component, works management should consider the valid design standards (eg ASME, DIN, BS) as a minimum requirement.

The above aspects should be particularly considered when designing components containing flammable, explosive or toxic gases or liquids above their boiling point.

7.3. Manufacture of components

Works management or the technology supplier should ensure that the manufacture of components important for the safety of the installation is carried out with appropriate quality assurance measures.

Works management or the technology supplier should select only experienced manufacturers for the manufacture of these components.

Works management or the technology supplier should arrange for inspection and control measures to be carried out, when appropriate, in the manufacturer's workshop by either qualified workers or third parties.

These inspection and control measures should be specified at an early planning stage. They should be valid for all important stages of the manufacturing process and documented accordingly.

7.4. Assembly of the installation

Works management or the technology supplier should:

(a) ensure that assembly of the installation on-site is carried out with appropriate quality assurance measures;

(b) ensure that safety-relevant work, such as welding, is carried out only by qualified workers;

(c) arrange for all on-site work on components important for the safety of the installation to be inspected by either qualified workers or third parties;

(d) decide whether repair is sufficient or replacement required when failures are detected during assembly;

(e) ensure that functional tests are carried out on components, control devices and safety devices important for the safety of the installation before start-up of the operation.

7.5. Process control

To keep an installation safely within the design limits, works management should provide an appropriate control system.

This control system should, where appropriate, make use of such features as:

- manual process control;
- automatic process control;
- automatic shut-down systems;
- safety systems;
- alarm systems.

Based on the above features, works management should establish an operational safety concept for a major hazard installation.

The operational safety concept should maintain the installation or the process in a safe condition by the sequence of:

(a) monitoring a process variable in order to identify abnormal conditions which require manual process control (monitoring system); *and then*

(b) initiating automatic process control when a limit value is exceeded (control system); *and then*

(c) taking automatic action to avoid a hazardous condition (protective system).

Process variables monitored and controlled by such systems should include temperature, pressure, flow rate, mixing ratio of chemical substances, rates of pressure or temperature change.

In order to operate such control systems, facilities should be made available by works management to monitor the process variables and active components of the installation, eg pumps, compressors and blowers, with regard to operation and to hazardous conditions such as excessive pressure.

In establishing an operational safety concept, special attention should be paid to different phases of operation such as start-up or shut-down.

7.6. Safety systems

All major hazard installations should be equipped by works management with safety systems, the form and design of which will depend on the hazards present in the installation.

To prevent deviations from permissible operating conditions, works management should provide the major hazard installation, as appropriate, with:

(a) sensors and controllers to monitor temperature, pressure and flow, and to initiate actions such as emergency cooling, etc;

(b) pressure-relief systems such as:

- safety valves; or
- bursting discs;

which where necessary should be connected to a:

- blow-down system;
- scrubber;
- flare; or
- containment system;

(c) emergency shut-down systems.

To prevent failure of safety-related components, such components should be specially equipped by works management for higher reliability, for example using 'diversity' (different systems doing the same job) or 'redundancy' (several identical systems performing the same task).

All safety-related utility supplies, such as electricity supply to control systems, compressed air for instruments or nitrogen supply as an inert gas, should be examined by works management to determine whether a second source, eg emergency generators or batteries, a buffer-storage tank or an extra set of pressure gas cylinders, is necessary in the event of a primary system failure.

To determine the existence and the cause of a malfunction and to enable the proper counteraction, works management should provide a major hazard installation with alarm systems which may be connected to sensors.

Over and above the safety systems which help to keep the installation in a safe condition, protective measures should be taken by works management to limit the consequences of an accident. Such measures may include:

(a) water-spray systems (to cool tanks or to extinguish a fire);

(b) water jets;

(c) steam-spray systems;

(d) collecting tanks and bunds;

(e) foam-generating systems;

(f) detector-activated systems.

To mitigate the consequences of an accident, an emergency plan (on-site and off-site) should be drawn up by works management and local authorities in consultation with workers and their representatives. The plan should include technical as well as organisational measures.

Measures to prevent human and organisational errors, which are a frequent cause of accidents, should be considered by works management as a key issue for the prevention of accidents.

The following examples should be used by works management as guidelines:

(a) use of differently sized connections on flexible hoses to prevent unintentional mixing or use of reactive or incompatible substances;

(b) prevention of materials mix-ups by means of proper marking, labelling, packaging, inspection on receipt, and analysis;

(c) interlocking of safety-related valves and switches to prevent unintended modes of operation;

(d) clear marking of switches, knobs and displays on control panels;

(e) proper communication devices for the workers;

(f) safeguarding against inadvertent switching actions.

7.7. Monitoring

To ensure the safety of a major hazard installation, a monitoring schedule should be prepared by works management for the condition of all safety related components and systems.

A monitoring schedule should include such tasks as:

(a) checking of safety-related operating conditions both in the control room and on-site;

(b) checking of safety-related components of the installation;

(c) monitoring of safety-related utilities (electricity, steam, coolant, compressed air, etc);

(d) monitoring corrosion of critical components.

7.8. Inspection, maintenance and repair

Taking into account the contributions of the workers familiar with the installation, works management should draw up a plan for the inspection, maintenance and repair of the major hazard installation.

A plan for on-site inspection should include a schedule, and the equipment and procedures to be adhered to during inspection work.

For repair work, strict procedures should be specified for carrying out any tasks involving hot work, opening of normally closed vessels or pipelines, or work which could compromise a safety system or which involves any change in design or component quality. These procedures should cover the qualifications required by personnel, quality requirements for the work to be performed and requirements for the supervision of repair work.

Requirements specified in national or internationally recognised standards or practices for inspection and repair work should be considered by works management as minimum requirements for major hazard installations.

A maintenance plan should be prepared by works management specifying the different maintenance intervals, qualifications required by personnel and the type of work to be carried out. All maintenance work and defects noted should be documented in accordance with the plan.

7.9. Management of change

All changes in technology, operations and equipment that would fall outside current design limits should be subject to the same review as for new installations.

Before authorising a change, works management should complete documentation of the proposed change, including:

- effects on safety;
- effects on equipment and operating procedures.

7.10. Training of workers

The overall safety arrangements at a major hazard installation should recognise that the human factor is critical to the safety of the installation. Therefore, works management should adequately train workers in the safe operation of the major hazard installation. For new installations, this training should take place before start-up. Necessary facilities for such training should be provided by works management.

The training should include, but should not be limited to, such topics as:

(a) broad understanding of the overall process used in the installation;

(b) the hazards of the process and the substances used, and precautions to be taken;

(c) process control and monitoring of all operating conditions, including those at start-up and shut-down;

(d) operating procedures, including those in the case of malfunctions or accidents;

(e) emergency procedure exercises;

(f) experience in similar installations elsewhere, including accidents and near misses.

Safety training for workers by works management should be a continuous process. Training sessions should be repeated at regular intervals under conditions as near to reality as possible. The effectiveness of safety training should be assessed and training programmes reviewed in co-operation with workers and their representatives.

7.11. Supervision

Works management should provide adequate supervision of all activities performed in a major hazard installation. Supervisors should have the necessary authority, competence and training to exercise their role properly.

7.12. Control of contract work

Special attention should be given to work performed by outside contractors or temporary workers. Works management should ensure that work performed by outside contractors or temporary workers meets the requirements detailed in all the provisions mentioned in this chapter, as appropriate.

8. Emergency planning

8.1. General

Emergency planning should be regarded by competent authorities, local authorities and works managements as an essential element of any major hazard control system.

Emergency plans for major hazard installations should cover the handling of emergencies both on-site and off-site.

Works managements should ensure that the necessary standards appropriate to the safety legislation in their country are being met. They should not regard emergency planning as a substitute for maintaining good standards inside the installation.

When making arrangements for emergency planning, the competent authorities and works managements should take into account the United Nations Environment Programme (UNEP) handbook, *Awareness and Preparedness for Emergencies at Local Level (APELL): A process for responding to technological accidents*, designed to assist decision-makers and technical personnel in improving community awareness of major hazard installations and planning for local emergencies.

8.2. Objectives

The objectives of emergency planning should be:

(a) to localise any emergencies that may arise and if possible eliminate them;

(b) to minimise the harmful effects of an emergency on people, property and the environment.

8.3. Identification and analysis of hazards

For the initial stage of both on-site and off-site emergency planning, works management should systematically identify and assess what accidents leading to an emergency could arise on its installations.

For both on-site and off-site emergency planning, this analysis should be based on those accidents which are more likely to occur, but other less likely events which would have severe consequences should also be considered.

The analysis of possible accidents by works management should indicate:

(a) the worst events considered;

(b) the route to those worst events;

(c) the time-scale to lesser events which might lead to the worst events;

(d) the size of lesser events if their development is halted;

(e) the relative likelihood of events;

(f) the consequences of each event.

Guidance on the harmful properties of hazardous substances should be obtained where necessary from the suppliers of those substances. In addition, the publications of the UNEP/ILO/WHO International Programme on Chemical Safety (IPCS) should, if necessary, be consulted to obtain practical advice on, for example, the safe storage, handling and disposal of chemicals.

8.4. On-site emergency planning

8.4.1. Formulation of the plan

Each major hazard installation should have an on-site emergency plan.

The on-site plan should be prepared by the works management and should be related to an estimate of the potential consequences of major accidents.

For very simple installations, the emergency plan may consist merely of putting the workers on stand-by and calling in the outside emergency services.

For complex installations, the plan should be much more substantial, taking account of each major hazard and its possible interaction with the others, and should include the following elements:

(a) assessment of the size and nature of the potential accidents and the relative likelihood of their occurrence;

(b) formulation of the plan and liaison with outside authorities, including the emergency services;

(c) procedures for raising the alarm and for communicating both within and outside the installation;

(d) appointment in particular of the site incident controller and the site main controller, and specification of their duties and responsibilities;

(e) the location and organisation of the emergency control centre;

(f) the actions of workers on-site during the emergency, including evacuation procedures;

(g) the actions of workers and others off-site during the emergency.

The plan should set out the way in which designated workers at the site of the accident can ask for supplementary action, both inside or outside the installation, at an appropriate time. In particular, the plan should include the provision for attempting to make safe the affected part of the installation, for example by shutting it down.

The plan should contain the full sequence of key workers to be called in from other parts of the installation or from off-site.

Works management should ensure that the requirements of the plan for emergency resources, both workers and equipment, are consistent with available resources which can be quickly assembled in the event of an emergency.

Works management should consider whether sufficient resources exist at the installation to carry out the plan for the various assessed accidents in conjunction with the emergency services.

Where the plan requires the assistance of the emergency services, works management should ascertain the time taken for these services to be fully operational on-site and then consider whether the workers can contain the accident during all of that period.

The plan should take account of such matters as absence of workers due to sickness and holidays, and periods of installation shut-down. It should be sufficient to apply to all foreseeable variations in manning.

8.4.2. Alarms and communication

Works management should arrange for the onset of any accident or emergency to be quickly communicated to all appropriate workers and personnel off-site.

Works management should inform all workers of the procedures for raising the alarm to ensure that the earliest possible action is taken to control the situation.

Works management should consider the need for emergency alarm systems, depending on the size of the installation.

Where an alarm system is installed, there should be an adequate number of points from which the alarm can be raised.

In areas where there is a high level of noise, works management should consider the installation of visual alarms to alert workers in those areas.

Works management should make available a reliable system for informing the emergency services as soon as the alarm is raised on-site. The details of the communication arrangements should be agreed between works management and the emergency services, and should also be included in the off-site emergency plan.

8.4.3. Appointment of key workers and definition of duties

As part of the emergency plan, works management should nominate a site incident controller (and a deputy if necessary) to take control of the handling of the accident.

The site incident controller should be responsible for:

(a) assessing the scale of the incident (both for internal and external emergency services);

(b) initiating the emergency procedures to secure the safety of workers and minimise damage to the installation and property;

(c) directing rescue and fire-fighting operations until (if necessary) the fire brigade arrives;

(d) arranging for a search for casualties;

(e) arranging the evacuation of non-essential workers to assembly areas;

(f) setting up a communications point with the emergency control centre;

(g) assuming the responsibilities of the site main controller until he or she arrives;

(h) providing advice and information, as requested, to the emergency services.

The site incident controller should be easily identifiable by means of distinctive clothing or headwear.

As part of the emergency plan, works management should nominate a site main controller (and a deputy if necessary) who will take overall control of the accident from the emergency control centre.

The site main controller should be responsible for:

(a) deciding whether a major emergency exists or is likely, requiring the emergency services and off-site emergency plan to be implemented;

(b) exercising direct operational control of the installation outside the affected area;

(c) continually reviewing and assessing possible developments to determine the most probable course of events;

(d) directing the shutting down of parts of the installation and their evacuation in consultation with the site incident controller and key workers;

(e) ensuring that any casualties are receiving adequate attention;

(f) liaising with chief officers of the fire and police services, local authorities and the government inspectorate;

(g) controlling traffic movement within the installation;

(h) arranging for a log of the emergency to be maintained;

(i) issuing authorised statements to the news media;

(j) controlling the rehabilitation of affected areas after the emergency.

Where the emergency plan identifies other key roles to be played by workers (eg first-aiders, atmospheric monitoring staff, casualty reception staff), works management should ensure that these workers are aware of the precise nature of their roles.

8.4.4. Emergency control centre

Works management should arrange for the on-site emergency plan to identify an emergency control centre from which the operations to handle the emergency are directed and co-ordinated, and should provide a suitable control centre consistent with the plan.

The control centre should be equipped to receive and transmit information and directions from and to the site incident controller and other areas of the installation, as well as outside.

Where applicable, the emergency control centre should contain, for example:

(a) an adequate number of both internal and external telephones;

(b) radio and other communication equipment;

(c) a plan of the installation showing:

- areas where there are large inventories of hazardous substances;
- sources of safety equipment;
- the fire-fighting system and additional sources of water;
- sewage and drainage systems;
- installation entrances and roadways;
- assembly points;
- the location of the installation in relation to the surrounding community;

(d) equipment for measuring and indicating wind speed and direction;

(e) personal protective and other rescue equipment;

(f) a complete list of workers;

(g) a list of key workers with addresses and telephone numbers;

(h) lists of other persons present on-site, such as contractors or visitors;

(i) a list of local authorities and emergency services with addresses and telephone numbers.

Works management should arrange for the emergency control centre to be sited in an area of minimum risk.

Works management should consider the identification of an alternative emergency control centre should the main centre be put out of action, for example, by a toxic gas cloud.

8.4.5. Action on-site

The primary purpose of the on-site emergency plan is to control and contain the accident and thereby prevent it from spreading to nearby parts of the installation, and to minimise casualties.

Works management should arrange for sufficient flexibility to be included in the emergency plan to enable appropriate action and decisions to be taken on the spot.

Works management should consider how the following aspects are covered in the emergency plan:

(a) evacuation of non-essential workers to predetermined assembly points through clearly marked escape routes;

(b) designation of someone to record all workers arriving at the assembly points so that the information can be passed to the emergency control centre;

(c) designation of someone in the emergency control centre to collate lists of workers arriving at the assembly points with those involved in the accident and then to check against the list of those thought to be on-site;

(d) arranging for the lists held in the emergency control centre to be updated as necessary with details of absences due to holidays and sickness, changes in persons present on-site, etc;

(e) arranging for records of workers, including names and addresses, to be kept in the emergency control centre and to be regularly updated;

(f) arranging for the authoritative release of information during any emergency of significant length, and appointing a senior manager to be the sole source of this information;

(g) procedures for rehabilitation at the end of the emergency, including instructions for re-entering the accident area.

8.4.6. Planning shut-down procedures

Works management should ensure that emergency plans for a complex installation take account of the interrelationship of its different parts, so that ordered and phased shut-downs can take place when necessary.

8.4.7. Rehearsing emergency procedures

Once the emergency plan is finalised, works management should ensure that it is made known to all workers and to external emergency services where applicable.

Works management should arrange for the emergency plan to be regularly tested, including the following elements:

(a) communications systems which would be in operation during an accident;

(b) evacuation procedures.

8.4.8. Plan appraisal and updating

In the process of developing a plan and its rehearsal, works management should involve workers familiar with the installation, including the safety team as appropriate.

Works management should arrange for emergency planning rehearsals and exercises to involve workers familiar with the installation and to be monitored by observers, eg senior emergency officers and government inspectors, who are independent of the installation.

After each exercise, works management should ensure that the plan is thoroughly reviewed to take account of omissions or shortcomings.

Works management should ensure that any changes in the installation or in hazardous substances on-site are reflected where necessary in changes to the emergency plan.

These changes should then be made known to all those with a role in handling the emergency.

8.5. Off-site emergency planning

8.5.1. General

The off-site emergency plan should be the responsibility of the local authority and works management, depending on local arrangements.

The plan should be based on those accidents identified by works management which could affect people and the environment outside the installation.

The plan should therefore follow logically from the assessment used as the basis for the on-site emergency plan.

It is important that the plan should have sufficient flexibility to deal with emergencies other than those specifically included in the plan.

8.5.2. Aspects to be included in an off-site emergency plan

The off-site emergency plan should include the following (as appropriate):

(a) *organisation* – details of command structures, warning systems, implementation procedures, emergency control centres, names of the emergency co-ordinating officer, the site main controller, their deputies and other key workers;

(b) *communications* – identification of personnel involved, communications centre, call signs, network, lists of telephone numbers;

(c) *specialised emergency equipment* – details of availability and location of heavy lifting gear, bulldozers, specified fire-fighting equipment, fire boats;

(d) *specialised knowledge* – details of specialist bodies, firms with specialised chemical expertise and laboratories;

(e) *voluntary aid organisations* – details of organisers, telephone numbers, size of resources;

(f) *chemical information* – details of the hazardous substances stored or processed in each major hazard installation and a summary of the risks associated with them;

(g) *meteorological information* – arrangements for obtaining details of weather conditions prevailing at the time of an accident, and weather forecasts;

(h) *humanitarian arrangements* – transport, evacuation centres, emergency feeding, treatment of the injured, first aid, ambulances, temporary mortuaries;

(i) *public information* – arrangements for dealing with the media and informing relatives of casualties, etc;

(j) *assessment* – arrangements for collecting information on the causes of the emergency, and for reviewing the effectiveness of all aspects of the emergency plan.

8.5.3. Role of the emergency co-ordinating officer

The off-site plan should identify an emergency co-ordinating officer and a deputy, if necessary, with the necessary authority to mobilise and co-ordinate the emergency services.

The emergency co-ordinating officer should take overall command of the off-site handling of the emergency.

The emergency co-ordinating officer should liaise closely with the site main controller throughout the emergency to receive regular briefing on the development of the accident on-site.

8.5.4. Role of works managements of major hazard installations

Where the responsibility for preparing the off-site emergency plan lies with works management:

(a) works management should ensure that the plan is known to all organisations and personnel with a role to play in handling the emergency;

(b) it should appoint the emergency co-ordinating officer;

(c) it should arrange for the off-site plan to be rehearsed and tested in conjunction with on-site exercises and to be updated from the experience gained at these rehearsals.

Where the responsibility for preparing the off-site emergency plan lies with the local authority, works management should establish a liaison with those preparing the plan and provide information to assist them in that task.

This information should include a description of possible on-site accidents with potential for off-site harm, together with their consequences and relative likelihood.

Technical advice should be provided by works management to familiarise outside organisations which may become involved in handling the emergency.

Works management should ensure that any changes in the installation or hazardous substances on-site which may affect the off-site plan are passed to those responsible for producing the off-site emergency plan.

8.5.5. Role of the local authorities

Where the duty for preparing the off-site plan lies with the local authorities, they should (as appropriate) develop any necessary administrative structures or arrangements and appoint an emergency planning officer to take charge of this task. In addition, they should appoint an emergency co-ordinating officer to take overall command of subsequent off-site emergencies.

The emergency planning officer should liaise with works management to obtain the information to provide the basis for the plan. This liaison should be maintained to keep the plan up to date. Where more than one major hazard installation is operating within any local authority, that authority should make appropriate arrangements for the co-ordination of the off-site emergency plans covering every installation, to produce where necessary an overall plan.

The emergency planning officer should ensure that all those organisations which will be involved in handling the emergency off-site are familiar with their roles and are able to fulfil them.

Local authorities should attempt to enlist the help of the media in the emergency planning process.

The emergency planning officer should arrange for the off-site plan to be rehearsed and tested in conjunction with on-site exercises and to be updated from the experience gained at these rehearsals.

Where a major accident could result in a major spill or environmental harm requiring attention and investigation, the emergency planning officer should identify those authorities who will carry out these tasks and inform them, as appropriate, of their role in the off-site plan.

8.5.6. Role of emergency services

The roles of the police, fire and health authorities and other emergency services should be consistent with the normal practice in each country, which may entail a redistribution of the roles listed below.

The police should take responsibility for protecting life and property, and controlling traffic movements during the emergency.

Depending on local arrangements, the police should also be responsible for tasks such as controlling bystanders, evacuating the public, identifying the dead, dealing with casualties, and informing the relatives of the dead and injured.

The control of a fire on-site should normally be the responsibility of a senior fire-brigade officer upon arrival at the site, in co-operation with works management.

Depending on local arrangements, the senior fire-brigade officer may have similar responsibilities for other major accidents such as explosions and toxic releases.

Fire authorities having major hazard installations in their area should, at an early date, familiarise themselves with the location on-site of all stores of flammable materials, water and foam supply points, and fire-fighting equipment.

Health authorities, including doctors, surgeons, hospitals, poison centres and ambulances, should have a vital role to play following a major accident.

Health authority services should form an integral part of an off-site emergency plan.

Health authorities should be familiar with the short- and long-term effects on people of a major accident arising from a major hazard installation in their area.

Where hazardous substances are stored or handled at major hazard installations in their area, health authorities should be familiar with the appropriate treatment for anyone affected by these substances.

Where accidents with off-site consequences may require medical equipment and facilities additional to those available in their area, health authorities should arrange a 'mutual aid' scheme to enable the assistance of neighbouring authorities to be obtained.

8.5.7. Role of the government safety authority or inspectorate

Depending on local arrangements, government inspectors should:

(a) check to ensure that works management has properly identified potential major accidents which could affect people and the environment outside the installation, and where appropriate has provided the information required by the local authorities;

(b) check that works management has prepared an on-site emergency plan and has provided information about the plan to the local authorities;

(c) check that the organisation responsible for producing the off-site plan has made adequate arrangements for handling emergencies of all types;

(d) check to ensure that the various elements of the emergency plan have been tested and rehearsed;

(e) be clear as to their expected role during the actual emergency, including advisory and monitoring duties;

(f) in the event of an emergency, advise works management and emergency coordinating officers of the suitability of an affected area for re-entry and reuse once the emergency has ended;

(g) consider whether parts of the installation or equipment should be secured for on-the-spot examination and subsequent testing;

(h) interview witnesses as soon as practicable after the emergency;

(i) institute any necessary action in the light of lessons learned from a major accident, including evaluating the effectiveness of the emergency plan.

8.5.8. Rehearsals and exercises

The organisation responsible for preparing the off-site plan should appropriately test its arrangements in conjunction with on-site exercises.

In particular, it should ensure that the various communication links needed for overall co-ordination are able to operate efficiently under emergency conditions.

After each rehearsal exercise, the organisation responsible for the plan should thoroughly review the exercise to correct shortcomings or omissions in the off-site plan. The effectiveness of the plan should also be reviewed following a major accident.

9. Information to the public concerning major hazard installations

9.1. General information

Competent authorities should make arrangements to provide information to the public living or working near a major hazard installation. These arrangements should require that works management make available such information in co-operation with the local authority for all existing installations, and for new installations before they start to operate.

This information should include:

(a) the designation of the installation as a major hazard installation;

(b) a broad description in simple terms of the major hazard activities in the installation, the hazardous substances used and how they are controlled;

(c) ways of recognising that an emergency is occurring (alarm system);

(d) the action that the public should take in the event of an emergency;

(e) the known effects to the public of a major accident;

(f) remedial treatment appropriate for anyone affected by a major accident.

Advice on the distance from the major hazard installation within which the public are to be informed should be obtained from the Group of Experts or elsewhere.

All different available forms of communicating this information should be considered in order to make these procedures as effective as possible, taking into account the different target groups (schools, hospitals, etc).

The general information should periodically be repeated and if necessary updated to allow for any movement of the population into and out of the locality.

Local authorities in co-operation with works managements should assess whether the general information has been effectively communicated and understood, and take appropriate action to revise it if necessary.

Arrangements for informing the public should allow for the existence of major hazard installations operating near a territory that comes under a different local authority or country from the one in which the installation is situated.

Provision should be made for people living near the installation but in the adjoining territory to be similarly informed.

9.2. Information during an emergency

Works managements should provide information to the public living or working near a major hazard installation, giving warning of the occurrence of a major accident as soon as possible after it has taken place.

This should be carried out according to the procedures detailed in the general information.

Works managements should regularly update this information during an emergency, for example with the co-operation of the media, particularly if it becomes necessary for the public to take action different from that given in earlier information.

9.3. Information after a major accident

Works managements should provide information for communication to the public who have been affected by a major accident, on the outcome of their investigation into the accident, and on the short- and long-term effects on the public and the environment.

After a major accident, works managements should review the general information in consultation with local authorities and the public to see if any revisions are necessary.

10. Siting and land-use planning

Competent authorities should establish arrangements to ensure that new major hazard installations are appropriately separated from people living or working nearby. These arrangements should take full account of both the relative likelihood of a major accident and its consequences, allowing for any special local factor.

Additionally, they should seek to ensure that these arrangements prevent inappropriate developments being built close to any hazard installations, especially where these developments will contain significant numbers of people.

Competent authorities should obtain specialist advice from a designated source within their country, such as the Group of Experts, to enable them to formulate a policy for the siting of new major hazard installations and the use of land near all such installations.

This policy with regard to proposed developments near major hazard installations should take account of the following factors concerning the proposed development, as appropriate:

- the proportion of time spent by individuals in the development (eg homes, shops, hotels);
- the size of the development in terms of number of users at any one time;
- ease of evacuation or other measures in the event of an emergency on-site;
- vulnerability of individuals using the development (eg children, disabled people, elderly people);
- physical features of the development (eg height of buildings, type of construction).

Competent authorities should, where appropriate, apply this policy to designate zones around major hazard installations, with clear guidelines as to what types of developments are appropriate for each zone.

This policy should seek to ensure that sensitive developments such as schools, hospitals and homes for the elderly are placed further away from the major hazard installation than developments such as factories and normal housing.

In addition, competent authorities should designate zones suitable for new major hazard installations which will depend on the type and maximum quantity of hazardous substances proposed for the new installations.

Competent authorities should examine all existing major hazard installations to determine whether their separation from nearby developments is consistent with their policy. Where it is not, they should consider whether it is appropriate to seek improvements.

11. Reporting to competent authorities

11.1. General

A major hazard control system should include the principal requirement for works managements to report in writing to the competent authorities within a specified time period. The requirement should include:

(a) notifying the existence of, or proposal for, a major hazard installation;

(b) reporting about the hazards of the major hazard installation and their control (safety report);

(c) immediate reporting of major accidents.

11.2. Objectives of the reporting system

Reporting to the competent authorities should be arranged in such a way that the information can be used:

(a) inside the installation:

- to create awareness of the hazards of the particular installation;
- to inform the workers concerned;
- to decide on the appropriate level of safety and the required safety provisions;

(b) outside the installation:

- to inform the competent authorities;
- to support decisions on land-use planning and siting;
- to assist the competent authorities in setting priorities for inspection of the major hazard installation;
- to give guidance in the preparation of the off-site emergency plan;
- to inform the public nearby.

11.3. The notification of major hazard installations

Works managements should notify the competent authorities of the existence of, or proposal for, major hazard installations. In the case of a new development, notification should precede the start of work on the installation. Notification should take place within the time period specified by the competent authorities.

This notification should include information on:

(a) works management;

(b) the installation;

(c) existing licences or permits;

(d) the hazardous substances, their names, maximum expected quantities and physical conditions.

Any notification for a new installation should take into account any foreseeable increase in the range or quantity of hazardous substances, to allow for the planned growth of the installation.

11.4. The safety report

11.4.1. General

The works management of a major hazard installation should provide or make available to the competent authorities a safety report containing all safety relevant information about the major hazard installation.

The preparation of the safety report should be carried out under the direct responsibility of works management. Attention should also be paid to the input from, and participation of, the workers familiar with the installation. For specific items, the assistance of external consultants may be appropriate.

The safety report should be arranged in such a way that it gives information about the installation, its hazards and their control. It should:

(a) identify the nature and quantities of hazardous substances used in the installation;

(b) give an account of the arrangements for safe operation of the installation, for control of abnormal conditions that could lead to a major accident and for emergency procedures at the site;

(c) identify the type, relative likelihood and consequences of major accidents that might occur;

(d) demonstrate that works management has identified the major hazard potential of the installation and has provided appropriate safety measures.

The safety report should contain sufficient information to be understood without previous knowledge of the particular installation.

A safety report should contain the following elements:

(a) description of the installation, the processes and the hazardous substances used;

(b) description of the hazards, their control, and consequences to workers, the public and the environment of potential major accidents by means of systematic hazard analysis;

(c) description of the organisation of the installation and the management of its safety;

(d) description of the emergency provisions in order to mitigate the consequences of major accidents.

11.4.2. Description of the installation, processes and hazardous substances

The description of the installation should give safety-relevant information on:

(a) the installation and the surroundings;

(b) the design parameters;

(c) protection zones;

(d) area classification;

(e) the equipment and materials used.

The description of the processes should give information about:

(a) the technical purpose of the installation;

(b) the basic principles of the process;

(c) process conditions, including the static and dynamic process parameters and safety-relevant data;

(d) utility supplies;

(e) discharge, retention, recycling or disposal of liquids, gases and waste products.

A list of all hazardous substances should be given, including:

(a) raw materials and the final products;

(b) intermediate products and by-products;

(c) waste products;

(d) catalysts, additives, etc.

Information about the hazardous substances should include:

(a) the process stage in which the substances are involved;

(b) the quantity of substances used;

(c) safety-related physical and chemical data;

(d) toxicological data;

(e) environmental impact data.

11.4.3. Description of the hazards and their control

The description of the hazards of the installation should be based on a systematic hazard analysis, including:

(a) the identification of hazards;

(b) the analysis of hazards;

(c) the analysis of the consequences of major accidents.

Works management should, where appropriate, consider using a rapid ranking system in its installation in order:

(a) to have a rapid indication of the hazards of the various parts of the installation;

(b) to set priorities for more detailed examination.

The identification of hazards should lead to the selection of safety relevant items. This identification should preferably be based on a preliminary hazard analysis.

The analysis of hazards should concentrate on the safety-relevant items. This analysis should be based on a hazard and operability study or recognised equivalent.

The description of the safety-relevant components should include data on:

(a) function, type and extent of operating conditions;

(b) design criteria;

(c) controls and alarms;

(d) pressure-relief systems and valves;

(e) dump tanks, sprinkler systems and fire protection.

For particularly sensitive features, such as instrumentation, an additional reliability study should be considered, which should indicate whether sufficient precautions are taken to avoid major accidents.

In the safety report an analysis should be given of the consequences of an identified major accident. This information should particularly be related to:

(a) possible releases of energy in the form of a blast wave, including its effects on the surrounding area;

(b) possible thermal radiation in the case of fire;

(c) possible dispersion of released substances, particularly toxic chemicals, including its effects on the surrounding area.

11.4.4. Description of the organisation

The safety report should contain information about the organisation of the installation and the management of its safety. Descriptions should be given of:

(a) management structure;

(b) general safety policy within the -installation;

(c) duties and responsibilities of works management and workers;

(d) consultation procedures with workers and their representatives;

(e) safety and operational procedures.

An organisational diagram should be included with a description of the position and line responsibilities of the various production and supporting departments, such as operations, safety, engineering, maintenance, and so on.

The allocation and delegation of responsibility for plant safety within works management should be described. The role and duties of workers, works management and safety departments should be detailed.

A description should be given of the procedures for safety consultation with workers. It should indicate whether a works council or a safety committee is involved in the safety consultation, and how the safety department and medical service function within this framework.

Information should be given about:

(a) the education and qualifications required of workers in particular jobs in the installation;

(b) the training of workers.

All procedures that are relevant to the safe operation of the installation should be described. These procedures should either be given in the report or a reference made to information available in the installation. The procedures described in the report should include:

(a) installation design and modification;

(b) start-up, operation and shut-down of the installation;

(c) inspection, maintenance and repair;

(d) communication and follow-up of accidents;

(e) internal safety audits;

(f) management of change.

11.4.5. Description of emergency provisions

Emergency provisions should be described in the safety report, including both organisational and technical aspects.

The organisational aspects should include:

(a) instructions and procedures in case of an emergency;

(b) communication within the installation and with third parties;

(c) the relationship between internal and external emergency services;

(d) practical training on, and rehearsals of, the emergency plan.

The technical emergency measures, which should be described in the report, include:

(a) alarm systems;

(b) emergency shut-down systems;

(c) fire-fighting equipment;

(d) evacuation plans;

(e) personal protective equipment, etc.

11.4.6. Handling and evaluation of safety reports

The competent authorities should check the safety report for completeness and accuracy and consider whether additional safety provisions are required.

Both works management and the competent authorities should use the information in the report to evaluate the safety precautions.

The evaluation of safety reports should be carried out by the competent authorities according to national guidelines drawn up either by the Group of Experts or elsewhere.

Evaluation should include a systematic study of the major hazard potential of the installation, including domino and missile effects.

Evaluation should cover:

(a) all handling operations, including internal transport;

(b) the consequences of process instability and major changes in process variables;

(c) the consequences of the location of one hazardous substance in relation to another;

(d) the consequences of common mode failure, eg sudden total loss of power;

(e) the consequences of the identified major accidents in relation to off-site neighbouring populations.

Where appropriate, competent authorities should consider the assistance of external consultants for the evaluation of major hazard installations, particularly where the off-site consequences of a major accident would be very serious.

11.5. Updating of safety reports

Works management should regularly update their safety report within a time period specified by the competent authorities.

Safety reports should be updated immediately in the event of significant modification to the installation.

Updated reports should take account of new important information about the hazards of the substances used and the process.

Details of minor changes taking place in every installation should be kept on file inside the plant by the plant personnel. On a regular basis, at least every five years, the safety report should be updated completely.

11.6. Reporting of accidents

11.6.1. Immediate report

The works management should report major accidents immediately to the competent authorities.

The report should include, as far as it is available, information necessary for an initial evaluation, such as:

(a) the nature of the accident;

(b) the substances involved;

(c) an indication of the possible acute effects on persons and the environment, and data needed to assess these effects;

(d) the initial measures taken.

The report should give information in order to let the competent authorities and, where appropriate, the local authorities decide whether urgent action is necessary off-site and whether the off-site emergency plan should be put into operation.

The competent authorities should provide a standard form for the immediate reporting of major accidents.

11.6.2. Complete report

Works management should later provide further information in the form of a complete report to the competent authorities.

This complete report of a major accident should contain:

(a) an analysis of the causes and contributing factors of the accident;

(b) the steps taken to mitigate the effects, acute as well as long term;

(c) the provisions made to prevent a recurrence of the accident;

(d) lessons learnt for the safety of the installation;

(e) all available data useful for assessing possible long-term effects on workers, the public and the environment.

The competent authorities should make information on the accident available to works managements and competent authorities elsewhere.

12. Implementation of a major hazard control system

12.1. General

The competent authorities should establish by policy, regulation or legislation a time schedule for the implementation of the various elements of a major hazard control system.

The speed of implementation of a major hazard control system should depend on:

(a) resources available locally and nationally for the different components of the control system;

(b) the number of major hazard installations in the country.

Priorities should be set by the competent authorities for the staged implementation of the major hazard control system. Care should be taken not to attempt too much in the short term where local resources are limited.

Where sufficient national and local resources are available, the competent authorities should arrange for any new major hazard installation to come within the full major hazard control system. Existing major hazard installations should be allowed a time period by the competent authorities to meet the various requirements of the system.

12.2. Identification of major hazard installations

The competent authorities should draw up a definition of a major hazard installation. This definition, based on a list of hazardous substances with their threshold quantities, should be clear and unambiguous.

The competent authorities should confirm this definition as part of major hazard legislation to enable both existing and proposed new major hazard installations to be identified.

As a start of the identification, the competent authorities should consider whether existing major hazard installations can be identified by non-statutory means, using tentative criteria.

12.3. Establishment of a Group of Experts

For countries setting up a major hazard control system for the first time, the competent authorities should consider the establishment of a Group of Experts.

The Group should consist mainly of trained engineers, chemists and physicists and should have the task of advising the competent authorities, works managements, trade unions, local authorities, government inspectorates, and so on, on all aspects of a major hazard control system.

Where appropriate, competent authorities should consider seconding experts from industry, trade unions, universities, research and technology institutes and consultancies to assist in this task.

Competent authorities should ensure that the chosen experts work as a group, in order that individual experiences can be shared by the group.

12.4. On-site emergency planning

Competent authorities should ensure that all major hazard installations have an on-site emergency plan.

Works management should make the necessary arrangements to draw up an on-site emergency plan. This plan should be based on the consequences of potential major accidents.

Works management should ensure that it has sufficient workers and safety management available to meet the requirements of the on-site emergency plan.

Works management should ensure that the on-site emergency plan is tested and rehearsed to identify any weaknesses in the plan, and that such weaknesses are quickly corrected.

12.5. Off-site emergency planning

Competent authorities should clarify, by means of policy, regulation or legislation, whether the works managements or the local authorities have the responsibility for preparing the off-site emergency plan.

Where the responsibility lies with the local authorities, works managements should assist them with the necessary technical information.

The off-site emergency plan should be based on information about the potential consequences of major accidents off-site.

The off-site emergency plan should be consistent with the on-site emergency plan.

All parties having a role in the off-site emergency plan should be advised as to their responsibilities by the party responsible for the plan.

The off-site emergency plan should specifically address whether those living near the installation should remain indoors or be evacuated, and what action is necessary in either case.

The organisation responsible for the plan should ensure that the plan is tested and rehearsed to identify any weaknesses, and that such weaknesses are quickly corrected in the modified plan.

12.6. Siting and land-use planning

The siting of major hazard installations and the use of land surrounding the installations should be regarded by the competent authorities as a fundamental element of the major hazard control system.

Competent authorities should establish criteria for the appropriate separation of installations from people living and working nearby.

If required, advice on such criteria should be obtained from the Group of Experts.

As a first priority, competent authorities should establish an appropriate siting policy for all new major hazard installations.

Where the separation from nearby developments is less than that indicated under the siting policy, the government inspectorate should urgently consider the need for additional safety control on-site.

12.7. Training of government inspectors

Competent authorities should take account of the key role that its government inspectors are likely to hold in any major hazard control system.

Competent authorities should take relevant measures to provide appropriate training to government inspectors and to establish minimum academic and professional qualifications enabling them to carry out their duties within the major hazard control system, which may include:

(a) identification of major hazard installations;

(b) licensing of, or issuing permits for, the installations;

(c) inspection of the installations;

(d) evaluation of safety reports from works managements;

(e) advising about off-site emergency planning.

Competent authorities should consider using the Group of Experts to assist in the training of government inspectors.

Alternative sources of training, which should also be considered by competent authorities, include:

(a) joint participation in industry safety training courses;

(b) fellowships under the supervision of experienced inspectors either in the country or abroad (if appropriate);

(c) professional meetings and seminars about major hazards;

(d) periodical literature and reports about major hazard control developments in other countries with established control systems.

12.8. Preparation of checklists

Checklists should be considered both by the competent authorities and by works managements as an effective way of transferring experience to less experienced users.

Where appropriate, checklists should be considered for:

(a) properties of hazardous substances;

(b) detailed design requirements;

(c) inspection systems;

(d) internal audit systems;

(e) management control systems;

(f) guidance on the contents of safety reports;

(g) reporting of major accidents;

(h) evaluation of hazards;

(i) preparing emergency plans both on-site and off-site;

(j) siting and layout of the installation;

(k) accident investigation.

Checklists should be kept up to date in order to be effective.

12.9. Inspection of installations by government inspectors

Competent authorities should make arrangements for major hazard installations to be regularly inspected by government inspectors.

The initial inspection programme should be drawn up based on the details provided at the time of notification. Subsequent inspections should take into account the findings from the examination of the safety report, and the results of previous inspections.

Government inspectors should set priorities for an inspection programme at each installation based on a sample inspection of one plant component to represent the standard of safety of all similar components.

Government inspectors should confirm by inspection which parts of the major hazard installation contain hazardous substances in sufficient quantity to cause a major accident.

Government inspectors should, through their inspections, make sufficient checks on the actions taken by works managements to satisfy themselves as to the competence of the latter to operate the plant safely and to maintain control in the event of an accident.

Government inspectors should keep a record of all inspections carried out, together with actions required of works managements, in order to ensure continuity where there is a change of inspectors.

Government inspectors should initiate action to remedy any significant defects discovered during the inspection.

12.10. Inspection of installations by specialists

The role of specialists, including electrical, mechanical, civil and chemical engineers, should be to provide support for general government inspectors.

Competent authorities should consider the need for specialists in their country according to the resources available.

The work of specialists should include, for example:

(a) advising the general government inspectors on the selection of sample components to be inspected inside the major hazard installation;

(b) inspecting pressure vessels for design, operation and maintenance to approved standards and regulations;

(c) checking computer-controlled major hazard installations for software accuracy and reliability;

(d) checking the procedures for modifying installations in order to maintain the initial integrity of the plant after modification;

(e) checking the design and maintenance procedures for pipelines carrying hazardous materials.

Specialists should be aware of the world-wide experience of accidents involving their particular discipline and should be able to advise general government inspectors and works managements accordingly.

12.11. Actions following the evaluation of safety reports

Evaluation in conjunction with the safety report of the installation should provide both works managements and government inspectors with a basis for:

(a) deciding if a new process should be allowed to proceed;

(b) evaluating the adequacy of the layout of a new installation or process;

(c) evaluating the adequacy of hardware and software control arrangements, eg automatic shut-off valves;

(d) formulating an on-site emergency plan and providing information for an off-site emergency plan;

(e) evaluating the separation proposed between the installation and the neighbourhood;

(f) deciding the extent to which the public nearby should be informed about the major hazard installation.

13

Insurance, Losses and Risk Management

In this chapter:

- Introduction.
- Insurance.
- Losses.
- Risk management.
- Appendix 1: British Insurance Brokers Association Jargon Buster.
- Appendix 2: Your insurance needs.
- Appendix 3: Protection against fire.

Introduction 13.1

This chapter provides disaster-risk managers, including business continuity practitioners and others charged with the responsibility of managing risk and crises, an introductory overview of insurance, claims and risk management. The primary focus is for businesses, but generally the information is applicable to most organisations, including local authorities. This chapter covers three topics:

- *Insurance* – includes an overview of the historical background and key roles of insurance, the principles involved, what constitutes an insurable risk, types of cover available, the functions of brokers and underwriters etc. Appendix 1 (at **13.50** below) consists of a 'jargon buster' provided by the British Insurance Brokers' Association (BIBA), which should assist if faced with unfamiliar language, and Appendix 2 (at **13.51** below) 'Your insurance needs', by the Association of British Insurers (ABI) highlights the types of insurance most businesses will consider buying.
- *Losses* – examines the procedures of insurance and duties of insureds and insurers in the event of a loss incident, the disaster-insurance process, damage mitigation

and the roles of loss adjusters, claims handlers and disaster restoration companies – see **13.37** below.

- *Risk management* – defines risk management and gives an overview of the risk management process – see **13.45** below. This is critical to understanding and contextualising insurance. This section introduces loss prevention, a part of risk management, which is sometimes referred to as risk reduction. The terms will be used synonymously in this instance with the primary concern of reducing the uncertainty of loss associated with disaster-risk. Appendix 3 (at **13.63** below) 'Protection against fire', covers loss prevention advice issued by the ABI.

Relevant themes 13.2

All three topics referred to in **13.1** above cover enormous subject areas and, therefore, this chapter can only introduce some of the themes relevant to disaster-risk managers. The chapter will not attempt to cover the regulatory environment or the legal aspects because risk management is an approach that can be applied to a broad range of organisational structures including manufacturers, financial service providers and governments. It will, however, cover the basic common law doctrines or principles of insurance. The legal aspect is covered in **CHAPTER 14: THE LAW RELATING TO EMERGENCIES AND DISASTERS**.

Key groups 13.3

The disaster-insurance process includes assessing risk exposure, buying insurance, incurring a loss incident (disaster), managing the crisis, mitigating the impact and recovering. This process engages a number of key groups:

- brokers (insurance intermediaries);
- underwriters (insurers);
- emergency response services (fire, ambulance, police);
- loss adjusters;
- claims handlers and damage restoration companies.

Financial loss 13.4

Insurance is a composite risk treatment – a risk transfer mechanism that reduces the adversity of the financial impact of hazards by:

- directly indemnifying insured against financial loss;
- indirectly reducing financial vulnerability to disasters; and,
- ensuring adherence to risk management procedures including loss prevention measures.

It is a method of spreading, over time and over a wider body of individuals/organisations, the financial losses arising from the occurrence of some types of uncertain events in a given period of time (van Oppen 'The Role of Insurance in Disaster Reduction' (2001) Research Paper Collection). According to Merkin (Merkin R *Insurance Handbook* (1st edn, 1994) Tolley), the most widely quoted definition of insurance is that of Channell J in *Prudential Insurance Company v IRC [1904] 2 KB 658*. In that case three elements of insurance were identified:

- it is a contract under which, in consideration for payments (generally called premiums), the assured secures some benefit on the occurrence of an event;
- the event must be uncertain;
- the event must be adverse to the insured, in that he possesses an insurable interest in the subject matter insured.

Principle of insurance **13.5**

Defining insurance is useful as some doctrines of the common law only apply to insurance contracts, most importantly the rules relating to insurable interest and utmost good faith, which will be covered at **13.11** and **13.12** below. In addition, regulation and taxation issues apply specifically to insurance, rather than other risk financing strategies such as savings. However, what is most important to remember is that insurance is a composite risk treatment that goes hand-in-hand with other risk treatments, including risk avoidance, risk retention and loss prevention/risk reduction. 'The principle of insurance always remains the same: the many policyholders that do not suffer loss pay for the few that do. If properly managed insurance can increase the capacity of an enterprise to deal with Disasters' (see van Oppen (2001)).

'Risk' is a term which incorporates 'hazard × value × vulnerability', where:

- hazard = a source of danger, a peril (eg volcano, earthquake, avalanche, faulty signal, defective building etc);
- value = anything of human value, including life, property and/or livelihood; and
- vulnerability = extent of exposure to hazard impact.

What constitutes a Disaster? **13.6**

'A Disaster may be defined as the realisation of a hazard that severely impacts things of human value' (Smith K *Environmental Hazards, Assessing Risks and Reducing Disasters* (1996) Routledge). The hazard impact may be direct (such as the immediate material damage in the aftermath of a storm) and/or indirect (otherwise termed 'consequential', such as business interruption). The hazard impact may also be tangible (ie financially measurable) and/or intangible (ie cannot be satisfactorily assessed in monetary terms, eg grief/stress or loss of a national treasure) (van Oppen (2001)).

On a world scale a Disaster may be considered to be a mass covariant risk (ie one that severely impacts large numbers of people, property and livelihoods) that is fundamental in nature (ie arises from causes outside the control of any one individual or group). However, within the scope of this book, we are mainly interested in events that exceed the ability of the affected enterprise to cope using only its own resources, such as a major fire. There need be no application of a numerical scale on what constitutes a Disaster as this will depend on the resilience of the enterprise impacted, for example, a large business can withstand a larger loss than a small business. Importantly, a hazardous event on its own, such as an earthquake, is not a Disaster *per se*. The Disaster is the occurrence of the hazardous event and the impact or effect of that hazardous event on things of human value.

Insurance

Historical background 13.7

Historically, insurance has developed for centuries. Indeed the practice of 'bottomry' dates back to about 1800 BC when the Code of Hammurabi included 282 clauses devoted to the subject. Bottomry was a loan or mortgage taken out by the owner of a ship to finance the ship's voyage. No premium in the modern sense was paid. If the ship was lost, the loan did not have to be repaid. The principle extended to life assurance as well. The debts of a soldier who died in battle were forgiven and did not have to be repaid. The realisation that risk bearing affected productive activity was appreciated by the Roman Emperor Claudius (10 BC–AD 54), who, eager to boost the corn trade, made himself a one-man, premium-free insurance company by taking personal responsibility for storm losses incurred by Roman merchants. This could be likened to intervention by governments post-disaster to restore the status quo (Berstein P L, *Against the Gods: The remarkable story of risk* (1996) Wiley & Sons, p 92).

The rise of trade in the Middle Ages accelerated the growth of finance and insurance. Bruges established a Chamber of Assurance in 1310. The earliest remaining example of an insurance policy (polizza) dates back to 1347 from Genoa (Hodgin R, *Insurance Law* (1998) Cavendish). Farm produce insurance became important in Italy to compensate farmers subjected to risks such as drought, flood or pestilence outside of their control. The Monte dei Paschi bank was established in Siena in 1473 to serve as an intermediary for agricultural co-operatives to insure one another. The first English policy appears to date from 1547. In 1601 Francis Bacon introduced a Bill in Parliament to regulate insurance policies. However, the full development of insurance as a commercial concept and subsequent development of English insurance law – largely attributable to the judgments of Lord Mansfield – was not achieved until the second half of the eighteenth century (Hodgin (1998), ch 1).

Insurance categories 13.8

Insurance is generally categorised into:

- *Life* – namely, long-term business and reinsurance including life and annuity; marriage and birth; linked long term; permanent health; tontines; capital redemption; pension fund management; collective insurance and social insurance.
- *Non life* – namely general business and reinsurance including accident, sickness, land vehicles, railway rolling stock, aircraft, ships, goods in transit, fire and natural forces, damage to property, motor vehicle liability, aircraft liability, liability for ships, general liability, credit, suretyship, miscellaneous financial loss, legal expense and assistance.

The insurance industry has £1,000 billion invested in company shares and other assets on behalf of millions of savers in the UK and elsewhere. It accounts for over 20% of investments in the London stock market (see the Association of British Insurers' (ABI) web site http://www.abi.org.uk/). There are approximately 800 companies authorised to conduct insurance business in the United Kingdom, and 4,000 in the European Union (see Hodgin (1998)).

Insurance benefits 13.9

Successful insurance implementation may lead to significant social and economic benefits (Ray P K, *A practical guide to multi-risk crop insurance for developing countries* (1991) Science Publishers Inc, USA) that may in turn improve the insureds capacity to cope with hazardous events that in turn will reduce the Disaster. The benefits will vary, being tangible, intangible, direct and indirect in nature.

Insurance:

- stabilises incomes for insureds by removing the burden of unexpected payouts for losses;
- improves insured's position in relation to obtaining credit. For example, most lenders require the item for which the money was borrowed and/or the life/health of the borrower to be insured. An example of this is a life endowment policy to cover a mortgage;
- improves capital formation by giving an incentive for greater risk-taking;
- stimulates production by increased investment. For example, an insured person need not save for adversity but may pay a relatively smaller premium allowing for increased investment in his/her productive activity;
- steadies and even reduces the cost of production by stabilising risks in the long run. This occurs because the producer is indemnified after a loss, which enables them to continue production. If the producer is uninsured, their loss may prevent future production, as they will go out of business;

- steadies and even stimulates consumption through stability of income. This may occur because the confidence of the insured increases with the security of insurance. Less money is needed in savings;
- may relieve the State, to a large extent, of the irregular financial burden of providing relief and distress loans to the community in case of Disaster. This will particularly occur where macro-insurance schemes are in place, such as the natural catastrophe government insurance 'CAT NAT' scheme in France.

Over and above the direct benefits of insurance (eg having a loss paid), it is most important to recognise the indirect benefits (eg income stability, ability to obtain credit, consumer confidence etc) to both the insureds and the local economy. These indirect benefits, just as much as the direct benefits, improve the insured's ability to prepare for and mitigate the impact of Disasters (van Oppen (2001)).

Principles of insurance **13.10**

To understand insurance it is necessary to have a grasp of a number of principles or doctrines that make up the foundation of insurance practice. These principles have been developed over several hundreds of years with many being upheld by courts or codified by Acts of Parliament. The Chartered Insurance Institute textbooks provide a good starting point for anyone studying insurance and much of this section is attributable to Dickson and Steele (see Dickson G C A, and Steele J T, *Introduction to Insurance* (2nd edn, 1984); Steele J T, *Principles and Practice of Insurance* (1989) Chartered Institute, Study Course 040; and Dickson G C A, *Risk and Insurance* (1991) Chartered Institute, Study Text 510).

Insurable interest **13.11**

There must be insurable interest in the subject matter of insurance. The subject matter of insurance can be any form of property or an event that may result in a loss of a legal right or creation of a legal liability. In this way, the subject matter of insurance of a fire policy can be buildings, stock or machinery; under a liability policy it can be a person's legal liability for injury or damage; with a life assurance policy it is the life of the assured. The insurable interest is in the pecuniary interest of the insured in that building, injury, life etc. Insurable interest constitutes 'the legal right to insure arising out of a financial relationship, recognised at law between the insured and the subject matter of insurance'.

Although statute requires insurable interest, there is no statutory definition of insurable interest with the exception of marine insurance policies where insurable interest has been codified in the *Marine Insurance Act 1906.* Definitions have otherwise been developed by case law, including Lawrence J in *Lucena v Craufurd (1808) 1 Taunt 325* for a definition which makes it clear that more than one person may have insurable interest in a single subject matter and *Macaura v Northern Assurance Co Ltd [1925] AC 619* that held insurable interest was necessary.

Insurance has been likened to a wager but there are some very important differences as the comparison in Figure 1 below shows (Steele (1989)).

Figure 1: Insurance/wagering contract comparison

Insurance contract	*Wagering contract*
Insurable interest in the subject matter of insurance essential.	The interests are limited to the stake to be won or lost and, as such, they are not recognised at law.
The insured is immune from loss and his identity is known before the event.	Either party may win or lose and the loser cannot be identified until after the event.
Full disclosure (utmost good faith) is required by both parties to the contract.	Full disclosure is not required by either party.
In most cases an indemnity only is secured.	The stakes are not paid by way of indemnity. Payment is made without suffering loss beforehand.
The contract is enforceable at law.	Neither party has any legal remedy.

Utmost good faith 13.12

It is usual to complete a proposal form when applying for insurance cover. This will enable the insurer (underwriter) to assess the risk, set the terms and conditions, quote a premium or decline to quote. The information will include:

- the name and address;
- the nature of insured's business;
- the loss record and details pertaining to the risk to be insured.

Most commercial contracts are subject to the doctrine of *caveat emptor* (let the buyer beware). Although various statutes protect the consumer, it is fundamentally the responsibility of each party to the contract to ensure they make a good or reasonable bargain. In most cases each party can examine the item or service which is the subject matter. Provided one does not mislead the other party and answers the questions truthfully, there is no question of the other party avoiding the contract. There is no need to disclose information that is not asked for.

Disclosure 13.13

Insurance is different. Buyers of insurance are subject to the legal doctrine of *uberrima fides* (utmost good faith) which means they are legally obliged to offer all information pertinent to the risk being insured by the insurer. In general terms this means ensuring that the proposal form is fully answered and all relevant facts concerning the business are disclosed to the insurer. Failure to disclose all relevant facts fully, whether specifically asked or not, might entitle the insurer to treat the policy as void (ABI (2002)).

This is because the nature of the subject matter of insurance, and the circumstances pertaining to it, are facts particularly within the knowledge of the insured. The insurers are not generally aware of these facts unless the insured tells them. For this reason the law intervenes to make the situation more equitable by imposing a duty of utmost good faith on both parties to an insurance contract. The duty has been defined as

> 'a positive duty to voluntarily disclose, accurately and fully, all facts material to the risk being posed, whether asked for them or not . . . Every circumstance is material which would influence the judgement of a prudent insurer in fixing the premium or determining whether he will take the risk' (*Marine Insurance Act 1906, s 18*(2)).

Reciprocal duty 13.14

There is also a reciprocal duty: the underwriter may not mislead expressly or withhold information that leads the insured into a less favourable contract, for example, by not advising that a security system entitles an insured to a discount (Steele (1989)). Waring *et al* argue that 'save for' *caveat emptor*, the power in the relationship between insurer and insured is unevenly distributed in favour of the insurer' (Waring A, and Glendon A I, *Managing Risk: Critical issue for survival and success in the 21st century* (1998), Thomson Learning, pp 102/103). In his opinion it is unwise to assume that an insurance claim will be met at all. Insurers have a duty to ensure not only that a claim is valid but also that the terms and conditions of the policy have been met by the insured. This endorses the generally accepted view that it is crucial for organisations to select their insurances with prudence.

The insurer may require a survey of the premises by an insurance surveyor to accurately assess the risk before cover is agreed. The survey will record any hazardous features and make recommendations for loss prevention. It may be necessary to comply with these recommendations before cover is issued. When a proposal is accepted, the insured will receive a policy document which sets out details of the contract including the scope of the cover and the exclusions and conditions. With motor and employers' liability insurance, the law requires evidence of cover to be provided by a certificate of insurance in a prescribed form (ABI, 2002).

Causation 13.15

The insurer is liable only for those losses proximately caused by an insured peril (*Marine Insurance Act 1906, s 55*(2)(*a*)), and it is for the insured to prove that this loss was so caused (Merkin R, *Insurance Handbook*, First Edition (1994)).

> 'Proximate cause means the active, efficient cause that sets in motion a train of events which brings about a result, without the intervention of any force started and working actively from a new and independent source' (*Pawsey & Co v Scottish Union and National Insurance Co* (*The Times*, 17 October 1908)).

The proximate cause is not the first cause, nor the last cause: it is the dominant cause (*Leyland Shipping Co v Norwich Union Fire Insurance Society Ltd [1918] AC 350*) or the efficient or operative cause (*P Samuel & Co Ltd v Dumas [1924] AC 431*). Consider the 'domino effect', the proximate cause was the knocking over of the first domino. If an observer had intervened and stopped domino three from hitting domino four – and then changed his/her mind and set four off against the next, there would have been an intervention in the train of events and so the first domino would no longer be the proximate cause. The point of the doctrine is practical and it is to explain who should pay for the loss. If a policy is an all risks policy, such that the insured's responsibility is simply to demonstrate that he/she has suffered a loss, the insurer can escape liability by demonstrating that the loss was proximately caused by an uninsured peril.

Indemnity **13.16**

With the notable exception of life and accident policies, most contracts of insurance are subject to the common law principle of indemnity, under which the insured is entitled to recover at most a sum equivalent to the amount which he/she has lost by reason of the happening of an uncertain event. In other words, when an insured suffers a loss he/she should be indemnified by the insurer, which means that the insurer should provide financial compensation in an attempt to place the insured in the same pecuniary position after the loss, as he/she enjoyed immediately before it. There is a link between indemnity and insurable interest as it is the insured's interest in the subject matter of insurance that is, in fact, insured.

In the event of loss, indemnity may be provided in various ways including:

- replacement;
- cash;
- reinstatement; or
- repair.

The option as to which method is to be employed is normally given to the insurers by the wording of the policy. In the vast majority of cases, claims are settled to the satisfaction of insureds and most insurers will assist insureds in compensating them in the most favourable way, but they will obviously not want to increase their costs. A Loss adjuster will be involved at this stage for most Disaster situations (see **13.40** below).

Under insurance **13.17**

It should be noted that the maximum amount recoverable under any policy is limited by the sum insured or the limit of indemnity. Indemnity very often exceeds the sum insured especially where policies have not been updated for a few years. Buyers of insurance should ensure that their insurances provide sufficient cover. It should also be noted that where there is under-insurance the insurers are only receiving a premium for a proportion of the entire value at risk and any settlement would take this into account. This is known as 'average' and will be calculated with the following simple formula:

(sum insured ÷ full value) × loss

Hence under-insurance is likely to have an extremely detrimental effect on a business and could potentially bankrupt it.

Insurer's right of subrogation **13.18**

Having indemnified the insured (in insurance contracts of indemnity) the insurer is entitled to 'step into the shoes' of the insured and may commence proceedings in the insured's name against the third party that caused the insured's loss. The insurer may then keep the sum recovered. This is known as the insurer's right of subrogation: its primary purpose is to prevent the insured from recovering more than indemnity.

Characteristics of insurable risks **13.19**

Before engaging in the terms and conditions and premium rating, underwriters should determine the acceptability of risk based on some fundamental risk criteria. Those seeking insurance for risk activity should consider their activity within these parameters. Insurers may well develop ways of insuring risks that traditionally fall outside this guide, but they will generally have to protect themselves against unpredictable losses in this event. There are essentially six characteristics of what constitutes an insurable risk.

Risk 1: Capable of financial measurement **13.20**

The risk must be capable of financial measurement (tangible). Insurance pays monetary compensation for loss only. Sentimental value is not relevant to insured loss. Underwriters calculate premiums based on historical data and, to a lesser extent, risk exposure forecasts. Insurance market conditions, affected by investment returns and claims experience *inter alia*, will also significantly influence insurance pricing. New types of risk such as e-commerce and changes in the socio-political environment may cause a deficiency in historical data for underwriters to estimate future losses, which in turn may lead them to decline participation on those risks. This type of market failure may lead to government intervention; for example, the UK Government led the insurance industry in setting up Pool Reinsurance Company Limited, commonly known as Pool Re, to insure terrorism risk.

There should be a large enough number of similar risks. Insurance works on the principle that those that do not incur losses pay for those that do incur losses. Losses will be calculated based on the Law of Large Numbers which explains that the larger the sample of similar events the less the variation will be away from the mean outcome. For example, in repeatedly throwing a dice, 'the average of a large number of throws will be more likely than the average of a small number of throws to differ from the true average by less than some stated amount' (see Berstein (1996), p 122).

The mass covariant (ie wide spread) nature of region-wide Disasters may mean that a majority of policyholders incur losses at the same time. This is not desirable for insurance, but may be overcome by the risk spread achieved by multi-national insurers and/or reinsurance. Reinsurance is the practice of buying insurance against losses to an insurance portfolio of risks – usually against catastrophic losses. However, international reinsurance is expensive, which is one of the reasons that frustrates insurance as a development solution to Disasters like mass flooding in less developed countries (LDCs), such as Bangladesh, whereas Tokyo can exchange earthquake insurance with Los Angeles. On a local scale, houses built on flood plains (a common UK problem due to poor planning) provide a poor risk, as they are all likely to flood at the same time.

Risk 2: Policy holders must be 'average' **13.21**

Policyholders must be 'average'. Sometimes only those with above average risk buy insurance. This is known as 'adverse selection'. If only those who are highly vulnerable to Disasters buy insurance – the claims will be high and consequently the cover will either be very expensive or the scheme will go bust. Insurers need to protect themselves against adverse selection. For example, house owners with houses in flood plains may all seek cover, while those with little or no risk of flood may decide not to insure.

Risk 3: The risk will be 'particular' **13.22**

The risk will preferably be 'particular'. A risk that is particular in nature originates from an individual event and its impact is local, for example, theft of property or a building fire. Contrarily, 'fundamental' risks arise from causes outside the control of any one individual or group, for example, war, hurricanes, tsunamis, inflation. They may be insurable, but usually with considerable limitations in cover.

Risk 4: The loss must be 'fortuitous' **13.23**

The loss must be 'fortuitous' (a chance event) as far as the insured is concerned. Insurers need to guard against 'moral hazard', which is concerned with the attitude and conduct of people (see Steele (1989)). The insurer will want to avoid insuring risks in which the insured has undue influence over the occurrence of an insured event. It is common in these instances for him/her to take increased risks, follow bad practice or adopt a lackadaisical attitude to insured property which leads to an increase in claims. The worst form of moral hazard, however, comes from those that misrepresent their risk and fraudulently claim. Upward premium adjustments are insufficient compensation for poor moral hazard: education is really the only solution.

Some Disasters are not fortuitous and are, therefore, uninsurable. For example, war, famine (usually) or an active volcano. Flooding in some parts of Bangladesh or even on a flood plain in the UK is more or less a certainty and would, therefore, be deemed uninsurable.

Risk 5: The premium must be reasonable **13.24**

The premium must be reasonable. Premium needs to be economically affordable to the insured and financially sufficient to cover losses, overheads and make a profit for the insurer (or meet insurance fund criteria if subsidised). Some Disaster types, such as terrorism, may be too infrequent and too severe when they do occur for insurers to set reasonable premiums. In this instance governments may feel compelled to intervene as noted earlier.

Buying insurance **13.25**

There are various options available to buyers of insurance. The usual route is to seek advice from an insurance broker. Insurance brokers are full-time specialists offering advice and help in arranging the most appropriate cover available. By law, any individual or firm using the title 'insurance broker' must be registered with the Insurance Brokers' Registration Council, the statutory body with responsibility for ensuring that the requirements of the *Insurance Brokers (Registration) Act* 1977 are met. The requirements include financial controls and professional qualifications.

Insurance intermediaries selling general insurance who are not able to use the title 'insurance broker' may be 'company agents' (representing a maximum of six insurance companies) or 'independent intermediaries' with no limit on the number of companies with whom they can deal. Both independent intermediaries and company agents selling non-life insurance must follow the Association of British Insurers (ABI) Code of Practice for the Selling of General Insurance. Code compliance by independent intermediaries is monitored by Pricewaterhouse-Coopers on behalf of ABI and insurers (ABI (2002)).

Independent intermediaries are responsible for the advice which they give and must have insurance cover against any professional errors they may make. The insurance company concerned is responsible for the actions of a company agent. Both independent intermediaries and company agents must display in their offices a declaration of their status.

Lloyd's **13.26**

The insurance market consists of Lloyd's and insurance companies. Lloyd's was founded in 1688 as an insurance market place (not an insurance company). It is made up of a number of syndicates which are financed by 'Names' and 'Corporate Capital'. There are about 10,000 Names offering their private wealth. In the early 1990s this figure was about 32,000. Severe losses and legal wrangling led to the need for 'Corporate Capital' – a new phenomenon to Lloyd's – which is provided by some 200 limited companies (see Hodgin (1998), ch 4). Lloyd's syndicates do not deal directly with the public and enquiries must be made through a Lloyd's broker. Insurance companies, on the other hand, may sell to the public direct or through insurance brokers, intermediaries or agents. Many companies have a network of branch offices from which quotations and advice can be obtained. Both may transact all kinds of insurance business or may specialise in certain types (ABI (2002)).

See **13.52** below for advice from the ABI on classes of insurance that owners of small businesses should consider.

Risk assessment, underwriting and rating 13.27

The role of the underwriter is to accept insurance business on behalf of the insurer. A Lloyd's underwriter will do exactly the same job, but for a Lloyd's syndicate. The job entails determining the terms and conditions of cover and the price which will reflect the level of risk exposure the case brings to the fund by way of potential loss. Generally, the aim is to accept sufficient premium into the fund to pay for the losses.

Figures 2 and 3 explain premium allocation and claims payment. Figure 2 shows how premium may be allocated. It should be noted that approximately 10% would go to the broker in most cases in the form of brokerage or commission for providing the business and servicing the insured's needs. An estimated 10% will be anticipated for profit to satisfy the shareholders that have invested in the insurance business, which

Figure 2: Insurance premium

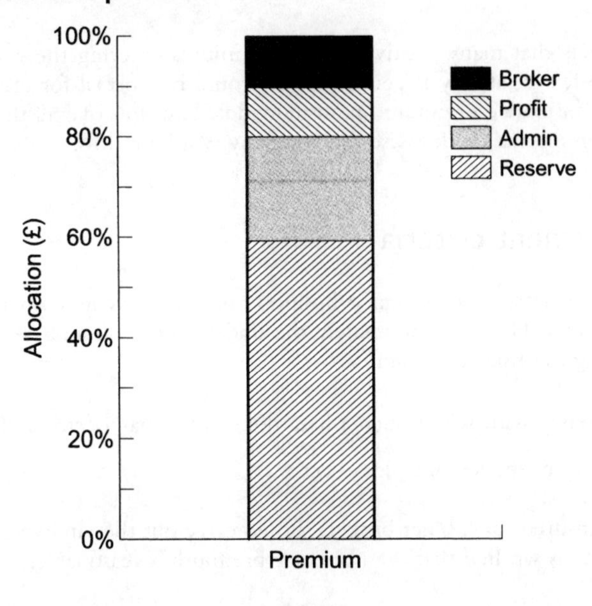

Figure 3: Insurance fund

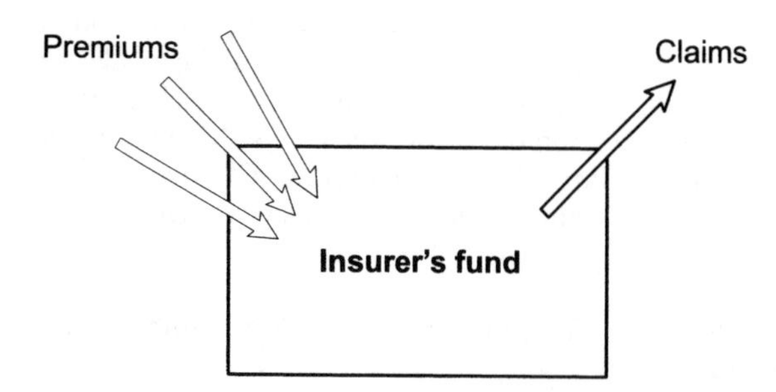

enables the company to have sufficient collateral to underwrite risks. This margin will vary enormously depending on the nature of the risk underwritten. The underwriter will be seeking, however, to maintain a steady profit rather than highs and lows. This is not an easy task as the insurance industry is notoriously cyclical with hard markets following soft markets where price highs follow price lows. These cycles tend to have a three or more year cycle.

A further 20% or more will be used up in administering the underwriting process – paying salaries, rents, stationary, and data processing computers *inter alia*. That leaves a mere 60% in reserve to pay for claims. These reserves will take advantage of other investments to maximise profits between receiving the premiums and paying out the claims – a gap of some considerable time for some classes of business. The investment of premium funds is strictly regulated to prevent insurers taking high risks with policyholders potential compensation money. Interest rates and the performance of the stock markets among other investment opportunities will impact investment returns.

Figure 3 depicts that many relatively small premiums entering the reserve account go to pay the few relatively larger claims. It would be normal for claims payments to exceed the insurance premium on an individual basis but overall they should not exceed total premiums – otherwise the company will be in loss.

Risk assessment criteria **13.28**

There are many classes of insurance business and each has its own particular risk assessment criteria. However, ultimately the underwriter will address the following issues regarding the risks identified:

(a) the frequency with which such a hazardous event may occur; and

(b) the severity of the loss if it does occur.

Similarly the insured (or his/her broker) should carry out the same exercise to assess whether he/she is satisfied that the insurance premium is reasonable.

Frequency **13.29**

The uncertainty of a risk can be thought of in terms of three elements (Brown and Churchill (1999)):

- ***if*** the risk event will occur
- ***when*** the risk event will occur; and
- ***how*** often the risk event might occur.

Severity **13.30**

A loss may be a one time payment (eg replacement of stolen property) or ongoing (eg loss of future earnings due to long-term disability). The relative severity will depend on the finances of the person (or group). The loss of a bicycle to an old age pensioner who uses it to get to the shops may be equivalent to several months' savings and lead to the additional cost of a bus fare. The same loss to a bank manager may be irrelevant to his transport needs and equivalent to one day's labour.

Frequency and severity may be ranked out of (say) five, and scored on a matrix as illustrated in Figure 4 below (van Oppen (2001)).

Figure 4: Frequency/severity matrix

Severity						
5	5	6	7	8	9	10
4	4	5	6	7	8	9
3	3	4	5	6	7	8
2	2	3	4	5	6	7
1	1	2	3	4	5	6
0	0	1	2	3	4	5
	0	**1**	**2**	**3**	**4**	**5** Frequency

Frequency:

Score out of 1–5 the historical frequency of the natural hazard event, where (for example):

5 = Regularly (eg annually for storm or flood/every 25 years for quake).

4 = Often (eg every other year for storm or flood/every 50 years for quake).

3 = Periodically (eg every five or so years for storm or flood/every 100 years for quake).

2 = Occasionally (eg every ten years or so for storm or flood/every 200 years for quake).

1 = Rarely (eg historical records show hazard occurrence has occurred).

0 = Never.

Severity:

Score out of 1–5 the most likely impact of a natural hazard event (eg a tropical storm), where (for example):

5 = Catastrophic impact.

4 = Serious impact.

3 = Moderate impact requiring outside assistance.

2 = Low-grade impact manageable in-house.

1 = Incidental damage with no significant operational impact.

0 = No remarkable impact.

It should be noted that by scoring the impact and not the hazard itself, the score takes on an element of 'vulnerability to' the hazard. This method, as opposed to scoring the hazard itself, has more value, because a hazard may be of varying significance to different 'values at risk'. For example, an earthquake may be more devastating to a brick building than to a wooden one. In this case an earthquake may be scored as a five for the brick building and, say, a two or three for the wooden one (van Oppen (2001)).

Allocating price **13.31**

Generally this sort of quantification greatly assists decision-making and may assist the underwriter in allocating the most equitable price that reflects the attributes of the risk. Usually he or she will start off with a base price, which will reflect the market price or the average price. This will consist of a rate calculated on the sum insured, employees' wages or the turnover (gross revenue) of the business for liability risks. This rate can then be increased or reduced to reflect the risk assessment. For example, discounts may be given for good physical features, such as sprinkler systems. Premiums may be flat, which means that they are not variable; or may be adjustable – based on a variable, such as the wage expenditure or gross turnover.

Limit of liability **13.32**

The sum insured is the maximum liability of the insurer. With some exceptions, it is not the amount that the insurer promises to pay in the event of loss (as loss may only be partial), nor is it an admission of the value of the property insured. With the exception of employer's liability insurances, which usually give unlimited indemnity, it is usual for liability policies to have a limit of liability (ie a maximum amount payable within the policy terms). This limit may be based on 'any one occurrence' during the policy period, and have an aggregate limit for the period. On this basis, for example, if a product is found to cause damage to a third party, the policy will pay out for each occurrence (ie each case) up to the limit of liability. But if it is aggregated, then the policy will cease paying out claims once the aggregate limit is exhausted. Thus, if a policy limit reads '£1,000,000 any one occurrence and in the aggregate', the policy may pay out up to ten claims of £100,000 or one claim of £1,000,000, but once the aggregate limit is paid, the policy is not bound to pay anymore.

The premium **13.33**

The premium, once paid, applies to the entire duration of the policy, and the insured cannot be required to pay a supplement should the insurer find itself facing an undue number of claims. Mutual insurances, on the other hand, rely on a 'call' from the participants to retain adequate funds for claims payments. The insurer determines the amount of premium at the outset and its payment is generally stipulated in the conditions of cover. Non-payment may deem the contract as void or enable the insurer to sue for the unpaid premium. It is common for the insurer to stipulate that the insurance cover will commence as soon as the premium is received.

Insurance premium tax **13.34**

Most types of general insurance, with the exception of life insurance, pensions and income protection insurances, are subject to insurance premium tax (IPT). The premium quoted by the insurer will include the tax where necessary. Insureds are not able to set IPT paid by them against VAT on the goods and services they sell to others. In the event that the insured sells cars, electrical or domestic appliances, or holidays, and they also arrange insurance for their customers, then in certain circumstances they may be liable to pay IPT themselves. It is recommended that an accountant, tax adviser or HM Customs and Excise be consulted (ABI (2002)).

Conditions in insurance policies 13.35

Insurance agreements consist of four main types of clause (*Tolley's Insurance Handbook*, First Edition (1994) Merkin) as follows:

- the insuring clause, which sets out the insured perils;
- the excluded perils clause;
- promises or warranties, made by the insured as to the situation or use of the subject matter;
- other terms, which are often interchangeably classed as 'terms' or 'conditions'.

It is important to note the difference between them as this has significant legal connotations.

Renewals 13.36

Large proportions of insurance policies are operated on a monthly payment (direct debit) basis by the insured. However, whether premiums are paid monthly or at inception, the insured will be asked at the end of the policy period either to pay the renewal premium or to complete a further bank mandate authorising the continuation of the direct debit. The insured will usually be granted 'days of grace' if they are late in taking up the offer. An indemnity policy is periodic and comes to an end at the end of the stated period. Renewal by the insured creates a totally new contract. The grant of days of grace by the insurer for payment of the insurance premium does not, therefore, amount to an extension of the old policy, but is an offer to the insured of fresh cover should the premium be paid in the days of grace. The insurer is not necessarily holding the insured covered for this period. It is worth checking the renewal terms, as there may be changes to reflect loss trends or other risk altering factors.

Losses 13.37

So far this chapter has introduced the underwriting process – the principles of insurance, the characteristics of insurable risk and the role of brokers and underwriters. Assuming that all the guidelines have been followed and successful insurance has been implemented, it shall now be assumed that a loss has occurred and the insured needs guiding through the process of making a claim, managing the crisis, mitigating the impact and recovering. The process may be viewed in Figure 5. Attention will be paid to the roles of loss adjusters, claims handlers and disaster restoration companies.

Claims duties and procedures 13.38

There are certain implied or unwritten duties and certain express duties imposed on the insured in the event of an occurrence likely to lead to a claim. The law

Figure 5: The loss process

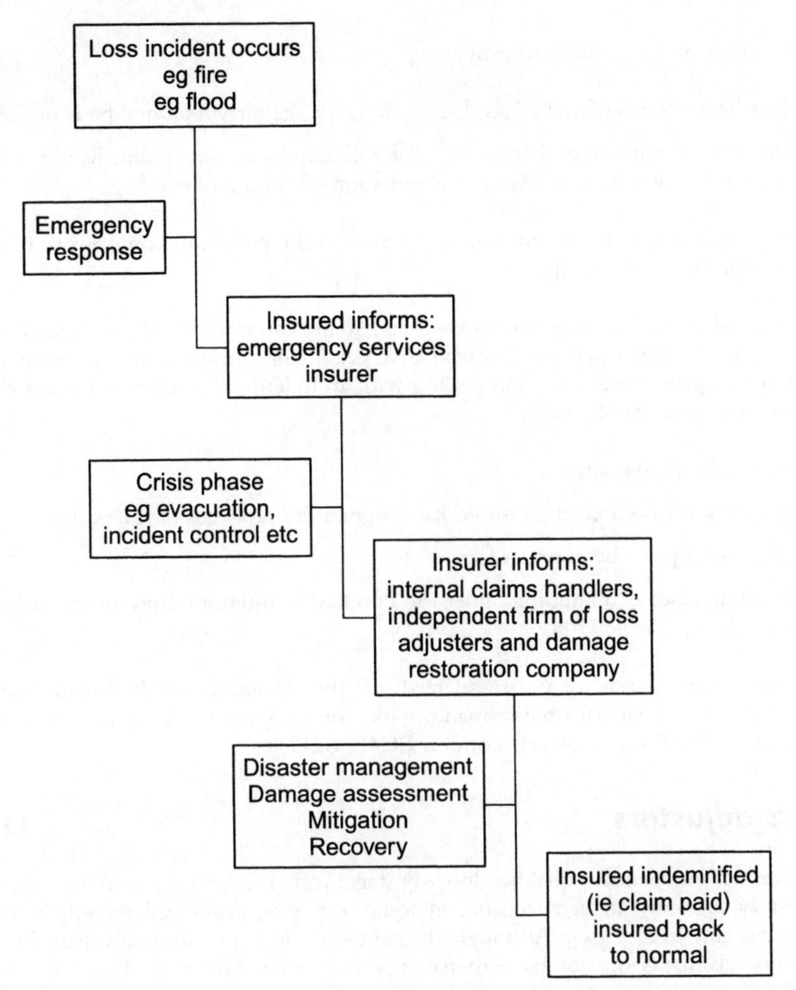

requires that the insured should act as if he were uninsured and take all reasonable steps to minimise his or her loss. The emergency services must not be hindered in any way.

Almost all policies stipulate that the insurer should be notified immediately of any event which could give rise to a claim and to provide full particulars within a given period of days. It is essential that the insurer is notified as soon as possible in order that a full investigation can be made. In Disaster situations it is in the insured's interest to have the assistance of the insurer's claims handling service and the appointed loss adjuster (see **13.40** below) and, in some cases, appointed disaster restoration experts. At common law, insurers have the right, jointly with the insured, to save the subject matter of insurance.

Onus to prove 13.39

The onus is on the insured to prove:

- that he/she has suffered a loss due to an event against which he/she is insured;
- the value or amount of that loss (except with employers' and public liability claims where the insurers take over the negotiations from the outset).

If the insurer wishes to rely on some exclusion in the policy, the onus is on them to prove that the exclusion applies.

Claims will be paid to the insured except in liability cases, when the payment is made directly to the third party and solicitors. Occasionally, insurers may exercise their option to repair or replace when dealing with an indemnity. If payment is not made to the insured it may be made to:

- his legal representative;
- any person to whom the insured has assigned the proceeds of the policy;
- to another party by court order;
- to a purchaser of a building under the Purchaser's Interest Clause of the Standard Fire Policy (Steele (1989)).

Disputes over claims may be referred to the Financial Ombudsman Bureau at http://www.financial-ombudsman.org.uk/ or the General Insurance Standards Council at 110 Cannon Street, London EC4N 6EU.

Loss adjusters 13.40

Insurers use independent qualified loss adjusters for their experience and the expertise necessary to carry out detailed and, in some instances, prolonged investigations of complex and large losses. Although the adjuster's fees are invariably paid by the insurers (claimants do not have to pay any fee to loss adjusters), he or she is an impartial professional person and makes his or her judgement on the amount to be paid in settlement solely on the basis of established market practice. It is his or her task to negotiate a settlement which is within the terms of the policy and equitable to both insured and insurer.

Should the loss adjuster not be an expert in a particular discipline which is necessary or desirable to pursue the negotiations, he or she will consult or employ such an expert. Chartered loss adjusters must operate under a Royal Charter to a code of conduct.

In addition to a thorough knowledge of insurance and of the specialist area in which they work (eg office buildings), they can advise both the insurance company and the policyholder on repair and replacement techniques. After discussions with the policyholder, the loss adjuster's report to the insurance company enables the company to process the claim without delay. Each loss adjuster works for many insurance companies who rely on the skill and impartiality of the loss adjuster.

The loss adjuster will visit the policyholder, soon after submission of the claim form, to discuss the circumstances of the claim. The loss adjuster will check:

- that the loss or damage falls within the terms of the insurance policy;
- that the sums insured on the policy are adequate; and
- that the amounts being claimed are fair and reasonable.

Loss adjusters are able to advise on loss prevention techniques, including security and safety, to avoid a further incident.

The loss adjuster will submit a report of his or her findings including recommendations to the insurance company. The company is then in possession of the required facts to indemnify the insured.

Policyholder information **13.41**

As with all insurance claims, it is helpful if the policyholder can provide full and prompt information. It will help the loss adjuster in his work if policyholders can provide:

- estimates for repairs;
- the cost of any 'emergency' repairs that were needed;
- valuations, receipts and other proof of ownership and value;
- the police crime reference number;
- any badly damaged property – for example burnt or soaked – should be stored in a garage or outbuilding so the loss adjuster can inspect it.

For more information on loss adjusting refer to the Chartered Institute of Loss Adjusters (CILA) on their web site at http://www.cila.co.uk/.

Loss assessors **13.42**

In motor insurance, a loss assessor is an engineer, but in other classes a loss assessor is a person who, in return for a fee (usually a percentage of the amount claimed), acts for the claimant in negotiating the claim. Loss assessors should not be confused with loss adjusters. Loss assessors are not impartial. It has been observed that some less scrupulous loss assessors have been known to follow the emergency services to incidents and persuade the potentially vulnerable claimant to instruct their services to handle the claim.

Claims handlers **13.43**

In the event of a claim advice being received by an insurer, they will issue it with a file and a reference number. Thereafter they will appoint the loss adjuster if necessary or deal with third parties or their solicitors for liability claims. A claims handler or manager will ultimately settle the claim.

Damage restoration companies 13.44

According to Doug Blanks, founder of Ark and General, Britain's largest damage restoration and cleaning company:

> '... the insurer ultimately wants two things of the loss adjuster: to make sure the claim is reasonable; and to make sure the insured is retained. As a rule of thumb a loss adjuster will endorse the spending of £1.00 to save the insurer £1.20'.

For example, if restoration of a painting costs £100 to put it back in its original condition, it must be worth at least £120 to make worthwhile. Otherwise the insurer would be better off paying out a total loss.

Damage restoration companies are appointed by the insurer, often recommended by the loss adjuster, and they are experts in mitigating the impact of Disasters – this may occur before the fire is out or the tide has subsided. They may offer specialist advice in business continuity and recovery.

Risk Management

Risk management defined 13.45

The insurance industry has witnessed a gradual move towards a more pro-active preventative strategy to risk management (Waring *et al* (1998)).

> 'Risk management is the systematic, positive identification and treatment of risks posing a threat to values (or resources) of human concern. It is a collective term encompassing all aspects of planning for and responding to risks – both pre-event and post event. Risk management includes preventative measures and continuity, contingency and crisis planning. Essentially risk management is about decision making which will depend on all the variables of the decision making environment (including issues such as risk perception and political, societal, social and economic conditions)' (van Oppen (2001)).

The risk management process is illustrated in Figure 6 at **13.46** below.

Risk management process 13.46

Figure 6 below is adapted from Irukwu (1991), but forms the commonly accepted approach to risk management. Its stages are quite straightforward and are explained below.

Figure 6: The risk management process

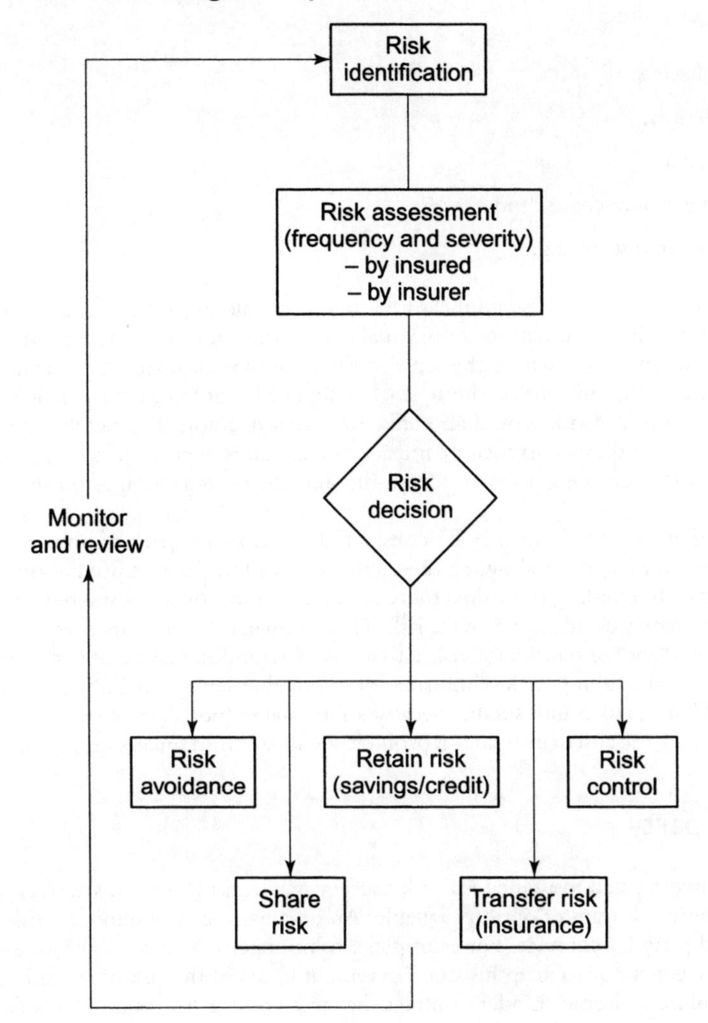

As noted above in **13.27 RISK ASSESSMENT, UNDERWRITING AND RATING**, a risk assessment involves evaluating the three core parts:

- hazard;
- values; and
- vulnerability.

Hence, in the first instance it is necessary to identify what the hazards are, what is at risk (the values) and how vulnerable the exposed values are (ie proximity and sensitivity). There are many risk identification techniques available (see Dickson (1991)), including:

- site surveys;
- check lists;
- flow charts;
- organisational charts; and
- simple brainstorming.

Having identified the risks and measured their probable impact in terms of frequency and severity, the organisation should make a decision founded on the principles of cost/benefit analysis – where the 'cost' is the downside of taking a particular course of action and the 'benefit' is the upside. It should be noted that a person's attitude to or perception of risk would also affect the final decision. Ultimately, the strategy seeks to reduce the occurrence or impact of hazardous events, thereby preventing a disaster and/or enabling a return to pre-disaster situations as soon as possible.

Some activities carried out may be considered too risky in terms of human or financial impact, for example, allowing the public to have fun rides on fire fighting units at a town fair. If avoiding the risk activity is not an option or is considered adverse, an organisation may decide to retain the risk. This action may be involuntarily because they cannot avoid and/or transfer the risk activity. Risk retention may be the only option for many reasons including: lack of information about alternatives (including communication problems); risk is uninsurable because of its quality (not all risks are insurable – see **13.19** above); the cost of insurance is prohibitive; and/or insurance is simply unavailable.

Third party **13.47**

As already discussed insurance is a risk transfer mechanism (see **13.4** above), but it is not the only risk transfer option available. An organisation may transfer a risk activity to a third party by contract. For example, a fishmonger may pay a cold store with an electricity generator to keep his stock overnight to avoid the risk of perishing due to power failure at home. Under contract, he may recover the value of his fish if the cold store does not perform their part of the bargain and the fish unduly perish. In this instance, the fishmonger may alternatively decide to keep the risk of perishing within his domain and not get a third party to take care of them, but transfer the risk by insuring it. The fishmonger will effectively be exchanging a known cost (premium) for an unknown cost (loss) at some future date during an agreed period.

Loss control **13.48**

Loss control covers two distinct areas in risk management:

- loss prevention, otherwise termed risk reduction; and
- damage limitation.

As the language implies, loss prevention is concerned with preventative and mitigating activities carried out pre-loss. **APPENDIX 3: PROTECTION AGAINST FIRE** (see **13.63** below) provides an example of advice issued by the ABI about pre-loss activities that should reduce the probability of fire.

It is mandatory to comply with certain conditions set by law, such as providing protective clothing to employees operating in hazardous environments. Loss prevention goes beyond the 'hard' risk controls and includes 'soft' risk prevention techniques, such as:

- management supervision;
- employee training and awareness; business continuity planning; and
- monitoring the risk management process for responsibilities.

Once an incident has occurred, such as a fire or a flood, plans should be activated to minimise or limit the damage. Pre-loss planning to this extent should include consideration for action such as evacuation, training to use portable fire appliances and computer back up etc.

Post-loss measures will include the implementation of the Disaster Recovery Plans, including the swift involvement of the insurance company's loss adjuster and, if applicable, damage restoration company.

Monitoring and reviewing **13.49**

Above all else, the process of risk management requires systematic monitoring and reviewing to ensure that the insurances are kept up to date and the decisions remain pertinent to the risk exposure which will reflect the changing nature of the activities carried out. The risk management strategy should be 'owned' by the appropriate manager and the operational tasks should be clearly allocated so that employees are aware of, trained in and encouraged to prioritise their loss prevention responsibilities. In the event of poor performance they may be held accountable and disciplined accordingly. Insurers of large companies use sophisticated models to assess the risk features relating to the holistic management of risk – where holistic risk management encompasses all areas of strategic, operational and environmental risk.

APPENDIX 1

British Insurance Brokers Association Jargon Buster **13.50**

The following is based on the Jargon Buster provided by the British Insurance Brokers Association (BIBA). The full version is available on their website (http://www.biba.org.uk/).

The insurance industry uses quite a number of technical terms, usually for reasons of precise meaning, which are not necessarily easily understood by the layman. This appendix clarifies the meaning of key words and phrases found in insurance documents.

Glossary of insurance terms

Act of God: *Nugent v Smith (1876) 1 CPD 423* 'Natural causes directly and exclusively without human intervention and that could not have been prevented by any amount of foresight and pains and care reasonably to have been expected'.

Adjuster: One who investigates and assesses claims on behalf of insurers (claims adjuster or loss adjuster).

Aggregate limit of indemnity: The maximum amount an insurer will pay under a policy in respect of all accumulated claims arising within a specified period of insurance.

All risks: Term used to describe insurance against loss of or damage to property arising from any fortuitous cause except those that are specifically excluded.

Assurance: A term interchangeable with insurance but generally used in connection with life cover as assurance implies the certainty of an event and insurance the probability.

Claims: Injury or loss to the insured arising so as to cause liability to the insurer under a policy it has issued.

Common law: The common law consists of the ancient customs and usages of the land, which have been recognised by the courts and given the force of law. It is, in itself, a complex system of law, both civil and criminal, although it is greatly modified and extended by statute law and equity. It is unwritten, and has come down in the recorded judgments of judges who for hundreds of years have interpreted it.

Consequential loss: Insurance of loss following direct damage, eg loss of profits; loss of use insurance.

Cover note: A document issued to the insured confirming details of the insurance cover placed. Some cover notes are a legal requirement, eg motor.

Deductible: The specified amount a loss must exceed before a claim is payable. Only the amount which is in excess of the deductible is recoverable.

Endorsement: Documentary evidence of a change in the wording of, or cover offered by, an existing policy or qualification of wording if the policy is written on restricted terms.

Excess: The first portion of a loss or claim which is borne by the insured. An excess can be either voluntary to obtain premium benefit or imposed for underwriting reasons.

Exclusion: A provision in a policy that excludes the insurer's liability in certain circumstances or for specified types of loss.

Inception date: The date from which, under the terms of a policy, an insurer is deemed to be at risk.

Insurable value: The value of the insurable interest which the insured has in the insured occurrence or event. It is the amount to be paid out by the insurer (assuming full insurance) in the event of total loss or destruction of the item insured.

Insured: The person whose property is insured or in whose favour the policy is issued.

Insurer: An insurance company or Lloyd's underwriter who, in return for a consideration (a premium) agrees to make good in a manner laid down in the policy any loss or damage suffered by the person paying the premium as a result of some accident or occurrence.

Limit: The insurer's maximum liability under an insurance, which may be expressed 'per accident', 'per event', 'per occurrence', 'per annum', etc.

Loss: Another term for a claim.

Material fact: Any fact which would influence the insurer in accepting or declining a risk or in fixing the premium or terms and conditions of the contract is material and must be disclosed by a proposer.

Name: Another term for an underwriting member of Lloyd's.

Negligence: Perhaps the most common form of tort. In *Blyth v Birmingham Waterworks Co. [1843–60] All ER 478* it was defined as 'the omission to do something which a reasonable man guided by those considerations which ordinarily regulate the conduct of human affairs would do, or doing something which a prudent and reasonable man would not do'. Gives rise to civil liability.

Non-disclosure: The failure by the insured or his broker to disclose a material fact or circumstance to the underwriter before acceptance of the risk.

Peril: A contingency, of fortuitous happening, which may be covered or excluded by a policy of insurance.

Period of risk: The period during which the insurer can incur liability under the terms of the policy.

Policy: A document detailing the terms and conditions applicable to an insurance contract and constituting legal evidence of the agreement to insure. It is issued by an insurer or his representative for the first period of risk. On renewal a new policy may well not be issued although the same conditions would apply, and the current wording would be evidence by the renewal receipt.

Policyholder: The person in whose name the policy is issued. (See also insured.)

Premium: The consideration paid for a contract of insurance.

Quote: A statement by an insurer of the premium he will require for a particular insurance.

Reinstatement: Making good. Where insured property is damaged, it is usual for settlement to be effected through the payment of a sum of money, but a policy may give either the insured or insurer the option to restore or rebuild instead.

Salvage: A recovery of all or part of the value of an insured item on which a claim has been paid. The insurer will normally dispose of the item and apply the proceeds to reduce the cost of the claim.

Schedule: The part of a policy containing information peculiar to that particular risk. The greater part of a policy is likely to be identical for all risks within a class of business covered by the same insurer.

Statute law: Presently the most important source of law is statute law, otherwise known as Acts of Parliament; which may create entirely new law, overrule, modify, or extend existing principles of common law and equity, and repeal or modify existing statute law.

Subject to survey: Phrase used by an insurer to signify provisional acceptance of an insurance pending inspection by a surveyor whose report is necessary to determine the rate and conditions applicable.

Sum insured: The maximum amount payable in the event of a claim under contract of insurance.

Third party: A person claiming against an insured. In insurance terminology the first party is the insurer and the second party is the insured.

Warranty: A very strict condition in a policy imposed by an insurer. A breach entitles the insurer to deny liability.

Wear and tear: This is the amount deducted from claims payments to allow for any depreciation in the property insured which is caused by its usage.

APPENDIX 2

Your insurance needs 13.51

Based on The Association of British Insurers's (ABI) website (www.abi.org.uk) (ABI (2002)), the guidance notes reproduced below were intended for small businesses, which should provide most readers with an overview of the types of cover to consider.

Property **13.52**

Buildings and contents can be insured against fire, lightning, explosion of gas and boilers used for domestic purposes without the addition of 'special perils' such as explosion, riot, malicious damage, storm, flood, impact by aircraft, road and rail vehicles, escape of water from tanks or pipes and sprinkler leakage.

'All risks' insurance gives wider cover including any accidental damage or loss not specifically excluded. However, 'all risks' will not cover wear and tear, electrical or mechanical breakdown and gradual deterioration which will be specifically stated in the policy document.

Business premises should be insured for their full rebuilding cost (including professional fees and the cost of site clearance) and not just for their market value. Expert advice may be needed to calculate the rebuilding cost, which often differs significantly from market value.

Stock should be insured for its cost price without any addition for profit. Provision can be made for seasonal stock fluctuations. Plant and business equipment can be insured on either a 'replacement as new' or on an 'indemnity' basis. If indemnity is chosen, wear and tear will be taken into account when settling any claims.

Engineering **13.53**

Engineering insurance provides cover against electrical or mechanical breakdown for most machinery, including computers. By law, a qualified person must regularly inspect many items of plant such as boilers, lifts and lifting machinery. Insurers can arrange to provide this service.

Theft **13.54**

Contents are usually covered against theft providing there has been forcible and violent entry to, or exit from, the premises. Damage to the building resulting from theft or attempted theft will also normally be covered. Theft by employees is usually not covered – cover against employee dishonesty can be arranged by a fidelity guarantee policy – see Other Risks at **13.60** below.

Money **13.55**

Money insurance is on an all-risks basis and covers cash, cheques, postage stamps and certain other negotiable documents.

Different limits will apply to money on the premises in and out of business hours, in safes, at the homes of directors or employees and in transit. There may be requirements in the policy relating to safe keys and the method of transit.

Personal assault cover may be included, which will provide compensation for the insured or his/her employees following injury during theft or attempted theft of money.

Goods in transit 13.56

Goods in transit insurance covers goods against loss or damage while in the insured's vehicle or when sent by carrier. The sum insured may be a limit for each vehicle or any one consignment.

Business Interruption (BI) 13.57

Even minor damage to insured property could seriously disrupt business leading to loss of income and extra expenses.

Business interruption insurance will compensate for the short-fall in gross profit together with paying any increased working costs and extra accountants' fees incurred.

When arranging this insurance it is necessary to estimate the maximum time needed to get the business working normally following the most serious damage. The insurers will ask for an estimate of the anticipated gross profit. If an auditor later certifies an actual figure materially lower than this estimate, a return of premium is normally given.

Motor vehicles 13.58

By law it is compulsory to insure legal liability for injury to others and damage to third party property arising from the use of vehicles on the road – third party insurance.

Most business policies are comprehensive or third party, fire and theft. Comprehensive cover includes damage to the insured vehicle. The third party section of a commercial vehicle policy is unlimited in respect of injury to others but, for damage to others' property, usually limited up to £1m – with private cars it is unlimited.

Insurers require full details of the types of vehicle and their usage, details of goods and samples carried, details of drivers and, where applicable, the maximum number of fare-paying passengers.

Where more than five vehicles are owned, a fleet policy may be arranged. The claims experience of the fleet will provide the main rating factor in assessing the cost of the policy.

Legal expenses 13.59

The cost of taking or defending legal action places a considerable financial strain on businesses. Legal expenses insurance (LEI) covers legal costs such as solicitors' fees and expenses, the cost of barristers and expert witnesses, court costs and opponent's costs if awarded against the insured in civil cases.

Types of dispute normally covered include contractual and employment matters, for example claims for unfair dismissal, racial or sexual discrimination, the recovery of bad debts, and disputes with a landlord other than those relating to rent or service charge (particularly relevant for small businesses when dealing with a large corporation or local authority landlord).

The policy cover will also include the cost of employing specialist accountants and lawyers to protect the insured's rights, for example, if their business is subject to investigations on tax or VAT matters. Most policies will offer a confidential telephone legal advisory service to answer commercial legal enquiries.

The vast majority of LEI policies are sold 'before the event', to cover the consequences of an event which has not yet occurred. However, there is a very small part of the market which is sold 'after the event', to cover legal expenses in a case where the disputed event has already happened, although the expenses of the court case resulting from it have yet to be incurred. It is likely to be available only where the chances of winning the case are high, in which case the insurer is likely to be able to recover the costs from the other side.

Other risks **13.60**

The following insurances may be provided under a 'package' business policy. Separate individual policies can also be issued.

- Credit Insurance – gives cover against the risk of debtors becoming insolvent and unable to pay.
- Book Debts – cover against loss of money arising from accidental damage or theft of books of account.
- Fidelity Guarantee – provides cover against loss of money or stock arising from dishonesty by the insured's employees.
- Frozen Food – provides cover against loss of frozen food in deep-freeze units caused by breakdown or damage to the unit or failure of the electricity supply.
- Glass – provides cover for the replacement of glass following malicious or accidental damage.
- Travel Insurance – individual or group travel insurance policies can be arranged providing cover during business journeys abroad which includes medical and legal expenses, personal accident and loss of baggage.

Legal liabilities **13.61**

Running a business creates considerable legal responsibilities towards the insured's employees, the public and customers. Injury to employees and members of the public could result in directors being legally liable to pay damages if they or their employees have been negligent or are found in breach of a statutory duty.

An employer's statutory duties are set out principally in the *Health and Safety at Work etc Act 1974* and associated Regulations. It is necessary to identify the safety and health needs of the insured's particular business activities by carrying out a risk assessment and action should be taken to remove or control the risks in line with the appropriate legislative requirements. Good management is an ongoing process. Any changes in workplace activities should lead to a reassessment of hazards present.

Advice on good health and safety management is available from the Health & Safety Executive and local authority environmental health departments. Further, liability insurers or brokers will be able to provide assistance with risk assessments and action plans.

Liability insurance will pay the amounts of any court awards or damages, claimants' costs and expenses where the insured or their employees are held legally liable, subject to any policy limits.

The main liabilities faced in business are:

- employers' liability;
- public liability;
- motor vehicle liability; and
- product liability.

Health, life insurance and pensions 13.62

It is worth considering the consequences of serious injury, illness or death to the owner or employees – and the effects this could have on the business. Policies providing benefit for serious illness, disability or death of employees and retirement benefits can be arranged on either an individual or group basis. There are valuable tax concessions for both employers and employees for retirement benefits schemes. There are usually tax concessions for employers who arrange insurance for their employees against death, disability or sickness. Accountants or insurance advisers should be able to provide further details. The options will include:

- Personal Accident and Sickness Insurance.
- Income Protection Insurance.
- Private Medical Insurance.
- Life Assurance.
- Pensions.

APPENDIX 3

Protection against fire **13.63**

The following advice is taken from 'Insurance Advice for Small Businesses' on the Association of British Insurers (ABI) website (http://www.abi.org.uk/).

Introduction **13.64**

Fire causes more damage to business properties than any other hazard. All businesses should take precautions and implement working procedures to reduce the risk of fire damage. Many businesses use substances and produce waste, which are potential sources of fire. Highly combustible materials can ignite and assist fire to spread very quickly. Even non-combustible goods can become a fire hazard if packaged in combustible materials such as cardboard, polythene or wood.

Construction **13.65**

Materials which encourage the spread of fire should not be used in the construction of the building, including ceilings, linings and partitions.

Buildings divided into fire-resisting compartments are less vulnerable to the spread of fire.

House-keeping **13.66**

'House-keeping' refers to the attitude of a business towards management control and supervision, tidiness, maintenance and safety. These are important in reducing fire hazards.

Remember:

- the owner or manager should take responsibility for fire safety;
- accumulated waste creates an unnecessary fire hazard as do blocked passageways. Waste should be cleared and removed from the site at frequent intervals;
- when a building is unoccupied all combustible material should be removed and services disconnected;
- check with your insurer before draining any sprinkler and de-activating the alarm;
- inflammable liquids and gases should be stored in a suitably constructed and secure detached building;
- combustible goods should not be stored alongside buildings or in yards;

- there should be restrictions on smoking, to reduce the fire hazards;
- a final inspection of the premises should be made before closing.

Heating, lighting and power 13.67

If the heating system is potentially dangerous, insurance cover may be refused. 'Safe' methods of heating are hot water radiators fed from a boiler segregated in a brick compartment or fixed electric fan heaters incorporating fan failure cut-outs. Your insurance company will be able to offer advice. Sufficient fixed power points and lighting sockets should always be available.

Electrical safety 13.68

Fires caused by electrical appliances are another danger.

Management should ensure that:

- an annual visual inspection is carried out and a more detailed inspection and test programme in conducted at least every five years or as recommended by the Institution of Electrical Engineers (IEE) Regulations;
- specialist advice is taken before installing or using electrical equipment where flammable material is used;
- staff are aware of new hazards when updating or renewing equipment. Overloading can occur where extra electrical equipment and machines are installed;
- residual Current Devices are considered when installing new electrical equipment;
- staff are trained in dealing with electrical fires. This should include isolation of electrical equipment: treatment of electrical shock and using the correct types of extinguishers or electrical fires.

Your insurance company will be able to offer advice.

Employees should:

- report promptly potential hazards such as damaged flex, stop using equipment that might be defective and not attempt to repair it;
- take care when using portable appliances, making sure that the leads will not be damaged. Also be careful when using heat producing appliances such as soldering irons;
- where possible, switch off and unplug appliances at the end of each day; ensure that ventilation grilles are always kept clear.

Detection and extinguishing equipment **13.69**

Staff should be trained in the use of extinguishers as well as the action to be taken in the event of fire. Insurers will normally give a reduction in premium if detection or extinguishing equipment, such as automatic sprinkler systems, are installed to certain standards.

Arson and malicious damage **13.70**

Arson and maliciously started fires are the greatest cause of fire damage.

14

The Law Relating to Emergencies and Disasters

In this chapter:

- Introduction.
- Statutes relevant to Emergency and Disaster management.
- Directors' responsibilities for health and safety.
- The 'Turnbull' guidelines.
- Criminal and civil law relevant to Emergency and Disaster management.
- The future: proposals for corporate killing.
- Public inquiries.
- Conclusions.

Introduction 14.1

Health and Safety law can trace its history back two hundred years. In 1802 legislation was passed known as the *Act for the preservation of the Health and Morals of Apprentices and others employed in Cotton and other Mills, and Cotton and other Factories* (Geo III, c73). 35 years later there was the landmark decision in *Priestly v Fowler (1837) 3 M & W1.* This case held that an employer owed a duty of care to his employee and if this was breached causing injury to the employee, the employee had a right to sue for compensation.

In contrast, Disaster and Emergency management, to the extent that it is a separate field, only developed legislation from the early part of the twentieth century. Even so, the pace of updates and new law has been rather infrequent.

In terms of emergencies, legislation was introduced after World War II, which focused primarily on arrangements for dealing with hostile attacks. However, the

Government accepted it no longer provided an adequate framework. John Prescott, Deputy Prime Minister, announced a review of emergency planning arrangements in the wake of the fuel crisis and severe flooding in the autumn and winter of 2000. In 2004, the Civil Contingencies Act was passed by the Parliament. Amongst other things, this repealed the Civil Defence Act 1948 and the Emergency Powers Act 1920.

In the last few decades there has been the development of legislation relating to disasters. This began in the wake of the chemical plant explosion at Flixborough, England (see **6.1** above), which killed 28 people and, two years later, the accidental leak from a pressing plant of toxic dioxin fumes in Seveso, Italy, which, although it did not lead to any immediate fatalities, caused contamination of some ten square miles of land and vegetation (see **8.9** above). This legislation has been led by the European Union.

Disasters have also largely been the main catalyst for the development of 'corporate manslaughter' law and the push for a new offence of 'corporate killing' when a company's management failures result in deaths.

Statutes Relevant to Emergency and Disaster Management

Geneva Convention No IV – Protection of Civilians in Time of War 14.2

This is one of the Geneva Conventions of 1949. It sets out how civilians are to be treated by a hostile power in time of war. It is legally binding in international law.

Geneva mandate on disaster reduction 1999 14.3

The 1990s were the International Decade for Natural Disaster Reduction (IDNDR). At the conclusion of the IDNDR forum in 1999, the participants issued this mandate detailing what they wished to achieve in the future and, as such, it is a document of intent. The thrust of the mandate is to ensure that risk management and disaster reduction become essential elements of government policies.

Seveso II Directive 14.4

Since World War II there has been a significant increase worldwide of accidents involving dangerous substances, particularly in the chemical industry. It is no coincidence that this increase has occurred as the world has become more industrialised.

The Seveso accident mentioned in **14.1** above has given its name to Directives that have been passed by the EEC which set out the basic principles of major accident hazard controls (see also **6.4** above).

In 1982 *Council Directive 82/501/EEC* on the Major Accident Hazards of Certain Industrial Activities was adopted. This has become known as the Seveso Directive. There were two amendments to the Directive (*Directive 87/216/EEC* in 1987 and *Directive 88/610/EEC* in 1988) aimed at broadening the scope of the Directive, in particular to include the storage of dangerous substances. These amendments followed the disasters at the Union Carbide factory at Bhopal, India in 1984 (where a leak of methyl isocyanate caused more than 2,500 deaths – see further **8.3** above) and at the Sandoz warehouse in Basle, Switzerland in 1986 (where fire fighting water contaminated with chemicals caused extensive pollution of the Rhine).

The Seveso Directive required a review by the Commission. As a consequence of this and certain resolutions by Member States, in 1996, the Directive was fully replaced when *Council Directive 96/82/EC* was adopted. This is known as the Seveso II Directive. This gave Member States up to two years to bring into force national laws, regulations and administrative provisions to comply with the Directive.

The original Seveso Directive was brought into UK law by the *Control of Industrial Major Accident Regulations 1984* (SI 1984 No 1902) (*CIMAH*) and Seveso II by the *Control of Major Accident Hazards Regulations 1999* (SI 1999 No 743) (*COMAH*) (these Regulations are dealt with in detail in CHAPTER 6: EMERGENCY PLANNING FOR COMAH AND NON-COMAH SITES), to replace the earlier *CIMAH Regulations*.

Aim of the Directive **14.5**

The aim of the Seveso II Directive is:

- to prevent major accident hazards involving dangerous substances; and
- if an accident does occur, to limit the consequences of this type of accident, not only to people but also to the environment;
- with a view to ensuring high levels of protection throughout the European Community in a consistent and effective manner.

The Seveso II Directive applies to establishments where dangerous substances are present. It covers both industrial activities and the storage of dangerous chemicals. It provides for three levels of proportionate controls. Essentially the more dangerous the substances, the greater the controls needed. It should be noted there are some important exclusions from the Directive, which include nuclear safety, military establishments, the transport of dangerous substances by rail, road, sea etc and the transport of dangerous substances by pipelines outside the establishments covered by the Directive.

The Directive contains general and specific obligations on operators and the Member State's authorities. The provisions fall into two categories:

- the prevention of major accidents; and
- the limitation of their consequences.

All operators covered by the Directive have to send notification to the 'competent authority' of the Member State containing information including its type of operation and details of the dangerous substances stored. Also, the operators are required to have a Major Accident Plan Prevention Policy. For operators of top-tier establishments (ie that store large quantities of dangerous substances), they need to establish a Safety Report, a Safety Management System and an Emergency Plan.

Further, there is a requirement upon Member States to ensure the aim of the Directive is taken into account in their land use policies. This objective is to be pursued through controls on the siting of new establishments, modification to existing establishments and new developments such as transport links, locations frequented by the public and residential areas that are in the vicinity of existing establishments.

Member States have to ensure that information on safety measures and on procedures in the event of a major accident is supplied to the public that will be affected if an accident occurs. Member States also have an obligation to report major accidents to the Commission.

Revision of the Directive **14.6**

In September 2003 the European Commission welcomed the agreement for the first revision of the Seveso II Directive (see also **6.27** above). The Directive is strengthened in a number of areas, including information for the public, training for emergencies and the involvement of subcontracted personnel. The new Directive also requires industrial operators to produce risk maps showing areas that might be affected by a major accident and Member States to provide the Commission with minimum data on all Seveso sites within their territory. The revision has come about following the analysis of the accidents at a firework manufacturer's premises in Enscede, in the Netherlands, and a mine in Baie Mare, in Romania (see **8.15** above).

Civil Contingencies Act **14.7**

Following a fuel crisis and severe flooding in the autumn and winter of 2000 and the outbreak of Foot and Mouth Disease in 2001, the Deputy Prime Minister announced a review of emergency planning arrangements. During that review, the responsibility for Emergency Planning passed from the Home Office to the Cabinet Office. Following the review, a Civil Contingencies Bill was introduced to the Parliament on 7 January 2004. The Bill received Royal Assent on 18 November 2004 and became known as the Civil Contingencies Act, 2004.

The Act, together with the accompanying regulations and non-legislative measures, such as the two manuals, Emergency Preparedness and Responding to Emergencies, are designed to provide a single framework for civil protection in the UK capable of meeting the challenges of the twenty-first century.

A comprehensive outline of the Act, together with associated Regulations and Guidance is set out in Annex A to this chapter. Suffice to say at this stage that the Act is divided into two parts:

1. Part 1 sets out local arrangements for civil protection, establishing a statutory framework of roles and responsibilities for local responders. Local responders are divided into two categories:

   (i) Category 1, which primarily consists of the emergency services, local government authorities, health and the Environment Agency;

   (ii) Category 2, which includes the utilities, organisations associated with transportation and the Health and Safety Executive.

2. Part 2 establishes a modern framework for the use of emergency powers that might be necessary to deal with the effects of the most serious emergencies. It replaces the Emergency Powers Act, 1920.

In relation to Part 1 of the Act there are certain exceptions to its application throughout the UK. For instance, whilst it specifies how those bodies that have a UK-wide responsibility, such as the British Transport Police, the Health and Safety Executive and the Maritime and Coastguard Agency, shall carry out their duties in Scotland, civil protection is largely devolved to Scotland. However, the Scottish Parliament has consented to Part 1 of the Act being extended to Scotland. As a result, the powers conferred on Ministers under Part 1 of the Act (power to make regulations and guidance, etc.) are, in relation to devolved matters in Scotland, exercisable by Scottish Ministers. The Scottish Ministers and UK Ministers must consult each other when exercising their legislative powers under Part 1.

In Wales, UK Ministers will make legislation and issue guidance in relation to responders in Wales. However, the Act requires the UK Ministers to obtain the consent of the Assembly before taking action in relation to a responder in Wales, which falls within devolved competence.

In Northern Ireland, different administrative arrangements at the local level make it impossible for Part 1 to apply to Northern Ireland in the same way as it applies in the rest of the UK, although it does apply to the Police Service of Northern Ireland, the Maritime and Coastguard Agency and telecommunication providers because the functions exercised by these bodies have not been devolved. The Northern Ireland Administration is currently developing the Northern Ireland Civil Contingencies Framework to ensure that those responders who perform duties that have been devolved to the Administration have similar responsibilities to those outlined in the Civil Contingencies Act, 2004.

Whilst Part 2 of the Act ensures that devolved administrations will, whenever possible, be consulted if emergency powers are to be used in their territory, the use of emergency powers remains with the UK Parliament at Westminster. The Act allows for emergency powers to be used in Scotland, Wales or Northern Ireland alone for the first time.

The Civil Contingencies Act 2004 (Contingency Planning) Regulations 2005 — 14.8

The Civil Contingencies Act 2004 (Contingency Planning) Regulations 2005 refer more specifically to the duties imposed on certain persons and bodies and the manner in which those duties are to be performed. Again these have been incorporated into the table at Annex A.

Health and Safety at Work Act 1974 — 14.9

The purpose of the *Health and Safety at Work Act 1974* (*HASAWA 1974*) is contained in *section 1* which states that it is 'securing the health, safety and welfare of persons at work', protecting the public, effective control over dangerous substances/explosives and the effective control of emission of noxious substances into the atmosphere. Implicit at least is the intention that the *HASAWA 1974* will influence actions that contribute to Disaster and Emergency situations, whether industrial, chemical or environmental.

The Act sets out general duties upon employers. The main duties are to ensure 'so far as is reasonably practicable' that employees (*section 2*) and non-employees, eg members of the public, (*section 3*) are not exposed to risks to their health and safety from the employer's undertaking (ie business).

The test for what is reasonably practicable was set out in the case of *Edwards v National Coal Board [1949] 1 AER 743*. This case established that the risk must be balanced against the 'sacrifice', whether in money, time or trouble, needed to avert or mitigate the risk. By carrying out this exercise, the employer can determine what measures are reasonable to take. This is effectively an implied requirement for a risk assessment.

HASAWA 1974, s 14 gives expressed powers to the Health and Safety Commission to investigate and to hold inquiries in relation to accidents and dangerous occurrences.

Management of Health and Safety at Work Regulations 1999 — 14.10

Regulation 3 of the *Management of Health and Safety at Work Regulation 1999* (*SI 1999 No 3242*) (*MHSW Regulations*) is the general requirement upon employers to carry out risk assessments of health and safety hazards resulting from their undertaking to which employees and non-employees are exposed. The purpose of the risk assessment is to identify precautionary and preventative measures that need to be taken.

The *MHSW Regulations* have an Approved Code of Practice (ACOP). This provides useful high-level guidance on the principles of risk assessment and what constitutes a 'suitable and sufficient' risk assessment.

A risk assessment considers the likelihood of harm occurring and the potential severity of that harm if it does occur. Thus it is important to appreciate that risk is the product of both elements. It may well be the likelihood of harm is small, but measures will need to be put in place if the potential harm is severe. This was emphasised by Mr Justice Clark in his sentencing remarks in the Health and Safety prosecution following the collapse of a walkway to a ferry at Port Ramsgate in 1994, which killed six people. He said:

> '. . . if thought had been given to its responsibilities especially having regard to the provision of the [*HASAWA*], Port Ramsgate could have appreciated that there were potential risks, albeit, perhaps very small risks . . . Further, once it was appreciated that there were potential risks, it would have been appreciated that such risks should have been guarded against because of the catastrophic consequences if anything went wrong'.

(*R v Fartygsentreprenader AB, Fartyskonstructioner AB, Port Ramsgate Ltd and Lloyd's Register of Shipping* (28 February 1997) unreported.)

It should be noted in *R v The Board of Trustees of the Science Museum [1993] ICR 876*, the Court of Appeal held that the term 'risk' denotes the possibility of danger rather than actual danger.

Regulation 4 of the *MHSW Regulations* sets out the principles of implementing preventative and precautionary measures. *Regulation 5* requires employers to have appropriate arrangements for the effective planning, organisation, monitoring and review of these measures.

Regulations 8 and *9* deal with issues more specific to Disaster and Emergency management.

Regulation 8 concerns procedures for serious and imminent danger. *Regulation 8(1)* requires employers to:

- have in place appropriate procedures to be followed if there is a serious and imminent danger to people working in his or her undertaking;
- nominate a sufficient number of 'competent persons' to carry out those procedures as they relate to evacuation;
- ensure none of his or her employees enter any areas which are restricted on health and safety grounds unless the employee has received adequate health and safety training.

In relation to these procedures, *Regulation 8(2)* says:

- workers exposed to serious and imminent danger are to be informed, so far as is reasonably practicable, of the nature of the hazard and what steps are being taken to protect them;
- these workers are to be enabled to stop work and immediately proceed to a place of safety;
- except in exceptional circumstances, the workers are to be prevented from resuming work where the serious and imminent danger still exists.

Clearly, in relation to setting up these procedures and operating them, risk assessment will play an important role. Further, they will also need to take account of any other relevant legislation (eg *COMAH Regulations*)

Regulation 9 of the *MHSW Regulations* requires employers to ensure that appropriate external contacts are in place to make sure there are effective provisions for first aid, emergency medical care and rescue work for incidents that require urgent action.

Reporting of Injuries, Diseases and Dangerous Occurrences Regulations 1995 — 14.11

The *Reporting of Injuries, Diseases and Dangerous Occurrences Regulations 1995* (*SI 1995 No 3163*) (*RIDDOR*), imposes a duty on employers to report to the enforcing authority, usually the Health & Safety Executive (HSE), certain types of accidents. These include the following:

- fatal accidents;
- major injury accidents; and
- dangerous occurrences.

All of the above must be reported immediately (normally by telephone) to the enforcing authority. This is then followed by a written report on Form F2508 within ten days.

Health and Safety (First Aid) Regulations 1981 — 14.12

The *Health and Safety* (*First Aid*) *Regulations 1981* (*SI 1981 No 917*) do not explicitly mention first aid arrangements necessary in the event of an emergency or disaster. However, the Regulations will apply in all types of accidents and major incidents. The Regulations require there to be adequate medical equipment, competent and trained first aiders or appointed persons and information on first aid facilities and equipment to staff/others.

The Construction (Health Safety and Welfare) Regulations 1996 — 14.13

Regulation 20 of the *Construction* (*Health Safety and Welfare*) *Regulations 1996* (*SI 1996 No 1592*) requires employers to make emergency arrangements to cope with foreseeable emergency situations. *Regulation 20* is primarily aimed at construction sites covered by the *Construction* (*Design and Management*) *Regulations 1994* (*SI 1994 No 3140*)(*CDM Regulations*), which require employers to have a Health and Safety Plan. The plan should factor in emergency arrangements and procedures in the event of Major Incidents on site. Note that the regulations are due to be replaced sometime in 2006.

The Confined Spaces Regulations 1997 — 14.14

Regulation 5 of the *Confined Spaces Regulations 1997* (*SI 1997 No 1713*) requires that where it is necessary for persons to work in confined spaces, employers must have suitable and sufficient arrangements for their rescue in an emergency. Note that the changes mentioned in **14.13**, may also affect *The Confined Spaces Regulations 1997*.

Carriage of Dangerous Goods by Road Regulations 1996 — 14.15

The *Carriage of Dangerous Goods by Road Regulations 1996* (*SI 1996 No 2095*) impose a duty on an operator of a 'container', 'vehicle' or 'tank' to provide information to other operators engaged/contracted regarding the handling of emergencies or accident situations. Similar Regulations cover the carriage of dangerous goods by rail (*SI 1996 No 2089*).

The Environment Act 1995 — 14.16

Although the *Environment Act 1995* does not explicitly deal with Disaster and Emergency management issues, its main objective is the mitigation or prevention of such events.

Terrorism Legislation — 14.17

There are two main pieces of legislation that gives the Government its main powers in relation to terrorism. These are the *Terrorism Act 2000* and the *Anti-terrorism, Crime and Security Act 2001*.

Terrorism Act 2000 — 14.18

The *Terrorism Act 2000* replaced the 'temporary' *Prevention of Terrorism Act 1996* that had been brought in originally in response to 'terrorism connected with the affairs of Northern Ireland'. It was passed by Parliament on 20 July 2000 and came into force on 19 February 2001.

The Act defines 'terrorism' as action or the mere threat of action that is made for advancing a cause, that attempts to influence Government and falls into one of several categories that include serious damage to property or disruption of an electronic system.

The main measures of the Act are:

(a) it outlaws certain terrorist groups and makes it illegal for them to operate in the UK;

(b) it gives the police enhanced powers to investigate terrorism;

(c) it creates criminal offences such as inciting terrorist acts, seeking or providing training for terrorist purposes and providing instruction or training in the use of firearms, explosives or chemical, biological or nuclear weapons.

The Act has sometimes been used in controversial circumstances. For example, the police used it against protesters at the Defence Systems and Equipment International arms fair in September 2003. The use of the Act in this way was upheld by the High Court on 13 October 2003.

Anti-Terrorism, Crime and Security Act 2001 14.19

This *Anti-Terrorism, Crime and Security Act 2001* was enacted in the wake of the terrorist attack in New York in September 2001. It is an extension of the *Terrorism Act 2000*. Some of the specific measures are aimed at:

- preventing terrorists from abusing immigration procedures;
- strengthening the protection and security of aviation and civil nuclear sites and the security of dangerous substances held in labs and universities; and
- cutting off terrorists from their funds by allowing assets to be frozen at the start of an investigation.

This Act also has a provision for the detention of foreign terrorist suspects. The Privy Counsellor Review Committee's report on the Act presented to Parliament in December 2003 'strongly recommend that the powers which allow foreign nationals to be detained potentially indefinitely should be replaced as a matter of urgency' with provisions that do 'not require a derogation from the European Convention on Human Rights'.

Directors' Responsibilities for Health and Safety 14.20

The approach of the Board of a company to health and safety is important to the Disaster and Emergency management of that company.

In recent years there has been guidance (which is not legally binding) as to what is expected of companies: this is contained in a specific guidance for directors on health and safety and in what has become known as the 'Turnbull' guidelines. However, as Lord Cullen wrote in his Report following the Public Inquiry into the Piper Alpha Disaster concerning the offshore gas explosion in 1988 (published November 1990) at paragraph 21.2: 'There is, of course, nothing new in the idea that safety requires to be managed'.

Guidance upon directors' responsibilities for health and safety 14.21

In 2001 the Health and Safety Commission issued guidance for board members of all types of organisations in both the public and private sectors entitled *Directors' Responsibilities for Health and Safety*. It sets out five action points:

- The Board needs to accept formally and publicly its collective role in providing health and safety leadership in its organisation.
- Each member of the Board needs to accept his/her individual role in providing health and safety leadership for their organisation.
- The Board needs to ensure that all Board decisions reflect its health and safety intentions, as articulated in the organisation's health and safety policy statement
- The Board needs to recognise its role in engaging the active participation of workers in improving health and safety.
- The Board needs to ensure that it is kept informed of, and alert to, relevant health and safety risk management issues. The guide recommends that one of the Board members be appointed as the Health and Safety Director.

The guide states that by appointing a Health and Safety Director the company will have a board member who can ensure health and safety risk management issues are properly addressed. However, some are concerned that there is a danger company directors will appoint a Board member to be a scapegoat for any serious management failures that occur.

The 'Turnbull' Guidelines 14.22

In 1999 the Institute of Chartered Accountants for England and Wales published the Turnbull Report that provides complementary guidance to the Stock Exchange's *Combined Code on Corporate Governance* (1998). Turnbull requires listed UK companies to have a system of internal control (IC) so that the Board can identify and control its exposure to significant risks to the business. Although it is aimed at stock market listed companies, its relevance is far wider.

The Turnbull guidelines require an IC to help safeguard shareholders' investments against all risks to the business, which includes health and safety risks as well as financial risks.

The health and safety risks include:

- operational risk;
- catastrophic accident;
- multiple health claims and legal risk.

Where the risks are substantial, Turnbull requires the board to assess, control and, if necessary, report on them. These risks can occur from human error (particularly of frontline workers), staff not adhering to control measures, poor decision making by management and the occurrence of unforeseeable circumstances.

The Board of a company has responsibility for the IC. When considering the risks, the Board will rely upon information and assurances provided by its senior and other management.

The Board is required to carry out an annual assessment of the IC and make a statement in its annual report and accounts. The assessment should include any incidents where there were significant failings in control measures or weaknesses as well as whether these have resulted in unforeseen outcomes.

Criminal and Civil Law Relevant to Emergency and Disaster Management

Criminal investigations where deaths have occurred 14.23

In 1998 the Association of Chief Police Officers (ACPO), Health and Safety Commission (HSC) and Crown Prosecution Service (CPS) collaborated to develop a protocol that outlined an agreed procedure for liaison between them where there is a work-related death in England and Wales. It is entitled 'Work-related deaths: a protocol for liaison'. A revised version (to which the British Transport Police and the Local Government Association are also signatories) was published in March 2003 and can be viewed on the HSE's website www.hse.gov.uk.

The main essence of the protocol is that if there is a suspicion of deliberate intent, gross negligence or recklessness then the police will lead the investigation to ascertain whether there is evidence to bring manslaughter or other serious charges. In these circumstances the HSE will provide technical assistance. If there is sufficient evidence, the offences are prosecuted by the CPS.

If the police decide there are no grounds for such suspicions or there is insufficient evidence to bring manslaughter or other charges, then the HSE will continue with the investigation to determine whether there should be Health and Safety prosecutions. If during the HSE's continued investigation, evidence is found which may support manslaughter or other serious charges, then the matter will pass back to the police for reconsideration.

The HSE will not normally take a decision on whether to prosecute until after the conclusion of the inquests into those who have died. If there are verdicts of 'unlawful killing' then the matter is automatically referred back to the police for further consideration. The HSE not only investigates Health and Safety offences but also prosecutes them.

Arguably, a direct consequence of the protocol has been the increased number of prosecutions against companies and individuals following a work-related death incident.

Health and safety prosecutions **14.24**

Health and Safety prosecutions are brought under the *HASAWA 1974*. Most prosecutions against companies are for breaches of *sections 2* and *3* of the Act. If an organisation is convicted in the Crown Court, there is no limit upon the fine that can be imposed.

In *R v F Howe and Son (Engineers) Ltd [1999] IRLR 434* the Court of Appeal provided guidelines on the sentencing of companies. It was said that any fine should reflect not only the gravity of the offence but also the means of the defendant.

In July 1999 at the Old Bailey, Great Western Trains were fined £1.5 m for a breach of *section 3* of the Act in connection with the Southall Train Crash of 1997 where seven people died. In January of the same year, in an Environmental, not Health and Safety, prosecution, Milford Haven Port Authority was fined £4 m at Cardiff Crown Court in relation to the oil spillage from the *Sea Express* that ran aground in 1996.

However, despite some large fines in high profile cases, in a press release on 5 November 2003 Timothy Walker, Director General of the HSE, complained that fines were still too low. He said:

> 'It is incomprehensible that fines for especially serious big company breaches in health and safety are only a small percentage of those fines handed down for breaches of financial services in similarly large firms. I understand that financial service breaches can affect people's wealth and well-being, but breaches in health and safety can, and do, result in loss of limbs, livelihoods and lives.'

The level of 'negligence' required for a conviction is not that high. Mr Justice Wright observed in June 1991 when sentencing British Rail for breaches of *sections 2* and *3* in connection with the Clapham Junction train crash of 1988:

> '. . . the charges that British Rail face today do not involve any connotation or allegation of recklessness. The allegation is no more than . . . a failure to maintain and observe the high standards required by the Health and Safety at Work legislation . . . '

Individual prosecutions will normally be for a breach of *HASAWA 1974, s 7*, which requires employees to take reasonable care for the health and safety of themselves and others who may be affected by their conduct. This section is somewhat of a grey area of law since few cases have been reported in the law reports.

However, there are basically two types of prosecutions:

- those of employees that can be described as 'front line' workers; and
- those of employees that are at management level (normally lower or middle management).

For a 'front line' employee to be prosecuted his/her actions normally have to be bordering on the reckless. For a manager to be prosecuted the negligence does not have to be so bad. A manager can be prosecuted if he or she has failed to carry out his/her job (or has violated procedures) so that health and safety standards are as a consequence significantly lowered. People convicted of this offence are usually required to pay a modest fine.

Directors, managers or senior officers of an organisation can be prosecuted as individuals pursuant to *HASAWA 1974, s 37* if an organisation's breach is alleged to have been committed with their consent or connivance or have been attributed to any neglect on their part. In the past such prosecutions have been rare and if a conviction achieved, the fine has amounted to a few thousand pounds. However, since the publication in January 2002 of the HSC's new Enforcement Policy Statement (which can be viewed on the HSE's website www.has.gov.uk), these types of prosecutions have increased. Further, an application to disqualify a convicted director under the *Company Directors Disqualification Act 1986* is now more commonly sought by the prosecution.

In recent times the HSE has appeared more willing to prosecute than before. Not all agree with this approach. On 28 November 2003 the *Guardian* newspaper reported on the resignation of the former director of railways at the HSE, Alan Osborne. In his resignation statement the paper reports him as saying that the HSE was failing to get its point across in enforcing safety standards. He referred to the failed prosecution of the Metropolitan police in 2003 for lax safety culture after the death of an officer who fell through a fragile roof. Of this prosecution Mr Osborne is quoted as having written:

> 'The prosecution of the Metropolitan police demonstrated just how out of touch the [HSE] has become'.

It is important to note that in any disaster causing deaths, it is not necessary for a successful prosecution to show the conduct of an organisation or individual that caused those deaths, only the that conduct resulted in the creation of a risk that harm might be caused.

For example, BP was fined a total of £1 million at Falkirk Sheriff Court, Scotland, on 18 January 2002 following two incidents at the company's Grange-mouth refinery in June 2000, one where a catalytic converter caught fire and the other where there was a steam rupture. The company pleaded guilty to breaches of *HASAWA 1974, s 2* and *s 3*. It was noted at court that only good fortune avoided fatalities and serious injuries.

At present the courts only have power to impose custodial sentences in a few health and safety offences. The Government's and HSC's strategy document *Revitalising Heath and Safety* (June 2000) recommended higher fines and making imprisonment more widely available for health and safety breaches.

Manslaughter **14.25**

The form of manslaughter charge against an individual in relation to a disaster causing deaths is likely to be based on *gross negligence*. The leading case on this area of law is *R v Adamako [1995] 1 AC 171.*

The case concerned an anaesthetist in charge during an eye operation. He failed to notice that an endotracheal tube (to allow the patient to breathe normally) had become disconnected for a period of about six minutes. As a consequence the patient died. The defence conceded at the trial that the doctor had been negligent, but denied this negligence was so bad that it should be deemed a criminal offence.

Adamako establishes that to convict someone of gross negligence manslaughter the jury must be satisfied that:

- the defendant owed a duty of care to the deceased; and
- he/she was in breach of that duty; and
- the breach of duty was a *substantial* cause of death (ie it has to be more than trivial, it has to be a significant); and
- the breach was so grossly negligent that the defendant can be deemed to have had such disregard for the life of the deceased that it should be seen as criminal and deserving of punishment by the State.

In many ways, this is a civil case in a criminal arena. It has to be proved that the defendant's negligence was one of the causes of the death as in a civil case. The differences are:

- the standard of proof, ie beyond reasonable doubt in a criminal case (as opposed to on the balance of probabilities in a civil case);
- the negligence must be a *substantial* cause and not just a cause;
- the negligence must be *gross*.

A person convicted of gross negligence manslaughter is likely to receive a custodial sentence of around 18 months to two years.

Corporate manslaughter **14.26**

In 1987 the *Herald of Free Enterprise* ferry capsized at Zebrugge killing 192 people. The immediate cause was the failure to close the bow doors of the ferry before leaving port. However, there was no effective reporting system to check the bow doors.

The prosecution of P & O Ferries, the operator of the ferry, confirmed it was possible to charge a company with corporate manslaughter. In this case, not only was the company charged, but also seven directors were charged with manslaughter (see *R v P & O European Ferries (Dover) Ltd [1991] 93 Cr App R72*).

However, the trial Judge threw out the case. He did not find that any of the directors were culpable. He ruled the prosecution had failed to satisfy the 'doctrine of identification'. This is the principle of showing that the company's conduct, which led to the deaths, is the conduct of a senior person in the company. In other words, for a company to be guilty of corporate manslaughter, a senior person, such as a director, also has to be guilty of manslaughter.

A similar result arose from the case brought against five executives working for Network Rail, formerly known as Railtrack, and its contractor, Balfour Beatty, following the Hatfield train crash in October 2000 that resulted in the death of four people and caused injuries to 102. In this case, engineers from Balfour Beatty had identified a faulty rail on the section of track, where the train came off the rails, 21 months prior to the accident. A replacement rail had been delivered but it was left alongside the damaged rail for 6 months. No speed restrictions were imposed in the area of the faulty rail; consequently when the accident happened the train was travelling at 115 mph. As he dismissed a charge of corporate manslaughter against the five executives, Mr Justice MacKay made it clear that the law, as it stood, left him no alternative, concluding his remarks with the comment that 'This case continues to underline a long and pressing need for the long-delayed reform of the law in this area of unlawful killing. There are thankfully signs this reform is now in sight.'

Following the decision in the Hatfield case, the Crown Prosecution Service decided not to bring manslaughter charges against Network Rail and its contractors, on this occasion Jarvis, following another train accident, this time at Potters Bar in May 2002 in which 7 people died and 70 were injured, because 'there was no realistic prospect of securing a conviction for manslaughter against either an individual or a corporation' (*Daily Telegraph, 18 October 2005, p. 2*). On this occasion, a set of points outside the station broke when an express train sped over them.

As a result, the whole debate on corporate manslaughter was renewed (see **14.31**)

The 'directing mind' **14.27**

Great Western Trains (GWT) were charged with corporate manslaughter following the Southall train crash. The crash occurred when one of GWT's train drivers passed a red signal and collided with a goods train. A safety system on the train, known as the automatic warning system (AWS) designed to stop drivers missing red signals was not working. No one on the Board of Directors of GWT was charged with manslaughter. The trial Judge dismissed the charge of corporate manslaughter because the prosecution had not identified the 'directing mind'. He said:

> 'In my judgement it is still necessary to look for... a directing mind and identify where gross negligence is that fixes the company with criminal responsibility... Accordingly I conclude that the doctrine of identification

> which is both clear, certain and established is the relevant doctrine by which a corporation may be fixed for manslaughter by gross negligence . . . '.

The Court of Appeal agreed that this was the correct position – see the Attorney General's appeal to the Court of Appeal on 15 February 2000 – *Attorney General's Reference No 2/1999.*

A person identified as the directing mind is likely to be prosecuted on the basis of gross negligence manslaughter. He/she will be sentenced like any other individual found guilty of this type of manslaughter. In relation to the company there is no limit as to the amount of fine that can be imposed.

There have been only a few convictions for corporate manslaughter, all against small companies where it has been easier to substantiate the directing mind. An example is *R v OLL Ltd and Kite* (8 February 1996, unreported). This case concerned a canoeing accident in 1993 in which four students died. A teacher, a group of eight students and two instructors from OLL Ltd tried to canoe across the sea from Lyme Regis to Charmouth, a distance of about one and a half miles. The teacher got into difficulties and one of the instructors went over to assist. The other instructor proceeded with the students. The group drifted out to sea where the students drowned.

Mr Kite was one of two directors of the company. He had the primary responsibility for devising, instituting, enforcing and maintaining an appropriate safety policy. The jury found he failed in his responsibility and this was a substantial cause of the deaths.

The company was fined £60,000. Mr Kite was sentenced to three years imprisonment, but this was reduced to two years on appeal.

Dealing with risk **14.28**

The issue of how a company and its Board deal with risk is likely to become more relevant to these cases. An illustration of this is the prosecution of Euromin and its director Mr Martell following the death of Simon Jones in April 1998.

Simon Jones (aged 24) died on his first day at work at Shoreham Docks. While unloading cargo from a ship the jaws of a grab closed around his neck. Ultimately, the case was unsuccessful with the company and its director being acquitted of manslaughter at the Old Bailey in November 2001. It should be noted that the manslaughter investigation did not start until six weeks after the incident, which as a consequence, meant there were evidential problems.

However, the case is significant for the judicial review by the deceased's family of the original decision by the CPS not to prosecute (*R v DPP ex p Jones [2000] IRLR 373*). At the judicial review the Director of Public Prosecutions argued that there was no evidence that the managing director had been warned of the dangers of the system of work being operated. The Judge said:

> 'The point can be put in this way. A clear and potential . . . major criticism [of the managing director and the company] is that they set up the unsafe

> system . . . it was unsafe because, *inter alia*, it arguably presented a danger of death – the danger that in fact had eventuated. Such a system requires detailed precautions to be taken to ensure the incipient dangers did not in fact eventuate'.

Another way of looking at this is to say the company should have carried out a risk assessment and that failure to do so was a cause of the death. It would be for the jury to consider whether that was a substantial cause and if the failure amounted to gross negligence.

Causation **14.29**

The difficulty with proceeding against large organisations is the problem of linking any failures by Board members or senior management with the incident. Often the initial cause of an incident is an error by a frontline worker. According to the HSE, 80% of accidents involve some form of human error. If an error has been made by a frontline worker, then the argument advanced is that this breaks the 'chain of causation'.

Important to the issue of causation is the way incidents are analysed. According to HSG 65, *Successful Health and Safety Management* (p 9), accidents, ill health and incidents seldom are random events and generally arise from failures of control, involving multiple contributory elements. It goes on to say the immediate cause may be human or technical failure, but they usually arise from organisational failings that are the responsibility of management. See also **CHAPTER 11: HUMAN ERROR AND HUMAN FACTORS**.

Reducing error and influencing behaviour (HSG 48), the HSE's *Guide to Human Error*, states (p 4):

> 'Many accidents are blamed on the actions or omissions of an individual who was directly involved in operational or maintenance work. This typical but short-sighted response ignores the fundamental failures which led to the accident. These are typically rooted deeper in the organisation's design, management and decision-making functions'.

The question is whether in these cases causation can be established from a legal point of view. It is in this area that matters are perhaps developing.

In the environmental case *Empress Car Co (Abertillery) Ltd v National Rivers Authority [1998] 1 All ER 481* there was an argument as to whether the act of an unknown person broke the chain of causation. The company was prosecuted for polluting a river contrary to the *Water Resources Act 1991, s 85(1)*. The House of Lords said what had to be determined was whether the event, the act of the unknown person, was ordinary (if so the company was guilty and the chain of causation was not broken) or extraordinary (if so the company was not guilty).

In 2000, Lord Cullen chaired the public inquiry into the causes of the Ladbroke train crash of October 1999 that killed 29 passengers and the drivers of both trains involved in the collision. The driver of the Thames Turbo train, Michael Hodder, passed signal SN109 at red. This signal had been passed on eight previous occasions in just over five years and each time the incident was recorded as 'driver error'. Three days after the crash, Her Majesty's Railway Inspectorate (part of the HSE) served a prohibition notice on Railtrack, the then operators of the railway infrastructure.

Prior to the start of the Inquiry the police announced there would be no manslaughter prosecutions. Lord Cullen, in his Report (Part 1 published in June 2001), declined to blame Mr Hodder for the crash. He concluded Mr Hodder passed the signal because of poor 'sighting' of the signal, both in itself and with other signals on the gantry on which it was situated.

After publication of the Report, lawyers for the bereaved and injured campaigned for a reconsideration of corporate manslaughter charges. Following advice concerning in part causation from a leading academic in this field of law, in May 2002 it was announced that the police investigation would be re-opened. Thames trains were fined £2 million and ordered to pay £75,000 costs for breaching health and safety regulations.

In 2003 corporate manslaughter charges were commenced against Network Rail (the successor to Railtrack) and Balfour Beatty, with senior managers being charged as 'directing minds', in relation to the Hatfield train derailment in which four people died. No frontline workers have been charged. The initial cause of the derailment was a broken rail. Following a seven month trial, five rail executives and Balfour Beatty were cleared of manslaughter charges but the company admitted breaches of the safety standards.

On 10 February 2004 Barrow Borough Council were summonsed for corporate manslaughter relating to the deaths of seven people from the legionnaire's disease outbreak in August 2002. A design services manager employed by the council is also being prosecuted for manslaughter as the directing mind of her employers. This is the first time a local authority has been prosecuted for corporate manslaughter. Barrow Council were cleared of manslaughter charges and the jury could not agree a verdict in the case of the design services manager.

The foregoing cases highlight the difficulty of securing convictions for corporate manslaughter under current legislation.

However, because of the difficulties of bringing corporate manslaughter charges there have been calls for a change in the law. This is discussed at **14.31** below.

Civil proceedings **14.30**

A civil claim is a legal action by one party, the claimant, against another party or parties, the defendant(s), to settle a 'dispute'.

Claims for compensation can arise out of negligence, ie where one party owes a duty of care to another that is breached and, as a consequence, the other party suffers injury or loss. A company cannot escape liability if the accident is due to the negligence of one of its employees as in legal terms it is 'vicariously liable' for that employee's acts and omissions.

However, even if a disaster is caused by a natural event, this does not automatically mean there is no legal liability. Following severe flooding in Cardiff in 1979, property owners successfully sued Cardiff City Council and South Glamorgan County Council for failing to provide adequate flood warnings. The case raised serious issues regarding statutory responsibilities for providing public warnings. Public agencies were advised that if they had no clear responsibility they were at risk of litigation if they used powers to assume a responsibility.

In a recent development it is possible that an enforcing authority cannot escape the possibility of some liability if there is a major accident. Thames Trains have sued the HSE for a contribution for any compensation it may have to pay as a result of the Ladbroke Grove train crash. The HSE issued an application to strike out the proceedings on the basis that it had no reasonable ground of succeeding. The application was dismissed and an appeal failed to overturn the decision. It was held that whilst the HSE owed no general duties arising out of the *HASAWA 1974* or by reason of the fact that the HSE was the safety regulatory body for the railways, it was arguable that it owed a 'common law' duty to the victims of the crash in respect of the particular facts of the incident.

In the Hillsborough Disaster in 1989, 96 people were crushed to death at a football match. In the enquiry into the disaster, Lord Justice Taylor concluded that the cause of the disaster was over crowding, but that the reason for it was the failure of police control.

The Hillsborough Disaster resulted in an important personal injury case that laid down guidelines when it was possible to recover compensation for Post Traumatic Stress Disorder (PTSD), if the claimant had not suffered physical injury (*Alcock v Chief Constable of South Yorkshire Police [1992] 1 AC 310*). Essentially, the claimant generally needs to witness a single horrific event and either, be present at the incident or see its immediate aftermath.

There may also be claims between different companies for compensation for damage to plant, machinery and materials etc, along with loss of revenue, which flow from the disaster. These disputes may involve contractual issues as well as issues of negligence. The claims may be the subject of court proceedings in the normal way or may be settled through the Alternative Dispute Resolution (ADR) process.

The Future: Proposals for Corporate Killing 14.31

In 1996, the Law Commission recommended the reform of the law by proposing the creation of a special offence of Corporate Killing. The Government pledged to do so in the Labour Party manifesto for the 1997 election and, in May 2002, the Home Office issued a consultation document setting out proposals for reforming the law on involuntary manslaughter. However, it was not until March 2005 that the Minister for the Home Office presented a Draft Bill for Reform of the Law relating to Corporate Manslaughter to Parliament. The Draft Bill provided for a further period of consultation, but at the time of submitting this revised chapter for publication, the Home Office has yet to move to the next stage.

Corporate manslaughter – the Draft Bill **14.32**

The Draft Bill laid before the Parliament in March 2005 set out a new, specific offence of corporate manslaughter. An organisation would be prosecuted for this if a gross failing by its senior managers to take reasonable care for the safety of its workers or members of the public caused a person's death. The new offence would apply to all companies and other types of incorporate bodies such as Government departments and other crown bodies and many in the public sector, including local authorities.

It describes a senior manager of an organisation as a person who plays a significant role in:

- the making of decisions about how the whole or a substantial part of its activities are to be managed or organised, or
- the actual managing or organising of the whole or a substantial part of those activities.

It is intended that the new offence would be based on failures in the way an organisation's activities are managed or organised, referred to as 'management failure'. The offence is designed to capture corporate failings in the management of risk and therefore applies to an organisation's senior managers, either individually or collectively, and not on failings that are purely local. It would therefore focus on the arrangements and practices for carrying out an organisation's work, rather than any immediate act by an employee (or potentially someone else) causing death.

As it currently stands under the Draft Bill, there are four elements to the new offence:

1. The organisation must have owed a duty of care to the victim that is connected with certain things done by the organisation.
2. The organisation must be in breach of that duty of care in the way its senior managers manage or organise a particular aspect of its activities. This introduces an element of 'senior management failure' into the offence.
3. This management failure must have caused the victim's death. The usual principles of causation in the criminal law will apply to determine this question.
4. The breach of duty must have been gross. In arriving at a decision as to whether there has been a gross breach of duty, the jury must take into consideration whether the evidence shows that the organisation failed to comply with any relevant health and safety legislation or guidance, and if so:
   - how serious was the failure to comply;
   - whether or not senior managers of the organisation:
     - knew, or ought to have known, that the organisation was failing to comply with that legislation or guidance;
     - were aware, or ought to have been aware, of the risk of death or serious harm posed by the failure to comply;
     - sought to cause the organisation to profit from that failure.

For a breach of duty to be gross there must be conduct that falls far below what could reasonably have been expected, which will inevitably mean that the breach will have involved a risk of death or serious injury.

The Draft Bill only applies to England and Wales. However, it does apply to ships registered under Part 2 of the Merchant Shipping Act 1995, to British-controlled aircraft, to British-controlled hovercraft and to offshore activities to which an Order in Council under section 10(1) of the Petroleum Act 1998 applies.

Proceedings can only be brought by the Director of Public Prosecutions and the penalty on conviction is an unlimited fine.

Corporate manslaughter – the debate 14.33

Currently, a corporate body can only be found guilty of the common law offence of gross negligence manslaughter if a 'directing mind' of that body is also guilty of gross negligence manslaughter (see **14.26–14.27**). Whilst building on aspects of this, the proposed new offence makes corporations and a range of government departments liable for the way in which the organisation's activities are run by its senior managers, rather than making liability contingent on the guilt of a particular individual.

On the face of it, anything that improves the accountability of senior managers for the safety of both employees and the public alike is to be welcomed. As such there are some positive points to be taken from the introduction of the Draft Bill. For instance:

- The Government has finally commenced the process of honouring the commitment made in 1997.
- The bill, if passed in its present form, will, for the first time, allow a charge of corporate manslaughter to be brought against senior managers.
- The court will have powers to order a convicted organisation to take remedial action.
- Crown immunity will be abolished for many government departments and agencies.

However, a variety of organisations, from firms of solicitors, such as Thompsons, to the Transport and General Workers Union, have criticised the Government's approach, claiming that the Draft Bill provides the bare minimum to justify calling it corporate manslaughter. For instance:

- It focuses only on corporate liability and completely fails to address individual liability.
- The definition of 'senior managers' is confusing and restrictive. As a result many companies may unjustifiably escape prosecution. For instance:
  - There may well be managers who have an important role in carrying out certain activities who are not categorised as senior managers, e.g. in the

absence of legally binding legislation as to where responsibility for health and safety issues lie, responsibility for health and safety could be given to a person who is not regarded as a senior manager.

– Also, in larger companies where there is often a complex management structure or on construction sites where there may be a number of companies operating, it may not be any easier to bring a charge of corporate manslaughter than it has in the past.

- Claiming that the level of fines under existing legislation is grossly inadequate, and therefore has little deterrent effect, the new Draft Bill fails to give courts the power to impose any greater penalties than can be imposed for existing offences. For example, organisations can already been fined an unlimited amount under the Health and Safety at Work Act 1974. In this respect, it is recommended that the courts should be able to link fines to the profitability of the company; the option of imprisonment should be retained; there should be an option for the court to disqualify directors, particularly if they have a poor safety record; in certain cases, probation orders should be available; and the court should be able to order a company to pay punitive levels of compensation to the families of deceased persons.
- Its relevance to companies from England and Wales operating overseas is limited, in the main, to ships, aircraft, hovercraft and offshore oil installations. It does not appear to include other forms of transport, such as coaches, tour operators, industrial premises and the like.

The need for reform **14.34**

Since the mid-1980s a number of disasters have evoked demands by the public for the use of the law of manslaughter because it was felt that senior managers of relevant companies had a responsibility for the failures that resulted in those disasters.

In the original consultation document in relation to corporate killing, the four disasters specifically mentioned in this category were:

1. The Clapham train crash (1988)
2. The capsizing of the Herald of Free Enterprise (1987)
3. The King's Cross Station fire (1987)
4. The Southall train crash (1997).

To this could be added a host of others. Amongst the most recent are:

- The Ladbroke Grove train crash (1999)
- The Hatfield train crash (2000)
- The Potters Bar train crash (2002).

In some of the above cases, prosecutions were brought against directors for manslaughter but, in each case, they failed because of weakness in the current legislation. In others, because of these failures, prosecutions were not even commenced because it was felt there was little chance of success. This has led to a perception by the public that the law in relation to corporate responsibility does not work. So clearly there is a need for reform on this count alone.

But there is arguably another reason for reform in addition to the apparent failure, in law, of the existing legislation. Many would argue that the main objective of any new legislation must be to achieve some degree of change in behaviour, which will result in a reduction in the number of deaths and serious injuries to employees during the normal course of their work and also to people who are being provided with some kind of service.

Many people doubt that the new Corporate Manslaughter Act, if it becomes law in its present form, will assist in changing the negative regard the public has for the law relating to corporate responsibility. Neither do they believe that it is likely to assist in a change of behaviour towards providing a safer environment.

The Government, on the other hand, has expressed its commitment to provide safer and more secure communities, at home and in the workplace, and has expressed its determination that the criminal justice system should command the confidence of the public. It remains to be seen who will be right.

Investigations 14.35

After a disaster or major accident has occurred, an investigation of some kind will inevitably follow. The types of investigation and the various laws that relate to them are detailed in **Chapter 16.**

Conclusions 14.36

With the changing risks and threats to society, whether from, for example changing climate conditions or from terrorist attacks, there is a need for new legislation to deal with modern day emergency planning.

In relation to Disaster management, the legislation is developing to require organisations to have more sophisticated risk management and control systems to prevent disasters and/or mitigate their effect. The measures required by the Seveso II Directive do not have the qualification 'so far as is reasonably practicable'. Unlike other UK legislation enacting EC Directives, neither does the measures required by the *COMAH Regulations* (although the ACOP to the regulations does mention this qualification). Perhaps, therefore, on a strict interpretation of the regulations a higher duty is required of all 'practicable' measures to be taken.

In terms of civil law, the boundary of an organisation's duty of care is continually broadening. This has implications for the insurance market. Insurers will no doubt make the level of premiums more and more dependant upon the organisation's risk management and control systems.

The issue of corporate governance is, at present, a highly topical and emotive one. It relates not only to financial matters (for example, with the Enron and WorldCom scandals) but also to health and safety ones. There is a public wish in the UK and elsewhere for organisations and those that run them to be held criminally culpable for major accidents when due to poor management. Although the corporate killing proposals are still yet to be turned into legislation, the police and CPS have shown they are still prepared to prosecute under the current law. However, whatever happens with the law of corporate manslaughter/killing, given the new enforcement policy of the HSE, more health and safety prosecutions of employers and directors/senior management with higher fines seem likely. From a prosecuting point of view, health and safety offences are easier to prove, as causation with the incident does not have to be established, only that a risk of harm occurred.

When a major accident occurs there will be a number of legal issues that arise. It is important to ensure there is communication between the various representatives dealing with them, otherwise there could be detrimental consequences. For example, an admission of negligence in a civil action might make it difficult to argue negligence issues in a related corporate manslaughter prosecution based on gross negligence.

Finally, an organisation should make sure that there are systems in place to deal appropriately and kindly with those injured and the relatives of those killed as a result of a disaster. A jury will give much weight to these sorts of matters. At the Public Inquiry into the Southall train crash a great deal of criticism was made by the lawyers representing the injured and bereaved of how their clients were treated following the incident by the companies involved. To act decently in such circumstances does not amount to an admission of guilt or civil liability but does provide good evidence of the attitude of the organisation.

Annex A

The Civil Contingencies Act 2004, associated Regulations and Guidance

Part or Section of the Act	The Civil Contingencies Act 2004	The Civil Contingencies Act (Contingency Planning) Regulations 2005	Statutory and Non-Statutory Guidance
	Came into force on the 14th November 2005 but the duty on local authorities to provide business continuity advise came into force on 15th May 2006. The Civil Contingencies Act 2004 was introduced in order to: • provide a holistic piece of legislation with a coherent, singular and unified approach to emergency response in the UK; • provide enabling legislation which can allow the Secretary of State or a Minister of the Crown to create regulations and emergency regulations based upon an assessment of 'urgency';	The Regulations came into force on the 14th November 2005. The Regulations address the 'how' issues whilst the Act is concerned with the 'what', 'who', 'when' and 'where'. The Regulation is the 'child' of the Act and exists because of the enabling powers of the Act. The Regulation has the following structure: Part 1: Introductory Part 2: General Part 3: Duty to assess risk of emergency occurring Part 4: Duty to maintain plans Part 5: Publication of plans and assessments	There are two documents issued by the Cabinet Office: 1. Emergency Preparedness: Guidance on Part 1 of the Civil Contingencies Act, its associated Regulations and non-statutory arrangements (Nov 2005) 2. Emergency Response and Recovery: Non-statutory Guidance to compliment Emergency Preparedness (Nov 2005). Both documents provide a 'road map', mirror sections of Part 1 of the Act and much of the Regulation but 'Emergency Response and Recovery' also covers Part 2 of the Act unlike the 'Emergency Preparedness' guidance. This is not legislation but 'Emergency Preparedness' is an 'approved code' type document having 'quasi-legal' effect, whilst 'Emergency Response and Recovery' is a 'best practice' type document. Both documents could be viewed as 'standard setting'.

	• reflect upon the lessons learnt from the UK experience in relation to Exotic Animal Diseases, e.g. Foot and Mouth as well as other serious threats such as terrorism, flooding and other natural hazards; • be both proactive law as well as providing advise on effective reaction to emergencies. The Act has the following structure: • Part 1: Local Arrangements for emergency response within counties of the UK as well as duties, functions and responsibilities of stakeholders (s1–18); • Part 2: Emergency Powers in relation to the Regions of England and Parts (Scotland, Northern Ireland and Wales) of the UK – creation, use, role and scope of emergency regulations (s19–31);	Part 6: Arrangements for warning and provisions of information and advise to the public Part 7: Advise and assistance to business and voluntary organisations Part 8: Information Part 9: London Part 10: Northern Ireland Schedule. The Regulations relate mostly on Part 1 of the Act. This is Subordinate Legislation.	'*Emergency Preparedness*' has the following structure: Chapter 1 Introduction 3 Chapter 2 Co-operation 10 Chapter 3 Information sharing 24 Chapter 4 Local responder risk assessment duty 34 Chapter 5 Emergency planning 47 Chapter 6 Business continuity management 74 Chapter 7 Communicating with the public 93 Chapter 8 Advice and assistance to business and voluntary organisations 109 Chapter 9 London 128 Chapter 10 Scotland 132 Chapter 11 Wales 136 Chapter 12 Northern Ireland 141 Chapter 13 Monitoring and enforcement 146 Chapter 14 The role of the voluntary sector 154 Chapter 15 Sectors not covered by the Act 160 Chapter 16 The role of the Minister 163 Chapter 17 Co-operation at the regional level in England 167 Chapter 18 Planning at the regional level in England 174 Annexes 178 Annex 2A Model terms of reference for the Local Resilience Forum 178 Annex 3A Formal procedures for requesting information 179 Annex 3B Information request proformas 181 Annex 4A Summary of the six-step local risk assessment process 183 Annex 4B Illustration of a Local Risk Assessment Guidance (LRAG) 186

	• Part 3: General Provisions – the 'birth certificate' part of the Act; commencement, title, scope, etc. (s32–36); • Schedule 1: Has 4 Parts relating to Category 1 and 2 Responders; • Schedule 2: Amendments and Repeals to Part 1 and 2; • Schedule 3: Repeals and Revocations. Duties under the Act can be Strict ('shall') or So Far As is Reasonably Practicable (SFRP) which involves weighing up risk/time issues against cost considerations of reducing the risk and achieving the goal in a time efficient way. This is Primary Legislation.		Annex 4C Example of an individual risk assessment 193 Annex 4D Likelihood and impact scoring scales 195 Annex 4E Community Risk Register 198 Annex 4F Risk rating matrix 199 Annex 5A Examples of generic and specific plans 201 Annex 5B Generic plan: emergency or major incident 203 Annex 5C Specific plan 204 Annex 5D Example of a plan maintenance matrix for a local authority 205 Annex 7A Communicating with the public: the national context 206 Annex 7B Lead responsibility for warning and informing the public 208 Annex 7C Checklist of suggested protocols 210 Annex 14A Examples of voluntary sector activities in support of statutory services 211 Annex 17A Model terms of reference for a Regional Resilience Forum 214 Glossary 215 Bibliography 228 It is Chapters 2 through to 8 that are the core focus of the guidance as they reflect the 7 core civil protection duties of responders: 1. Co-operation (all Responders) 2. Information sharing (all Responders) 3. Local responder risk assessment duty (Category 1 Responders)

			4. Emergency planning (Category 1 Responders) 5. Business Continuity Management (Category 1 Responder) 6. Communicating with the public (Category 1 Responder) 7. Advice and assistance to business and voluntary organisations (local authorities). Co-operation, Information Sharing and Risk Assessment are the three critical duties and functions which 'drive' Emergency Planning, BCM, communications with the public and advise to businesses. *'Emergency Response and Recovery'* has the following structure: Contents Chapter 1 Introduction 3 Chapter 2 Principles of effective response and recovery 6 Chapter 3 Responding agencies 11 Chapter 4 Management and co-ordination of local operations 20 Chapter 5 Care and treatment of people 34 Chapter 6 Information and the media 43 Chapter 7 The Government Offices for the English regions 51 Chapter 8 Regional Civil Contingencies Committees in England 54 Chapter 9 Response arrangements in Scotland 59

			  This document follows 8 key principles: 1. *continuity* of activities and operations; 2. *preparedness* in terms of roles and responsibilities; 3. *subsidiarity* – local decision making and high level co-ordination; 4. *direction* in terms of strategic aim, prioritisation and teamwork; 5. *integration* at local, regional and national levels to ensure a unified level of response; 6. *co-operation* between all responders, agencies and others with regards to information sharing;

			7. *communication* – efficient, accurate, authoritative and two way; 8. *anticipation* through dynamic Risk Assessment.
Part 1: Local Arrange-ments for Civil Protection			
Section 1: Meaning of Emergency	The Act says an 'Emergency' refers to: (a) an '*event or situation*' (which is then categorised by the section); (b) that which '*threatens*'; (c) '*serious damage*' (the section does not define the scope, extend and magnitude of damage); (d) to human welfare in the UK, the environment in the UK and the security of the UK (through war or terrorism).	The Regulations affirm the definition in Section 1 of the Act.	A. 'Emergency Preparedness', Paragraph 1.15, Page 5 of Chapter 1 says '*The definition of "emergency" is concerned with consequences, rather than with cause or source. Therefore, an emergency inside or outside the UK is covered by the definition, provided it has consequences inside the UK.*' The same document adds, '*The Act, the Regulations and this guidance consistently use the term emergency, but there is nothing in the legislation that prevents a responder from continuing to use the term "major incident" in its planning arrangements for response*' (Chapter 5 'Emergency Planning' page 59, op cit). Neither the Act nor the Regulations define how an 'Emergency' differs from a 'Disaster'. There is also some variance between the term 'Emergency' and the terminology used by the ILO Conventions on Major Accidents. In addition, the term 'Major Incident' is not mentioned in the Act nor the Regulations; which hitherto has been the preferred term of the emergency services. (See Chapter 5 of this book on how these terms differ).

Section 2: Duty to Assess, Plan and Advise	1. Subsection (2)(1): • s2(1)(a) and (b) are both Strict and applies to Responders defined by this Part of the Act. These subsections are concerned with assessing the risk of an emergency occurring and the preparedness of response. • s2(1)(c) is a SFRP duty and concerned with the 'business continuity' of Responders – can they continue to function when an emergency occurs? • s2(1)(d) is Strict and in effect is asking 'have we got a plan to cope with the emergency and to bring it under control?' • s2(1)(e) is Strict and broadly is asking Responders to adopt a 'dynamic' approach to Emergency situations – continue to assess and to ensure effective response and to be able to modify plans if required.	Much of Part 2–8 makes reference to s2(1) of the Act (Part 1 is the 'Introductory' part focusing on citation, interpretation of terms, etc.). Part 2: General 1. Reg 4(1) and 4(2) establishes a Strict duty on Category 1 Responders for a defined local area in England and Wales to co-operate with each other in order to perform their duties under s2(1) of the Act. Reg 4(3) refers to 'Local Resilience Forums' (LRF) as a mechanism for such co-operation. Reg 4(4) requires SFRP that such Category 1 Responders meet at least once every six months. Reg 4(5) requires Category 2 Responders to co-operate with Category 1 Responders for that local area and furthermore shall SFRP attend the LRF meetings or send a representative. There is a duty on Category 1 Responders to keep Category 2 Responders informed of LRF meetings.	B. 'Emergency Preparedness' guidance document provides information on the form and approach to establishing co-operation (Chapter 2) as required by S2(1) and Regs 4,8, 9, 10 and 11. A key focus is on the *modus operandi* of the LRFs. It is advised that the: • Chief Constable for that locality is the Chair (Category 1) • Senior Emergency Planning Officer as Secretary (Category 1) • Other Category 1 officers, e.g. county and district councils, chief fire and chief ambulance officer, etc. • Attendance also by Category 2 Responders, e.g. utilities, HSE, etc. • Others who may not be Category 1 and 2 Responders, e.g. Red Cross. A key requirement of the LRF will be 'Leadership', effective understanding of local risks of emergencies and rehearsal of each members roles and responsibilities as required under the Act and the Regulations.

	• s2(1)(f) is Strict and focuses on publishing any assessments and plans if it assists proactively and reactively in improving emergency response. • s2(1)(g) is Strict and is concerned with information to the public relating to the emergency. 2. Invariably, during an Emergency, situations can get 'serious' and Responders need to revisit the assessment, plan and resources to cope. This is the theme of subsections 2(2)(a) and 2(2)(b) – both are Strict. 3. s2(3) highlight the enabling powers of the Act (Strict) and the power to redefine duties of Responders. Likewise s2(4) has the same affect as s2(3) but relates to Scotland. 4. s2(5) are Strict and relate to the specifics of the enabling power under s2(3) and so relates to the Minister of the Crown (not Scottish) with regards to existing, additional, modified or extended duties of Responders. Although enabling regulations made under modified	2. Reg 5 is similar to Reg 4 but applies only to Scotland. In addition, the co-operating medium is referred to as the 'Strategic Coordinating Group' (SCG) and not the LRF. 3. Reg 6 extends the same logic of Reg 4 and 5 to Northern Ireland in broad terms. 4. Reg 7 focuses on protocols that can be entered into between Category 1 and 2 Responders as well as with Scottish Category 1 and 2 Responders. This could be of relevance to Responders on border areas on either of say England and Scotland. 5. Reg 8 is a 'division of labour' regulation and allows a Category 1 Responder to perform their duties under s2(1) of the Act with another Responder or to ask another Responder to perform such a duty of their behalf. The Regulations do not preclude the Category 1 Responder from jointly performing or delegating with say a Category 2 Responder.	C. The non-statutory guidance 'Emergency Response and Recovery', Chapter 3, identifies the roles, responsibilities and functions of those agencies that maybe involved in the emergency response. This includes the Category 1 and 2 Responders as well as others, e.g. HM Coroner, Armed Services, etc. D. Chapter 4 of the same non-statutory guidance focuses on 'Management and coordination of local operations'. It highlights: • The role and responsibilities of the three levels of command: Gold (Strategic, oversight, impacts and legislative consequences under CCA and Regulations); Silver (Tactical, to coordinate, liaison, line management, resource management); and Bronze (Operational, 'hands on', immediate actions). • Where multi-agency response is needed to an emergency then a Strategic Coordinating Group (SCG) may exist consisting of Gold level command from the emergency services and others. The SCG does not remove the LRF from its statutory duties. • The SCG may instigate a Recovery Working Group (RWG) or request one that already exists to bring key agencies together so that the recovery process can

	or extended duties of Responders. Although enabling regulations made under this subsection can confer additional 'functions' to e.g. a Scottish Minister. This means that whilst Scotland for example can create its own regulations they can be affected by any regulations by a Minister of the Crown. 5. s2(6) has broadly similar affect to s2(5) but strictly to Scotland.	6. Reg 9 says that 2(1)(a)–(f) of the Act are to be regarded as 'relevant civil protection duties'. Furthermore if there were two Category 1 Responders with these relevant civil protection duties in a particular area then Reg 9(3) and 9(4) allows for one to be lead-Category 1 Responder and the other the 'non-lead Category 1 Responder'. In addition, Reg 10 and 11 set out the terms and duties to co-operate between the lead and non – Lead Category 1 Responders. In short, they need to keep each other informed, share information and avoid duplication of effort and resource. 7. Q. Are Category 1 Responders bound to perform the duties under s2(1) of the Act for all emergencies in a given area? There are exemptions Reg 12 says, when situations arise that are within say the remit of the Control of Major Accident Hazards Regulations 1999 (as amended) or other pipeline, radiation, etc. regulations. Part 3: Duty to assess risk of emergency occurring	be speeded. The RWG would operate within a strategic framework and under the SCG's terms of reference. • The document also refers to the 'Inner and Outer Cordon' issues that arise during an emergency. E. Section 2(1)(c) is mirrored in Chapter 6 of 'Emergency Preparedness' guidance. The need for Responders to reflect upon their own business continuity is critical. Readers are referred to Chapter 1 of this book for approaches to Business Continuity Management (BCM). F. 'Emergency Preparedness', Chapter 4, focuses on 'Local Responder Risk Assessment Duty'. It makes the following recommendations: • Usage of a Risk = Impact × Likelihood matrix with a 5 × 5 scale (Impact can be 1 = Insignificant, 2 = Minor, 3 = Moderate, 4 = Significant and 5 = Catastrophic. Likelihood can be 1 = Negligible, 2 = Rare, 3 = Unlikely, 4 = Possible and 5 = Probable as defined in Annex 4D of the Chapter). • A '6 Steps' General Risk Assessment approach (Establish the Context – what, why, when and where; Identify the Hazards and Threats; Analyse Risks; Evaluate Risks; Accept Risk; Treat Risk.

		8. We know from s2(1)(a)–(b) that Responders have a duty to assess for the risk of an emergency occurring. Category 1 Responders under Reg 13 need only do so for the specific area they are responsible for. 9. Reg 14 allows for the responsible Minister for that part of the UK to issue guidance and an assessment to a Category 1 Responder as to the likelihood or extend of an emergency. 10. Reg 15 requires Category 1 Responders in England and Wales to maintain a Community Risk Register (CRR) of each assessment conducted as required by s2(1) as well as to share this information with other Category 1 Responders for that area (subject to national security and other considerations). Reg 16 extends this logic and requires Category 1 Responders in adjoining areas to share the CRR information with each other and/or with the Welsh Assembly or the Minister as appropriate. Reg 17 applies the same to Scotland. Reg 18 requires	• LRF to set up Risk Assessment Working Group (LRAW) and the lead assessors of this Group need to consider the likelihood of the hazards occurring within the next 5 years. • Local Risk Assessment Guidance (LRAG) to be provided by central government to Category 1 Responders. This provides a 'statement' on the assumptions and types of threats and hazards that could occur. This is not the same as the Specific Risk Assessment which needs to be localised and be 'dynamic' as hazards and threats in the area change. A standard LRAG will be a template with 5 columns – Column 1 'Type of Risk'; Column 2 'Risk Categories'; Column 3 'Outcome Description'; Column 4 'Likelihood Assessment, Lead Department & Assumptions'; Column 5 'Variations and Further Information' as defined in Annex 4B of the Chapter. • Annex 4E provides an outline of a Community Risk Register (CRR) as required under Reg 15. The CRR can have the following columns: Column 1: Risk Reference Column 2: Hazard or Threat Category Column 3: Hazard or Threat Sub-Category Column 4: Outcome Description Column 5: Likelihood

		the sharing of information between Category 1 Responders where such information is critical to the performance of another Category 1 Responder's function.	Column 6: Impact Column 7: Risk Rating Column 8: Capability Required Column 9: Controls Currently in Place Column 10: Additional Risk Treatment Required (with timescales) Column 11: Risk Priority Column 12: Lead Responsibility Column 13: Review Date
		Part 4: Duty to maintain plans 11. Reg 19 requires Category 1 Responders to ensure that when an assessment of an emergency occurring is conducted then such information needs to be taken into account when maintaining business continuity plans and emergency plans. This is a pragmatic requirement of ensuring that plans are not maintained in ignorance of assessments. In addition Category 1 Responders need to ensure that emergency plans take account of public information needs (Reg 20). 12. Reg 21 and 22 specify the types of plans; generic and specific plans as well as multi-agency plans (when more than one Category 1 Responder is involved or acting jointly). All of these plans relate to Category 1 Responders only.	G. Chapter 5 of 'Emergency Preparedness' provides guidance on Emergency Planning. This can be summarised as follows: 1. Emergency Planning is stated in the guidance as an 8 Step process: Step 1: 'Take direction from Risk Assessment' Step 2: 'Set Objectives' Step 3: 'Determine actions and responsibilities' Step 4: 'Agree and finalise' These 4 constitute the 'Consulting' part of the Planning process. Step 5: 'Issue and disseminate' Step 6: 'Train key staff' Step 7: 'Validate in exercises and in response' Step 8: 'Maintain, review and consider revision' These 4 constitute the 'Embedding' part of the Planning process.

		13. Reg 23 relates to the role of voluntary organisations and ensuring that Category 1 Responders are cognisant of such bodies. 14. Reg 24 focuses on 'how do we know an emergency has occurred?' Category 1 Responders in their plans need to ensure that a criteria exists which addresses this question and issues of 'can we cope with our existing deployment of resources?' 15. Reg 25 imposes a duty on Category 1 Responders to train and exercise its own staff and others it regards as necessary to ensure that each plan can work effectively if activated. 16. The relevant Minister for that part of the UK could require the plans maintained under the Act to be added to or modified and Category 1 Responders shall act accordingly (Reg 26). Part 5: Publication of plans and assessments	2. Emergencies can be classified as Single Location, Multiple Locations, Wide Area and Outside Area. Consequently plans need to reflect these locations. 3. The guidance also gives an outline of how an Emergency Plan could be structured and presented, with the following sections: • General Information relating to the Risk Assessment evidence upon which the Plan is based ('why the plan is needed') • Management, control and coordination ('how' and 'who' issues) • Activation, including alert and standby ('when' issues) • Action ('what' issues) • Annexes ('who' issues). 4. The guidance also states the '5 Steps' involved in developing such plans: i. Develop the Risk Profile ii. Set the Objectives of the Plan iii. Identify Tasks and Resources required iv. The Organisation ('who') v. The Arrangements ('what' and 'how').

		17. This is a short Part and says that Category 1 Responders in performing its duty under s2(1)(f) of the Act must not alarm the public unnecessarily (Reg 27). Part 6: Arrangements for warning and provisions of information and advise to the public 18. When a Category 1 Responder warns or informs the public about an emergency or pending emergency they need to reflect back to the relevant plan and ensure that the advise and information is consistent with that plan (Reg 28). Category 1 Responders may maintain general and specific arrangements to inform and warm (Reg 29) but again need to ensure the public is not unnecessarily alarmed (Reg 30). Category 1 Responders need to train and exercise themselves and others it deems appropriate to activate the plan including 'warning the public issues'. Reg 32 requires that there be a lead-Category 1 Responder who has the responsibility of warning and informing the public and non-Lead Category 1 Responders need to co-operate with the lead body. This is again to ensure a consistent and	H. Chapter 7 of the 'Emergency Preparedness' guidance provides advise on 'Communicating with the Public'. The Chapter focuses on some key issues on effective 'Public Communication': 1. It suggests a three-phase approach: • 'Public Awareness' before the emergency so that the public are informed and made aware of risk and emergency planning issues • 'Public Warning' during the emergency so that the public is alerted • 'Informing and Advising the Public' after the emergency from immediate through to long-term issues. 2. The Chapter then focuses on the method, mode and process of effective communications, e.g. using existing websites, summarising key content and making, e.g. schools, business chambers aware of key advise. Furthermore, the Regional Media Emergency Forums (RMEFs) can be a source for Category 1 Responders to liaise with, as such bodies usually consist of key media as well as local and central government officials.

		clear message is sent out to the public. Reg 33 goes one step further and suggests that such Lead Category 1 Responders need to make clear arrangements for contacting, informing and collaborating with other Category 1 Responders with respect to warning and informing the public. There will also be cases when a Category 1 Responder wishes to provide information and advise in a given local resilience area but it is not the lead body. In this case, Reg 34 requires such a non-lead body to consult and inform the lead body of its intent and message. Reg 35 says that Category 1 Responders also need to be cognisant of the advise being issued by say Food Standards Agency or the Met Office or indeed Category 2 Responders with regards to an emergency in a local resilience area. Is this duplicating the information we have? Should we as a Category 1 Responder use this other information? Does it carry more accuracy and is it easier to understand by the public?	I. There is further guidance in the non-statutory guidance in Chapter 6 'Information and the media' which highlights issues such as how Category 1 Responders can co-operate with the media during the emergency.

Section 3: Section 2 Supple-mental	As the title of this Section suggests it is a sort of 'addendum' to s2 above. 1. s3(1) and 3(2) are Strict and give Ministers of the Crown and those in Scotland the power to issue guidance to Responders. 2. s3(3) is Strict and requires Responders to comply with any regulations and guidance issues. 3. s3(4) and 3(5) introduces the two types of Responders under Category 1 and 2 (see Schedule 1 below).		
Section 4: Advice and Assistance to the Public	1. s4(1) is a Strict duty on Responders to provide business continuity advise to the commercial sector and the not-for-profit sector. 2. s4(2) and 4(3) are both Strict and are enabling powers of the Minister of the Crown or Scottish Minister. They may make regulations identifying the duties and how the Responders are meant to advise.	Part 7: Advise and assistance to business and voluntary organisations 19. This Part deals with the 'relevant responders', i.e. local authorities in England and Wales. It links with s4(1) of the Act. It should be noted that there are transitional arrangements in place as the full duty on local authorities to render business continuity advise comes into force on the 15th May 2006 so till then the Regulations says that do not regard the actions that local authorities have to perform under this Part as 'duties' but rather as 'powers' under the law to render advise and assistance (Reg 36 and 37).	J. Chapter 8 of the 'Emergency Preparedness' guidance focuses on 'Advise and assistance to business and voluntary organisations'. The key focus is on Business Continuity Forums and how Category 1 Responders and LRFs can coordinate and co-operate with their local business community and non-commercial organisations. A number of examples are provided by the guidance. A 10 Step approach is recommended: Step 1: 'Identify requirements of the legislation', i.e. s4 of the Act and Part 7 of the Regulations

	3. s4(4) and s4(5) are Strict and largely relate to the charges that can be made by Responders if called on by say the commercial sector to advise them ('consultancy service' situation). 4. s4(6)–4(8) are Strict and deal with the guidance that the Minister of the Crown or Scottish Minister and issue and the duty of the Responders to follow both the regulations and the guidance.	20. When giving advise on business continuity, the relevant responder needs to be aware of the Community Risk Register (CRR) as this will identify the risks of the emergency as well as other facts that will enable the relevant responder to give specific or generic business continuity advise and assistance (Reg 38). 21. Reg 39 and 40 says that relevant responders may issue business continuity advise and assistance to businesses and voluntary organisations only in their local area and further may recommend or seek the advise from an experienced and competent business continuity consultant. 22. Reg 41 requires relevant responders within a given resilience area to co-operate with each other. Other responders (general responders who are not local authorities) must also co-operate with relevant responders for that particular resilience area. In addition, Reg 42 says, general responders may co-operate with Scottish Category 1 Responders so as to enable the latter to fulfil its duties under s4(1). For example, a general	Step 2: 'Assess implications of the legislation: review existing work', e.g. conduct a Business Impact Analysis (BIA) Step 3: 'Identify audience for promotion strategy', e.g. business community with less than 250 staff in the area Step 4: 'Identify partnerships for promotion strategy', e.g., with the Business Chamber of Commerce Step 5: 'Identify resource requirements and availability', e.g. time, cost and money; budgets Step 6: 'Evaluate and review programme' – will the programme assist Category 1 Responder's to comply with their legal duties and will it improve awareness of the target audience? Step 7: 'Delivery (no later than the date when the duty comes fully into force)' Step 8: 'Identify means of delivery and timescales for programme', e.g who, when and how issues?

		responder, e.g. emergency service in Scotland or even on the English border with Scotland may co-operate with Scottish Environment Protection Agency (SEPA) which is a Category 1 Responder in Scotland. 23. Relevant responders need to check to see what other advise and assistance is being provided by other responders to businesses and voluntary organisations in their area and where needed avoid duplication (Reg 43). The relevant responder may also charge for advise and assistance 'requested' under s4(1) and Reg 39(3)(a), i.e. advise and assistance to 'businesses at large'.	Step 9: 'Assemble materials' which needs to be relevant and simple to meet the target audiences needs. Step 10: 'Formulate objectives and message with reference to the Community Risk Register and Local Resilience Forum partner needs'
Section 5: Civil Protection – General Measures	This section from 5(1)–5(5) is Strict and deals with the 'chain of command' issues. For instance that 'orders' issued by the Minister of Crown (or Scottish Minister) must be followed and that Responders can be asked to act or to perform some function to prevent an Emergency situation. In addition, that arrangements can be made to work with, delegate or collaborate between the Responders. Further, other devolved assemblies in the UK can be given additional functions.	No part of the Regulations refer to s5 of the Act other than Reg 26 again.	

Section 6: Disclosure of Infor-mation	This section is Strict and deals with the enabling powers of the Minister of Crown (or Scottish Minister) which can require the 'provider' and 'recipient' of emergency information to share such information. Guidance may be issued by the Minister if required and the duty holders are required to comply with the regulations if issued.	Part 8: Information The Act has five sections that require either the keeping, disclosure, sharing or sensitive use of information, namely sections 2, 4, 6, 9 and 12. Part 8 of the Regulations provide the rules with respect to sensitive information, its management and exchange of. This Part realises that during an emergency, e.g. terrorist event, there has to be the safe use, storage and exchange of information relating to the emergency. Reg 45 defines the meaning of 'sensitive information', i.e. disclosure of information which would affect national security, public safety, prejudice commercial interests of a person or comprise personal data conditions under relevant data protection laws. 25. Reg 46 says that a signed Certificate of a Minister of the Crown (or equivalent in other parts of the UK) would be sufficient evidence that disclosure of information would adversely affect national security. 26. Reg 47 allows requests for information from Category 1 and 2 Responders so long as that information assists in fulfilling a legal	K. Chapter 3 of 'Emergency Preparedness' guidance interprets the Act and Regulation. It reaffirms the importance of non-disclosure of sensitive information. The LRFs need to ensure that all members (and others that they may invite) are aware of their statutory duties to protect, preserve and non-disclose other than when required by those authorised by the legislation. This can be via using the Government's 'Protective Marking System' which grades sensitive information, e.g. 'Restricted' through to 'Top Secret'. Whilst not mentioned by the guidance, LRFs can refer to standards such as ISO 17799 (BS 7799) on Information Security as well as the Office of Government Commerce, UK Govt, which also has publications on Information Systems and Security.

		duty under s2(1)(a)–(d) and s4(1) of the Act or in connection with carrying out some other function relating to the emergency. Reg 48 deals with the format of the information request, e.g. in writing, etc. 27. Reg 49 relates to the duty of the responder who receives a request for information to supply that information, unless this is sensitive information in which case the responder must not disclose (but must give reasons to the requesting responder as to the non-disclosure). Also, for non-sensitive information the information must be supplied within a reasonable time and to the location requested (Reg 50). General responders cannot publish nor disclose such sensitive information to another unless authorisedby a certificate of a Minister of the Crown or by a relevant provision of these Regulations (Reg 51). Any sensitive information received by a general responder must only be used for purposes described by the Act or the Regulations (Reg 52). Reg 53 imposes duties on sensitive	

		information holders to ensure that there are secure means in place for the storage and authorised access to that information. Reg 54 relates to the Health and Safety at Work, etc Act 1974 where there are provisions for the confidentiality of information. However, such restrictions will not apply if the information is disclosed to another responder in pursuance of that responder's duties under the 2004 Act or these Regulations. 28. Part 9 relates to London (role, function and power of the London Fire and Emergency Planning Authority (LFPA) to maintain Community Risk Register, is nominated to maintain emergency plans for all London Category 1 Responders regarding pan-London emergencies as well as the effective training of such Category 1 Responders. Further all other Category 1 Responders must co-operate with the LFPA as the lead body). 29. Part 10 relates to Northern Ireland, 'Duty to have regard to the activities of other bodies involved in civil protection' and 'joint discharge of function'	

Section 7: Urgency	Sometimes a regulation or order cannot be immediately made or issued due to the procedures involved. If a situation necessitates, then the Minister of the Crown can issue a 'direction'. This could have the same effect as a regulation or order. Such a direction has a 'shelf life' of 21 days from the day of issuance. The duties under s7 are Strict other than s7(4)(b) which requires the Minister to revoke the direction SFRP.	The Regulations do not refer to 'directions' as defined in Section 7.	The non-statutory guidance, Annex 13A, lists the questions that need to be asked when making emergency regulations (necessity, why, advise of SCGs, etc.). 'Directions' issued could be developed on similar grounds to an emergency regulation.
Section 8: Urgency Scotland	As Section 7 and same SFRP with regards to s8(4)(b).	As above	As above
Section 9: Monitoring by Govern-ment	This is Strict and is an 'oversight' duty. It compels the Responders to provide information (maybe in a set format and by a set date) as to what they are doing, what they may have not done and to justify their actions. This section relates to Scotland as well.	The Regulations as stated above can compel the Category 1 Responder to submit details of emergency plans, information and CRR to the Minister.	This is explained on the 'Cooperation' chapter in the statutory guidance as discussed above.
Section 10: Enforce-ment	Q. What if a Responder fails to comply with its duties under say s2(1) above for example? This section deals with the High Court/Court of Session proceedings that can be brought by say a Minister of the Crown (or another body/Responder under Schedule 1)	The Regulations do not state the penalties for failing to perform the duties of the Category 1 or 2 Responders. The Act provides the advise on this in Section 10.	No details on Enforcement provided.

	against another Responder (be it a person or a body) that fails to act. A parallel is found in health and safety where for example the HSE can bring an enforcement action against say a local council. This section is using the same approach although this section deals only with civil sanctions (orders).		
Section 11: Scotland	As Section 10 but for Scotland only.	As above	As above
Section 11: Scotland	As Section 10 but for Scotland only.	As above	As above
Section 12: Provision of Infor-mation	This section says that any regulations or orders that deal with the provision and disclosure of information may give indication as to the 'information management' issues that invariably arise, e.g. storage and disposal, etc. This section is Strict. The section talks of 'information' rather than 'Emergency Information' so this may imply information both before, after, arising from, direct or indirect but related to the emergency situation.	See discussion above under Part 8 of the Regulations.	As mentioned above, whilst neither guidance makes reference to ISO 17799 (or BS 7799) this is a framework for information management issues.
Section 13: Amendment of Lists of Responders	An order can be issued by Minister of Crown or Scottish Minister to change, add to, remove or move the category of Responders and modify permanently or transitionally the function of Responders (in person or as a body).		

Section 14 (Scotland: Consultation), Section 15 (Scotland: Cross border collaboration) and Section 16 (National Assembly of Wales)	These sections deal with the Strict duties of the Minister of the Crown and that of Scotland to consult each other with regards to regulations and orders (s14). Section 15 is Strict and deals with the regulations and guidance that can be made by the Scottish Minister and the Minister of the Crown to require Responders in their jurisdiction to co-operate, share information and coordinate activities with other Responders, e.g. who may be in England (or vice-versa in Scotland). There is a Strict duty on Responders to comply with any such regulation or order. Section 16 deals with the Strict duty of the Minister of the Crown to consult with the Welsh Assembly on regulations, orders, directions, guidance, proceedings, etc. and that the Assembly shall be consulted before any such actions occur.	The Regulations dealt with 'cooperation' issues under Part 2 as above.	As above.
Section 17: Regulations and Orders	This section summarises that regulations and orders must be via statutory instrument so require the established rules on the making of such instruments to be followed. Such instruments can be made to take permanent or temporary effect and apply to a whole or a part of the country.		

Section 18: Interpretation	This section defines terms such as enactment, terrorism, war and function.		
Part 2 of the Act: Emergency Powers	Came into force on the 10th December 2004.		Chapter 13 and Annex 13 A of the non-statutory guidance provides a summary of Part 2 of the Act.
Section 19: Meaning of Emergency	At first sight, the meaning under section 19 may seem no different from Section 1. However, there are three important differences: 1 Section 19 relates to Emergencies 'in the United Kingdom or in Part or region' whilst section 1 relates to Emergencies 'in a place in the United Kingdom'. Thus, the definition in Part 2 is more 'regional' and 'national' whilst Part 1 is more 'local'. 2 Under Section 19 the Secretary of State can through an order amend or reclassify the types of events or situations that may be viewed as threatening human welfare. Whereas in Section 1, the Minister of the Crown or Scottish Minister can make this determination. 3 Any order made by the Secretary of State needs to be laid before each House of Parliament which is not the case under section 1.	The Regulations in Reg 3 states that the term 'emergency' is defined in accordance with s1(1) of the Act so not s19. This implies that the Regulations focus on Part 1 of the Act.	As above under discussion on Section 1.

Section 20: Power to make Emergency Regulations	This section deals with who can make 'emergency regulations'. Part 1 referred to 'regulations' whereas now we see the use of this term 'emergency regulations'. In short, the Secretary of State (Home Office) or the Prime Minister would be enabled to make such emergency regulations.		See Annex 13A of the non-statutory guidance.
Section 21: Conditions for making Emergency Regulations	1. There needs to be certain 'tests' applied before such emergency regulations can be made, namely: • the emergency has occurred, is occurring or could occur; • the emergency regulation is needed to ensure appropriate prevention, control or mitigation of some aspect of the Emergency; • that there is an urgent need otherwise prevention, control or mitigation could be affected. 2. In addition, there must be evidence that the existing legislation is not sufficient to address the Emergency and/or that existing legislation does not allow new provisions to be created to prevent, control or mitigate some aspect of that Emergency.	These three relate to Probability of occurrence, Management and Severity	See Annex 13A of the non-statutory guidance.

Section 22: Scope of Emergency Regula-tions	1. Emergency regulations can make provisions for anything necessary to prevent, control or mitigate the event or situation. The regulations can be such that they provide for the protection of human life through the continuity of the functions of the Houses of Parliament or the devolved assemblies. 2. s22(3) is very important in that it implies that an emergency regulation can make provisions to prevent, control or mitigate the event or situation – the kind that could be made by an Act of Parliament. These are listed in s22(3)(a)–(q) and cover a range of powers from confiscation of property, deployment of armed forces, prohibition of travel, etc., in fact anything needed to ensure that the Emergency at the local, regional or national level is effectively prevented, controlled or mitigated. 3. This section also says whoever makes such emergency regulations must take account of the need for proceedings (e.g. debate or review) by Parliament, High Court and Court of Session with regards to this regulation and provisions acted upon under it.		See Annex 13A of the non-statutory guidance.

Section 23: Limitations of Emergency Regulations	1. s23(1) states two 'tests' of making emergency regulations, namely the said regulations are 'appropriate' and 'proportionate'. 2. The regulations created can only result in summary convictions and so this sets limits to the powers and scope of the regulations. (This is quite common. Most regulations made under the enabling powers of the Health and Safety at Work, etc. Act 1974 are also limited to summary convictions.)		See Annex 13A of the non-statutory guidance.
Section 24: Regional and Emergency Coordinators	The emergency regulations made under the Act (so this section presumes that such regulations will come about) requires the Minister of the Crown (Strict Duty) to appoint an Emergency Coordinator for that part of the UK (other than England) and a Regional Nominated Coordinator (RNCs) for a specific region in the UK. Therefore, England has RNCs and other parts of the UK has Emergency Coordinators. Neither are servants or agents of the Crown and therefore subject to legal scrutiny for their actions, omissions and conduct.		The non-statutory guidance, Chapters 7–12, focuses on the organisation of government at a national and regional level to ensure the effective administration during an emergency (whilst Category Responders focus on the effective emergency management and response issues). The Coordinators referred to in s24 are not necessarily Responders and can be from the government sector.

Section 25: Establishment of Tribunals	If an emergency regulation was to establish a Tribunal then a senior Minister of the Crown must (Strict Duty) consult the Council of Tribunals. However, this can be overridden, for example, given the urgency of the situation. It can also be noted that if the Council was to present a report on the Tribunal being made reference to in the regulations then the Minister shall SFRP, lay such a report before Parliament. If the Minister did not consult the Council when making the emergency regulations then the Minister shall do so SFRP.		
Section 26: Duration	Emergency regulations shall have a 'shelf life' of 30 days or can lapse sooner.		
Section 27: Parliamentary Scrutiny	The senior Minister of the Crown shall SFRP lay the emergency regulations before Parliament. The regulations have a 'shelf life' of seven days from the time of laying. Within this time, both Houses need to pass a resolutions approving the emergency regulations otherwise they cease to take effect. Alternatively, if each House was to say the emergency regulations shall cease to take effect then they may specify a date when this is to happen or they may cease the following day from the passing of the resolution by both Houses. Each House can off-course recommend amendments to the emergency regulations and they shall take effect from the date specified or from the following day.		

Section 28: Parliamentary scrutiny: prorogation and adjournment	Let it be assumed that the emergency regulations are made on the 1st of a particular month. Let it also be assumed that five days pass from the date of making the emergency regulations, i.e. it is now the 5th of the month. If on the next day (the 6th) Parliament stands prorogued (discontinuing but not dissolving a session of Parliament) then by proclamation Her Majesty can require Parliament to meet 'on a specified day within that period'. There are two further scenarios: 1. What if it is the House of Commons in the above situation (and not Parliament as a whole) that is 'adjourned' (not the same as prorogued)? Then, the Speaker shall arrange for the Commons 'to meet on a day during that period'. 2. In the House of Lords when it stands adjourned, it is the Lord Chancellor who does the same (as the Speaker above). It can be noted that if Parliament knows in advance when the prorogation or the adjournment is to occur then the date of the meeting can be set in advance as can the location if the meeting is to occur elsewhere.		

Section 29: Consultation with devolved administra-tions	It is a Strict Duty on the senior Minister of the Crown to consult with the senior most Ministers from Scotland, Northern Ireland and Wales when such emergency regulations are to be made and which affect these parts. When there is an urgent situation then the senior Minister of the Crown does not need to consult.		
Section 30: Procedure and Section 31: Interpretation	Section 30 says that emergency regulations are to be made via statutory instruments and are subordinate legislation (in connection with the Human Rights Act 1998). Section 31 provides interpretation to terms used in the Act.		
Part 3 of the Act: General			
Sections 32–36	These Sections deal with: • Section 32: Amendment and repeals • Section 33: Money (Parliament to cover any expenditures incurred by the Minister in connection with this Act) • Section 34: Commencement of the Act • Section 35: Extent (Act applies to England, Wales, Scotland and Northern Ireland) • Section 36: Short Title: The Civil Contingencies Act 2004		

Schedule 1			
Part 1: Category 1 Responders (General)	1. Local Authorities, e.g. County Councils 2. Emergency Services (Chief Officer of Police including British Transport Police and Fire and Rescue) 3. Health, e.g. NHS, Primary Care Trust, Strategic Health Agency (due to amendment, the Health Protection Authority is now removed) 4. The Environment Agency 5. The Secretary of State 'in so far as his functions include responding to maritime and coastal emergencies (excluding the investigation of accidents'.		
Part 2: Category 1 Responders (Scotland)	1. Local Authorities 2. Emergency Services (Chief Constable of Police, Fire Authority, Scottish Ambulance Service Board) 3. A Health Board 4. Scottish Environment Protection Agency		
Part 3: Category 2 Responders (General)	1. Utilities, e.g. those holding a statutory license for electricity, gas, water, public electronic communications network 2. Transportation, e.g. London Underground, those that are licensed holders of railway assets, airport operator, harbour authority, Secretary of State with regards to highway authorities 3. Health and Safety Executive		
Part 4: Category 2 Responders (Scotland)	1. Utilities as above but with reference to Scotland 2. Transportation as above (but there is no mention of the highway authorities) but with reference to Scotland 3. Health, namely the Common Services Agency.		

Schedule 2			
Part 1: Amendment and Repeals Consequential on Part 1 (of the Act)	1. Civil Defence Acts 1939 and 1948 are both repealed 2. Civil Defence Acts (Northern Ireland) 1939 and 1950 are both repealed 3. Section 138 of Local Government Act 1972, subsection 1(A) is repealed (relating to likelihood of emergencies, necessary plans and the expenditure involved) 4. Civil Protection in Peacetime Act 1986 – repealed 5. The following are also amended: • Defence Contracts Act 1958 • Public Expenditure and Receipts Act 1968 • Road Traffic Act 1988 • Metropolitan Country Fire and Rescue Authorities is the new name of such authorities with omission to 'Civil Defence' in their previous title.		
Part 2: Amendment and Repeals Consequential on Part 2 (of the Act)	1. Emergency Powers Act 1920 – repealed 2. Emergency Powers Act (Northern Ireland) 1926 – repealed 3. Northern Ireland Act 1998 amended to reflect Civil Contingencies Act 2004 and delete reference to the 1926 Act		

Part 3: Minor Amendments	The following are amended: • Energy Act 1976 – inclusion of territorial control of energy in the UK, for example. • Highways Act 1980 – role of a police officer (whose rank maybe defined by regulation) in giving consent to traffic calming measures and works as well road construction work. • Road Traffic Regulation Act 1984 – orders can be made with regards to impact of 'danger or damage connected with terrorism in connection with road traffic which can only be made on the recommendations of a senior police officer for that given area. In addition, powers can be conferred onto constables as needed. Overall, the amendment relates to ensuring road traffic management is flexible during an emergency. • Roads (Scotland) Act 1984 – similar changes as to the Road Traffic Regulation Act 1984.		
Schedule 3: Repeals and Revocations	This is a summary of all Acts that have been repealed or amended which includes the above and others		

The Act, Regulations and guidance documents provide the starting point for duty holders. A key challenge for the authorities is to ensure not only that Responders across the parts and regions of the UK have a consistent approach but also that the volume of documentation and the administrative time involved can be reduced. The sources referred to above will no doubt evolve over time.

15

The Medical Response to Disaster

In this chapter:

- Introduction.
- Definition and classification of Disasters.
- Scene management.
- Medical organisation.
- Safety at the scene of a Major Incident.
- After the event.
- Conclusion.

Introduction 15.1

Hardly a day goes by without seeing on television or reading in the media about Disasters around the world. Human suffering is brought into our homes in graphic detail from war torn areas of the world or as a consequence of natural or human-made Disasters.

The events of 11 September 2001, with the destruction of the World Trade Centre in New York (see **3.39** above) and the attack on the Pentagon in Washington by a co-ordinated act of terrorism, gave the developed world new impetus in terms of planning for Major Incidents and Disasters. Prior to September 11th, people who talked of 'what if' incidents on the scale of the World Trade Centre were considered scaremongers. Now governments and Emergency planners around the world realise 'it could happen to them'. The events of 11 September 2001 also heightened awareness of the risks from chemical, biological and nuclear weapons to civilian populations where previously they were only perceived as largely military risks. In addition, the suicide bombers brought their particular brand of mindless carnage to parts of the world where it had never been witnessed previously.

In 2002, the worst floods for over a century swept across Europe claiming lives, and with historic buildings and priceless artefacts being lost for ever. Lives, too, were being lost in significant numbers in Disasters. In July 2002, 83 lives were lost and many injured when a Sukhoi SU–27 military aircraft crashed into the crowd at an air show in Lviv in the Ukraine. This event will enter the history books as the world's worst air show accident, eclipsing a similar accident at Ramstein in Germany in 1988 when 79 died and 530 were injured.

Sadly, as soon as another news story pushes the events of one Disaster out of the limelight they are all too readily lost from the corporate memory but not, of course, from the lives of those involved. The effects of a Disaster changes the lives of those involved forever, as does the way in which the events were handled by those present. Everything comes under scrutiny after the event, from the responding emergency services and the subsequent medical and psychological care of all involved, to the portrayal of events in the media along with the response of local and national governments.

The medical response to a Disaster is but one facet of its total management, and the ability of medical resources to respond to the challenge will vary around the world from country to country, and between a developed western country and poorer third world countries. This chapter will aim to review the accepted medical Major Incident response in the UK. It is recognised that terminology may differ around the world, as may some working practices, but it is hoped that this chapter will place medical management at the scene of a Major Incident or Disaster into a context that can be translated anywhere.

Planning for and Responding to Medical Emergencies in the UK 15.2

This chapter predominantly describes the medical response to a major incident but it would be remiss if, right at the outset, mention was not made of recent changes to the Health Service's management of the medical aspect of emergencies in the UK and also of major concerns that now appear on the horizon.

In England, the Department of Health has the responsibility to protect the people from danger insofar as health matters are concerned. The devolved administrations have a similar responsibility in Northern Ireland, Scotland and Wales.

The Health Protection Agency was set up under the Health Protection Act 2004. Under the Act, the functions of the Agency are 'to protect the community (or any part of the community) against infectious diseases and other dangers to health'. In addition, it also provides support to, and works in partnerships with, others who have health protection responsibilities and advises, through the Department of Health, all government departments and devolved administrations throughout the UK.

A number of important documents have been issued since the setting up of the Health Protection Agency. The principle ones, all of which can be downloaded from the Health Protection Agency's website (www.hpa.org.uk) are:

- *Beyond a Major Incident, issued on 13 December 2004.*
  This paper sets guidance and policy to assist the NHS to plan for a major incident of very serious proportions involving potentially large numbers of casualties, i.e. casualty numbers that are beyond the capacity created by the local implementation of major incident plans.
- *NHS Emergency Planning Guidance 2005, issued on 13 October 2005.*
  The guidance describes a set of principles to guide all BHS organisations in developing their ability to respond to a major incident or incidents and to manage recovery whether the incident or incidents has effects locally, regionally or nationally, within the requirement of the Civil Contingencies Act 2004.
- *UK Influenza pandemic contingency plan issued on 19 October 2005.*
  As its title implies, it outlines the Health Protection Agency's plan for responding to an influenza pandemic.

One other paper relating to the medical response is to be found on the UK Resilience website (www.ukresilience.info):

- *Guidance on Dealing with Fatalities in Emergencies, issued in May 2004.*
  It is primarily aimed at local responders and consolidates what local emergency service officers should consider when planning how to respond to an incident in which there are mass casualties.

Epidemics and Pandemics 15.3

Epidemics, a widespread outbreak disease occurring in a single community, population or country, are fairly frequent. Countries around the world, including the UK, are used to epidemics of 'ordinary flu', the last one occurring as recently as 1998. Of extreme concern to countries at the current time, however, is the danger of a pandemic. These occur on a much greater scale, spreading around the world and affecting many hundreds of thousands of people across many countries. They are, of course, not new. In 1918–1919, for instance, a pandemic known as Spanish influenza is estimated to have killed 40 million people worldwide, including approximately 280,000 in the UK.

At the time of writing, the main concern is avian flu, more commonly known as bird flu. On 8 December 2004, the World Health Organisation (WHO) announced:

> WHO and influenza experts worldwide are concerned that the recent appearance and widespread distribution of an avian influenza virus, influenza A/H5N1, has the potential to ignite the next pandemic.

Avian flu is a highly contagious disease of birds, caused by influenza A viruses. In birds, the viruses can present a range of symptoms from mild illness and low mortality to a highly contagious disease with a near 100% fatality rate. At the time of writing, the bird flu virus affecting poultry and some people in Asia is the highly pathogenic H5N1 strain of the virus. As the virus can remain viable in contaminated droppings for long periods, it can spread among birds, and from birds to other animals, through ingestion or inhalation.

On 1 March 2005, the Chief Medical Officer for England and Wales, Sir Liam Donaldson, announced:

> Wherever in the world a flu pandemic starts, perhaps with its epicentre in the Far East, we must assume we will be unable to prevent it reaching the UK. When it does, its impact will be severe in the number of illnesses and disruption to everyday life. The steps we are setting out today will help us to reduce the disease's impact on our population.

The UK's response, insofar as it relates to the Health Service, to such an outbreak is contained in the document, 'UK Influenza pandemic contingency plan'.

In consultation with the other Health Departments of the devolved administrations, the Department of Health (England) will appoint a UK National Influenza Pandemic Committee (UKNIPC). Should a pandemic lead to the possibility of introducing various controls in an effort to contain the disease, this will almost certainly have implications for other Government Departments. In such cases the Civil Contingencies Committee (CCC) with the support of the Civil Contingencies Secretariat (CCS) will become the central focus for such activity.

Scene Management 15.4

The scene of any large scale incident, whatever its cause, will initially be chaotic. It is the role of the emergency services to bring order to that chaos and they do it by their own command and co-ordination structures. In the UK the emergency services are primarily Police, Fire and Ambulance Services with HM Coastguard becoming involved with incidents at sea or on the shoreline. To these will be added medical resources, local authorities, public and private utilities and a range of voluntary organisations. In certain circumstances the Armed Forces may also become involved.

Emergency service priorities at a Major Incident 15.5

At the scene of any Major Incident the emergency services have a number of priorities which encompass all their actions and these are the:

- preservation of life;
- prevention of further loss of life;
- relief of suffering;
- protection of property;

- protection of the environment;
- restoration of normality;
- investigation of the cause.

In order to mount an appropriate response to a scene all emergency services need intelligence as to what they are going into. Similarly, the emergency service assets first on scene must send as much information back to their respective control rooms to ensure a proper response. There are a number of mnemonics that will help first responders to update their chains of command. The commonest are CHALET, ETHANE AND METHANE (see Figure 1 below). They all contain the same information and it is often quoted that METHANE is CHALET in a logical order. These mnemonics have been further refined and combined and some services now use SADCHALETS. The use of these mnemonics are of as much relevance to the medical services as they are to the other emergency services.

Figure 1: Mnemonics for the assessment of a Major Incident

CHALET	ETHANE	METHANE	SADCHALETS
C Casualties	**E** Exact location	**M** Major incident standby or declared	**S** Survey
H Hazards	**T** Type of incident	**E** Exact location	**A** Assess
A Access	**H** Hazards	**T** Type of incident	**D** Disseminate
L Location	**A** Access	**H** Hazards	**C** Casualties
E Emergency Services required	**N** Number of casualties	**A** Access	**H** Hazards
T Type of incident	**E** Emergency Services required	**N** Number of casualties	**A** Access
		E Emergency Services required	**L** Location
			E Emergency Services required
			T Type of incident
			S Safety of all personnel attending
			S Survey

Medical Organisation 15.6

For a Major Incident to be successfully managed it is essential that there is effective command and co-ordination of all personnel deployed to the incident. Each service will have their own Incident Officer and whilst the terminology used to describe them may vary around the country, and internationally, they must work very closely together.

Inter-relationship between Incident Officers 15.7

In the UK the police have overall responsibility for the co-ordination of a Major Incident scene and Figure 2 below illustrates the inter-relations between the Incident Officers of the various services.

Figure 2: The inter-relationship between Incident Officers

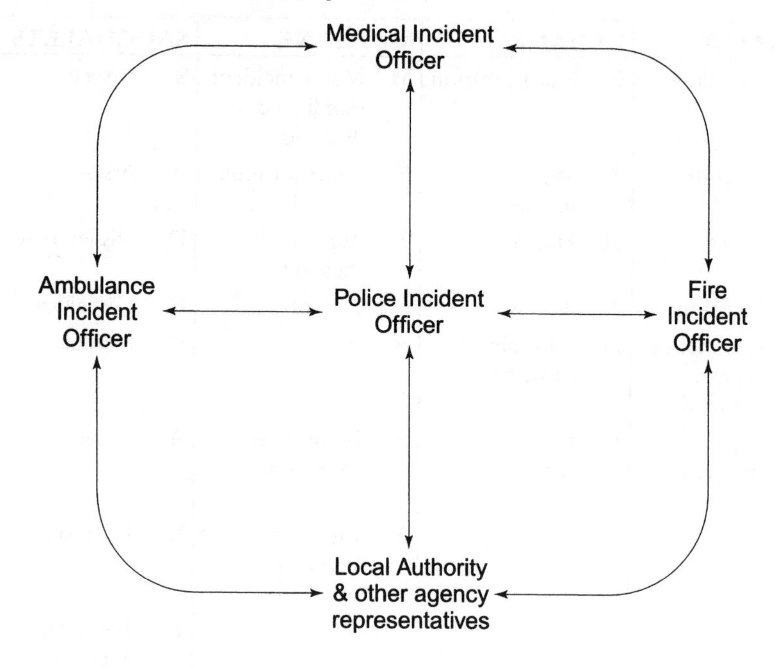

Reproduced with permission from Dr C J Eaton (author/publisher) *The Medical Aspects of Major Incident Management* (1999) British Association for Immediate Care.

In the UK, the Ambulance Service of the area in which the incident occurs is charged with the co-ordination of all the National Health Service assets committed to the scene. This response is led by the Ambulance Incident Officer who has overall responsibility for the functioning of the Ambulance Service. They work alongside the Medical Incident Officer who is responsible for the medical resources at the scene.

Gold, Silver and Bronze system of command and control 15.8

Within each service the Gold, Silver and Bronze system of command and control may be found. Gold is the Strategic level with Silver being the Tactical Commander at the scene. There is usually only one Silver Commander in any service and in an Ambulance Service it is the Ambulance Incident Officer, with the Medical Incident Officer as Silver of the Medical Services deployed to the scene. The Bronze level is Operational and there may be a number of Bronze Commanders covering different roles or areas of the operation within the incident scene. In Scotland, the terms Gold, Silver and Bronze are not used. Instead the three levels are referred to as Strategic, Tactical and Operational. This Gold, Silver, Bronze system is shown diagrammatically in Figure 3 below and it is a very useful concept that can be translated into many different scenarios.

Figure 3: The Gold, Silver and Bronze system of command and control

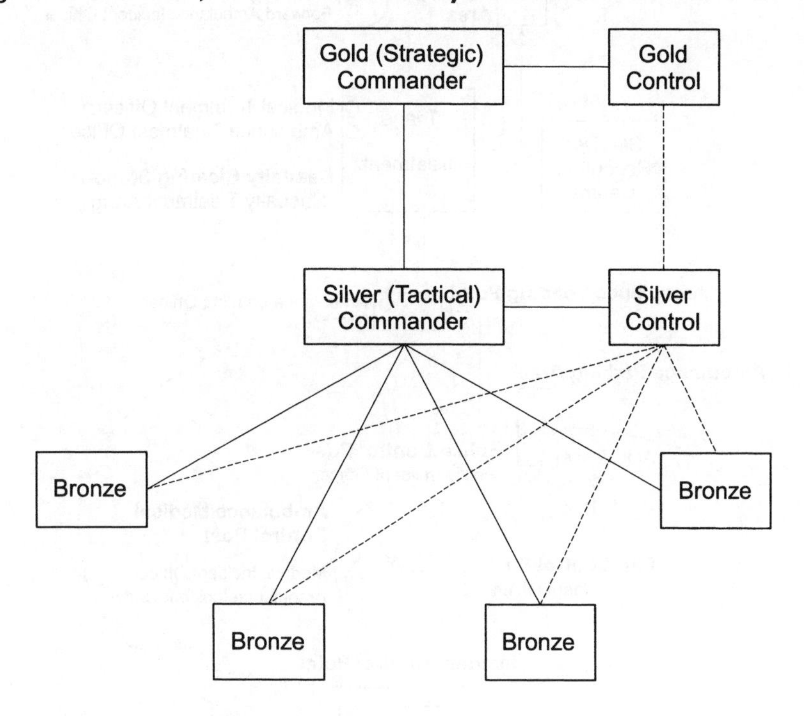

Ambulance and medical resources 15.9

The organisation of a Major Incident scene from a medical perspective is best shown diagrammatically as in Figure 4 below. Ambulance and Medical resources will be deployed to all the elements as indicated in the diagram, as they become available, by the Ambulance and Medical Incident Officers.

Figure 4: The Medical and Ambulance Service Organisation at a Major Incident

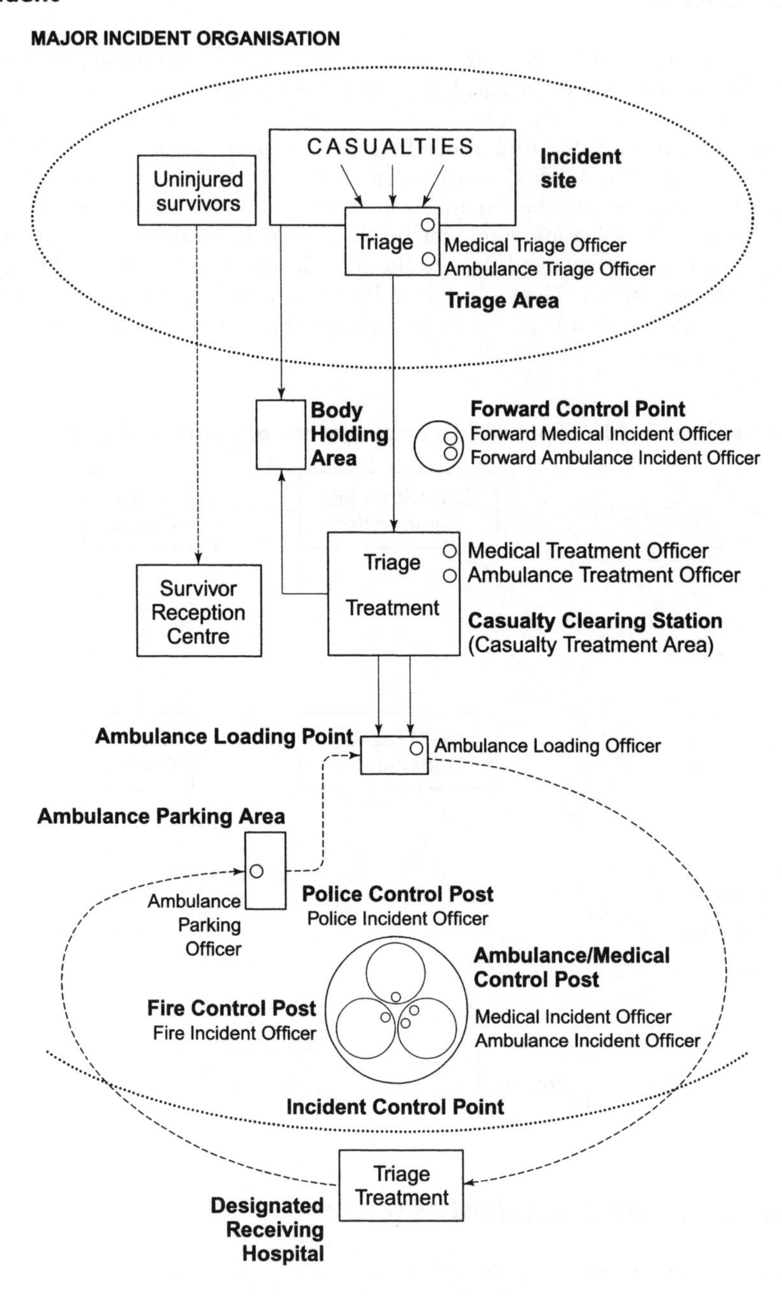

Reproduced with permission from Dr C J Eaton (author/publisher) *The Medical Aspects of Major Incident Management* (1999) British Association for Immediate Care.

Reproduced with permission from Dr C J Eaton (author/publisher) *The Medical Aspects of Major Incident Management* (1999) British Association for Immediate Care.

Safety at the Scene of a Major Incident 15.10

The scene of any incident, whether large or small, can be dangerous and all personnel deployed to a scene should have due regard for their own and their colleagues' safety. No member of the emergency services should expose themselves to unnecessary risk and become a casualty themselves. The Fire Service will have overall responsibility for the scene and, in liaison with the police, will establish the inner cordon and control access through it.

All ambulance and medical personnel on the scene must dress in appropriate personal protective equipment (PPE) that will include high visibility jackets bearing the designation and status of the individual. Appropriate head protection and footwear are essential. Appropriate standards have been issued concerning PPE for health care professionals in the UK. A higher level of PPE will be necessary to deal with chemical and biological risks and this will be available to appropriately trained healthcare personnel.

Decontamination 15.11

Until recently decontamination at an incident scene was usually only considered in chemical incidents. The risks and needs for mass decontamination were rarely addressed. The increase in terrorist activity has brought the possible requirement for decontamination very much to the fore in relation to chemical attacks with 'home made' agents and biological attacks with organisms such as anthrax (see **19.17** below).

Other risks are also present, such as man-made mineral fibres (MMMF), which can be found in modern aircraft, vehicles and railway carriage designs. These present a hazard both when burnt as toxic fumes or as the carbon-fibre dust that can be inhaled.

Any Major Incident plan must make provision for decontamination although the means and equipment available may differ around the UK and, indeed, around the world.

Decontamination will be achieved using water delivered through a shower unit and ideally warmed to body temperature. Victims' contaminated clothing must be bagged for safe disposal and some form of disposable replacement clothing will be required if they are not otherwise injured. Thought must also be given to the containment and ultimate disposal of the by now contaminated run off water and assistance with this can usually be provided by the Fire Services, local authorities or the Environment Agency.

Whilst equipment and procedures for the decontamination of a small number of people is relatively straightforward, mass decontamination of large numbers can present major challenges to all emergency services.

Triage 15.12

Triage is the process of sorting casualties according to the severity of their injuries and, in the mass casualty situation, it is to ensure that the most good is done for the greatest number. Where sufficient resources exist to manage an incident, the aim should also be to get the right casualty to the right hospital within the right time frame. There are a number of different triage classifications as shown in Figure 5 below, but they are all basically saying the same thing.

Figure 5: Various triage classification systems

Colour	*Priority*		*Clinical*	*Time*	
Red	Priority 1	P1	Critical	Immediate	T1
Yellow	Priority 2	P2	Serious	Urgent	T2
Green	Priority 3	P3	Minor	Delayed	T3
				Expectant	T4
White			Dead	Deceased	T0

The Red (Priority 1) casualties require immediate life saving interventions or they will not survive. The Yellow (Priority 2) require intervention but can wait until the priority 1 casualties are dealt with. The Green (Priority 3) casualties, often inappropriately called walking wounded, will form some 60% of the casualties in most Major Incidents. These three categories could be described simplistically as 'Can't wait; can wait; must wait' (see Figure 6 below).

The Expectant category shown in Figure 6 is only used in the overwhelming mass casualty situation and is for those casualties who are unlikely to survive, even if treated aggressively. In addition, there are the uninjured but involved who really constitute a separate category and, whilst they should not enter the casualty evacuation chain, they will need to be corralled at a Survivor Reception Centre so that appropriate documentation can be undertaken.

The key to effective triage at the scene of a Major Incident is that it should be performed by a senior and experienced clinician and that some form of triage labelling should be used to convey information to the receiving hospital. There are a number of different triage labels available but they should all be of high visibility, robust and waterproof, allow for serial observations of the patient and allow for the patient's category to change in either direction.

One recent development with the Cruciform triage card is the addition of a numbered wristband (see Figure 6f below). This latest version also has a number of other self-adhesive labels, all bearing the same number, which can be used for property or other items, thus significantly improving the whole area of casualty tracking at a Major Incident.

Figure 6a: Cruciform Triage Card

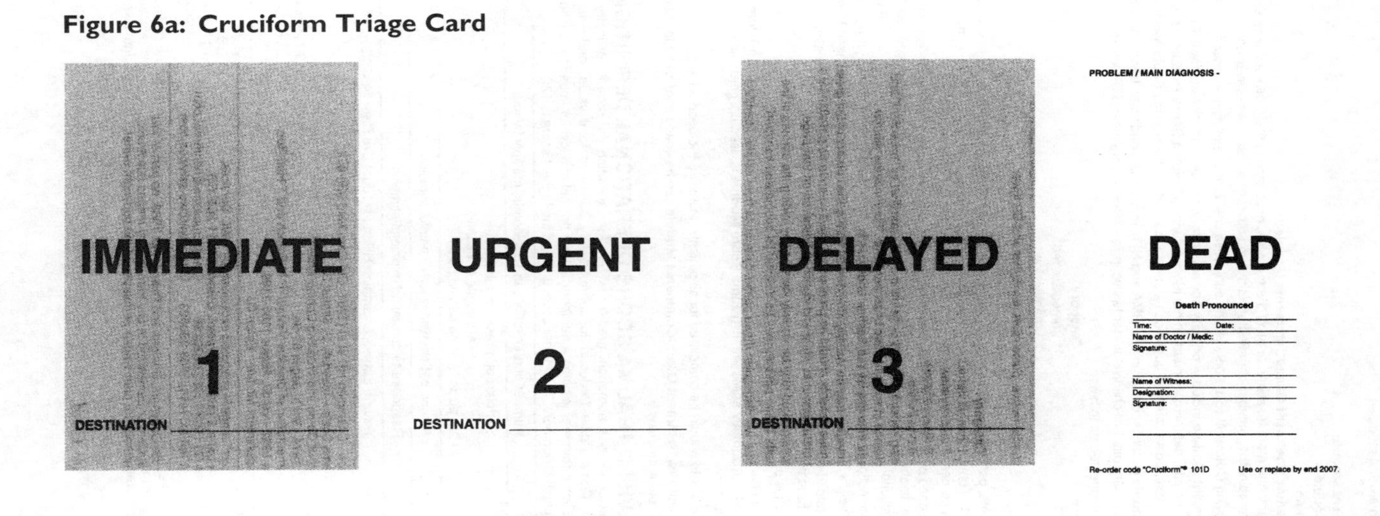

(Red) (Yellow) (Green) (White)

Note: The Cruciform Triage Card – the various elements of which are shown above and on the next five pages. Reproduced with kind permission of Carl Wallin of CWC Services.

Figure 6b: Cruciform Triage Card

Notes on Use

1. The card is divided into several modules.
   a) **Colour Module**
   b) **Casualty Assessment Module**
   c) **Casualty Details Module**
   d) **Trauma Score Module**
   e) **Additional Observation/Treatment Given/Comments Module**
2. Rescuers should use any one module, some, or all of the modules, according to their training, expertise, the time and amount of help available, the number of patients/victims and other circumstances prevailing.
3. The first assessment using **Colour Module** only will be subjective until further assistance arrives.
4. The **Casualty Assessment Module** includes primary and secondary surveys and management. Coma scale scoring started on this module, can be continued on the more extensive **Trauma Score Module** inside the card. Changes to the Sp O_2 levels, pulse rate and pupils can also be recorded on the **Trauma Score Module.**
5. The **Trauma Score Module** is only accurate in TRAUMA and NOT in medical conditions. It is not validated for children under five or the very old. For children under twelve, pulse rates are higher and BP lower.
6. Priority categories are suggested as follows:

Trauma Score	**Category**
1-10	1 immediate/red
11	2 urgent/yellow
12	3 delayed/green
0	4 dead/white

7. a) In large scale incidents, patients trauma scoring consistently at 3 or below may be classified as "expectant" and delayed on scene until patients with a greater chance of survival have been dealt with. They may be visually identified by folding the corners of the green page in to display red 'flashes' underneath.

   b) Patients considered dead by first aiders/paramedics may be visually identified by folding the corners of the green page in to display white "flashes" underneath. These patients should be pronounced dead as soon as possible by a leagally qualified professional who will alter the status colour to white and identify themselves on that page.

   c) Patients considered infectious, infested or contaminated may be visually identified by folding the corners of the green page in to show yellow "flashes" underneath. These patients may be delayed for specialised transport.

   Using the green page, folded as suggested above, will distinguish these patient categories from others, identified as green/delayed, who have no obvious significant injuries or notable conditions.

   NB. A normal scaore (12) is possible with significant injuries, eg fractures of long bones.
8. Triage levels are liable to change. Regular observations and repeated trauma scoring using the **Trauma Score Module** will allow this to be accurately assessed. If **Trauma Score Module** is used up, card can be sealed and another used for continuation.
9. The **Casualty Details Module** should be completed for all patients, including the dead, to assist identification.
10. The **Additional Observations/Treatment Given/Comments Module** can be used for ad hoc information not otherwise catered for in other modules.

TRIAGE REVISED TRAUMA SCORE OPERATIONAL DEFINITIONS

The Triage Revised Trauma Score is a numerical grading system for estimating the severity of injury. The score is composed of the Glasgow Coma Scale (reduced to approximately one third total value) and measurements of cardiopulmonary function. Each parameter is given a number (high for normal and low for impaired function). Severity of injury is estimated by summing the numbers. The lowest score is 3 and the highest score is 12.

Respiratory Rate	Number of respirations in 15 seconds, multiply by four
Systolic Blood Pressure	Systolic cuff pressure, either arm – ausculate or palpate No pulse – no carotid pulse
Best Verbal Response	Arouse patient with voice or painful stimulus
Best Motor Response	Response to command or painful stimulus
Acknowledgements	Trauma Score - Champion HR et al (1981) Crti Care Med 9(9)-672 Glasgow Coma Scale - Teesdale G, Jennet B "Assessment of Coma and impaired Conciousness" The Lancet 1974 - Vol II, pages 81-84 "Early Documentation of Disaster Victims" P M French and T Hamilton Anaesthesia 1982 - Vol 37, Pages 1186-1189. Casualty Assessment Module - Eaton CJ

"The Cruciform" range is produced exclusively by CWC Services.
17 Shavlock Lane, Crafthole, Torpoint, Cornwall PL11 3DF. GB.
Tel/Fax: +44 (0) 1503 230823/230796 Email: cruciform@btinternet.com
International Te/Fax: (+44) 1503 230823 URL: www.cwc-services.com

Notes for the use of Cruciform Triage Card.

Figure 6c: Cruciform Triage Card

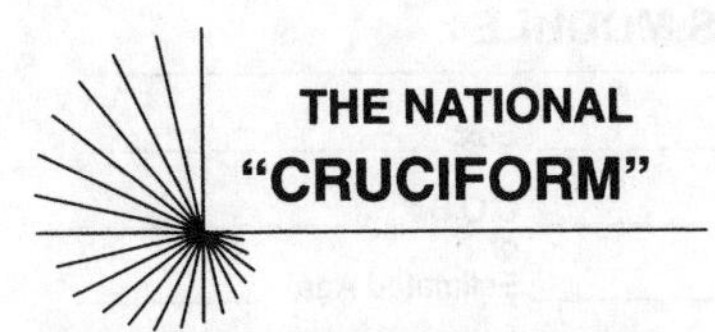

CASUALTY ASSESSMENT MODULE

PRIMARY SURVEY Time:			
Airway	☐ Clear	☐ Obstructed	
C. Spine	☐ Normal	☐ Possible injury	
Breathing	☐ Spontaneous	☐ Problem	
Circulation/ Haemorrhage	☐ External ☐ None/slight	☐ Possible interal ☐ Moderate	 ☐ Severe
Disability	☐ Alert	Responds to	☐ Verbal stimuli ☐ Pain ☐ Unresponsive

Exposure/Injuries

C# Closed Fracture
O# Open Fracture
B Burn (shade area)
F Foreign body
L Laceration
A Abrasion
E Eccymosis (bruising)
P Pain

PRIMARY MANAGEMENT			
Airway	☐ Oropharyngeal	☐ Nasal	☐ ET Tube
	☐ C/Thyrotomy	☐ Oxygen	☐ Suction
C. Spine	☐ C. Collar	☐ Ext Dev	☐ Long board
Breathing	☐ Ventilated	☐ Chest drain	
Circulation	Cannula size:	Rt............	Lt............
	IV Fluids	**Volume**	**Time**
	☐ H'manns/N. Saline	________	______
	☐ H'maccel/G'fusine	________	______

SECONDARY MANAGEMENT			
Analgesia	☐ N_2O/O_2	**Dose/Route**	**Time**
Drugs	☐ ________	________	______
(specify)	☐ ________	________	______
Splinting	☐ Frac Straps	☐ Traction: lbs/Kg ______	
	☐ Box	☐ Other (specify) ______	

SECONDARY SURVEY Time:	
Respiratory Rate	
Oxygen Saturation (Sp O_2)%	
Blood Pressure	
Pulse Rate	
Temp. / Core Temp.	/

Pupil: Size

1 Constricted
2 Normal
3 Dilated

Size R ☐
Size L ☐

Pupil: Reaction (✓or ✗) R ☐ L ☐

Eye Opening	Spontaneous	4	☐
	To voice	3	☐
	To pain	1	☐
	None	1	☐
Best Verbal Response	Orientation	5	☐
	Confused	4	☐
	Inappropriate words	3	☐
	Incomprehensible	2	☐
	None	1	☐
Motor Response	Obeys commands	6	☐
	Localises pain	5	☐
	Withdrawal (pain)	4	☐
	Flexion (pain)	3	☐
	Extension (pain)	2	☐
	None	1	☐
Coma Scale Score			

COMMENTS

Signed:
Status:

John Eaton 1998

The Casualty Assessment Module.

Figure 6d: Cruciform Triage Card$_2$

CASUALTY DETAILS MODULE

Surname: ______ Sex: ______

First names: ______ D.O.B.: ______
or
Address: ______ Estimated Age: ______

______ Telephone Number: ______

Next of Kin: ______ Telephone Number: ______

Type of Incident: ______ Date: ______

Incident Location: ______

Location of Casualty: ______

Position when found: ______

Time found: ______ Found by: ______

Particulars of Casualty (injuries may be shown on body outline overleaf)

Height: ______ Build: ______ Face: ______ Eyes: ______

Nose: ______ Teeth: ______ Complexion: ______ Hair: ______

Distinguishing Marks or Features: ______

Clothing or Dress: ______

State of Body – Normal / Slightly Dis. / Mutilated: ______

Ambulance call sign ______

Crew names ______

Police name: ______

Police number: ______

Location
Arrival time: ______
Location:
Depart time: ______
Destination hospital or other:

Destination
Arrival time: ______

Immediate care/GP/hospital doctor - Name/Call sign: ______

Work Address: ______

______ Telephone Number: ______

Nurse/First Aider/Medic/Other Rescuer - Name, Designation and Telephone Number(s)

The Casualty Details Module which comes as a carbonated tear-out sheet.

Figure 6e: Cruciform Triage Card

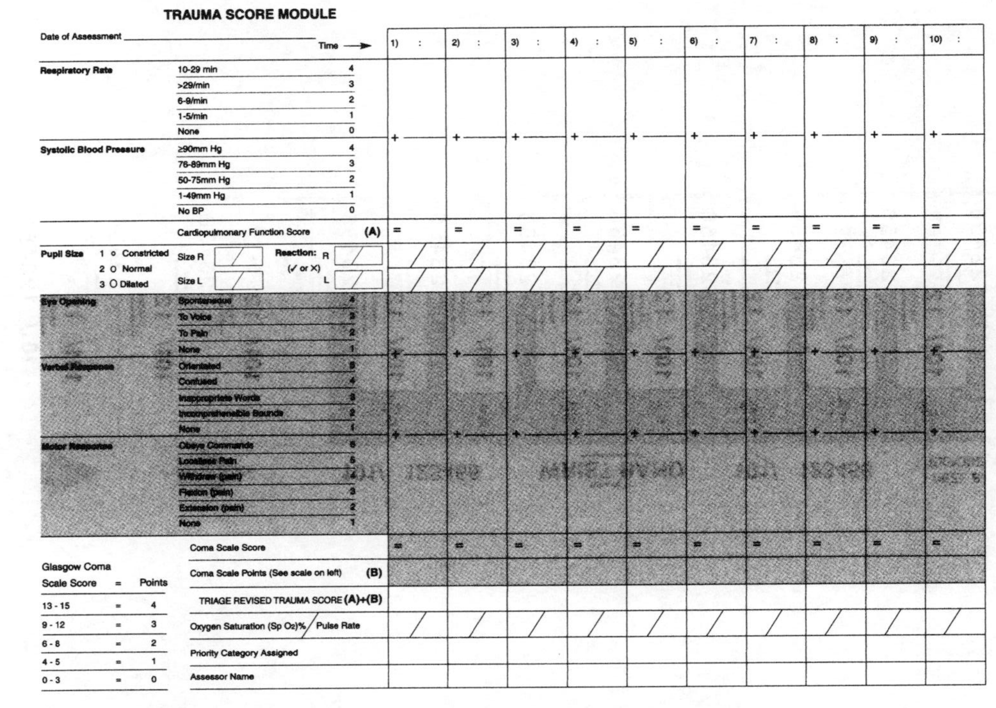

TRAUMA SCORE MODULE

Date of Assessment ____________

		Time →	1) :	2) :	3) :	4) :	5) :	6) :	7) :	8) :	9) :	10) :
Respiratory Rate	10-29 min	4										
	>29/min	3										
	6-9/min	2										
	1-5/min	1										
	None	0	+	+	+	+	+	+	+	+	+	+
Systolic Blood Pressure	≥90mm Hg	4										
	76-89mm Hg	3										
	50-75mm Hg	2										
	1-49mm Hg	1										
	No BP	0										
	Cardiopulmonary Function Score	(A)	=	=	=	=	=	=	=	=	=	=
Pupil Size 1 ∘ Constricted 2 O Normal 3 O Dilated	Size R / Size L	Reaction: R / L (✓ or ✗)	/	/	/	/	/	/	/	/	/	/
Eye Opening	Spontaneous	4										
	To Voice	3										
	To Pain	2										
	None	1	+	+	+	+	+	+	+	+	+	+
Verbal Response	Orientated	5										
	Confused	4										
	Inappropriate Words	3										
	Incomprehensible Sounds	2										
	None	1	+	+	+	+	+	+	+	+	+	+
Motor Response	Obeys Commands	6										
	Localises Pain	5										
	Withdraw (pain)	4										
	Flexion (pain)	3										
	Extension (pain)	2										
	None	1										
	Coma Scale Score		=	=	=	=	=	=	=	=	=	=
	Coma Scale Points (See scale on left)	(B)										
	TRIAGE REVISED TRAUMA SCORE	(A)+(B)										
	Oxygen Saturation (Sp O2)% / Pulse Rate		/	/	/	/	/	/	/	/	/	/
	Priority Category Assigned											
	Assessor Name											

Glasgow Coma Scale Score	=	Points
13 - 15	=	4
9 - 12	=	3
6 - 8	=	2
4 - 5	=	1
0 - 3	=	0

The Trauma Scoring Module which allows for the recording of serial readings of patient observations.

Figure 6f: Cruciform Triage Card

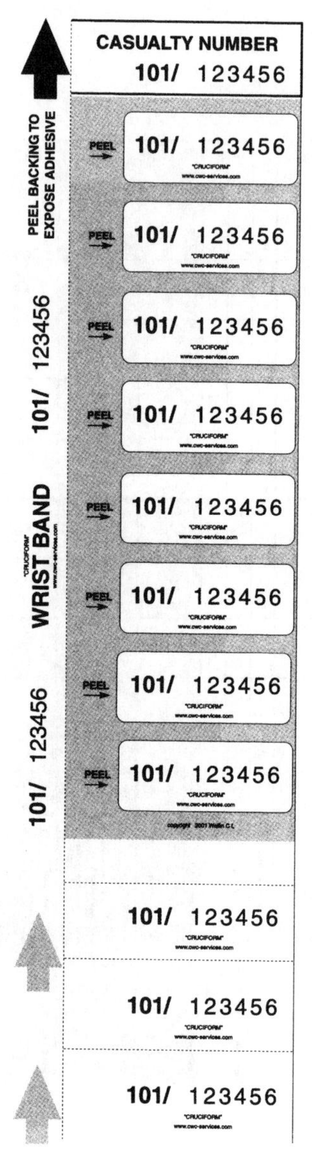

The patient wristband and additional labels for property, blood samples and the like, all showing the same identification number.

Casualty care **15.13**

The level of medical care that can be provided at the site of a Major Incident is very much a reflection of the numbers and skills of staff as well as the equipment levels available. The aims must be to follow best clinical practice within the constraints of the scene. The priorities for the clinical care of individual casualties are listed in Figure 7 below.

Figure 7: Priorities of casualty care

1. Airway – with cervical spine control where appropriate.
2. Breathing – with additional oxygen when available.
3. Circulation – with control of external bleeding.

For the medical management of any Major Incident or Disaster to be successful, the key is the availability of adequate equipment including appropriate drugs, and systems must be in place to deliver these to the scene. Medical Emergency planners must consider issues of temporary structures in which the professionals mobilised to the scene can work, in order to achieve the aim of triage namely that of doing the greatest good for the greatest number.

The organisation of casualty care at the scene of a Major Incident is based on the provision of a Casualty Clearing Station at which the medical, nursing and ambulance resources should work. Individual medical assets should only be deployed into the Disaster ground for the management of individual trapped casualties. Figure 8 below illustrates the organisation of a Casualty Clearing Station at which casualties will be stabilised ready for the journey to definitive hospital facilities.

Distribution of the casualty load 15.14

The size of the incident will define how many hospitals will be needed to cope with the casualty load. It is important that the casualties are spread amongst a number of receiving hospitals so that no one facility becomes overloaded. In the UK, the Ambulance Service, through the Ambulance Incident Officer at the scene, is responsible for the transportation of the casualties from the scene to hospital. The relevant Ambulance Service Control will establish the bed states and capacities of the various hospitals.

Within the UK most casualty transportation from a Major Incident is by road ambulance. However, when one starts to consider large-scale incidents, with long journey times to definitive care, other modes of transport such as rotary or fixed wing aircraft or rail could be considered.

It is of vital importance that the intended destination for each casualty is recorded as they leave the scene so that the information can be passed to the police casualty bureau, whose role it is to collate information both from the scene and from anxious relatives phoning in seeking news of missing family members.

Figure 8: Diagrammatic representation of a Casualty Clearing Station

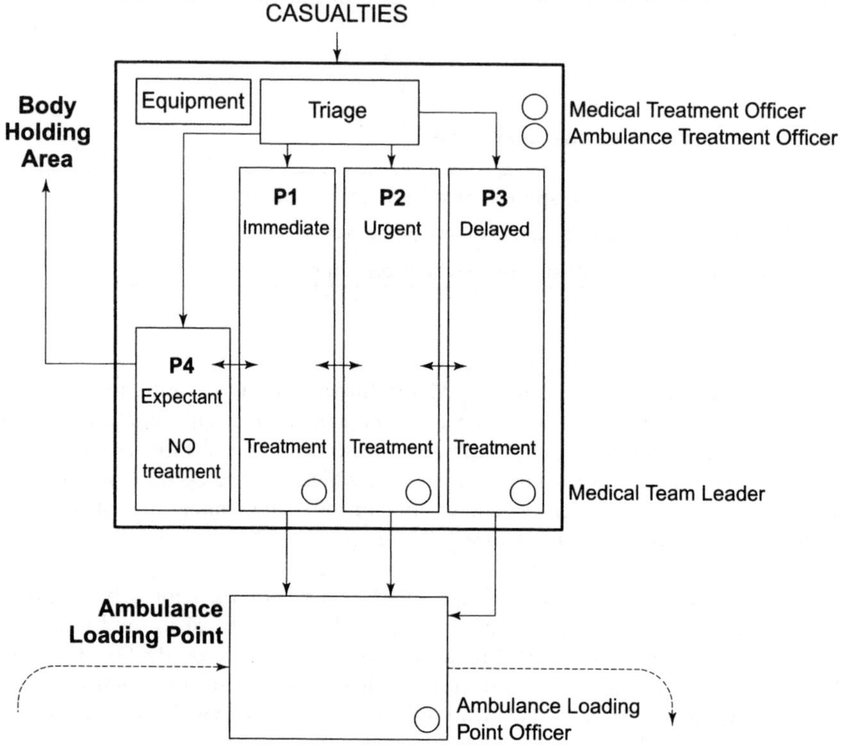

Reproduced with permission from Dr C J Eaton (author/publisher) *The Medical Aspects of Major Incident Management* (1999) British Association for Immediate Care.

Management of the dead 15.15

Part of the effective response to any Disaster or Major Incident situation must be the sensitive and efficient handling of the dead. The incident itself may well be a crime scene such as bomb blast and, therefore, the process by which fatalities are handled is an integral part of the investigation. In situations such as earthquakes and major floods, sanitary pressures may influence the way in which fatalities are managed.

The needs of relatives must be appreciated and the ethnic and religious beliefs of the community involved should also be considered. Although often seen as a totally separate issue, the management of the dead should be considered to be part of any medical response plan.

In any scene with mass fatalities it is important that the dead are labelled as such as early as possible so as to prevent multiple members of the emergency services being diverted from the living. The Medical Incident Officer should allocate a doctor as soon as one is available to work with the police confirming that victims are dead.

This should be done in the presence of a police officer, the time should be recorded and, as far as possible, victims should be left in situ to allow photography and other forensic investigation.

There may be incidents such as earthquakes where body decomposition, prevailing weather conditions and other factors require bodies to be buried as rapidly as possible to prevent the spread of disease. Where mass fatalities have occurred this may have to be done without the benefit of identification, but this is to be avoided if at all possible.

Body holding point **15.16**

In the more controlled Major Incident environment, the management of fatalities follows a more structured and organised path. Once recovered from the scene itself, the remains will be taken to a body holding point. This will usually be adjacent to the incident site, eg a building, temporary structure, or tent. From here the bodies will be taken to a temporary mortuary. This may be a pre-designated location such as an aircraft hangar where plans have already been drawn up by the local authority as to how the temporary mortuary will be set up and managed. It is here that police, coroner's officers, scenes of crime officers, pathologists and forensic odontologists will all work under the direction of HM Coroner for the area in which the accident happened. Post-mortem examinations will be carried out in the temporary mortuary to establish the cause of death of each victim. This can be of considerable importance in, for example, aircraft accidents where the question arises – was the death of the pilot natural causes or a terrorist act thus leading to the loss of the aircraft?

Within the temporary mortuary the identification process of the victims will also be undertaken. Sanitisation and embalming of the remains will be done so that they can be repatriated and returned to their next of kin once identified.

Identifying victims **15.17**

Identification of the victims also involves a range of specialists and many parameters can be used. Clothing, jewellery, tattoos, medical and dental records along with DNA profiling can all play a part in the identification process. A positive match in three parameters will allow an identity to be confirmed.

Private sector assistance **15.18**

Although the Police, acting on behalf of HM Coroner, have responsibility for the deceased in any Major Incident and may themselves have trained body recovery teams, private sector companies are sometimes contracted to provide specific aspects of the process such as the temporary mortuary, identification procedures, property handling and repatriation of foreign nationals.

After the Event

Operational debriefing **15.19**

A review of the medical response and staff debriefing is important and all levels of the response must be allowed to 'have their say'. Those in command appointments will be required to write formal reports and the requirements of any public or judicial inquiry will need to be considered. They should complete and file the reports and paperwork as soon as practical as public inquiries have been known to be held many years after the event.

It is also important that a joint emergency services debrief meeting is held as all incidents will generate problems. This is best held about a month after the event so that each individual organisation has completed its own review before coming together with other services.

Psychological issues **15.20**

Heads of agencies and organisations must consider the psychological and welfare needs of their personnel, many of whom will have worked long hours in very stressful situations. It is important simply for commanders to say 'thank you' to all their staff as early as possible after the stand down from the incident.

The chain of command during an incident should be able to pick up those members of staff who were not coping well so that they could be specifically followed up. More formal Critical Incident Stress Debriefing should be offered to all staff within 48 hours of the incident. There are a number of different models for such debriefings and any medical organisation should have thought how it was going to be delivered and by whom. This is often something that is contracted out to appropriate professional agencies to undertake.

Studies have shown that between 40–70% of those directly involved in a Disaster may experience symptoms during the following month. Not all of these will go on to develop full Post Traumatic Stress Disorder (PTSD) where symptoms must have been present for at least a month before such a diagnosis is normally made.

Studies have also shown that emergency service personnel are more likely to develop PTSD in three very distinct sets of circumstances:

- first, where they have been unable to do anything active to help the victims;
- second, where they have been involved with multiple fatalities, mutilated bodies or multiple child deaths; and
- the third set of circumstances is where the incident has been poorly managed and an individual's line management has been poor.

A detailed study of PTSD, its recognition and management is outside the scope of this chapter but all involved in medical response to any Major Incident or Disaster must be aware of the potential.

Community recovery **15.21**

Incidents such as earthquakes, floods and major fires can damage and destroy property and, indeed, whole communities may be wiped out or require evacuation. Temporary housing and transport infrastructure may need to be replaced. Safe water and sanitation may need to be established along with the provision of health care services in order to control the spread of disease. The mental care required for a community hit by Disaster may need to be in place for a significant period and often has to be enhanced at the anniversary of the incident.

Emergency planning **15.22**

There are lessons to be learnt from every incident no matter what its size or cause. These must be taken on board by the planners and incorporated into revisions of planning documents, validated during exercises and disseminated to all involved in the Emergency response. Thus, the Emergency management circle is closed.

Conclusion **15.23**

The management of Major Incidents or Disasters is one of the challenges of medicine. In the UK few doctors will ever experience more than one Major Incident in an entire career. It is important, therefore, to learn from the experiences of others and to expose oneself to current best practice training. If and when involved in an incident always be ready to share the experience and lessons with other professionals so that the corporate knowledge base is expanded for the benefit of all.

When one is confronted with the daunting experience of a Major Incident it will probably be unexpected. If one is in a command position the most important thing to do is to stop, take a deep breath and plan to bring order to the inevitable chaos by applying the principles outlined in this chapter. That, combined with collaborative working with the other entire emergency services and agencies, will ensure the best possible result.

Lives will be saved by a collaborative team approach and by the rapid delivery of the right skills in sufficient numbers to the scene. In the delivery of medical care remember the old adage 'evacuation is never urgent but resuscitation is'.

Community recovery

[illegible]

Emergency planning

[illegible]

Conclusion

[illegible]

[illegible]

[illegible]

16

Public Inquiries and other Investigations into Disasters and Major Emergencies

In this chapter:

- Introduction.
- Types of inquiry and investigations.
- Coroners' Inquest v public inquiries.
- Criminal proceedings v public inquiry.
- Technical investigation v criminal proceedings.
- Technical investigation v public inquiry.
- Delays and shortcomings.
- Conclusion.
- Appendix 1.

Introduction **16.1**

Following a major disaster several investigations, involving a number of different agencies, are likely to take place. A government report explains:

> 'Establishing want went wrong and how to avoid a recurrence is generally the business of technical investigations. For cases raising serious issues of wider concern, a public inquiry (in Scotland, a fatal accident inquiry) may be appropriate. Where there are signs that a criminal offence may have been committed a criminal investigation may lead to prosecution, and there may be the possibility of a private prosecution. Where death is involved, there will normally be a Coroner's Inquest in England and Wales, or an investigation by the Procurator Fiscal in Scotland. Regulatory authorities may have a role, in

considering disciplinary sanctions. Civil liability and questions of compensation will be dealt with in the civil courts.' (Department of Environment, Transport and the Regions. *Transport Safety, Part III.* Stationary Office, London, 1999, paragraph 3.02.)

Types of Inquiry and Investigations **16.2**

There is, therefore, the possibility of five types of inquiry or investigation in England and Wales:

- coroner's inquest;
- criminal investigation;
- organisational or Government Department internal inquiry;
- technical investigation; and
- public inquiry.

As discussed at **16.1** above, the procedures for Scotland are slightly different.

Coroner's inquest **16.3**

In the United Kingdom, the Office of the Coroner is one of the oldest legal appointments. It is the coroner's responsibility in England and Wales to enquire into the circumstances leading to an individual's death. In Scotland the responsibility rests with the Procurator Fiscal. If the death is not due to a natural cause, the coroner will hold an inquiry. This is known formally as an inquest. Coroners' enquiries and inquests are conducted in accordance with the *Coroners Act 1988* and the *Coroners Rules 1984* (*SI 1984/552*). All inquests are conducted in public.

An inquest is not a trial. It is a limited inquiry into the facts surrounding a death, ie who has died and how, when and where they died.

Following a major disaster in which a number of people have been killed, the coroner may sit with a jury, whose duty is to declare a verdict. The verdict is likely to be one of accidental death, misadventure, suicide, an open verdict (where the result of death cannot be reliably determined) or unlawful killing. However, if a verdict of unlawful killing is declared, a criminal prosecution would have to follow before a person could be convicted of and sentenced, for example, for the offence of manslaughter or, indeed, any other criminal offence.

If a criminal investigation is undertaken or an inquiry is ordered by a Minister under the Inquiries Act 2005, the coroner will delay the inquest until the investigation or the inquiry is complete. If a person is formally charged with causing the death or deaths, the coroner will delay the inquest until the trail is over. If a person is convicted of the manslaughter of those who died, or if the cause of death has been established by an inquiry, the coroner is unlikely to hold a full inquest.

If the Attorney-General considers that, in the interests of justice, it is desirable for the decision of a Coroner's Court to be quashed he may make application to the High Court under *s 13 of the Coroners Act 1988*, for a new inquest to be held.

Included in the reasons for a new inquest are fraud, rejection of evidence, irregularity of proceedings, insufficiency of inquiry and the discovery of new facts or evidence.

At the conclusion of the inquests in March 1991 into the deaths of those who died in the Hillsborough disaster (see **16.4** below), the jury returned verdicts of accidental death by a majority of nine to two. Many of the families of the victims were dissatisfied with the verdict and just over a year later, the Attorney-General was invited to use his powers under *s 13* of the *Coroners Act 1988* but declined to do so. In April 1993, six of the families were granted leave to move for judicial review to quash the original verdict of the inquests and seek a new inquest. A judicial review is a review of a law or an executive act by a government employee or agent for the violation of basic principles of justice. The court has the power to strike down that law, to overturn the executive act, or order a public official to act in a certain manner if satisfied that the law or act was unconstitutional or contrary to that practised in a free and democratic society. However, in November 1993, the Divisional Court dismissed their application.

Criminal investigation 16.4

When there is a suggestion that a criminal offence has been committed, a criminal investigation will be undertaken. A criminal investigation is the systematic collection of information for identifying, apprehending and convicting offenders against the criminal law. The collection of information must be systematic because there are legal rules that must be followed in order for that information to be admissible into a court of law.

The success or failure of an investigation largely depends both on the validity of the law under which the investigation is conducted and the thoroughness of that investigation. There is no doubt that, at the current time, the criminal law is unsatisfactory insofar as it relates to disasters and emergencies. Whilst it is relatively straight forward to bring a successful prosecution against a person who is on the front line, eg a train driver, it is far more difficult to acquire a criminal conviction against a senior representative of a company or a corporation. This is discussed in greater detail at **14.26–14.29** and **14.33–14.36**.

Following the Hillsborough disaster in 1989, in which 96 people died at the Liverpool versus Nottingham Forest football match, Lord Justice Taylor was appointed by the Home Secretary to:

> 'inquire into the events at Sheffield Wednesday football ground on 15 April 1989 and to make recommendations about the needs of crowd control and

> safety at sports events'. (Secretary of State for the Home Department. The Hillsborough Stadium Disaster 15 April 1989. Interim Report (Cm 765). HMSO, London, 1989).

Later that year, the Police Complaints Authority (PCA) (since 1 April 2004, the Independent Police Complaints Commission) directed that disciplinary charges of neglect of duty be preferred against Chief Superintendent David Duckenfield, the officer in overall command, and Superintendent Murray, who was effectively his deputy at the ground on the day of the match. However, before the charges could be heard, Duckenfield retired on the grounds of ill-health. Disciplinary charges can only be taken against a serving police officer; this meant the charges against Duckenfield were dropped. The PCA subsequently announced that it would be unfair to pursue what was in essence a joint charge against only one of the two officers and it also dropped the proceedings against Murray. In the autumn of 1990, the Director of Public Prosecutions (DPP) announced that no criminal proceedings would be taken against any of the police officers involved in the policing of the match at Hillsborough.

Criminal proceedings for manslaughter, by way of a private prosecution, were eventually brought by the Hillsborough Family Support Group in 2000, 11 years after the event, against Duckenfield and Murray who both faced two sample charges of killing fans. Superintendent Murray was found not guilty, but the jury failed to reach a verdict on the charges brought against Chief Superintendent Duckenfield. Normally this would result in a retrial. However, the trial judge, Mr Justice Hooper, stayed the two charges against Duckenfield on the basis that he would not receive a fair trial on the second occasion and forcing him to undergo another trail would clearly constitute oppression.

There will, of course, always be a criminal investigation following an incident that clearly involves crime such as an act of terrorism. Such was the case when Pan Am 103 crashed on Lockerbie in 1988 after a bomb had exploded on board the aircraft, killing all 259 passengers and crew and a further 11 people on the ground. The criminal investigation that followed eventually culminated, in January 2001, in the conviction of a Libyan citizen, Abdel Basset Ali al-Megrahi, for murder by three presiding Scottish judges at a court set up at Camp Zeist, a former US air base, in the Netherlands; another Libyan was acquitted. The court recommended that al-Megrahi serve a minimum of 20 years imprisonment.

Internal government department inquiry 16.5

Shortly before the end of the Foot and Mouth Crisis in Great Britain in 2001 (see **4.10** above), the Government announced three separate inquiries, but none of them amounted to the full public inquiry that many politicians and farmers had called upon the Department of the Environment, Food and Rural Affairs to convene. The first, chaired by Dr Iain Anderson, a former senior civil servant, was appointed to:

> 'conduct an inquiry into the Government's handling of the outbreak in order to draw out lessons and make recommendations'. (Ministry of Environment,

> Food and Rural Affairs. Foot and Mouth Disease 2001: Lessons to be Learned. Stationary Office, London, 2001, page 5.)

The second was a scientific review carried out by the Royal Society, led by Sir Brian Follett, which examined the transmission, prevention and control of epidemics. The third, chaired by Sir Don Curry, former head of the Meat and Livestock Commission, was set up to advise the Government on how to create a sustainable, competitive and diverse farming and food sector which would contribute to a thriving and sustainable rural economy.

In the context of this book, the only one that warrants further mention was the first. Dr Anderson's inquiry:

> 'was asked to start when the epidemic was judged to be over and then to report within six months.' (Ministry of Environment, Food and Rural Affairs. Foot and Mouth Disease 2001: Lessons to be Learned. Stationary Office, London, 2001, page 5.)

In his subsequent report, Dr Anderson described how he went about conducting the inquiry. He appointed a small secretariat, led by a senior civil servant with experience across government. The inquiry:

- received 576 written submission;
- visited many of the worst hit regions around the country, during which they met over 200 representatives of national and local organisations in round-table discussions and talked, in their homes or places of work, to many people who had been directly affected;
- invited over 100 people to be formerly interviewed; and
- met with officials in the Netherlands, France and the European Commission in Brussels.

In the final report, Dr Anderson made 81 recommendations as to how such a crisis might be avoided, but, if that proved impossible, how it could be managed in the future in order to bring about a more effective response.

Technical investigation 16.6

The purpose of a technical investigation is to establish what went wrong and how to avoid a recurrence. Various bodies concerned with, for example, the safety of transport can set up their own investigations.

Aircraft accidents, involving civil aircraft, in the United Kingdom are investigated under the *Civil Aviation (Investigation of Air Accidents and Incidents) Regulations 1996 (SI 1996/2798)*. These Regulations comply with the international requirements laid down in Annex 13 of the Chicago Convention. In November 1994, a Directive *Establishing the Fundamental Principles Governing the Investigation of Civil Aviation Accidents and Incidents* was adopted by the EU Council (*European Union Directive 94/56/EC*). The Directive, which basically incorporates the principles of Annex 13, ensures that all Member States have an 'independent' investigation body for aircraft accidents.

In a recent publication *Airline Insurance: An Introductory Guide* (Jim Bannister Developments Ltd, London, 2003), Ken Smart, the Chief Inspector of Air Accidents, set out the duties of the Air Accident Investigation Branch (AAIB) in Appendix 3. On average, there are about 350 aircraft accidents and other serious incidents at airports each year. Up until 1996, regulations covering the investigation of aircraft accidents allowed the Secretary of State for Transport to order a Public Inquiry if it is considered to be 'expedient in the public interest' to do so. This has been very rare, the last public inquiry being conducted into the disaster at Staines in 1972, when a British European Airways Trident aircraft crashed immediately after take-off, killing all 118 passengers and crew. Since then, there have been a number of serious incidents such as the fire on board an aircraft at Manchester Airport in 1985, killing 55 people, and the Kegworth crash in 1987 in which 47 people died and 79 were injured, most of them seriously. However, the investigation in each case was conducted by the Air Accident Investigation Branch.

With the exception of a public inquiry, it was the Chief Inspector of Accidents who has complete discretion to decide the extent and form of the investigation. If the accident or incident involves public transportation or general aviation facilities, a team of inspectors will generally carry out what is termed as a 'field investigation'. As a result of their initial enquiries the Chief Inspector will then decide whether or not to conduct a formal investigation under the Regulations. In only about 4% of all accidents and serious incidents does the Chief Inspector order a formal investigation, following which a report (known as the AAIB report) is submitted to the Secretary of State for Transport. All other field investigations result in a comprehensive report being submitted to the UK Civil Aviation Authority.

If any party disagrees with the findings of the report submitted to the Secretary of State for Transport, they are permitted to request a review board. This consists of a judge or senior Queens' Counsel sitting with independent technical assessors to review the AAIB accident report. The AAIB report is not amended by this procedure but the findings of the review board are published together with the AAIB report.

Similar procedures for the investigation of marine accidents exist under the *Merchant Shipping Act 1995*. The Marine Accident Investigation Branch (MAIB) conducts its investigations in accordance with the *Merchant Shipping (Accident Reporting and Investigation) Regulations 1999 (SI 1999/2567)*. The legislation applies to merchant ships, fishing vessels and, with some exceptions, pleasure craft. There is a Memorandum of Understanding between the Health and Safety Executive (HSE), the MAIB and the Maritime and Coastguard Agency as to which organisation will take the lead in investigations where they share a common interest, particularly at the ship/shore interface.

Following criticism of the way railway accidents were investigated, Part 1 of the *Railways and Transport Safety Act 2003* established the Rail Accident Investigation Branch (RAIB) along similar lines to that which already exists in the marine and aviation sectors. This followed recommendations by Lord Cullen in The Ladbroke Grove Rail Inquiry, Part 2 Report (HSE, London, 2001).

The following year, the EU issued a major Directive in respect of railways (EU Directive 2004/49/EC). The Directive specifies the need to have a Safety Authority (Articles 16–18), and an investigating body 'which shall be independent' from the railway authorities, including the Safety Authority, 'and from any party whose interests could conflict with the tasks entrusted to the investigating body'(Article 21). The Directive also describes, in broad terms, the procedures for the investigation (Article 22). In the UK, regulations in the form of the Railways (Accident Investigation and Reporting) Regulations 2005 describe how such an investigation will be conducted.

Finally, if the incident occurs in a chemical plant or similar facility, the investigation will almost certainly be undertaken by the Health and Safety Executive under the Health and Safety Act 1974. The rules for the conduct of such investigations are to be found in the Health and Safety Inquiries (Procedure) Regulations 1975.

During the last 20 years or so, even though it may not have been related to the cause of the accident, such reports have invariably commented on the performance of the emergency services and other agencies in response to the disaster.

Public inquiries 16.7

Until 1921 the public investigation of disasters, emergencies and crises tended to be undertaken by parliamentary committees of inquiry. Such inquiries suffered from two disadvantages. Firstly, they could not examine witnesses on oath. Indeed, in some cases, witnesses ignored requests from the committee. Secondly, political partisanship sometimes got in the way of the investigation. So, in 1921, the Government passed the *Tribunals of Inquiries (Evidence) Act.* Since then only 24 inquiries in total were appointed under this particular piece of legislation of which only 4 related to the type of disasters and emergencies referred to in this book, namely:

Table 1: Tribunals and Inquiries conducted under The Tribunals and Inquiries (Evidence) Act 1921 relating to disasters and emergencies

	Event	Chair	Date	Command Number
1.	The disaster at Aberfan in which 116 children and 29 adults were killed (see **11.2**)	Sir E Davies	1967	HC 553
2.	The events on Sunday 30 January 1972 (Bloody Sunday), which led to loss of life in connection with the procession in Londonderry on that day	Lord Widgery	1972	HC 220/72
3.	The shootings at Dunblane primary school on 13 March 1996	Lord Cullen	1996	3386
4.	A further inquiry into the events on 30 January 1972 (Bloody Sunday)	Lord Saville	1998	Not yet reported

(Source: British Tribunals of Inquiry: Legislative and Judicial Control of the Inquisitorial Process-Relevance to Australian Royal Commissions. Research Paper no. 5 2002–2002, Departmental of the Parliamentary Library of Australia.)

Note: The Tribunals of Inquiries (Evidence) Act has now been repealed by the Inquiries Act, 2005.

More recently, public inquiries into disasters and emergencies have tended to be held under the *Health and Safety at Work, etc Act 1974* (see also **14.39** above). *Section 14* of the Act enables the Health and Safety Commission to direct the HSE or authorise any other person to investigate and make a special report on any accident, occurrence, situation or other matter which the Commission thinks is necessary or expedient to investigate. Whilst it normally does, by virtue of *section 14(1)*, it is immaterial whether the HSE is or is not responsible for securing the enforcement of any of the statutory provisions relating to the accident, occurrence, situation or other matter. When it becomes what is broadly known as a public inquiry, the consent of the appropriate Secretary of State is sought (*section 14(2)(B)*). In such cases it will be 'held in accordance with regulations made for the purpose by the Secretary of State and shall be held in public except where or to the extent that the regulations provide otherwise' (*section 14(3)*). Under *section 14(4)(B)* the person holding the enquiry may be given the power to summon witnesses to give evidence or produce documents and also 'the power to take evidence on oath and administer oaths or require the making of declarations'.

The Commission may cause any report made by virtue of *section 14* of the Act 'or such part of it as the Commission thinks fit, to be made public at such time and in such a manner as the Commission thinks fit' (*section 14(5)*).

In Scotland, a Public Inquiry following an incident in which people have died is known as a Fatal Accident Inquiry.

The Tribunals of Inquiry (Evidence) Act 1921 **16.8**

The Tribunal of Inquiry (Evidence) Act 1921 could be summarised as follows:

- It allowed for a tribunal to be set up, with all the powers of the High Court as regards examination of witnesses and production of documents, in order to objectively investigate the facts relating to a matter of urgent public importance.
- A resolution of both Houses of Parliament was required to set up such an inquiry.
- It allowed the tribunal to enforce the attendance of witnesses and examine them on oath, affirmation, or otherwise.
- It allowed for the tribunal to compel the production of documents.
- Anyone who failed to attend on being summoned, or who refused to take the oath or produce a required document or take any other action which would, in a court of law, be a contempt of court, could be punished as if that person was guilty of contempt of court.
- Proceedings were normally in public unless in the opinion of the tribunal it was in the public interest to hold it, or a part of it, in private.

The Inquiries Act 2005 **16.9**

The Inquiries Act replaces over 30 different pieces of legislation on inquiries, including the Tribunal of Enquiries (Evidence) Act, 1921. Therefore, according to the Government, the Act, which came into force on 7 June 2005, creates a comprehensive new statutory framework for inquiries set up by ministers. It covers the setting up of inquiries, the appointment of people to conduct them, their procedures (the Act provides for the appropriate Minister to make regulations covering these (see **16.11**)) and powers, and the submission and publication of reports.

It is not intended that inquiries normally carried out under the various pieces of legislation that provide for technical investigations (see **16.6**) will be replaced but the Inquiries Act will be used to look into a past event, or series of events, that has caused – or is capable of causing – public concern. The Act will only be used for inquiries where the Minister decides that it should have statutory powers.

There is, according to the Government, a presumption that such inquiries will allow public access to the hearings. However, the Minister setting up the inquiry or the inquiry chairperson can place restrictions on public access if they can justify such a restriction in law, or it is in the public interest or it will better enable the inquiry to fulfil its terms of reference. However, it is unlikely that the Minister or the inquiry chairperson would place such restrictions on an inquiry set up to investigate a major disaster unless it resulted from a terrorist incident in which sources of intelligence were a factor.

In the setting up of an inquiry, the Minister concerned must, as soon as is reasonably practicable, make a statement, written or oral, to the Parliament.

The Act also:

- gives power to an inquiry chairperson to require the production of evidence for the establishing Minister or the chairperson, or both, and place restrictions on the public access to the inquiry where appropriate;
- places a duty on the chairperson to deliver a report to the commissioning Minister and set out what the final report may contain;
- gives powers to the ministers of the devolved legislatures in the UK to set up inquiries under the Act;
- contains provisions on inquiries established jointly by two or more ministers, including cross-border inquiries within the UK;
- creates offences and provides for the enforcement of inquiry orders by the High Court or Court of Session;
- gives members of an inquiry immunity from civil proceedings;
- makes arrangements for the payments of expenses of witnesses including legal representation where appropriate.

Clearly, the Government has been concerned over the spiralling cost of recent inquiries and almost certainly any new Regulations introduced by the Minister under the Act will place a requirement on the inquiry chairperson to have regard to the costs. For instance, the regulations are likely to introduce tighter control over legal representation and the taking of evidence. A new requirement to publish the costs of the inquiry will provide a greater level of public scrutiny.

Criticisms of the new legislation 16.10

Some criticism of the new legislation has been expressed both during the debate stage and since, namely:

- The new Inquiries Act appears to have brought about a fundamental shift in the manner in which the actions of government and other bodies can be investigated in the UK. Whereas, it was the Parliament that instigated an inquiry under the 1921 Act, it now appears that it is the Minister and, whilst he must inform the Parliament about various actions he intends to take or has taken, the Parliament cannot debate the action he has taken. Therefore, it is suggested that the Parliament's role has been reduced to that of the passive recipient.
- The Minister's powers to interfere at various stages during the process of the inquiry robs it of any independence, and notable people have expressed concern of this. For instance, in a letter to Baroness Ashton at the Department of Constitutional Affairs, Lord Saville, who chaired the most complex inquiry ever held in the UK, the Bloody Sunday Inquiry, expressed 'grave reservations' about the

Act, particularly about the restrictions that can be placed upon it by a Minister, saying:

> 'I take the view that this provision makes a very serious inroad into the independence of any inquiry and is likely to damage or destroy public confidence in the inquiry and its findings, especially in cases where the conduct of the authorities may be in question.'

He pointed out that any ministerial interference with 'a judge's ability to act impartially and independently of government would be unjustifiable' and went on to say that 'neither he nor his fellow judges on the BSI would be prepared to be appointed as a member of an inquiry that was subject to a provision of this kind'.

- Lord Norton, Professor of Government at the University of Hull, expressed similar sentiments during a parliamentary debate on the Inquiries Bill:

  > Given the powers vested in a Minister, one has to wonder who would accept appointment to serve on an inquiry if independence were not guaranteed. *(Hansard, House of Lords, 9 December 2004, col. 1002)*

In response, the Minister for Constitutional Affairs, Baroness Ashton, stated:

> 'The Act consolidates existing inquiries legislation, fills the gaps and codifies best practice from past inquiries. For the first time in statute the Act lays down all key stages of the inquiry process – from setting up the inquiry, through appointment of the panel to publication of reports.
>
> The Act does not, as has been suggested, radically shift emphasis towards control of inquiries by Ministers. Instead, it makes clear what the respective roles of the Minister and chairman are, thereby increasing transparency and accountability.' (*Investigatory Inquiries and the Inquiries Act 2005. Standard Note: SN/PC/2599 updated 28 June 2005. House of Commons, London, p. 3*)

Rules of procedure and guidance 16.11

The Act specifies that rules of procedure may be produced by the Lord Chancellor, the Scottish Ministers, the National Assembly for Wales and the First Minister and Deputy First Minister in Northern Ireland. In 2004, the Department of Constitutional Affairs issued a discussion paper, suggesting that it would, in due course, issue statutory provisions, which would be likely to cover matters of evidence, representation and procedure, handling inquiry records and assessing costs. At the same time, the paper pointed out that, whilst not making them part of the statutory provisions, there was likely to be a need to issue guidance which would take the form of practical advice to the inquiry team on how to organise and run the inquiry. This might cover such things as finding suitable accommodation and IT systems, the role of the secretary, solicitor and counsel, media handling, and arrangements for making reports available on the day of publication. The guidance is likely to be based on the experience of previous inquiries.

Terms of reference and independence **16.12**

Under the Inquiries Act, 2005, the terms of reference will be set by the Secretary of State in consultation with the person appointed to chair the inquiry. If it is held under the *Health and Safety at Work, etc Act 1974*, the terms of reference are likely to be drawn up by the Health and Safety Commission and approved by the appropriate Secretary of State. The appointed Chairperson of the Inquiry is also normally consulted before the terms of reference are announced. The terms of reference tend to be short and to the point. They are designed not to fetter the inquiry in establishing the truth. The Chairperson can subsequently apply for them to be amended but this is not usual given the fact that he or she has been consulted during the initial drafting.

Despite the reservations expressed earlier (see **16.10**), any inquiry set up either by a government department or by the HSE with government approval is expected to act independently. In 1993, the Court of Appeal, in *Crampton* v *Secretary of State for Health* (the Allitt Inquiry) laid down that inquiries 'should be searching, thorough and completely independent'.

Gathering of evidence **16.13**

The gathering of evidence is usually the responsibility of Government Solicitors under the supervision of the Presenting Counsel. If it is a particularly large or complex inquiry, other agencies such as the police or the HSE may be charged with the responsibility of undertaking preliminary investigations and collecting evidence on behalf of the Government Solicitors. An exception to this was the Hillsborough disaster. In this case, Lord Justice Taylor decided 'that the investigation of the disaster and the gathering of evidence . . . should be conducted by an independent police force' and the Chief Constable of West Midlands Constabulary was asked to undertake this task. Lord Justice Taylor emphasised that, on this occasion, the police were directly responsible to him for the conduct of the investigation. (para 5, page 1 Interim Report)

Representation **16.14**

The parties involved in an inquiry are normally legally represented. If there are parties with similar interests, the Chair may direct that a single legal team represent them, e.g. the injured and the bereaved. 'Substantial bodies' will normally be expected to pay their own costs of representation. For organisations, this will often be met by insurance. The Chair can recommend that the State meet or contribute to certain parties' costs if they would not otherwise be able to afford representation.

Procedure **16.15**

Currently, the actual procedures to be followed during the hearing of an inquiry are at the discretion of the Chair although this may change once the regulations and guidance which are due to accompany the Inquiries Act 2005 have been issued.

Normally Counsel to the Inquiry opens the hearing, outlines its purpose, details what occurred and the evidence that is due to be heard. The legal representative of each party (or the party themselves if not legally represented) will then make a short opening statement about the matter (a more lengthy written statement might also be submitted).

The witnesses are called by Counsel to the Inquiry and taken through their evidence. Then each party's lawyer has the opportunity to cross-examine that witness. Once the evidence has been heard, the parties put in written submissions to the Chair. The hearing concludes with the parties' representatives making a short oral statement outlining their written submissions.

Hearings **16.16**

In some cases, the Chairperson will invite written submissions to be made prior to or early on in the inquiry. Most of the evidence from those who were closest to the disaster is heard orally by the inquiry. In addition, where he or she considers that a written submission is particularly relevant, the Chairperson may invite that person or persons to appear before him or her to give oral evidence.

By virtue of the legal authority vested in the Chairperson, he or she can investigate anything that they consider to come within the Terms of Reference. Unless it is an important matter of state security, evidence relating to the investigation will be heard in public.

Witnesses **16.17**

In 1966, Lord Salmon laid down six principles relating to witnesses, namely:

1. Before any person becomes involved in an inquiry, the inquiry must be satisfied that there are circumstances which affect him or her and which it proposes to investigate.
2. Before any person who is involved in an inquiry is called as a witness, he or she should be informed in advance of allegations against him or her and the substance of the evidence in support of the allegations.
3. He or she should be given an adequate opportunity of preparing his or her case and of being assisted by a legal adviser, and his or her legal expenses should normally be met out of public funds.
4. The witness should have the opportunity of being examined by his or her own solicitor or counsel and of stating his or her case in public at the inquiry.
5. Any material witness he or she wishes to be called at the inquiry should be heard, if it is reasonably practicable.
6. The witness should have the opportunity of testing by cross-examination conducted by his or her own solicitor or counsel, any evidence which may effect him or her.

(Royal Commission on Tribunals of Inquiry 1966. Report of the Commission under the Chairmanship of the Right Honourable Lord Justice Salmon, HMSO, 1966.)

In 1993, in the case of *Crampton and others v Secretary of State for Health (9 July 1993)*, Lord Bingham said: 'the rationale of the six cardinal principles is undoubtedly sound and anyone conducting an inquiry of this kind is well advised to have regard to them'. However, Lord Bingham pointed out that: 'the Royal Commission Report itself (had) not been embodied in legislation and numerous inquiries have been . . . satisfactorily conducted, since 1966 without observing the letter of those principles'.

In a Public Inquiry unconnected with disasters and emergencies, Sir Richard Scott suggested a number of additions to Lord Salmon's principles, namely:

7. Adequate advance notice should be given to witnesses on the matters in respect of which they will be questioned.
8. Adverse and damaging allegations or proposed criticism made by other witnesses or by any other means should (if they are relevant) be drawn to the attention of the object of the allegations so that he or she can, if desired, respond to them.
9. Legal assistance should be available to those involved, both at the stage of giving evidence and at the stage of responding to criticisms.
10. Adversarial procedures, such as the right to cross-examine other witnesses, the right to have an examination-in-chief or re-examination conducted by a party's own lawyer or the right for interested parties to participate over and above the extent mentioned, in oral hearings, should not be incorporated into the procedures at an inquisitorial inquiry, except to the extent that fairness requires it or the particular circumstances warrant it.

(*Inquiry into the Export of Defence Equipment and Dual-Use to Iraq*. Chairman: Sir Richard Scott. HMSO, London, 1996, Chapter 1 of Section K).

Presenting Counsel can normally demand the attendance of witnesses. Reasonable expenses for travel, subsistence and loss of earnings can be paid out of public funds.

Appeal **16.18**

Although there is no appeal against the findings of an inquiry, the Inquiries Act 2005 does make provision for an application for judicial review of a decision made by the Minister in relation to an inquiry or by a member of an inquiry panel. In extreme circumstances, the Minister could order a new inquiry if it was thought that significant new material had come to light that would affect the original findings.

In 1989, Lord Justice Taylor carried out an investigation into the Hillsborough disaster (see **16.4** above). On 5 December 1996, Granada Television broadcast a programme on Hillsborough that suggested there was fresh evidence that had not been available to Lord Taylor and which called into question the verdict at the Coroner's inquests and other decisions; for instance that made by the Director of Public Prosecutions. As a result, the Hillsborough Family Support Group brought pressure to bare on the Government and, on 30 June 1997, the Home Secretary appointed the Rt Hon Lord Justice Stuart-Smith:

> 'To ascertain whether any evidence exists relating to the disaster at the Hillsborough Stadium on 15 April 1989 which was not available:
>
> (a) to the inquiry conducted by the late Lord Taylor; or
>
> (b) to the Director of Public Prosecutions or the Attorney General for the purpose of discharging their respective statutory responsibilities; or
>
> (c) to the Chief Officer of South Yorkshire Police in relation to police disciplinary matters;
>
> and in relation to (a) to advise whether any evidence not previously available is of such significance as to justify establishment by the Secretary of State for the Home Department of a further public inquiry; and in relation to (b) and (c) to draw to their attention any evidence not previously considered by them which may be relevant to their respective duties; and to advise whether there is any other action which should be taken in the public interest'. (Secretary of State for the Home Department. *Scrutiny of Evidence Relating to the Hillsborough Football Stadium Disaster.* The Stationary Office, London, 1998, Introduction, para 3.)

Stuart-Smith found that there were only two matters that Lord Taylor had not considered in his 'otherwise very full inquiry' and neither gave grounds 'for reopening the judicial inquiry or any other proceedings'. (Secretary of State for the Home Department. *Scrutiny of Evidence Relating to the Hillsborough Football Stadium Disaster.* The Stationary Office, London, 1998, Chapter 6, para 32.)

In 1972, a Tribunal was appointed under the *Tribunals and Inquiries (Evidence) Act 1921* to investigate the fatal shooting of 13 men and the wounding of a further 15 on the occasion of a protest march in Londonderry on 30 January 1972. The sole member and chairman of the Tribunal was Lord Chief Justice Widgery. The primary tasks of the Tribunal were to:

- establish the truth about what happened on Bloody Sunday;
- satisfy the public that the full facts has been established; and
- generally help restore public confidence in the integrity of government.

Ever since the Tribunal published its report in April 1972 it is claimed, particularly by nationalists in Northern Ireland, that there were substantial grounds for believing that Widgery failed to achieve any of these objectives. Twenty-six years later, in 1998, the UK Government announced another inquiry, this time under Lord Justice Saville, which has still to report (see **Table 1**).

Why is there a need for an investigation or inquiry? 16.19

In an unpublished paper (*Inquiries into Major Disasters*, Typescript, Cranfield University), Mervyn Jones listed three reasons as to why it is important to hold an investigation following a major disaster or major emergency.

Firstly, when such an event does occur, the truth is often distorted, sometimes deliberately, but generally unintentionally. Intentional distortion, at the very least, may be a subtle bias of interpretation that is articulated by a person or persons who may be motivated by individual, commercial or political considerations. At the very worst it can be an outright lie. Unintentional distortion of the truth can occur through ignorance of the facts and through rumour. The media can often contribute to this form of distortion, particularly when reporting what is generally called a 'breaking news' story. It is important, therefore, to establish the truth.

Secondly, it is important to publish the truth. In other words, when there has been intentional or unintentional distortion that has appeared in the public domain, it is important that people know precisely what did happen and what recommendations have been made to ensure that it does not happen again.

Finally, public inquiries are often instituted in response to clamour from the public, often led by concerned relatives of the deceased or those who have been injured, who are concerned that there may be a 'cover-up' of the truth, leading to a lack of accountability. Occasionally, the event may be of such magnitude and be subject to such far-reaching public concern that the Government, often fuelled by demands in the media, may decide that a public inquiry is in the 'public interest'.

What is the public interest? 16.20

Public Inquiries tend to be reserved for those disasters or emergencies that cause the greatest public outcry. In his report into the capsize of the Herald of Free Enterprise, Mr Justice Sheen suggested that:

> 'In every formal investigation it is of great importance that members of the public should feel confident that a searching investigation has been held, that nothing has been swept under the carpet and that no punches have been pulled'. (Department of Transport. *MV Herald of Free Enterprise*. HMSO, London, 1987, para 60.)

Lord Justice Clarke makes a similar comment in his report, *Thames Safety Inquiry* (Secretary of State for the Environment, Transport and the Regions. The Stationary Office, 2000), suggesting that:

> 'the purpose of a public inquiry is . . . to carry out a full, fair and fearless investigation into the relevant events and to expose the facts to public scrutiny'.

However, he takes the view that:

> 'No member of the public has a right to a public inquiry. Whether such an inquiry should be ordered will (as ever) depend upon all the circumstances of the case. It will depend upon where the public interest lies. The public interest is not of course the same as the interest of the public. The public may be interested in many things which it would not be in the public interest to investigate publicly'. (Secretary of State for the Environment, Transport and the Regions. Thames Safety Inquiry, The Stationary Office, London, 2000, para 5 10.)

Clarke suggests that a public inquiry has two purposes. Firstly, to ascertain the facts, and secondly, to learn lessons for the future. He continues:

> 'In the vast majority of cases the second is a very important ingredient... because it is to be hoped and indeed expected that the detailed examination of the causes of a particular casualty will yield valuable information from which lessons can be learned'. (Secretary of State for the Environment, Transport and the Regions. Thames Safety Inquiry, The Stationary Office, London, 2000, para 5.3.)

In order to meet with Clarke's suggestion, a public inquiry should seek to establish:

- exactly what went wrong;
- why it went wrong;
- whether the disaster could have been avoided had appropriate and reasonable action been taken; and
- make recommendations of the action that needs to be taken in order to avoid a recurrence of such a disaster.

But, says Clarke, it is not in the public interest that a public inquiry should seek

> 'to establish civil liability or to consider whether a crime has been committed. The former is the role of the civil courts and may involve many questions of fact and law which it would not be appropriate to debate at an inquiry. The latter is the role of the police, the CPS and the DPP'.

He concludes that the 'role of an inquiry is simply to find the facts', although he does recognise 'that those facts may form the basis of civil liability or indeed of an allegation that a crime has been committed'. (Department of Transport. *MV Herald of Free Enterprise*. HMSO, London, 1987, para 5.5.)

Inquiries and the railways **16.21**

There have been numerous inquiries into disasters on the railways over the last 20 years. Four are mentioned here because in each case the event resulted in a public inquiry.

The King's Cross Underground Fire, 1987 **16.22**

Five days after a serious fire at King's Cross Underground Station on 18 November 1987, in which 31 people died and many were injured (see **11.21** above), Desmond Fennell, QC, was appointed by the Secretary of State for Transport to hold a formal investigation under *section* 7 of the *Regulation of Railways Act 1871* into the causes and circumstances of the fire. According to Fennell, the terms of *section* 7 required him to have regard to three particular matters:

- the causes of the accident;
- the circumstances attending the accident; and
- any observations or recommendations arising out of the investigation.

Fennell made it clear at the outset that the inquiry:

> 'was to be an investigation and not litigation: it was not a lawsuit in which one party wins and another party loses. It was quite different from the ordinary criminal process which is accusatorial in character. The Investigation was inquisitorial. It was an exercise designed to establish the cause of the disaster and to make recommendations which will make a recurrence less likely. Those who died deserve nothing less'. (Department of Transport. *Investigation into the King's Cross Underground Fire*. HMSO, London, para 5, page 22.)

The inquiry was divided into two parts. Part 1 was mainly devoted to eyewitness evidence, both oral and written, and concluded with expert evidence as to the mechanics of the flashover. Part 2 was devoted principally to the human and physical state of affairs in place at the underground station on the night of the disaster and within London Transport at the time. There was also further extensive scientific evidence. The final report made 157 recommendations grouped under three headings:

- most important;
- important; and
- necessary.

The Southall Railway Accident **16.23**

Within hours of a collision between two trains on 19 September 1997 at Southall, in which seven people died and 139 were injured (see also **11.29** above), Professor John Uff, QC, was asked by the HSE to chair an inquiry 'to determine why the accident happened, and in particular to ascertain the cause or causes, to identify any lessons which have relevance for those with responsibilities for securing railway safety and make recommendations'. (HSE *The Southall Rail Accident Inquiry Report*, HSE Books, Norwich, 2000, page a).

Although the inquiry proceedings began in December 1997, with a formal opening in February 1998, it was suspended after the driver of one of the trains, Larry Harrison, was charged with manslaughter in April 1998. The company for whom he worked, Great Western Trains, also faced seven charges of corporate manslaughter. In July 1999, at the Old Bailey, the prosecution offered no evidence against Harrison because it was felt that he had been so psychologically affected that there was no chance of him being able to present a defence. Mr Justice Scott Baker then threw out the seven charges of corporate manslaughter against Great Western Trains because of weaknesses in the law (see **14.26–14.28** above). The Law Commission, in report 237, advised the Government that the law of corporate manslaughter should be changed in 1997. As at the beginning of 2006 it remains unchanged. However, Great Western Trains did plead guilty to a charge under health and safety legislation and were fined a record £1.5 million. The health and safety inquiry under Professor Uff eventually resumed on 20 September 1999 and closing submissions were made on 20 December 1999.

Professor Uff made 93 recommendations, dividing them between those that should be implemented within six months, those that should be implemented within 12 months and those that should be implemented within two years.

Ladbroke Grove **16.24**

Meanwhile, on 5 October 1999, a further major rail accident, in which 31 people, including the two train drivers died and over 400 people were injured, occurred between two trains at Ladbroke Grove, London (see also **11.7** above).

On this occasion, with the consent of the Deputy Prime Minister, who, at that time, had a responsibility for transport matters, Lord Cullen was appointed by the Health and Safety Commission (HSC) to conduct a public inquiry under *section 14(2)(b)* of the *Health and Safety at Work, etc Act 1974*:

> 'To inquire into and draw lessons from, the accident near Paddington Station on 5 October 1999, taking account of the findings of the HSE's investigations into immediate causes.
>
> 'To consider general experience derived from relevant accidents on the railway since the Hidden inquiry, with a view to drawing conclusions about:
>
> - factors which affect safety management
> - the appropriateness of the current regulatory regime.
>
> 'In the light of the above, to make recommendations for improving safety on the future railway'. (HSE. *The Ladbroke Grove Rail Inquiry*. HSE Books, Suffolk, 2001.)

The Hidden inquiry related to a railway accident just outside Clapham Station in 1988 (see **11.23** above).

The joint Cullen and Uff inquiry **16.25**

When it became obvious that there were a number of issues which were common to both Cullen's and Uff's inquiries, the HSC, again with the consent of the Deputy

Prime Minister, set up a third inquiry, chaired jointly by Lord Cullen and Professor Uff, to look specifically into train protection and warning systems, the future application of automatic train protection systems; and signals passed at danger (SPAD) prevention measures. These issues were dealt with under the Southall and Ladbroke Grove Joint Inquiry into Train Protection Systems, etc (HSE Books, Suffolk, 2001).

Changes in public inquiries **16.26**

The changes in the complexity of public inquiries over the years can be traced by an examination of a series of investigations that were undertaken in the twentieth century in response to disasters or serious incidents at football grounds. When 25 people were killed and another 517 were injured at Ibrox Stadium, Glasgow during a match between Scotland and England in 1902 there was no inquiry of any kind other than a very cursory investigation by the police. Eighteen years later, when over 1,000 people required medical treatment at the first FA Cup Final to be played at Wembley Stadium – fortunately no-one was killed – the Home Secretary asked the Right Honourable Edward Shortt, KC to chair a committee of six:

> 'to inquire into the arrangements made to deal with abnormally large attendances on special occasions, especially attendances at athletic grounds, and to consider, in view of the increased facility and rapidity of transport, what further steps, if any, should be taken to ensure the safety of the public'.

The match had taken place on 28 April 1923 but the committee was not appointed until 13 June and it is interesting to note how some bodies treated this inquiry. For example, the Football Association refused to comply with a request for information, as did a number of clubs in the First Division (Secretary of State for the Home Department. *Report on the Departmental Committee on Crowds.* HMSO, London, 1924.)

In 1946, following the death of 33 people at an FA Cup 6th Round tie between Bolton Wanderers and Stoke City, the Home Secretary appointed R Moelwyn Hughes, KC 'to conduct an enquiry into the circumstances of the disaster at the Bolton Wanderers' football ground on 9 March 1946, and to 'report to me thereon'. Although the inquiry heard 64 witnesses the report itself was only 12 pages in length and, compared with later reports, made few recommendations (Secretary of State for the Home Department. *Enquiry into the Disaster at the Bolton Wanderers' Football Ground on 9 March 1946.* HMSO, London, 1946).

Following two disasters on the same day – at Bradford 55 people died in a fire and at Birmingham, a young man was crushed to death when a wall at the ground collapsed – Mr Justice Popperwell was appointed:

> 'to inquire, with particular reference to the events at Bradford City and Birmingham football grounds on 11 May, into the operation of the *Safety of Sports Grounds Act 1975*; and to recommend what, if any, further steps should be taken, including any that may be necessary under additional powers, to improve both crowd safety and control at sports grounds'.

Popperwell submitted two reports. The interim report made recommendations as to the immediate action that needed to be taken to avoid a repetition. The final report reviewed previous inquiry reports, dealt with the working of the *Safety at Sports Grounds Act 1975* and the need for further regulation to protect the public from fire and the hazards of faulty construction, and also looked at crowd control in general. It made a number of recommendations for new offences and for some increased powers for the police in relation to sports grounds and hooliganism (Secretary of State for the Home Department. *Committee of Inquiry into the Crowd Safety and Control at Sports Grounds*. HMSO, London, 1985 (Interim Report); 1986 (Final Report)).

Finally, Lord Justice Taylor, like Popperwell, submitted two reports. The interim report dealt primarily with the events at Sheffield on the day and apportioned blame. The final report looked at, amongst other things, safety at sports grounds and football hooliganism in general. (Secretary of State for the Home Department. The Hillsborough Stadium Disaster 15 April 1989, HMSO, London, 1989 (Interim Report); 1990 (Final Report).

The key lesson to come from an examination of these four reports is that had all the recommendations of the two earlier reports been implemented, there is an argument to suggest that neither the Bradford fire nor Hillsborough should have occurred.

Criticisms of the current system of inquiries and investigations 16.27

During the recent debate to decide whether a rail accident investigation branch should be set up, the Association of Personal Injury Lawyers (*Establishing a Rail Accident Investigation Branch: A Response by the Association of Personal Injury Lawyers*, October 2001) expressed some of their concerns as follows:

- there was a lack of co-ordination of the many possible investigations and legal proceedings. Consequently, there is much duplication of effort in the recording of evidence;
- criminal prosecutions have delayed the progress of investigations into the causes of an accident;
- confusion and delay sometimes surround whether a public inquiry is to be held; and
- there is no compulsion on anyone to implement the recommendations of any inquiry.

Duplication of effort 16.28

Dealing with each of these in turn, critics of the current system of investigating major disasters or serious incidents claim there is considerable overlap, even conflict, between these various types of inquiries. The new Inquiries Act 2005 is unlikely to alleviate the validity of these criticisms.

Coroners' Inquest v Public Inquiry 16.29

This criticism was addressed, in part, insofar as it affects public inquiries and coroners' inquests, by the publication, in March 1997, of the report of the Interdepartmental Working Group on Disasters and Inquests. The report recommended that, when a fatal disaster occurred and a judge was appointed to chair a public inquiry to establish the facts surrounding the events, provision should be made to obviate the need for an inquest 'if the inquiry was likely to establish adequately the cause of the deaths arising from the disaster'.

The recommendation was accepted by the Government (see Home Office Circular No. 59/1999) and the change in procedures was implemented following its inclusion in the *Access to Justice Act 1999*. Introducing a new *section 17A* into the *Coroners Act 1988*, the new provision generally requires a coroner to adjourn an inquest once he has been informed by the Lord Chancellor that a judge has been appointed to conduct an inquiry into the events surrounding the death or deaths. A coroner can only resume the inquest under exceptional circumstances but:

> 'he may not do so for 28 days following the publication of the findings of the public inquiry or, if the Lord Chancellor so notifies the coroner, for 28 days after the conclusion of the public inquiry. Where the inquest is resumed, there will be no obligation on the coroner to summon a jury, although there will be a discretion to do so. If a jury is summoned, the inquest must proceed as if it were a fresh inquest'.

The new rules came into force on 1 January 2000.

Criminal Proceedings v Public Inquiry 16.30

Under the present law it is generally impractical to have a criminal investigation running in parallel with a public inquiry because there is always a risk that the reporting of any information disclosed in the latter may prejudice criminal proceedings. This conflict can sometimes lead to the adjournment of a public inquiry pending the outcome of criminal proceedings, as it did in the Southall railway accident case. This causes frustration, particularly to the relatives of the deceased and those who have been injured, who are both keen to know why an accident happened and to see someone brought to justice if they have been negligent. That frustration is magnified if, as in the case of Southall, the criminal prosecution fails (see **16.23**).

In theory, criminal proceedings tend to be adversarial whilst inquiries are meant to be inquisitorial. However, because the outcome of an inquiry can strongly influence civil proceedings, the conduct of questioning in inquiries sometimes borders on the adversarial.

However, even if the prosecuting authorities have indicated there is not to be a prosecution, a witness could still incriminate himself/herself when giving evidence at the inquiry. For this reason, the Chair, if authorised by the Attorney General, can give an undertaking that no evidence given by an individual or disclosed in a document produced by them can be used in a criminal prosecution against them. However, as shown in the Ladbroke Grove train crash, this does not prevent a police investigation being re-opened.

Technical Investigation v Criminal Proceedings **16.31**

Whilst it is essential that nothing occurs which could prejudice any criminal proceedings, it is also essential that the conclusions of any accident investigation as to the cause are published as early as possible in the interests of preventing further accidents. Of late, the HSE have published reports that are purely technical in nature despite the fact there is an ongoing criminal investigation.

Technical Investigation v Public Inquiry **16.32**

The demand for a public inquiry is not as great when deaths are caused as a result of aircraft accidents as they are following other types of human-made disaster. For example, as has already been indicated, there were a greater number of deaths following both the fire on board an aircraft at Manchester Airport, 55, and the Kegworth aircraft accident, 47, than in any of the railway accidents in recent years. Lord Justice Clarke maintains that this is because the AAIB have 'acquired an international reputation for full and fair investigation such that its reports were accepted by those concerned'. (*Thames Safety Inquiry* (2000), para 6.6.)

Delay in Announcing a Public Inquiry **16.33**

In some areas, there has been a tendency to conduct investigations or hold inquiries in private. According to Professor Ian Kennedy, who chaired a public inquiry unrelated to disasters and emergencies, there is a misplaced feeling amongst some people that 'enquiring in private is more conducive to getting at the real truth'. He states:

> 'Holding an inquiry in private is more likely to inflame than protect the feelings of those affected by the inquiry, not the least because of the notion of secrecy and exclusion which it fosters. Furthermore, the public's confidence in the organisation or service under review, or indeed in Government as a whole, is unlikely to be enhanced, if they, and particularly the press, are excluded'. (*Learning from Bristol: the Report of the Public Inquiry into Children's Heart Surgery at the Bristol Royal Infirmary* 1984–1995. Command Paper: CM 5207. HMSO, London, 2001, Chapter 2, para 6.)

In August 1989, a large dredger, Bowbelle, collided with a pleasure cruiser, the Marchioness, on the River Thames shortly after midnight. As a result, the Marchioness sank with the loss of 51 lives. Following this incident, the Marine Accident Investigation Branch (MAIB) launched an investigation (Marine Accident Investigation Branch. *Report of the Chief Inspector of Marine Accidents into the Collission Between the Passenger Launch Marchioness and the MV Bowbelle with Loss of Life on the River Thames on 20 August 1989.* HMSO, London, 1989). Friends and relatives of the victims, together with those who had survived – there were 132 people on board at the time of the collision – formed the Marchioness Action Group and pressed the then Conservative Government for a public inquiry into the disaster. Instead, in 1990, the Secretary of State for Transport ordered a general inquiry into river safety under the chairmanship of John W Hayes (Department of Transport. *Report of the Enquiry into River Safety.* HMSO, London, 1992). Earlier and later transport accidents, notably a number of train crashes, eg Clapham, Southall and Ladbroke Grove, had resulted in fewer deaths and yet public inquiries had been held.

Over the years, the Marchioness Action Group continued to press for a public inquiry and in 2000, the Secretary of State for the Environment, Transport and the Regions announced that one would be set up following a recommendation by Lord Justice Clarke in the final report of the Thames Safety Inquiry (The Stationary Office, London, 2000).

Clarke's recommendation was not based on the emergence of new evidence but on the fact that:

- the MAIB investigation was private, therefore, that the evidence considered by the inspectors had not been publicly scrutinised (para 35);
- the public was particularly disadvantaged by the removal of names, marginal references to evidential sources and appendices in the process of amending the original inspector's report into the published Chief Inspector's report (para 34).

Clarke went on to explain that the holding of a public inquiry should not depend solely upon the number of people who were injured or who lost their lives, although this may be an important factor. What is important 'are the overall circumstances of the case'. In the Marchioness case 'the sense of shock felt by the public and the feeling that such an accident should not be possible on the Thames in the conditions which prevailed on 20 August 1989 are important factors' (para 6.4). Clarke concluded that 'the collision between the Bowbelle and the Marchioness was a suitable case for a public inquiry' and one should have been set up in 1989 immediately following the disaster. (para 6.7.)

No Compulsion to Implement Recommendations 16.34

Even though a public inquiry is set up by or in consultation with the appropriate government department, neither the Government, nor anyone else, is compelled to implement the recommendations made by that inquiry. Indeed, they may choose not to do so for financial or political reasons. The Association of Personal Injury Lawyers pointed out that although 'each inquiry produces a long list of recommendations designed to prevent a further recurrence of the accident' there was 'little or no public scrutiny' as to whether or not these recommendations had been implemented. A prime example of this occurred over the introduction of an automatic train protection (ATP) system. In his report into the Clapham Train Crash in 1988, Sir Anthony Hidden, QC, recommended the provision within five years across the UK network of ATP, with a high priority being given to those lines on which there was a high density of traffic. The then Transport Secretary, Cecil Parkinson, was unambiguous. He said:

> 'We have made it clear that the cost of implementing these recommendations will be met. And that British Rail in carrying through won't be hampered in any way by a shortage of money'. (Changing policy on safety. BBC News, 7 October 1999.)

However, such a system had still not been implemented at the time of the Southall and Ladbroke Grove railway disasters. As Gwyneth Dunwoody, MP, Chairman of the House of Commons Transport Select Committee said immediately following the issue of the Cullen/Uff report:

> 'The Government which thought it was a good idea, then ran away from it when they discovered how much money they were going to have to cough up'.

In 1999, the Department of Transport announced that the cost of introducing a full ATP system would be about £17 million per life saved. Even introducing a simpler train protection warning system (TPWS) would cost between £4.6 and £6.5 million per life saved. TPWS would have averted the Ladbroke Grove disaster.

Shortcomings in Public Inquiries 16.35

Professor Kennedy points out that when something has gone wrong the call for a public inquiry is becoming increasingly common. He claims that 'this would appear to be because there is no clear criteria or guidance, for the Government or the public, which analyses what inquiries are for and about, when they are justified, and whether and why they should be in public or private'. He also called for an urgent need for the development of guidance. (Chapter 2, para 7.)

There have been occasions when no public inquiry, or fatal accident report, has been set up following a major disaster. For example, no public inquiry has ever been held in response to the Lockerbie disaster. Before the trial of Abdel Basset Ali al-Megrahi, there had been considerable speculation as to the events leading up to the blowing up of Pan Am 103. During the proceedings, the court heard evidence from 230 witnesses over a period of 84 days resulting in 10,232 pages of transcript. And yet, insofar as the relatives of the victims were concerned, the trial left them still demanding a public inquiry because it left crucial questions unanswered.

On 19 December 2002, in answer to a question in the House of Lords, the Leader of the House, Baroness Amos, suggested that the previous investigations had been extensive. They included a fatal accident inquiry, an inquiry by the US President's Commission on Aviation Security and Terrorism, a Transport Select Committee inquiry, various internal Department of Transport reviews of aviation security, and the trial, conviction and unsuccessful appeal of al-Megrahi. She stated:

> 'We have concluded that, given the absence of any significant new information, the fact that the key issues have already been extensively explored, and action taken, including substantial changes to airport procedures, it is most unlikely that any further form of inquiry would unearth further lessons to be learned, 14 years after the event, which had not already been identified by earlier investigations. The Government have therefore decided not to initiate any further form of review on Lockerbie'. (*Hansard*, 19 December 2002, Col WA141.)

However, this is not the way it is seen by litigation expert Philip Rodney of the Scottish law firm, Burness. He explains:

> 'A trial is conducted using an adversarial approach. The court does not play an active role in determining the factual truth. It acts as an arbiter on the basis of the evidence presented to it by the opposing parties. It can only make its decisions based on the evidence brought to its attention by the protagonists. At a public inquiry, however, an inquisitorial approach is taken. It takes a positive role in determining the truth and will seek the assistance of witnesses as it considers necessary'. (BBC News Online, 31 January 2001.)

Whilst not necessarily setting out to apportion blame, public inquiries now invariably end up doing so. As a result, individuals and, indeed, organisations, may feel under threat of being made legally accountable, in civil law (damages) and criminal law (manslaughter) for what has occurred. This may lead to defensiveness even by those who have nothing to fear. Defensiveness can be a severe obstruction to uncovering the truth.

If the scale of the disaster has provoked a huge emotional response, eg where many deaths have occurred, public approbation, particularly in the media, is likely to arise. This can be extremely damaging and long lasting. Therefore, exposure to critical comment at a public hearing or in a subsequent report can seriously damage an individual's or an organisation's professional reputation.

The last word is left to Lord Justice Clarke from his own personal experience. Pointing out that he had attended 'a number of formal investigations as counsel in the 1970s and 1980s, which had lasted a long time and cost the public purse a lot of money'; he went on to say that this 'was at least in part because parties (and their counsel) were inclined to use them to further their own interests or perceived interests in the litigation which is almost always contemplated after a serious casualty' (Secretary of State for the Environment, Transport and the Regions, Thames Safety Inquiry, The Stationary Office, London, 2000, para.5.6).

Conclusion **16.36**

Investigations into major disasters are usually very complex. Although the primary cause, eg a train driver going through a signal at danger, are relatively straightforward to establish, discovering the underlying causes of the accident and the lines of accountability of individuals and organisations can take months and, in some cases, a year or more. Generally, as the law stands at the moment, the final report may have to await the completion of legal deliberations, eg by the police and the Crown Prosecution Service. However, in the interests of ensuring that there is no immediate repeat of the disaster, the findings of any technical investigation, either by the HSE or by one of the three accident investigation branches (see **16.6**) must be made public as soon as possible.

APPENDIX I

Selected list of Inquiries and Investigations that make reference to the reasons why and the way public inquiries should be conducted

Formal Investigation into the MV Herald of Free Enterprise
Chairman: Mrs Justice Sheen – Wreck Commissioner.
Report of Court No. 8074.
Presented to the Secretary of State for Transport.
Printed by Her Majesty's Stationary Office, London, 1987.

Inquiry into the Export of Defence Equipment and Dual-Use to Iraq
Chairman: Sir Richard Scott.
Presented to Parliament by the President of The Board of Trade.
Printed by Her Majesty's Stationary Office, London, 1996.

Thames Safety Inquiry: The Final Report
Chairman: Lord Justice Clarke.
Command: 4558
Presented to Parliament by the Secretary of State for the Department of Environment, Transport and the Regions.
Printed by The Stationary Office, London, 2000.

Learning from Bristol: the report of the public inquiry into children's heart surgery at the Bristol Royal Infirmary 1984–1995
Chairman: Professor Ian Kenedy.
Command: CM5207.
Presented to Parliament by the Secretary of State for the Department of Health.
Printed by The Stationary Office, London, 2001.

17

Recovery – Creating Crisis Resilient Communication and Information Systems

In this chapter:

- Introduction.
- A stitch in time . . .
- Developing crisis resilient systems.
- Data recovery following a disaster.
- Role of ISO17799 in business continuity and disaster recovery management.
- Case study.
- Conclusions.
- Recommendations.
- Commonly used terminology.

Introduction 17.1

The integration of IT systems and Internet enabled technologies in business processes have changed the very nature in which business is conducted. However, the evident process efficiencies and cost benefits are also coupled with new and varied risks arising due to ever evolving technology and the open nature of the Internet.

Technology-driven business transformation, integrated systems and globally spread operations have given rise to a virtually extended enterprise, which coupled with the need for 24 × 7 operations and zero tolerance for downtime, make it essential for organisations to increasingly invest in business continuity measures.

Figure 1: Key drivers for business continuity planning

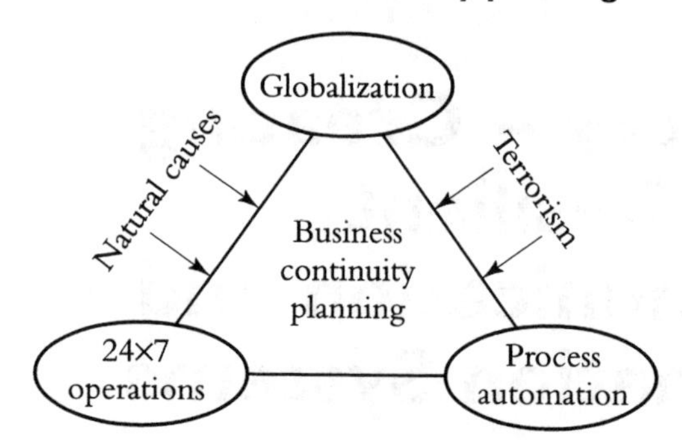

While the key drivers for creating disaster recovery measures remain high integration of IT, geographically spread organisations and 24 × 7 operations; acts of terrorism too have necessitated investment in disaster resilient communication and information systems.

The key question for businesses is no longer 'how do I respond in the event of a disaster?' instead, businesses increasingly need to ask 'how do I manage my risk so that I am always there for my customers and stakeholders?'

The answer and the challenge is to implement a strategy that takes into account the risk in totality, ensuring the availability, reliability and recoverability of information systems while balancing the cost of risk management against the opportunity cost of not taking appropriate action.

A Stitch in Time ... 17.2

Historically, Business Continuity (BC) and Disaster Recovery (DR) have been the concerns of the financial services sector, where the primary objective was to ensure the survival of the business. However, unprecedented changes in the way business is conducted have resulted in emerging business continuity concerns, which have forcefully evolved the nature of BC and DR options available today.

Recovery options available a decade ago probably translated to a site with computer equipment and office areas. Today, recovery site vendors not only offer work-area for end-users, but also provide for recovery of distributed client/server and telecommunications networks.

Traditionally, BC and DR have been considered a function of the organisations' IT or IS department, left solely at the discretion and responsibility of the CIO or IT manager and not requiring any senior business management involvement. While recent years have witnessed a shift in this mindset, organisations still assume that BC and DR supports or recovers only those functions or systems that form an integral part of the IT/IS department.

The key to the effective development, implementation and on-going maintenance of business continuity and disaster recovery measures lies in adopting an enterprise-wide framework, which is a culmination of people, processes and technology.

Table I: Components of effective business continuity framework

Dimensions	Desired components
People	• Management commitment • User awareness and participation
Processes	■ Holistic approach covering all critical business operations and associated IT systems. ■ A comprehensive DR plan: (a) integrating recovery efforts of various organisational teams and functions (b) considering the inter-linkages and inter-dependence with various business units and other external parties
Technology	• Fault-tolerant/fail-safe technologies • Data availability and management • Inter-operable and compatible technology components • Built-in early-warning signals

Management Commitment **17.3**

A successful BC and DR initiative essentially requires sponsorship and commitment from the senior management. Lack of management commitment and time are the key reasons for ineffective DR planning.

Senior management representatives need to be acquainted with the potential risks emanating from, and associated with, an IT implementation and their potential business impact. In particular, management must critically examine the impact on its bottom-line if there is a disruption in business operations due to unavailability of the supporting IT systems. The responsibility for proactively undertaking a BC and DR initiative and assigning further accountability lies with the senior management.

Further, various acts of law and regulations – such as banking regulations, privacy and data security acts, state legislation (both of the parent country and the trading country) and requirements from insurance underwriters – necessitate senior management involvement. Also, certain acts hold senior management financially liable for protecting and preserving the assets and resources of their organisation.

Developing Crisis Resilient Systems 17.4

Figure 2 below provides a diagrammatic representation of the various phases which an organisation may follow for developing BC and DR plans.

Figure 2: Stages of disaster recovery planning

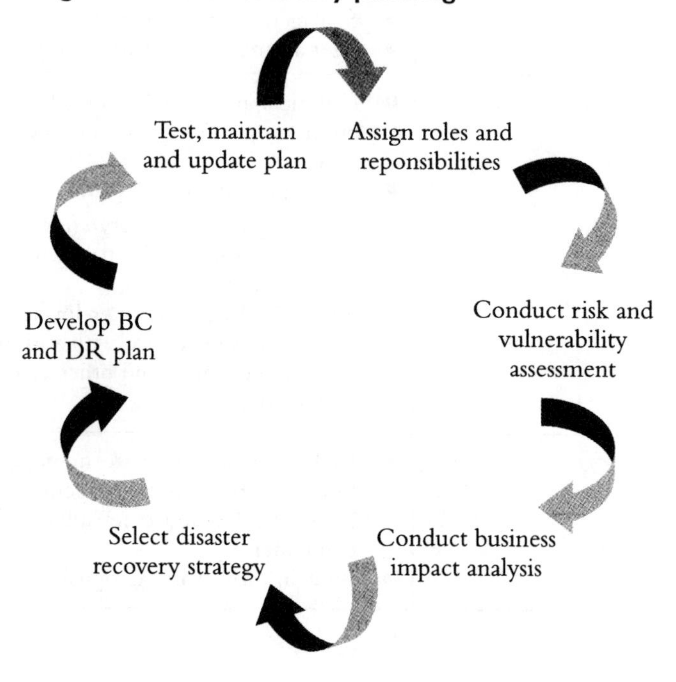

Assigning roles and responsibilities 17.5

One of the responsibilities, which the senior management must undertake towards the BC and DR plan development process is assigning accountability, roles and ownership to personnel across the enterprise for development and, thereafter, maintenance of the plans.

The management must select a team of personnel from within the organisation to steer the BC and DR development plan. Personnel representing the various business functions or departments must be selected to form part of this 'core team'. Even if an organisation chooses to undertake the assistance of external consultants for the BC and DR plan development, the core team must include representatives of various business functions from within the organisation. These personnel provide the important business knowledge and understanding which guides the DR strategy.

It is important to understand that while IT/IS personnel may form an integral part of the core team, it is essential that the ownership for the plans lies with business process owners, who are the prime users of the information systems supporting business operations. This will assist in identifying the critical business processes and ensuring that an appropriate recovery strategy is selected. Having personnel from various functions of the organisation as part of the core team brings in a holistic approach to the development, ensures enterprise-wide participation and co-operation, while ensuring that the BC and DR planning exercise is not treated as just another IT project.

Employee co-operation and participation is essential for the BC and DR development process. When undertaking such an exercise, the management must communicate frequently with the employees (through e-mails, notices, etc), informing them of the formation of the core team, its responsibilities and tasks. The management should also facilitate adequate training in the area of Business Continuity Planning (BCP) and Disaster Recovery Planning (DRP) for the core team to ensure effective contribution towards the development process and to equip them to undertake BC and DR awareness workshops for other employees.

Risk and vulnerability assessment 17.6

As a preliminary step, the core team should list the organisation's business processes and identify all the IT and other systems (including telecommunications and other external systems or vendors) associated with them. It should be noted that while BC and DR plans cover all the business operations, the critical business processes are given priority for immediate recovery following a disruption. Once the IT systems and the business processes dependent upon them have been identified, a detailed risk and vulnerability assessment needs to be carried out. The objectives of the exercise being to:

- identify potential threats and control weaknesses;
- assess the likelihood of their occurrence; and
- assess capability of existing mitigating controls and identify areas of improvement.

One of the important outcomes of conducting risk and vulnerability assessment is to strengthen weak controls and implement mitigating controls, wherever possible to minimise risk of failure due to a disruption. A DR plan would address all the residual risks.

Defining disaster scenarios 17.7

At this stage, all potential threats, the likelihood of these threats materialising and the resultant risks to the organisation's business operations are analysed. In particular, the core team must identify all material threats including internal and external threats, to the areas shown in Table 2 below.

Table 2: Considerations of threat assessment and sample threats

Areas	Considerations
Office work-space	• vital records retention • skills management • housekeeping policies • storage of hazardous materials • insurance policy and procedures
Data centre and communications infrastructure	• confidentiality, integrity and availability of data • data backup and storage • intrusion detection systems • documentation storage and management • data centre security – policy and procedures • physical access control mechanism • logical access control mechanism • air conditioning systems • equipment malfunction/failure • storage of hazardous materials
Building area	• equipment/machinery malfunction/failure • emergency response and evacuation procedures – vicinity to nearest hospitals, fire stations, police stations, etc • lack/failure of intrusion alarms, fire and smoke detection, warning and control • malfunction/failure of power controls • building air conditioning systems • storage of hazardous materials • breach/lack of perimeter protection and access controls • emergency response procedures • elevator controls
Surrounding area	• frequent flooding • civil disturbance targets • proximity to commercial or military airport • high crime area • proximity to nuclear power plant
Regional area	• seismology issues • meteorology issues

The core team must consider each of the sample risk areas defined above to identify potential disaster/disruption scenarios. Having identified potential disaster/disruption scenarios, the core team should document them in a risk assessment matrix. The matrix should list each threat, its likelihood of occurrence, the resulting risks and the existing controls for reducing or eliminating the threat. The risk assessment matrix forms the basis for the development of the BC and DR strategy.

The core team should also identify the immediate areas of improvement and assign responsibilities for the bridging risk and control gaps to appropriate business process owners and/or IT personnel.

Business Impact Analysis (BIA) 17.8

Having carried out the risk assessment, an organisation needs to evaluate the impact of each disruption/threat on its critical IT systems and associated business processes. During this phase, the impact of security threats and vulnerabilities are analysed and the critical IT systems and business processes that will need to be supported on a priority basis during an unplanned disruption are identified.

For each business process, an assessment of the business activities performed and the degree of reliance and importance of those activities towards delivery of final product/service, assist in determining the criticality of the process.

The objective of conducting BIA is to identify the affect of disruption on individual business locations, business units or processes in monetary and/or operational terms. An organisation must also assess the financial, service/operational, legal and/or reputation related business implications that may arise due to its inability to deliver a service or other work product in the event of a disruption. This will help prioritise which systems should be recovered first.

The next step is to identify the maximum acceptable timeframe for which the organisation can afford to go on without its critical business processes and related IT systems, without facing unacceptable financial, legal and/or reputation loss. The maximum time to recover critical business processes and IT systems is determined by the risk acceptance capability of the management and its ability to invest on implementing fail-safe and recovery mechanisms. The minimum time for which a business process may be unavailable following a disruption will be determined by its dependence on IT systems, other business processes and external vendors/parties. For determining the least recovery time accurately, the internal and external work inflows and outflows associated with each identified business process, the source or destination of the workflow and the associated IT systems are required to be studied in detail.

It is critical for any organisation to determine the least recovery time for its critical business processes following a disruption, since this will assist the management in selecting a suitable recovery strategy and also assist in undertaking appropriate and informed service level commitments with its customers, while also warranting similar Service Level Agreements (SLAs) from its vendors and service providers.

While determining the target recovery timeframe, which is dependent on both the least recovery time and the maximum acceptable downtime, the organisation must focus on inter-linkages among the various functions/departments and also the reliance of other locations on availability/non-availability of the IT systems and processes. The core team must ensure that the target recovery timeframe for IT systems and business processes is clearly and realistically defined.

Having identified the target recovery timeframe or the Recovery Time Objective (RTO – The maximum time a business can afford to go without a business process or information systems), the next step towards developing a BC and DR strategy is to identify the recovery resource requirements for both business processes and IT systems, such as:

- internal and external personnel resources;
- office equipment;
- software (application programs, computer platforms);
- computing equipment;
- documents;
- utilities (including power, etc); and
- voice and data communications requirements; etc.

There is a possibility that, depending on the time and cost of arranging for the required recovery resources, the management may decide to redefine its RTO and accept certain costs. The management must balance the cost of process and system unavailability against the cost of resources required for restoring the system.

Figure 3: Cost of disruption versus the cost to recover

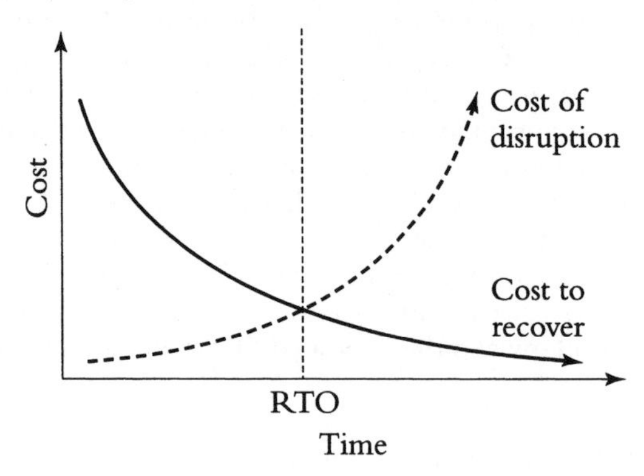

As depicted in Figure 3 above, the point where the two lines meet defines how long an organisation may afford to go on with system disruptions.

At the end of this phase, an organisation should have obtained an analysis of the availability requirements of IT systems and associated business processes, in the order of their criticality and the maximum recovery timeframe.

Selecting business recovery strategy 17.9

The objective of this phase is to identify various alternative recovery options and assess their appropriateness in meeting the organisation's recovery needs. The core team must identify multiple recovery options and present the same to the senior management for evaluation and final selection of a suitable recovery option. At this time, the core team may also seek assistance from external service providers. The following aspects need to be considered while evaluating various recovery options:

- Recovery time frame/recovery time objective (RTO);
- Recovery site requirements – many aspects are required to be considered when deciding on a recovery site. The first question is whether there is a need to recover the operations at an alternate site (off-site). If yes, then depending upon the requirements a suitable alternate site should be identified. The typical categories of alternate sites are mirrored-site, hot-site, warm-site and cold-site (in their decreasing order of preparedness). These details have been discussed in the subsequent sections of this chapter;
- Alternate hardware/software requirements;
- Documentation requirements;
- Personnel requirements (both internal and external);
- Back-up storage; and
- Approximate costs, etc.

Also the recovery options should address the requirements for:

- business critical processes; and
- the IT and communications infrastructure supporting the same.

The management should consider various feasible options ranging from the minimum recovery time option, which may involve significant investments both in terms of financial costs and resources required; to higher recovery time option, which may require lower financial and resource investments. The strategy should include a selection of methods that complement one another to provide recovery capabilities for the entire gamut of threats identified. Various recovery options, which may be considered, include implementation of cold, warm, or hot sites, data mirroring options, reciprocal agreements, SLAs with the equipment vendors, redundant arrays of independent disks (RAID) technologies, automatic fail-over mechanisms, uninterruptible power supply (UPS), etc.

The management needs to identify the best-fit recovery strategy considering cost and time factors. One technique that may be helpful is determining the annual loss exposure (ALE). ALE may be defined as the total cost or financial loss that an organisation incurs over a year due to a particular threat or risk.

ALE for a particular risk = (Annual frequency of occurrence of the threat) × (Monetary loss attributed to each occurrence)

The current ALE is compared to the ALE after a control is put in place and the difference provides the annual cost, which will be incurred on implementing the control. The management may use ALE to guide its recovery strategy decision. However, the following shortcomings of the ALE technique must also be considered:

- estimates of annual frequency of occurrence of a threat may be subjective;
- estimates of the costs of each impact may be subjective;
- the time difference between each incident/impact is not considered; and
- the fact that the implementation of one control may benefit more than one threat area is not accounted for.

The finalised recovery strategy forms the basis for the development of the BC and DR plan.

BC and DR Plan Development 17.10

The BC and DR plan should be typically in two parts – recovery procedures for onsite and offsite. The BC and DR plan for onsite recovery will address recovery procedures for the business processes and IT systems at the primary business site. On the other hand, offsite BC and DR plans will document recovery procedures to be followed in order to revive business processes and IT systems at a backup/disaster recovery site, which is different from the primary business site.

The BC and DR plan should include:

- procedure for invocation of the BC and DR plan;
- recovery team and responsibilities;
- recovery site – onsite/offsite;
- escalation procedures and contact lists;
- evacuation procedures;
- recovery procedures for IT systems;
- DRP testing procedures; and
- DRP administration/maintenance procedures.

Invocation of the BC and DR plans 17.11

The conditions for activating the recovery plans should be identified and detailed in the BC and DR plans. This includes procedures for notifying the recovery teams and other personnel once the disruption takes place, undertaking an initial assessment of the damage and the activation of the recovery plan.

A disruption may take place with or without prior notice. For example, advanced notice is often available in the event of a hurricane or storm or there may be information of a possible virus affecting systems. However, there may also be unexpected equipment failure or a criminal act. Notification procedures should be documented in the plan for both types of situation. The procedures should describe the methods (eg e-mail, telephone, radio, television announcements, etc) to be used for notifying recovery and other personnel (eg staff, families, customers, vendors, etc) during business and non-business hours. Notification procedures to be followed in case specific individuals cannot be contacted should also be included.

Personnel should be identified and assigned the responsibility for damage assessment following a disruption. It should be the responsibility of the damage assessment team (the subsequent section details various other teams, generally formulated to recover from a disaster) to determine the nature and extent of the damage. The team should assess the cause of the disruption, areas/systems affected, type of damage and the estimated time to restoration.

When damage assessment is complete, the appropriate recovery and support teams should be notified.

The recovery plans should be activated based on predefined activation criteria. Activation criteria may be formalised based on:

- safety of personnel and/or extent of damage to the business premises;
- extent of damage to IT systems (physical, operational or financial);
- criticality of the system to the organisation's business; and
- expected duration of disruption.

Recovery team and responsibilities **17.12**

The recovery plan document should specify the roles and responsibilities, which individuals must partake during the recovery process. At a minimum, individuals responsible for notification of disaster, damage assessment, declaration of disaster, plan activation and the recovery team should be identified. Alternatives should be nominated as required for the recovery team members. Each recovery team should comprise of a team leader, team members and backup team leader. Examples of recovery teams include:

- crisis management team – responsible for providing management support and guidance for recovery efforts following a disaster;
- IS support team – responsible for restoring IS infrastructure and operations at the recovery site;

- business function recovery teams – responsible for ensuring that recovery resources are available to business units and performing critical business functions at the recovery location;
- logistics support team – responsible for managing voice communications and other logistics issues related to the recovery effort; and
- restoration team – responsible for salvage operation and facilitates the transition to a permanent site.

The roles and responsibilities of the recovery teams should include leading the personnel to safety following a disaster. The responsibilities and the procedures to be followed by the recovery teams should be carefully detailed in the recovery plans. It is also essential that each recovery team is adequately trained for the procedures it must carry out following a disruption. Some of the other tasks, which may be assigned to recovery teams, for example, include:

- notifying internal and external business partners of the disruption/disaster;
- arranging for necessary office supplies and work space;
- obtaining and installing necessary hardware/software components;
- restoring critical operating system and application software;
- restoring data from backup media;
- testing system functionality including security controls;
- connecting systems to network or other external systems; and
- operating alternate equipment successfully.

Recovery site – onsite/offsite 17.13

The plan should detail the procedures, which describe the actions to be undertaken to move the critical business activities and systems to the alternative temporary or offsite locations. Further, the procedures detailed should be such that they ensure that the business processes are brought back into operation within the required recovery timeframe. Similarly, the plan also needs to describe the activities that need to be carried out for business resumption, ie to return to normal business operations.

Escalation procedures and contact lists 17.14

It is critical that escalation procedures for informing concerned individuals, internal personnel, media and general public are detailed. An effective method for escalation is to document in the form of an inverted tree. This technique involves listing the first individual to be contacted at the top and in the event of unavailability to move downwards. It also involves assigning notification duties to specific individuals, who in turn are responsible for notifying other recovery personnel. Contact details, including official and residential addresses must be maintained and updated to remain current. Arrangements should also be made for public relations management, ie the statement to be issued to public, vendors, internal users, etc need to be created as part of the plan and the individuals responsible for activating the same assigned.

Figure 4: Sample calling tree

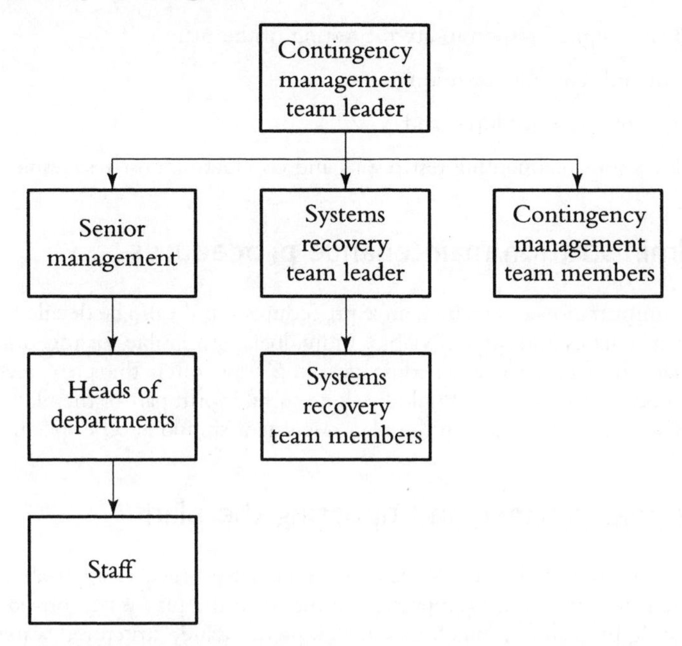

Evacuation procedures **17.15**

The plan should detail the evacuation procedures, which need to be followed in the event of an emergency. These include actions to be taken following an incident that jeopardises human life. The plan should detail the procedures for contacting the appropriate public authorities such as the hospitals, police, fire service, local government, etc.

Recovery procedures for IT systems **17.16**

The recovery procedures for IT systems including hardware, software, application systems, communications, infrastructure, etc. need to be detailed clearly in the form of a checklist, to ensure that systems can be recovered in the required timeframe with accuracy. To prevent any difficulty or confusion at the time of recovery, it essential that recovery procedures be detailed sequentially for each IT system. It is also essential that all relevant system documentation is stored in backup along with the system software.

Plan testing procedures 17.17

The plan should elaborate the procedures to be followed for testing the BC and DR plan. For example, the plan should include:

- procedures to be followed for validating the plan;
- individuals assigned responsibility for testing of the plan;
- time plan and schedule for testing;
- test performance procedures; and
- procedures for documenting test results and conducting post-test results analysis.

Plan administration/maintenance procedures 17.18

The plan administration and maintenance procedures should also be detailed to ensure that the plan remains current and viable. Individuals responsible for the same should be identified. A maintenance schedule detailing how often the plan needs to be tested and updated and how it should be done should form part of the plan. Further, procedures for the distribution of the plan document should also be specified.

Testing, maintenance and updating the plan 17.19

It is important to test the newly developed plan to assess its applicability when implemented. It is also equally important to maintain the plan on an on-going basis. Various testing procedures, which may be adopted, include structured walkthrough, component testing and full interruption test.

Structured walkthrough 17.20

Also referred to as a 'table-top' exercise, the structured walkthrough is a paper evaluation of the DRP designed to expose errors or omissions without incurring the level of planning and expenses associated with performing an operations test. In the structured walkthrough, a disaster scenario is established, and recovery teams assemble in a conference room and walkthrough their recovery actions.

A scenario will be made available in advance to allow the recovery team members to review their recovery actions in response to the test scenario. At the end of the structured walkthrough any changes to the plan that are found to be necessary are documented and implemented.

Component testing 17.21

Component tests are actual physical exercises designed to assess the readiness and effectiveness of discrete plan elements and recovery activities. The isolation of key recovery activities allows recovery team members to focus their efforts while limiting testing expense and resources. This testing is effective for identifying and resolving issues that may adversely affect the successful completion of a full operations test. Component tests include:

- evacuation tests;
- emergency notification test;
- backup restoration test;
- business application recovery test; and
- remote or dial-in access test.

Full interruption test 17.22

The full interruption test requires extensive planning and preparation and should not be performed until most, if not all, of the plan components have been tested. This test requires the simulated recovery of critical business application systems and communication components. It is the closest exercise to an actual disaster scenario. Although, a full interruption test requires weeks of planning and considerable co-ordination of personnel and resources, the exercise provides a level of confidence about the ability to recover in an actual event.

The success of a BC and DR plan is, to an extent, dependent upon the understanding and awareness of the end users. The management should conduct awareness workshops for end users and introduce them to their expected roles and duties in the event of a disruption and the measures they should incorporate within their day-to-day work to reduce the risk of an incident.

The BC and DR plans should be maintained by regular reviews and updates to ensure their continuing effectiveness. Procedures should also be included within the organisation's change management process to address updating of business continuity and disaster recovery plans. Examples of situations that may necessitate updating plans include:

- acquisition of new equipment;
- upgrade of systems;
- changes in personnel contacts;
- changes in business strategy;
- changes in location, facilities and resources;
- changes in requirements of customers, contractors, suppliers, insurers, legislation, etc;
- addition, removal or modification of business processes;
- variation in risks (operational and financial); and
- environmental factors.

Plan maintenance and update should also include a strategy for plan distribution and control.

It is important to store backup copies of the plan. A copy of the plan should also be stored at the recovery site. The core team leader or the BC and DR co-ordinator may also consider storing a copy of the plan at his or her home. Every recovery team member should receive a copy of the common sections and the sections relevant to their role. Also, the sections common for all organisation personnel (such as steps to be taken in case of disruption to reach for safety, how to report a disruption or intrusion, etc) should be published on the internal bulletin board or Intranet.

It is also necessary to work out the control mechanism to ensure all old copies of updated plans are kept out of circulation. Maintaining control of the recovery plans is extremely important for security purposes.

Critical Success Factors (CSFs) for BC and DR management 17.23

Critical success factors for BC and DR management include:

- senior management takes ownership and assigns responsibility and accountability to an effective team;
- effective personnel awareness programs – educating personnel about potential disasters and training personnel to react appropriately during an incident, including evacuation and contact procedures;
- disaster scenario planning covers incidents of unauthorised access and internal as well as external intrusion attempts along with scenarios of natural disasters, fire break-outs, power and telecommunication outages, political and terrorist threats;
- BC and DR measures include disruption mitigation processes at suppliers, strategic partners, external services providers, infrastructure service providers (such as mail and phone), transportation services, etc;
- ensure adequate storage, backup and archival systems are available for data stored both in physical as well as electronic form;
- provide for backup personnel and skill sets; and
- test, maintain and update.

Data Recovery Following a Disaster 17.24

Data recovery (preceded only by human/personnel safety) and, as illustrated in Figure 6 below, forms one of the most important aspects of any business continuity and disaster recovery strategy. The data backup strategy selected and implemented by an organisation has a direct impact on its data recovery strategy and capabilities. The management should finalise an appropriate data backup strategy based on the criticality of the IT systems for the continued working of the business process and the cost of each possible data recovery strategy.

Thus, an effective business impact analysis (how long can business operations continue without the supporting IT systems and underlying data) and a cost-benefit analysis (speedier recovery solutions may cost more than the cost of business process unavailability) needs to be undertaken.

The two key parameters to be considered carefully while designing a suitable data backup (recovery) strategy are recovery time objective (RTO) and recovery point objective (RPO).

- recovery time objective (RTO) attempts to determine – how long the business can afford to be without a particular IT system; and
- recovery point objective (RPO) indicates – when the systems and data are being recovered, how much data the business can afford to lose or recreate after recovering from the backups.

The shorter RTO and RPO will require developing a faster responding plan and more frequent data backups, thereby attracting higher investments; and vice versa. Therefore, an appropriate determination of these parameters can significantly impact the organisation's capability to recovery data in a faster and effective manner, while optimising the investment in data backup solutions.

Data recovery options 17.25

The recoverability of a business or the data recovery process followed after a disaster would depend upon the data backup mechanism implemented by the business. The evolving nature of business processes over the years has brought about significant changes in the business continuity and disaster recovery requirements of businesses, as is indicated in Figure 5 below.

All data that an organisation's application systems hold is not equally essential for continuing mission critical operations. When finalising data recovery objectives and the corresponding data recovery options, it is imperative that an organisation segregates its business data based on the value it associates with it (ie impact of system/data being unavailable for an extended period), the time required to get that data back after a failure (impact of restoring and re-entering data lost following a disruption) and the tolerance for data loss after an event (impact of data being irreplaceable following a disruption).

How an organisation segments data and prioritises the recovery of its business applications is highly dependent on individual business requirements. While one organisation may determine that e-mail is a mission critical application, another may consider it least important. For example, for a consulting organisation, in the event of a disruption or system failure, it may be acceptable for client files to be inaccessible for say 48 hours (RTO). However, since the consultants input data directly into the systems and do not enter the same on paper first, near zero data loss is acceptable

Figure 5: Changing disaster recovery and business continuity requirements

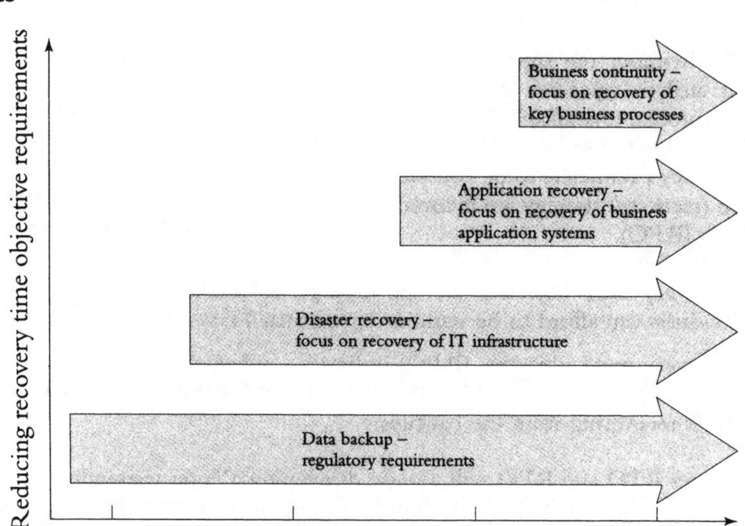

(RPO). Similarly, for an insurance agency, whose business is dependent on being available to its customers when a disaster strikes, it is important that its systems are back online processing customer claims in say four hours (RTO). However, since the agents maintain a paper trail for all transactions, re-entry of a small amount of data is acceptable, say in eight hours (RPO).

The following describes some of the data recovery mechanisms in their order of least time to recovery and decreasing cost of adoption.

Mirrored site **17.26**

Mirrored sites are fully redundant facilities with full, real-time information mirroring. Mirrored sites are identical to the primary site in all technical respects. These sites provide the highest degree of availability because the data is processed and stored at the primary and alternate site simultaneously. These sites typically are designed, built, operated and maintained by the organisation.

Hot site **17.27**

This may be described as an alternate business facility, which may be maintained either by the organisation itself or by agreement with a vendor, which is fully equipped with the resources required to recover business functions affected by the occurrence of a disaster. Hot sites may vary in type of facilities offered (such as data processing, communication, or any other critical business functions needing duplication). Location and size of the hot site is proportional to the equipment and resources needed for ensuring complete business recovery. Such a facility may generally enable an organisation to 'go live' or resume business process almost immediately following a disaster.

Warm site 17.28

An alternate processing site which is only partially (as compared to hot site, which is fully equipped) equipped with resources such as hardware, communications interfaces, electricity and environmental conditioning capable of providing backup operating support for critical business operations in the event of an unexpected disruption. Such a facility may generally enable an organisation to resume business operations within 6 to 12 hours following the disaster.

Cold site 17.29

An alternate business facility that offers only the environmental conditions such as air-conditioning and raised flooring for conducting business operations in the event of an unexpected disruption. Such a facility may either be maintained by the organisation itself or by agreement with a vendor. The required equipment and resources required for resuming critical operations must be set-up after a disaster has occurred. This option may generally enable an organisation to be operational within two to four days following a disaster.

There are various other options available such as a co-location site or a co-operative site, where more than one organisation may depend upon the data recovery facilities offered by a vendor. However, one of the most critical aspects to be considered when outsourcing data recovery is the disaster recovery/business continuity capabilities of the vendor.

So, is data backup the solution for data recovery? Don't be too sure! Gone are the days when manufacturers of data storage media wrote 'life time warranty'. Today, 'just like the Rolls Royce which never broke-down but only failed to proceed at times', virtually nothing is guaranteed to work, when required.

So, does that mean that having a mirrored hot site does not really guarantee data recovery? Well, yes if you think of it as a one time effort and expenditure. The golden rule is to ensure that backup data is up-to-date and the backup media is periodically tested to ensure data recovery when required. Never under-estimate the power of a backup media device, it may fail to function or have a sector go bad just when you need it and you could have a bigger disaster at hand.

Data recovery technologies 17.30

Over the years, disaster recovery technologies too have evolved to meet the industry's increasing requirements. Today, vendors offer dedicated backup and recovery devices (eg tape libraries, RAID arrays, etc); software (eg network management software, server load balancing software, etc) and services (including remote offsite backup centres, online offsite backup maintenance, etc) enabling companies to maintain business operations despite disruptions.

Wide range of data backup solutions are available for data residing on both servers and desktops/laptops. Some factors to be considered when selecting a backup solution include:

- equipment interoperability – to the extent possible, the backup device should be hardware independent and compatible with operating systems and applications in use;
- storage volume – the backup solution should provide adequate storage capacity;
- media reliability – typically, the cost of the backup media depends on the reliability it offers. The backup media should provide the reliability, which is commensurate with the business requirements;
- media life – each type of media has a different use and storage life beyond which the media cannot be relied on for effective data recovery;
- performance and speed – different type of backup solutions offer varying performance. The two key parameters, which define performance, are write speed and retrieve speed. Businesses having to backup large volumes of data will require backup solutions, which offer faster write and retrieve speeds; and
- backup software – the ease of use and compatibility with existing systems of the backup application should be considered.

Some of the backup options and media available for desktop/laptop, as well as server backups, include:

- floppy diskette – available with most desktop computers and represent one of the cheapest backup solutions. It has very low storage capacity and extremely low reliability. It can only be used for backing up a very small volume of end-user data and is not a preferred mode of backup any longer;
- magnetic tape – this is a relatively low cost option but offers more reliability than floppy diskettes. It offers a high-capacity backup solution. Tape drives are automated and require a third-party backup application or backup capabilities in the operating system;
- removable cartridge – portable or external backup devices offer a faster and more flexible backup option. The cost and reliability of these devices are comparable to tapes. The portable devices come with special drivers and application to facilitate data backups;
- compact disk (CD) – rewritable CDs are low-cost storage media and have a higher storage capacity than floppy diskettes. To write to a CD, a rewritable CD (CD-RW) drive and the appropriate software is required;
- network storage – data stored on networked PCs can be backed up to a networked disk or a networked storage device, which may be a server. PC users may backup

data to allocated space or a network backup system may automatically backup PC data;

- replication or synchronisation – data replication or synchronisation is a common backup method for portable computers. Handheld computers or laptops may be connected to a PC and replicate the desired data from the portable system to the desktop computer; and
- online backup – data backups are scheduled to a remote location over the Internet. Data can be encrypted for transmission.

Though off-site storage of backup media enables the system to be recovered, data added to or modified since the last backup could be lost during a disruption or disaster. To avoid this potential data loss, a backup strategy may need to be complemented by redundancy solutions, such as disk mirroring, redundant array of independent disks (RAID) and load balancing.

Redundant Array of Independent Disks 17.31

The RAID technology provides disk redundancy and fault tolerance for data storage and decreases mean time between failures (MTBF). RAID increases performance and reliability by spreading data storage across multiple disk drives, rather than a single disk. RAID can be implemented through hardware or software. In either case, the solution appears to the operating system as a single logical hard drive. With a RAID system, hot swappable drives can be used – that is, disk drives can be swapped without shutting down the system when a disk drive fails. RAID technology uses three data redundancy techniques: mirroring; parity and striping.

- Mirroring – with this technique, the system writes the data simultaneously to separate hard drives or drive arrays. The advantages of mirroring are minimal downtime, simple data recovery and increased performance in reading from the disk. If one hard drive or disk array fails, the system can operate from the working hard drive or disk array, or the system can use one disk to process a read request and another disk for a different processing request. The disadvantage of mirroring is that system performance may be compromised due to the simultaneous writing functions.
- Parity – parity refers to a technique of determining whether data has been lost or overwritten. Parity has a lower fault tolerance than mirroring. The advantage of parity is that data can be protected without having to store a copy of the data, as is required with mirroring.
- Striping – in striping, a data element is broken into multiple pieces and a piece is distributed to each hard drive. Data transfer performance is increased using striping because the drives may access each data piece simultaneously. Stripping is carried out at data byte level or data block level.

Currently, there are six fundamental RAID levels available, with each level providing a different configuration. Further solutions may be developed by combining two or more RAID techniques.

- RAID-0 is the simplest RAID level, relying solely on striping. RAID-0 has a higher performance in read/write speeds than the other levels, but it does not provide data redundancy. Thus, RAID-0 is not recommended as a data recovery solution.
- RAID-1 uses mirroring to create and store identical copies on two drives. RAID-1 is simple and inexpensive to implement, however 50% of storage space is lost because of data duplication.
- RAID-2 uses bit-level striping, however the solution is not often employed because it is expensive and difficult to implement.
- RAID-3 uses byte-level striping with dedicated parity. RAID-3 is an effective solution for applications handling large files, however fault tolerance for the parity information is not provided because that parity data is stored on one drive.
- RAID-4 is similar to RAID-3, but it uses block-level rather than byte-level striping. The advantage of this technique is that the block size can be changed to meet the application's needs. With RAID-4, the storage space of one disk drive is lost.
- RAID-5 uses block-level striping and distributed parity. This solution removes the bottleneck caused by saving parity data to a single disk in RAID-3 and RAID-4. In RAID-5, parity is written across all drives along with the data. Separating the parity information block from the actual data block provides fault tolerance. If one drive fails, the data from the failed drive can be rebuilt from the data stored on the other drives in the array. With RAID-5, the storage space of one disk drive is lost.

Load balancing 17.32

Server load balancing increases server and application availability. Through load balancing, traffic can be distributed dynamically across groups of servers running a common application so that no one server is overwhelmed. With this technique, a group of servers appears as a single server to the network. Load balancing systems monitor each server to determine the best path to route traffic to increase performance and availability so that one server is not overwhelmed with traffic. Load balancing can be implemented among servers within a site or among servers in different sites. Using load balancing among different sites can enable the application to continue to operate as long as one or more sites remain operational. Thus, load balancing could be a viable contingency measure depending on system availability requirements.

Electronic vaulting and remote journaling 17.33

Electronic vaulting and remote journaling techniques provide for data backups to be made to remote storage media over communication links. This technology compliments the regular backups since data processed in between the regular backups is also stored.

The electronic vault or the remote storage site could use optical disks, magnetic disks, mass storage devices or an automated tape library as the storage devices. With this technology, data is transmitted to the electronic vault as changes occur on the servers between regular backups. These transmissions between backups are sometimes also referred to as electronic journaling.

With remote journaling, transaction logs or journals are transmitted to a remote location. If the server needed to be recovered, the logs or journals could be used to recover transactions, applications or database changes that occurred after the last server backup.

Remote journaling and electronic vaulting require a dedicated off-site location to receive the transmissions. The site can be the organisation's or a third party hot site, off-site storage site or another suitable location.

Storage area network (SAN) 17.34

The storage area network (SAN) technology is built upon the concept of storage virtualisation, which is described as the process of combining multiple physical storage devices into a logical, virtual storage device that can be centrally managed and is presented to the network applications, operating systems and users as a single storage pool. Benefits of storage virtualisation are that storage devices can be added without requiring network downtime and storage volumes from an incapacitated server or a storage device can be reassigned.

A SAN is a high-speed, high-performance network that enables computers with different operating systems to communicate with one storage device. A SAN can be local or remote (within a limited distance) and usually communicates with the server over a fibre channel. The SAN solution moves data storage off the LAN, thus enabling backup data to be streamed to high-speed tape drives, which does not affect network resources as distributed and centralised backup architecture does. SAN moves away from the client/server architecture and towards the data-centric architecture.

Figure 6 below maps the relative availability of various recovery solutions (adapted from *NIST Guide – Contingency Planning Guide for Information Technology Systems*). High availability is measured in terms of minutes of lost data or server downtime; low availability implies that server recovery could require days to be completed.

Market analysis 17.35

There are various prominent participants in the disaster recovery and business continuity products and services market. Infrastructure vendors, such as Brocade, which provides fibre channel solutions, supply the underlying technologies necessary for business continuity. Solutions such as CA's BrightStor Enterprise Backup or EMC's Snap View are applications that specifically cater for the needs of backup and restoration, enabling faster and more intelligent restores or recoveries.

Figure 6

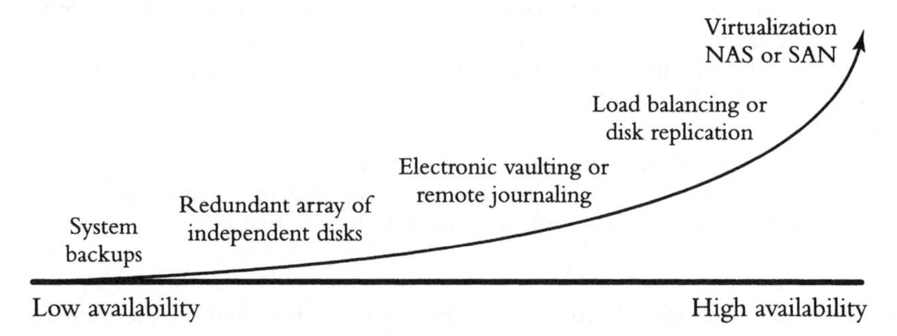

Vendors such as Sybase offer applications that specifically work towards improving system resilience, like load-balancing software. These applications include products such as the Sybase OpenSwitch that supplies a level of transparency between client connections and database servers. IBM offers an extensive suite of application continuity services. SchlumbergerSema's also offers a backup and recovery solution, whereby a full variety of hot or cold, dedicated or shared servers and datacentre space is made available.

Sungard offers a backup and restore solution, whereby server systems or mainframes can be restored onto alternative devices, if the primary device is incapacitated. Sungard also offers high availability services, which aim to provide a proactive solution that minimises business disruption. The service is based on concurrent systems running within a Sungard data centre that can be used in the event of a failure and, depending on the solution, the recovery time scale can be small.

Data recovery – best practices 17.36

There is no silver bullet when it comes to data recovery. Your data recovery strategy and its success largely depend on your prior planning. Here are some of the DOs and Don'ts, while recovering data following a disaster:

Dos	Don'ts
Always back up your data.	Don't assume that the backups will always work for you. Be prepared for the worst.
Encourage users to backup data.	Don't allow users to store critical data on personal desktops and laptops.
Restore data from backup media periodically to test for data recoverability.	Never upgrade a system without performing a complete backup and restore process.

Dos	**Don'ts**
When it comes to sophisticated technology, always involve an expert at early stage.	In the event of disaster (data corruption), don't panic, analyse and think through the problem before attempting to quickly fix it.
Prioritise your data recovery efforts.	Don't try to recover everything, you have limited time.
In most cases, the data is intact, just the master index is corrupted, follow a methodical approach.	Don't give up before trying every option. It is highly likely that you may recover the data as long as the media is available.
Perform a cost-benefit analysis – the cost of recovering all data may exceed the value.	Don't focus just on data, look at the business value of data.
Remain informed of the storage solution vendor's business and keep updated drivers and utilities.	When you are in a hole, don't dig deeper. Consult the vendor.

Information lifecycle management **17.37**

When discussing data recovery and storage measures, it is essential to touch upon the concept of information lifecycle management (ILM). The conception behind ILM is that all information has certain value attached with it, and this value does not remain constant over time. Typically, as time passes information becomes historical in nature and its current requirement declines. That is, not all information needs to be available and accessibly fast at all times. Information of high value needs to be available and accessible faster, but faster access means higher costs of storage and retrieval.

The ILM theory states that as data ages, it should be moved across media of lower and lower cost (and higher and higher access times) until eventually (if the occasion arises) the information being of no value may be destroyed. For example, over time and relevance information may be transferred from being on-line to disk drives to tape libraries.

ILM deals not only with data storage but also with maintaining and managing information throughout its lifetime, so that relevant information is available when needed. The most critical element of implementing ILM is on going data evaluation and segregation. For example, data may be segregated as:

- archival or cold data – these data may typically be saved to tape and are used primarily for record retention. These may be data that needs to be accessed infrequently.

- primary or warm data – these may be the files, databases and records that an organisation generates and uses in the day-to-day running of its business. This may be data that needs to be accessed continuously;
- temporary or hot data – this may be data which is temporary in nature (stored for little time). For example, data stored in multiple locations on servers to enhance access and delivery times for enterprise users or to reduce loads on a website.

To meet the requirements of business, organisations move away from traditional tape based backups and adopt technologies such as RAID, disk mirroring or disk replication to create multiple copies of data. These options enhance performance and accessibility while providing protection against hardware malfunction, since systems don't need to be taken offline for backups and data is stored on multiple media. Data may also be stored at both a primary and remote site.

However, a data backup and recovery strategy involving disk-to-disk backup solutions alone without the use of tape backups does not offer comprehensive security, since data is accessible or online at all times.

As the data is available online, these solutions do not provide protection against:

- data terrorists or hackers/intruders; and
- natural or man-made disasters.

Tape based or off-line backups provide protection against intrusion attacks fundamentally because they are offline and since they can be moved to a geographically different location, risks of natural disasters affecting both primary and remote sites are reduced. Another advantage is that tape backups can be stored in safe and controlled environmental conditions, eg fire proof safes, etc. Since tape backups also assist in data archival, tape backups combined with the disk-to-disk backups enable complete information lifecycle management.

Role of ISO 17799 in business continuity and disaster recovery management **17.38**

The ISO 17799 standard provides an open framework (independent of any technical platform) for the development, implementation, maintenance and measurement of enterprise-wide information and security management practices (covering confidentiality, integrity and availability of information), so as to provide confidence in the operations of the organisation. In particular:

- ISO 17799 assists organisations in developing their own information security framework;
- the standard relates to all information, regardless of the media on which it is stored or where it is located;
- it provides guidance and recommendations for both managers and employees responsible for initiating, implementing and maintaining information security; and

- the standard provides a comprehensive set of security controls representing best information security practices. These security controls are neutral to the nature of industry or size of the business and may be applied selectively depending on the needs of the business.

The scope of ISO 17799 17.39

The ISO 17799 standard owes its origins to the BS 7799 standard, which was developed by the British Standards Institute (BSI) in 1995 (and updated in 1999 and 2002) and covers the following ten domains to provide for enterprise-wide information security.

- Security policy – adopt a security process, which outlines the organisation's security requirements/expectations and highlights the management's commitment to security.
- Security organisation – plan a management structure for undertaking security management responsibility and accountability in the event of any security incident.
- Asset classification and control – conduct a detailed assessment and inventory of the organisation's information infrastructure and assets to enable appropriate security measures. *You can't be secure when you don't know what needs to be secured*!
- Personnel security – people are a key component of an organisation's security. Security screening at the time of recruitment, outlining security requirements in job responsibilities, confidentiality agreements and incident reporting mechanisms are some essentials.
- Physical and environmental security – often on low priority, not having adequate physical access controls including restricted access to premises and sensitive areas, backup power supply and equipment maintenance is similar to having an intruder alarm system but leaving your keys in the door.
- Communications and operations management – instil security consciousness among employees, deploy anti-virus mechanisms and routinely monitor logs.
- Access control – protect the network (LAN and WAN), systems and information assets from external as well as internal intrusions.
- Systems development and maintenance – security must form part of every system development/maintenance process.
- Business continuity management – formal business continuity plan (BCP) and disaster recovery plan (DRP) need to be developed to protect the organisation and its information assets against potential natural and man-made disasters. Moreover, the standard requires that a risk-based approach must be followed to develop BCP/DRP, and these strategies should be an outcome of careful consideration of risks and business impact due to non-availability of key information systems.
- Compliance – comply with all applicable regulatory and legal requirements with respect to information management such as the *Data Protection Act*, the Health Insurance Portability and Accountability Act (HIPAA), the Gramm Leach Bliley (GLB) Act, Copyright Act or any particular local statutory requirements.

How ISO 17799 assists in managing disasters **17.40**

As detailed above in **17.39**, business continuity management forms an integral part of ISO 17799 implementation. Further, since ISO 17799 advocates enterprise-wide security, business continuity of operations and processes is ensured.

When an organisation adopts ISO 17799, it is implementing strong security controls framework based on the industry best practices and hence builds resilience towards impending disasters.

Being an industry accepted security standard, becoming ISO 17799 compliant assists an organisation to:

- project the image of a highly security conscious and aware organisation;
- demonstrate robust and mature information security management capabilities; and
- enable mutual trust between networked sites and trading partners.

However, the single most important benefit which ISO 17799 compliance provides is 'increased management assurance' in its systems and capabilities of proactively managing a disaster and recovering from the same in acceptable time frame. Being ISO 17799 compliant also provides third party assurance in the organisation's security framework and its ability to recover from a disaster.

Is ISO 17799 really needed for preventing disasters? – A true story! **17.41**

> 'A major UK financial company took pride in its disaster recovery planning and had in place what it considered to be a comprehensive and fully tested plan of the highest quality. Indeed, the plan itself had been fully tested only days prior to the fateful incident.
>
> On a quiet Sunday afternoon tranquillity was disturbed by a large explosion in their main office complex in the centre of a major city. Not a bomb or a terrorist incident, but a serious gas explosion.
>
> The company confidently swung the business continuity plan into full effect, almost as quickly as the media hit town, to instantly discover something that the plan, as good as it was, had overlooked! The streets were strewn with paper from the office containing a wide variety of confidential customer information. Sensitive data was lying around for any passer by or observer to simply pick up and read.' (ISO 17799 News Letter, January 2004)

Despite having a disaster recovery plan, one simple security lapse turned costly for this company, as this aspect of the incident was much reported. The reason, of course, was that the company didn't have any enterprise-wide security in place. Had the company taken care of smaller issues such as 'a clean desk policy', 'secure filing of confidential data' and 'employee responsibility' as detailed in sections 'Asset allocation and control' and 'Personnel security' of the ISO 17799, such a 'disaster' could have been averted.

Case Study

The following case study has been adapted from the article authored by Abhinav Singh and printed in the January 2004 issue of the *Network Magazine* (www.networkmagazineindia.com/200401/casestudy01.shtml).

A large IT services company offering software development (both onsite and offshore) services, IT consulting services and system integration services has over 15 development centres in India, which are spread across five different cities. The company also has marketing and sales offices across multiple locations in India, Japan, the UK, and the US. Apart from this, the company has near-shore development centres in Canada, Japan, UK and the US.

With such a vast scope of operations and its need for servicing overseas customers, it became imperative for the company to formulate a DR strategy, which encompassed the critical business processes and systems across the organisation.

The company is also among the first few in India to achieve a BS 7799–2: 2002. The company has formulated a comprehensive DR policy aligned with the BS 7799 framework.

DR Framework **17.42**

The DR policy developed by the company essentially addresses procedures for managing availability risks arising due to unavailability or inaccessibility of people and the flow of processes and technology operations in case of a disaster.

The company's DR plan includes procedures for undertaking command and control of the organisation in the event of a disruption, recovery procedures and resources required, vital records management, and infrastructure (ie facilities, telecom and IT) requirements and recovery.

The company undertook a detailed business impact analysis and risk assessment which assisted the management in defining the scope of its recovery strategy and prioritising the restoration of data. Critical business processes were also identified as part of the BIA and risk assessment exercise. Technology measures such as the fallback storage and the server configuration required for restoration of a given process and data were also facilitated by the exercise.

Responding to disaster **17.43**

The company's development centres across the world are well-equipped to recover and protect vital data in case of a disaster. There is a Chief Recovery ➡

Officer, a Damage Assessment Team and an Emergency Response Team at the strategic centres of the company. The task of a damage assessment team is to classify the disaster and invoke the recovery process based on the level of disaster.

Besides these teams, the company has identified critical personnel across all functions and business processes. All of them have visas to travel across countries in case of a disaster. Online project management and configuration management systems assist the company in moving work seamlessly across locations. The aim is to quickly respond to emergencies, minimise the impact of the disaster, analyse the impact and prevent its reoccurrence through well-defined and well-documented policies and processes.

Recovery sites 17.44

Data and application systems in use have been prioritised on the basis of their importance to the company's business. Three recovery sites have been created (the hot site, warm site, and cold site) to enable recovery of data and systems (spread across locations) based on their criticality for continuing operations.

A hot site is kept operationally ready with specific hardware platforms and fallback connectivity for immediate availability in the event of disasters. WAN links, WAN equipment, application systems (such as ERP, CRM and SCM software), and employee self-service servers are included in its category. The recovery time varies from immediate recovery to a maximum of eight hours.

The warm site is a partially equipped alternate site with backup tapes and basic connectivity infrastructure ready for use. The required servers and desktops have been are pre-identified and can be installed upon notification of a disaster. Project servers, Intranet servers and quality function servers fall under this category. The recovery time is from 48 to 72 hours.

The cold site is an alternate site with basic facilities like space, power and air-conditioning. Equipment and links are installed there to duplicate critical business functions. However, the restoration time may be from 48 hours to a week.

Incremental backup is taken for mission critical applications every eight hours and may take longer for less mission critical applications. A full backup is taken at the end of every week.

Backups 17.45

The company maintains a combination of servers for DR. There are multiple links through routes like fibre and satellite for DR. The combined inter-office connectivity results in a very thick network pipe. ➡

Multiple Internet links have been leased from a variety of service providers for building in adequate redundancy and fault tolerance. The switch over of communication links has mostly been automated.

Tape backups are carefully stored and those containing important data are sent to the remote disaster recovery location. Tape backup is taken using Veritas' storage management software for servers and Altrinis for laptops.

DR centre **17.46**

The company has identified one of its data centres as the location for its DR centre and has installed a well-equipped facility there to run the operations of the entire company in a worst-case scenario.

Ten personnel have been deployed for maintaining the DR centre 24 × 7. The DR centre also provides storage facility for tapes that are stored in a fireproof area. All the data being processed by any of the company's facilities worldwide is replicated to this centre.

In case of a major disaster, the infrastructure at the DR centre has the capability to go live within four hours of any strategic centre going down. Every six months, the company undertakes a DR restoration exercise where for a whole week all the systems and processes of the company are shut down across the world and are run from its DR centre.

Conclusions **17.47**

The ever increasing integration of IT in business has brought about a sea of change in the attitudes and understanding of the management. The business management no longer questions the value of IT systems and the data contained in them.

However, business managers expect that the systems and data will be available as and when required and that the responsibility for ensuring the same lies with the IT personnel.

Business management, in conjunction with their IT personnel, need to proactively create crisis resilient systems and plan for disaster recovery measures, without which every business is at risk not only from natural and manmade disasters, but also from unexpected human errors, viruses, etc.

Business continuity and disaster recovery planning assists an organisation in ensuring enterprise-wide availability, reliability and recoverability of its systems and business processes.

Availability 17.48

BC and DR management assists in the identification and ongoing maintenance of the availability requirements for the organisation's IT infrastructure (including hardware, software, network and communication systems, remote sites and facilities management).

Figure 7: Benefits of implementing business continuity and disaster recovery solutions

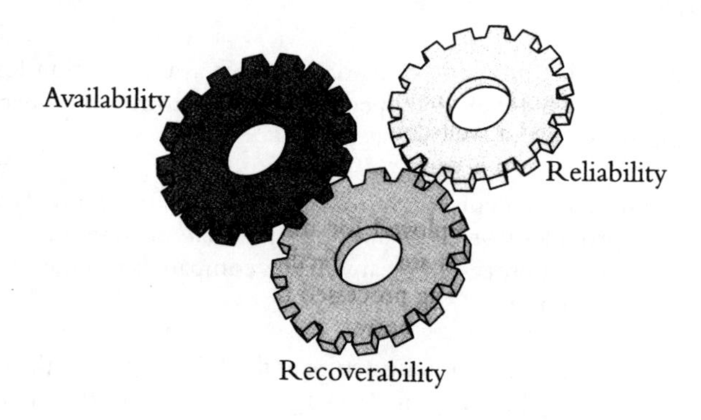

Reliability 17.49

BC and DR management assists in improving the overall operational reliability of IT infrastructure. It also assist in:

- extending the useful life of assets;
- improving quality assurance systems;
- better management of third-party services; and
- minimising software malfunction and reducing unplanned down time to facilitate 24 × 7 operations.

Recoverability 17.50

BC and DR management assists in minimising downtime of key processes in the event of a disruption. In particular, it facilitates:

- identification and reduction of potential risks; and
- development of emergency response procedures, including a BCP and DRP.

Recommendations **17.51**

Business continuity and disaster recovery solutions are like the nation's defence forces – you can't start building an army when the enemy attacks.

Expect the unexpected **17.52**

Think the unthinkable! Think outside the box! *Think like the enemy!*

- Prepare for diverse man-made disasters (including physical terrorist attacks and computer hackers).
- Meet the increasing availability expectations of customers (both internal and external).
- Think beyond technology and information systems (include people and processes).
- Broaden the scope of continuity planning (include service providers, customers and the public).
- Undertake proactive planning (BC and DR plans must be anticipatory in nature).
- Regulatory compliance (consider the legal requirements such as the Data Protection Act, HIPAA, GLB Act, Copyright Act, Sarbanes-Oxley Act, etc).

Self Assessment Checklist **17.53**

The following self assessment checklist providence a ready reckoner for an organisation to evaluate the effectiveness of its BC and DR strategy.

- Is our business continuity strategy event-driven or risk-driven and stakeholder-focused?
- How critical is information availability to our success?
- Are capabilities for managing business continuity aligned with organisational strategy?
- Who are our stakeholders and what is their tolerance for unplanned downtime?
- Does the risk management program address people, processes and technology, as well as the extended enterprise?
- Does the business continuity strategy eliminate single points of failure?
- How do we reinforce key management disciplines to ensure a reliable service delivery to all stakeholders?
- Does the risk management program support real-time service monitoring and reporting with predictive capabilities for critical infrastructure?
- How do we optimise the value of information flowing across the value chain?
- Does management have timely, independent assurance that its business continuity capabilities are adequate?

Figure 8: Key strategies

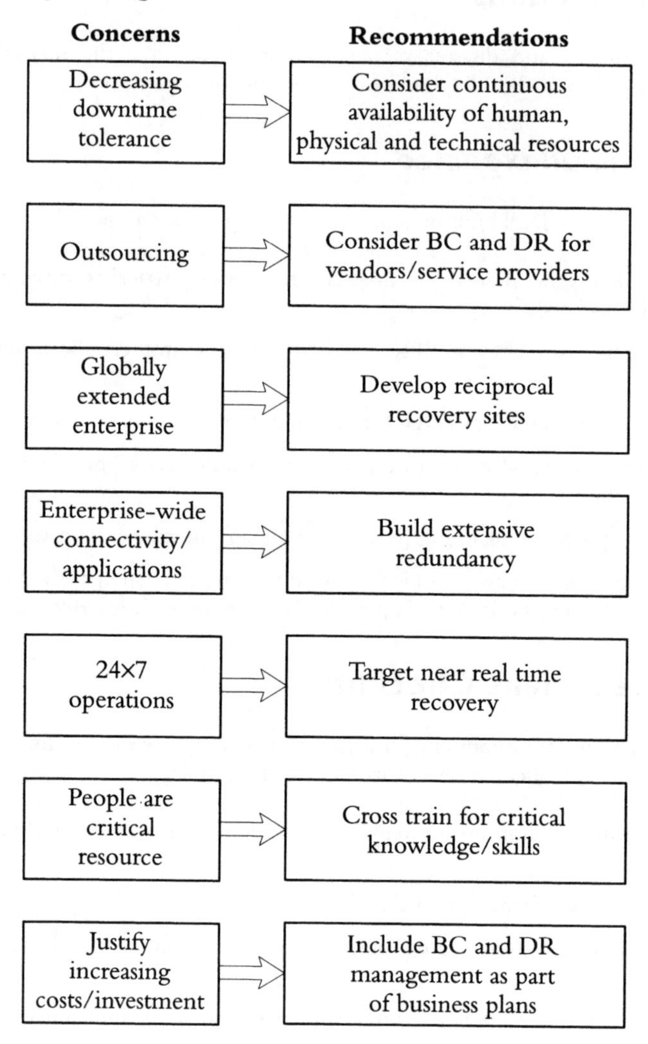

How do you Measure the Performance of your BCP/DRP 17.54

There are various ways by which businesses measure the performance and effectiveness of their business continuity and disaster recovery plans. According to a Business Continuity Study (*A Review of the Factors Influencing Business Continuity in the Next Millennium*) conducted by KPMG, most businesses rely on the results of actual tests for measuring the effectiveness of BCP. The other methods include service level monitoring, audit findings, performance reviews and comparison with industry norms.

Figure 9: Measures of BCP/DRP performance

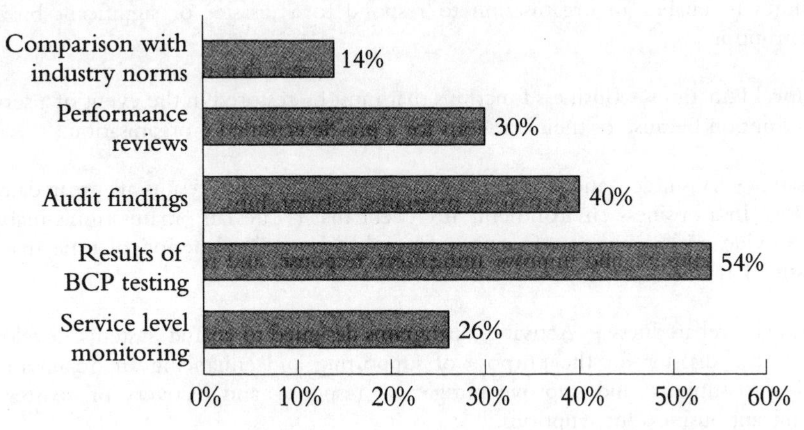

Commonly Used Terminology **17.55**

Alternate Site – A location other than the main facility that can be used to conduct business functions.

Business Continuity Planning (BCP) – The process of developing advance arrangements and procedures within an organisation for the purpose of responding to a disaster or significant business interruption so that critical business functions continue without interruption or essential change.

Business Impact Analysis – The process of ascertaining the impact of a disaster or service interruption upon an organisation in a quantifiable manner, such as, number of days and financial loss.

Business Resumption Planning – The process of advance planning and initiation of arrangements and procedures for execution when critical business functions for an organisation are disrupted as a result of disaster or other significant interruption.

Cold Site – An alternate operating facility equipped with air conditioning and electrical wiring. The addition of resources and equipment would need to be installed to process critical business operations of an organisation, such as, communications, UPS systems, business and computer operations. Utilising a cold site requires consideration of necessary time for equipment delivery, installation and testing before they are ready for use. Also known as a shell site.

Command Operations Centre – A facility that is separated from the main facility that is equipped with adequate communications equipment that is utilised by the management team to initiate and manage recovery efforts, and communications with individuals from the media and business units. It is a temporary facility that is typically used until alternate sites are functional.

Contingency Planning – The process of developing advance arrangements and procedures to enable an organisation to respond to a disaster or significant business interruption.

Critical Functions – Business functions that must be restored in the event of a service interruption because of their criticality to the operation of an organisation.

Disaster – A sudden and unplanned calamitous event that results in great damage or loss. In a business environment, any event that creates an organisation's inability to provide critical business functions for a pre-determined period of time may be classified as a disaster.

Disaster Preparedness – Activities, programs, relationships, and systems developed prior to a disaster for the purpose of supporting and enhancing an organisation's ability to support, and improve mitigation, response, and recovery of disasters or significant business interruptions.

Disaster Recovery – Activities and programs designed to enable an organisation to return to an acceptable condition following a disaster or significant business interruption.

Disaster Recovery Plan – An approved set of arrangements and procedures to enable an organisation to respond to a disaster and resume critical business functions within a pre-determined period of time. A disaster recovery plan is one of the deliverables of a business continuity planning process.

Disaster Recovery Planning – The process of developing advance arrangements and procedures that enable an organisation to respond to a disaster, and resume critical business functions within a pre-determined period of time, minimise the amount of loss suffered in such an event, and consider requirements for the replacement or restoration of impacted facilities in a speedy manner. Term is often used interchangeably with business continuity planning.

Hot Site – An alternate facility with the necessary equipment and resources necessary to recover critical business functions. The types of hot sites vary depending upon requirements. Examples include hot sites equipped for computer processing, communications equipment and electrical power.

Local Area Network (LAN) – A short-distance network used to connect terminals, computers and other peripherals within one or more buildings. A LAN does not utilise telecommunication providers to link devices together though it may have a gateway that provides access to non-LAN attached devices through public carriers.

Off Site Storage – An alternate facility other than the primary facility where duplicated vital records and documentation may be stored for use during a recovery from disaster.

Reciprocal Agreement – An agreement between two organisations with basically the same equipment that allows the other to process data for them in the event of a disaster or significant business interruption.

Recovery Point Objective (RPO) – The point in time for which data must be restored in order to resume transaction processing.

Recovery Time Objective (RTO) – The maximum acceptable time duration that can elapse before the lack of a business function will severely impact an organisation.

Warm Site – A partially equipped alternate site. A warm site will typically have some resources and equipment not found in cold sites but not to the level of a hot site.

Wide Area Network (WAN) – A network that links metropolitan, campus, or local area networks together across greater distances, and usually linked together by common carrier lines.

18

Rescue Equipment and Response

In this chapter:

- Historical overview of search and rescue equipment.
- Search and rescue framework for the UK.
- Search and rescue executive organisations.
- Common operational features of search and rescue.
- Special aspects of search and recovery.
- Specific search and rescue problems.
- Conclusion – lessons for practitioners.

Historical Overview of Search and Rescue Equipment 18.1

The following quote was made by Sir Eyre Massey-Shaw, Chief Officer of the London Fire Brigade, 1861–1892:

> 'A fireman, to be successful, must enter buildings. He must get in below, above, on every side; from opposite houses, over brick walls, over side walls, through panels of doors, through windows, through skylights, through loopholes cut by himself in the gates, the walls, the roof. He must know how to reach the attic from the basement by ladders placed on half-burned stairs, and the basement from the attic by rope made fast on a chimney. His whole success depends on his getting in and remaining there, and he must always carry his appliances with him, as without them he is of no use'.

Other than specialist rescue teams, the fire brigades have traditionally provided the basis for search and rescue worldwide. Equipment has evolved since the earliest beginnings to support the firefighter's dedicated role – extinguishment of fires and performance of humanitarian services. In former times, firemen used hand pumps to deliver water, such hot and demanding labour that they had perforce to be continually supplied with beer to keep them pumping and avoid dehydration and to provide them with the necessary energy. Wheeled pumping units were either manually or horse drawn from a network of fire stations where ladders were also kept. Wheeled escapes were developed which could reach to the fourth floor of a building and provide a stable staircase for evacuation; so successful the design that the London fire brigade sold off its escapes to surrounding shire brigades only in the 1980s.

Developing technology 18.2

Rescue technology has developed steadily and continues to evolve to support the changing needs of modern times – basically bigger and more rapidly developing fires and taller and more difficult buildings. Today's fire appliance carries a wide range of tools and equipment. Crews are supplied with dedicated protective clothing (fire gear) and breathing apparatus. High performance pumps deliver typically 4,500 litres per minute at several bars pressure using a hydrant or open water supply and can throw a powerful jet many tens of metres, laying down easily a ton of water every minute. For smaller fires, an on-board high-pressure hose reel is the preferred choice, using combination of jet and spray to great effect in killing fire with minimal water use (less water damage and run-off pollution). The longest ladder carried on a front line appliance is 13.5 m long and can easily reach the fourth floor of a building – it is the modern equivalent of the wheeled escapes.

Although a front-line fire appliance can deal with most emergencies including an increasing role in extrication of persons from vehicles, specialist backup is available in the form of turntable ladders and hydraulic platforms, able to reach to 30 m or more, thus the Fire Services are a formidable rescue force. Even so, for specific high-risk situations – oil refineries, airports etc – a range of specific rapid response vehicles have been developed, able to rapidly knock down the most severe fires, involving large inventories of highly flammable fuels and chemicals.

Specialist equipment 18.3

Basically fire brigade rescue equipment using manual pumps and escape ladders, evolved into its modern, highly sophisticated range of fire appliances – pumps, ladders, turntable ladders and hydraulic platforms. Specialist equipment now encompasses air lifting bags and hydraulic jacks, cutting and spreading gear along with a range of 'small gear'. Chemical protection suits are available to deal with chemical and biological incidents and decontamination procedures are defined and practiced.

The Fire Service role has basically developed within the community it serves and its focus is usually local, the emphasis is on providing technical competence and support to community emergencies. However, there are a number of specialist rescue services and organisations whose work is of a wholly different nature and which have evolved in a similar, yet different manner.

Air and sea rescue service 18.4

Rescue at sea and the Royal National Lifeboat Institution (RNLI) has a special place in the hearts of the population and the history of the lifeboats is well known. Every sea rescue is a drama in which the rescuers may put their lives on the line to recover those in distress – all the way from Grace Darling, who rowed out in an open boat to save the lives of a shipwrecked crew, through to a stricken ferry ship with 800 persons aboard in the North Sea (June 2002). Perhaps the heyday of the air-sea rescue services was during World War II when survivors from ditched aircraft were recovered using fast launches with aircraft support. Britain's most easterly airfield was RAF Beccles from which a variety of aircraft were deployed to save many lives – notably the Warwick, a cousin of the Wellington bomber which was adapted to carry lifeboats. The need to succour and rescue survivors was set against a background of water so cold that survival could be measured in minutes, invariably bad weather, lumpy seas and enemy action, all of which combined to make rescue exceedingly hazardous. The recovery of 33% of all downed crews by the year 1941 is a tribute to the air-sea rescue service.

Growth of hazardous sports 18.5

In more recent times, specialist volunteer rescue organisations have developed as a result of the growth of hazardous sports – acquiring and developing their own specialist equipment according to their needs. It is worthy of comment that much of the specialised safety gear designed for working at height was developed by the caving and mountaineering community. Cave, fell and mountain rescue teams become involved in searching operations as much as recovery of casualties, although this aspect too may be critical and requires specialist techniques of extrication and recovery to safety. These teams frequently become involved in recovery of persons injured or missing as well as those endangered in their sporting activity and also extrication of animals.

Search and Rescue Framework for the UK 18.6

The UK Government is a signatory to various conventions which commit it to provide search and rescue (SAR) infrastructure, namely:

- Convention on the High Seas 1958;
- Safety of Life at Sea (SOLAS) 1974;
- Maritime Search and Rescue 1979;
- Civil Aviation 1944.

These same structures also cover the UK responsibility which is divided into maritime, aeronautical (the Department for Transport (DT)) and inland (including inland waters) (police). The responsibility includes assessing the adequacy of the services to respond and ensuring that the infrastructure to deliver effective SAR is in place and kept in a suitable state of readiness.

The basic functional capabilities are:

- to receive details of aircraft, ships or persons in distress;
- to communicate between the various agencies involved in search and rescue;
- to deliver survivors to safety.

Thus, basically this involves an amalgam of government departments, the statutory emergency services (police, fire, ambulance), the military and voluntary and charitable organisations. At the time of writing, a management framework is being scoped and developed under the auspices of the DT, which has the overall strategic responsibility, particularly for maritime and aeronautical safety.

The Maritime and Coastguard Agency (MCA) deals with maritime, cliff, mud, shoreline SAR, pollution control and salvage. It discharges its functions via HM Coast Guard, the RNLI and in co-operation with other appropriate SAR authorities. Support from masters of vessels at sea, offshore oil and gas operators, the lighthouses and competent harbour authorities (who have service boats and tenders available).

The Civil Aviation Division (CAD) in co-ordination with the MCA and the Ministry of Defence (MoD) handles civilian aircraft incidents.

The MoD provides both military and civil SAR facilities and undertakes exercises and training and maintains an aeronautical Rescue Co-ordination Centre.

The Cabinet Office also has a role in ensuring quality in the provision.

Search and Rescue Executive Organisations 18.7

There are a number of search and rescue (SAR) executive organisations. These are summarised below.

Police 18.8

Responsibility for the co-ordination of land-based (above the high-water mark) and inland waters SAR rests with the Home Office and is exercised through the Police Service. This responsibility has been delegated by the Police Service to the MCA on certain high-risk inland waters such as Lough Neagh, Lough Erne, the Caledonian Canal, Loch Ness, the Norfolk and Suffolk Broads and various other estuaries including the River Thames.

The police co-ordination strategy works at operational, tactical and strategic levels of command (bronze, silver, gold) enabling a phased response and management control.

Fire Service 18.9

The Fire Service provides technical support, especially in fire matters. Whilst most brigade activity is local, they have the capability to mobilize rapidly and support emergencies anywhere in the UK. Although divided into 62 autonomous brigades, structures are in place to co-operate with neighbouring brigades. They may also attend maritime incidents and some brigades are specially trained for this role. Policy on SAR matters is set at Fire Authority level so there are local variations in provision. Thirteen of the 62 fire brigades also contribute towards the UK Fire Services Search and Rescue Team (UKFSSART), which has the capability to respond to both national and international SAR missions.

Ambulance service **18.10**

The ambulance service, which may be complimented by Helicopter Emergency Medical Service or air ambulance, has statutory duty to respond to traumatic and medical emergencies. There are 34 separate services that must provide a standard of service, which includes response time. Co-ordination and mutual support between services is via FM radio using an emergency reserve channel. Paramedic staff are trained to national standards and have a wide range of appropriate skills – but these normally do not extend to technical extrication or recovery which is under the remit of the Fire Service.

HM Coast Guard **18.11**

Her Majesty's Coastguard (HMCG) is responsible for the initiation and co-ordination of civil maritime SAR. This includes the mobilisation, organisation and tasking of adequate resources to respond effectively to persons either in distress at sea and to persons at risk through injury or death on the cliffs and shoreline of the UK. For this latter task, the HMCG has a number of Coastguard Rescue Teams (CRTs) that specialise in cliff and mud rescue. A volunteer auxiliary coastguard service comprising a number of teams support HNCG and are particularly concerned with local issues.

The Maritime Coastguard Agency exercises its responsibilities through 19 regionally based Maritime Rescue Co-ordination Centres (MRCCs). SAR helicopter capability is delivered through a combination of MCA, Royal Air Force (RAF) and Royal Navy (RN) assets dispersed at bases around the UK. The MCA has four bases, at Lee-on-Solent, Portland, Sumburgh and Stornoway. The RAF has six bases, at Boulmer, Chivenor, Leconfield, Lossiemouth, Wattisham and Valley. The RN has two bases at Culrose and Prestwick. These 12 bases have crews on standby at 15 minutes readiness by day and 45 minutes by night throughout the year.

Working closely with the emergency services and the MCA, all aerial SAR resources are controlled from the Aeronautical Rescue Coordination Centre (ARCC) at Kinloss in Scotland where a Nimrod, a long-range maritime patrol aircraft, is always on standby.

The coastguard keeps continuous radio watch on marine wavebands covering up to 150 nautical miles offshore, including international VHF and MF distress frequencies as well as telephone watch. A computerised command and control system enables sophisticated incident control including direction finding and drift prediction so that search coverage plans can be made.

Additionally, HMCG broadcasts maritime safety information, most notably weather forecasts (via the Met Office) but also navigational hazards and MoD information. Medical link calls are also relayed (the code calling for assistance is PAN PAN (ie not the well known MAYDAY to which all vessels nearby have a legal duty to respond immediately). Medico indicates the medical nature of the emergency.

Global marine distress is managed by MRCC Falmouth which is linked via satellite communications to various centres, especially ARCC RAF Kinloss. The EPIRB registry is held there – this is a listing of all the Emergency Position Indicating Radio Beacons so that ships may be identified in distress.

Royal National Lifeboat Institution 18.12

The Royal National Lifeboat Institution (RNLI) deals with maritime SAR through its fleet of all-weather lifeboats. Co-ordination is centred on RNLI headquarters at Poole which is continuously manned and provides liaison between local lifeboat stations and other SAR authorities. There are 232 lifeboat stations, 130 all-weather lifeboats (over ten metres in length and having ten hours duration at full speed) and the remainder have in-shore lifeboats only (to which launching restrictions are applied and have limited duration of three hours at speed). The service provides 24-hour cover capable of reaching anywhere offshore up to 50 nautical miles within 2.5 hours and for this uses a range of vessels from 4.9-metre inflatables through to 17-metre Severn Class all-weather lifeboats.

All lifeboats are provided with VHF radio and Global Positioning Systems (GPS) and for the all-weather boats these are linked to a computerized chart system. These lifeboats also have MF and HF radiotelephones and may have other radio facilities.

Mountain Rescue Council 18.13

The Mountain Rescue Council (England and Wales) acts as liaison body for 47 independent operational teams in eight regional associations; in Scotland there are 22 civilian teams plus various police and RAF teams affiliated to the Mountain Rescue Committee, however, the organisation is not regionally structured.

Cave rescue 18.14

Cave rescue (UK wide) is represented by the British Cave Rescue Council. 15 cave rescue organizations cover the main geographical regions.

Northern Ireland maintains a joint mountain, cliff and cave rescue co-ordinating committee.

Search and Rescue Dog Teams 18.15

The National Search and Rescue Dog Association (NASRDA) supports specialist dog teams, commonly drawn from members of mountain rescue teams. The Royal

Air Force, too, has mountain rescue teams. In addition, the UK Fire Service has an Urban Search and Rescue Dog Team giving it the capability to respond to both national and international SAR missions.

Beach Lifeguard Units 18.16

Beach Lifeguard Units are provided in a number of places and support the duties imposed on local authorities (LAs) with respect to safety on beaches and cliffs. These are trained by the Royal Lifesaving Society or the Surf Lifesaving Society and usually operate under contract to the LA and in liaison with HMCG.

Summary 18.17

Thus there is a complex and diverse range of organisations and structures available to provide SAR cover for virtually any foreseeable scenario. There is much highly specialised knowledge and equipment in existence and constantly under development. SAR teams and personnel are highly skilled and take considerable pride in themselves and their achievements.

Common Operational Features of Search and Rescue 18.18

Whilst SAR by its very nature can involve a very wide scenario range, nonetheless there are defined stages through which any incident will pass.

Mobilisation 18.19

Assembling a competent team. For inland incidents, mobilisation is usually initiated by the police; offshore incident by HMCG or the CAD.

There needs to be enough manpower made available and this is built up as the incident severity is assessed – using first attendance information to gauge whether this will be operational, tactical or strategic intervention. Fire brigade response, for example, is codified according to the number of appliances sent (eg 'make pumps four') – often it is the manpower rather than the appliances which are needed. The statutory SAR organisations are inherently more structured than those of the voluntary bodies, however, these latter also have robust call-out systems. There needs to be sufficient numbers of rescuers (phased callout) to operate the incident without wasting manpower or conversely overstretching and exhausting the first responders, which is often a danger. A relief system needs to be linked to the callout arrangements.

Medical and technical specialists will be required – location and availability of specialists can be a critical aspect. Rescue organisations use database systems and a warden system to optimise use of general and special manpower.

Location of equipment should be pre-planned. Whilst the fire brigade has a powerful asset register and can readily arrange specialist equipment such as heavy cranes, the voluntary organisations basically rely on rescuers carrying their own personal kit and making their own way to the scene. Stores of equipment must be established and be easily accessed. As an example, the West Brecon Cave Rescue Team maintains a modified Land Rover and a rescue store at its headquarters.

Supporting services 18.20

Fire, police, and the ambulance service will attend any incident as required but also when it is protracted, the welfare of the participants becomes an issue as, too, those of persons displaced by events. Red Cross, Salvation Army and Women's Voluntary Services are welcomed for provision of 'comforts':

> '... emerging into a scene which had turned into a crowded car park from a damp and forlorn field containing just two vehicles when we went underground 12 hours earlier – I made straight for the WVS mobile canteen'.

Command and communications 18.21

A forward communications centre should be established immediately with all necessary links to the police. Many rescues from mountain, fell and cave are initiated by the party to which misfortune has happened. Self-help, where practical, is important in maintaining morale and reducing the final time to recover casualties, but also to minimise the scale of the callout. Walking wounded can often be aided all the way to safety, however, it is essential to declare an Emergency and call for assistance via the police. Besides the obvious reasons that a second accident could occur or the weather or the casualty's condition could deteriorate unexpectedly, there is an insurance issue involved – the casualty's life insurance might be compromised! And the clock only starts running when the alert has been raised.

> 'We were 400 feet vertically almost directly below the rescue HQ, where the controller was sitting in front of a roaring fire. Our rescuer had popped out of a small hole with the words "I knew we would find you here" and then used his molephone to report us safely discovered.'

Accounting for rescuers by name and number is a key function of the forward command centre. Length of time involved (relieving exhausted rescuers) has already been mentioned. Special skills and equipment should also be noted. A rescue logbook should be initiated as soon as operations begin. It is vital that all information is accurately recorded *at the time* for legal reasons and reasons of debriefing so that lessons may be learned. The fire brigades use a tally system to log persons in and out of an incident.

The importance of the logbook 18.22

A caver in the Gouffre Berger went missing on his way out. He had not logged himself out at the logbook at the entrance tent, which, alas was not continuously manned at that time. Base camp was an hour's walk away over karst terrain which is notoriously dangerous because it contains many deep fissures and, it was possible, that being at night, exhausted and alone, he could have fallen into one. His body was not discovered for five months. In fact he was still in the cave. He had gone the wrong way up a narrowing passage and whilst traversing a tight rift, dislodged a boulder on to his chest, which would have killed him almost immediately.

Identifying casualties **18.23**

Accounting for persons involved (likelihood and numbers of injured or dead) can be extremely difficult, for example, the passenger manifests of commercial aircraft are notoriously inaccurate. In the case of a disaster such as the WTC collapse on September 11, 2001 it is impossible. Identification of some missing and unaccounted for at the Ladbroke Grove rail disaster was only made by observing apparently abandoned cars in a car park. Until all persons have been accounted for, the incident cannot be formally closed. Techniques for identification using DNA tissue typing is a major breakthrough in accounting for missing persons, however, the traditional formal identification of the dead by their relatives is still the most used. Thus, setting up a temporary mortuary will be needed where there are multiple fatalities. It is the police role to contact relatives of persons for this most harrowing aspect of Disaster Management.

Searching **18.24**

A major feature of mountain, cave and cliff rescues and the maritime equivalent, is locating missing persons. The mountains and the seas can be very big places and searches invariably seem to take place in hostile conditions.

> 'It was minus 1 degree Celsius but with a 60 knot wind. I was next last along the ridge and became suddenly very tired. I said to the man behind me that I was just going to have a little rest, forty winks. It was so seductive…' (radio operator, Morlais Mountain Rescue Team).

Intelligence is crucial to a successful search – knowledge of where last seen or heard from, probable destination or direction taken, and the likely mental, as well as physical, state of lost persons.

> 'We found him in a car park in the New Forest. I knew he would go there because he loved the place. He had the hosepipe and the rag in the car with him. Luckily, we were in time.'

The West Brecon Cave Rescue Team have successfully trialled route cards. Many caves and mines are in remote locations and finding the entrance can be problematic on featureless moorland or in dense forestry, especially at night. The route cards are waterproofed and described using permanent geographical features and they can be used with or without the backup of GPS – the latter may not be accurate enough to find cave entrances, some of which are physically small. Mountain rescuers depend heavily on radio communications and use specialist equipment, despite which there can be problems of dead spots. Helicopter support is available to support searching and subsequent recovery and so communication and co-ordination is an essential requisite to success. Communications are maintained by light portable radio sets and heavier vehicle and base sets.

In addition most teams are equipped with radios capable of communicating directly with MoD Helicopters and the Coast Guard, the use of which require a DTI VHF Radio Operators Certificate of Competency. Included in the radio equipment specifications are additional facilities according to the operational requirements of individual teams. These include speech links with, for example, MoD helicopters, area ambulance control stations and Police control points.

Radio links **18.25**

Underground communications in mines and caves poses further problems. Whilst radiolocation systems have been developed, surface contact has to be more or less vertically set up using loop antennas, so that persons may have to be dispatched to a grid reference in order to establish a radio link (the latest version, Heyphone is currently being rolled out to cave and mountain rescue organizations). The alternative is to run a cable and whilst single wire telephones are available it takes time and effort to set up (some caves have permanent lines installed).

Thermal imaging cameras **18.26**

Thermal imaging cameras have been used to locate persons by body heat but the technique is of limited application although they are useful in collapses or tangled wreckage and sometimes in thick undergrowth. Fire brigades have access to them – they work in the same way as night-sights used by the military.

Supporting casualties **18.27**

Whilst prompt recovery of casualties is always a primary consideration, support of the injured or unwell is just as essential. Cutting a person free of a vehicle, only to let them die of shock and hypothermia while waiting for an ambulance is not a good outcome.

First aid and medical support needs to be given throughout a recovery to prevent physical deterioration and maintain the morale of the casualty. Being fastened immobile into a rescue stretcher can become uncomfortable and also the casualty may become quite rapidly hypothermic and deteriorate while the team are concerned with the technical aspects of the rescue. For these reasons it is important to have a team member, preferably paramedic trained to act as patient manager.

Medical advice is also available by radio and it is in this way that the well being of many casualties is determined and monitored.

Medical equipment **18.28**

The innovations in medical equipment have been many and varied over recent years; some developed primarily for cave and mountain rescue use, some adopted from ambulance paramedic practice. The Little Dragon warm air breathing apparatus for the treatment of hypothermia generates warm, moist air for the patient to inhale, either independently or assisted. Other equipment in this field includes oxygen equipment with both automatic and on-demand supply, Entonox analgesics gas, intra-venous infusion sets (drip sets), and comprehensive resuscitation sets/kits. For splinting of limbs etc, a variety of splints are used, from inflatable splints to the more rigid, yet adaptable Kramer wire splint and exotic variants permitting traction to be applied. The monitoring of bodily function such as pulse, blood pressure, temperature, and electrocardiograph recordings are being carried out using portable electronic equipment.

Recovery

Carrying equipment **18.29**

The primary piece of equipment for mountain and cave rescue is a specially designed stretcher and a number of these have been developed from the original Neil-Robertson concept (in appearance and function they are rather like a large cricket pad with carrying handles). The Bell Mk3 and MacInnes are the mainstay of the mountain rescue teams, designed to give the casualty protection and at the same time permit manoeuvring in difficult terrain. The standard or split model like the Bell comes in two halves, each with its own integral carrying system; each half incorporates locking tubular retractable handles and weighs 22 kg when assembled. The MacInnes stretcher is extensively used in Scotland as well as throughout the world. It has very effective skids for lowering on snow slopes and also the capability to have a wheel attached for ease of movement over suitable terrain in the long glens.

Worthy of mention is the Alphin, a folding one-piece stretcher made by Troll Safety Equipment. It has a polycarbonate bed and short spinal protection strip. It is narrow and is excellent for constricted spaces, hence its popularity with cave rescuers, industry and the fire brigade. It handles well in awkward situations and is particularly versatile when raising or lowering – particularly horizontally. However, many of the UK Cave Rescue Teams now prefer to use the recently developed React stretcher as the main recovery method.

Specialist equipment **18.30**

Other specialist carrying equipment is the casualty bag – waterproof with a fibre pile inner with zip access to enable monitoring of the casualty without having to undo everything. Also a Vacuum Mattress is now available – a full body splinting device designed to effectively immobilise a casualty. It is the preferred option for neck or spinal injuries, being adaptable and also very useful in multiple injury patients.

Much of the equipment used in rescue is exactly that which is used in pursuit of sporting activity, eg rock and ice climbing and caving. Such items as karabiners, slings (tape and perlon), belay items (nuts and chocks, friends, pitons, ice screws, dead men belay plates, etc), ropes – both nylon and terylene, dynamic (lifelines) and pre-stretched (hauling, lowering and fixed rope situations), harnesses, helmets ice axes and hammers, crampons, etc. The list is all but endless.

Inventory and maintenance **18.31**

The inventory has to be stocked, stored and maintained by the rescue organisations and replaced when time expired, worn or damaged – this makes the operations expensive so that many of the organisations are charities and partake in fund raising activity.

Rescue is complete when a casualty reaches a medical centre. The rescuers must pass over identification details along with relevant medical information (eg analgesics administered).

Debriefing **18.32**

After any incident there should be a full debriefing and write-up. Logbooks, telephone logs, statements etc, must be made available and the report may be used as evidence in criminal or civil proceedings or to inform the coroner. It should, therefore, be as accurate and comprehensive as possible and signed off by the incident controller. The debrief should consider such matters as efficacy of callout arrangements, efficiency of command, control and communications and liaison between services, the effectiveness of techniques used and lessons to be learned.

Special Aspects of Search and Recovery

Searching in wreckage

Rescue dogs – National Search and Rescue Dog Association (NSARDA) **18.33**

The forerunners of modern search dogs are popularly considered to be the dogs of the monastery at the summit of the Great St Bernard Pass. The monks there were frequently called to assist travellers lost or caught in avalanches, operating a rescue service then, just as the people living in mountainous regions throughout the world are doing to this day. The famous 'St Bernards' were introduced in the seventeenth century, initially as guard dogs. Later, as they began to accompany the monks on rescue missions, the dogs' unerring sense of direction enabled the recovery of missing persons, even, it is claimed, to persons covered by snow. As far back as World War I, dogs were trained for use by the Red Cross to locate injured personnel on the battlefield during lulls in the fighting. They were trained to 'find' by use of the airborne scent from a human body.

Search dogs are trained to find missing people by following scent which is carried on the air. This is a very efficient method of searching large areas quickly and does not require items of clothing or effects of the missing person. Dog teams can be quickly deployed by helicopter to remote areas where they can quickly begin to start searching, whilst other search resources are being marshalled. At the present day within the NSARDA there are in excess of 90 qualified search dogs on 24/7 callout.

Dogs work equally well in the dark and use their senses of smell and hearing to their fullest under these conditions. It is calculated that a dog is equivalent to about 20 searchers in good conditions and many more in poor conditions. In ideal circumstances, a dog can pick up a human scent from about 500 metres.

Wide range of incidents where dogs are deployed 18.34

Search dogs are employed in a wide range of incidents. When operating under the control of a mountain rescue search co-ordinator, the handler and dog will follow instructions and integrate their abilities with that of other searchers. However, increasingly the handlers and dogs are responding to situations that do not involve other mountain rescue elements. In particular, when searching with police officers for possible crime victims, the handler and dog will rely strongly on their own team skills.

Search dogs are usually summoned by the police control room via a SARDA call-out co-ordinator who, after discussing the details will establish the necessary resource required.

Provision is commonly made for dogs and handlers to be deployed by aircraft to cut down on walk-in times with the dogs quickly becoming accustomed to being winched into and out of helicopters.

Scenting capabilities 18.35

The Fire Services are also making use of a dog's astonishing scenting capability and a number of firedog teams are now deployed in the UK to assist in forensic analysis of fires. They are able to detect traces of accelerants, such as petrol that would otherwise go undetected. Fire investigation dogs have a valuable contribution to make in the continuing fight against arson in the same way that trained dog teams are commonly used to tackle other criminal activities – the detector dog is well known in the high profile war against drug smuggling, with search dogs routinely deployed in airports and seaports to sniff luggage and passengers for traces of narcotics. Similarly, explosives search dogs are regularly used by the security forces. Smugglers or those attempting to conceal illicit materials go to extraordinary means in attempting to outwit the most sensitive detector device available to modern science – the dog's nose – and they usually fail.

Specific Search and Rescue Problems

Extrication from vehicles 18.36

Modern safety features are sophisticated and basically designed to keep occupants inside a vehicle, and to absorb the kinetic energy of a crash. These include crumple zones and shock absorbing bumpers, anti-burst door latches, side impact bars to keep the door from collapsing, and seat/shoulder belts and headrests to prevent occupants from being thrown around in a collision.

Shock absorbers 18.37

Shock absorbing bumpers have evolved from an original friction fit design which absorbed energy by plastic deformation, to mouldings which crumple upon impact, through to gas or oil filled shock absorbers, which at low speeds absorb the impact, then return to their original positions. There is a danger that shock-absorbing bumpers can become loaded and spring free dangerously during the extrication. Crumple zones help absorb some energy upon impact, however, distortion can change the structural integrity of the vehicle causing problems with stability and, at the same time, make it harder to open the doors which could become jammed.

Side impact beams, developed to prevent the door from collapsing into the passenger compartment, have also made it more difficult to gain access.

Door latch designs 18.38

Door latch designs are very variable in design, but they all have the same function – to prevent the door from opening. Safety standards require that latches be manufactured with high shear strength to prevent the possibility of a door opening during impact and allowing an occupant to fall out. This feature also makes it harder for rescuers to gain access through a door.

Head restraints 18.39

Head restraints are installed in vehicles to prevent whiplash; however, few people know how to adjust them properly. Improper use creates the potential for more extreme or extensive occupant injuries and, therefore, more complex casualty extrication operations.

Also, seat belts include an inertia-reel system to prevent an occupant from moving forward into the steering column and/or windscreen upon impact but may cause severe bruising and possible internal injuries instead.

Vehicle chassis design 18.40

Vehicle chassis design has evolved significantly over time. The original full-frame chassis, found on virtually all vehicles at one time and still on some today, gave strong anchor and stabilisation points. Next came the 'unibody' with stub frames that held the engine and differential in place, resulting in fewer anchor points. The full unibody vehicle also creates stabilisation problems due, in part, to false (plastic) contact points.

With the introduction of space frame vehicles, a new hazard has emerged: glass removal jeopardising the integrity of the vehicle. This can compromise the safety of both rescuers and occupants due to changing weight distribution and hence instability developing. In some situations the windscreen can become distorted, with the risk of explosive energy release when it breaks, destabilising the vehicle and showering shards of glass. Roof removal can also represent a hazard in so far as the roof and glass are part of the structural strength of the vehicle.

Undeployed air bags **18.41**

Undeployed air bags are a serious danger, and in some models there may be as many as 18 of them. The hazards associated with such varied numbers and locations in various makes and models are obvious. Activation of an airbag causes a sudden uncontrolled release of energy. Hazard factors include the travel distance of the bag, whether it is tethered or not, and what is in the path of travel or release area.

To mitigate this hazard, the rescuer must first assess if the hazard still exists. If so, the power source must then be located and isolated. This in itself can be a challenge because of the locations of batteries in today's vehicles. Even if the power source is eliminated, the bag may still deploy, as systems are designed with capacitors which hold enough charge to activate the bags; depending on the vehicle, charges can be held for up to 30 minutes.

Modern airbags **18.42**

Any vehicles manufactured since the year 2000 are equipped with 'smart' air bags, which have a first and second stage deployment system. With this design, there is the possibility that one of the two stages may not have been activated even if the bag has been deployed. Following the removal of a roof, there is potential for an airbag to vent up through the post in which it is positioned.

Note: Rescuers should always assume that undeployed air bags are live, and stay clear of the release area if at all possible.

To enhance the effectiveness of air bags, some manufacturers are installing seat belt pretensioners. Most units are activated mechanically and extrication efforts can set off the unit. For safety of the rescuer, it is vital to release the belt as soon as possible and not to work between the straps.

To aid in the recognition of these hazards, manufacturers' manuals are available with current information about battery, sensor, pretensioner, air bag and other hazard locations and should be consulted if at all possible.

Fuel-related hazards 18.43

Fuel-related hazards were associated primarily with petrol or diesel: its properties, leaks, spills and the containers. Today, rescuers must be familiar with a great variety of fuels used in and stored on vehicles: propane (LPG), natural gas (LNG), methane, and even a hybrid using one of the above fuels in combination with high voltage batteries.

The hazards of LPG and LNG are their properties and the fact that they are stored under pressure. Since they are stored as liquids, upon release to the atmosphere they vaporize rapidly, with the potential to cause cold injuries and, at the same time, create a flammable atmosphere. Pressurised tanks are likely to BLEVE (Boiling Liquid Expanding Vapour Explosion – an extremely violent event) if the vehicle catches alight.

Hybrid vehicles pose the normal fuel hazards but in addition, the batteries may leak corrosive fluids, cause electric shocks (up to 300 volts) and cause incendive sparking. The system has a service plug that disables it, however, the battery terminals and cabling will still present a risk.

Materials 18.44

Plastic components make it more difficult to gain a purchase point and also create a greater possibility of tool slippage during the extrication. When using hydraulic cutters, there is a tendency for the cutter to twist and not cut straight. This causes the blades of the cutter to spread apart and break, leading to flying shrapnel. When a reciprocating saw is used, the plastic has a tendency to melt around the blade, sealing itself and gumming up the blade. In most cases, removing the plastic panel and then performing the cutting operation is the safest way.

Performing cutting or spreading operations in areas containing micro-alloys which are extremely tough, could damage the tool, sending debris flying.

Struts have become commonplace in modern vehicles. The major concern associated with them is the means of attachment. Most struts are attached with plastic clips that can be damaged upon impact, leaving a loaded strut in place until disturbed during the extrication operation. When released, the stored pressure can propel the strut through metal and flesh.

The presence of fire at a road traffic accident does not mitigate these hazards and, in some cases, may even increase the danger. Some components may even release their energy after the fire is extinguished. Struts, driveshafts, bumpers, shock absorbers and tyres all have the potential to injure.

Stabilization, always a problem with larger vehicles on their sides, has brought back the use of high-lift jacks, and spurred manufacturers to develop new devices for the purpose of stabilisation. Air lifting bags and hydraulic jacks, whilst very versatile in operation are liable to leak and should always be backed up by physical supports.

Aircraft accidents **18.45**

Most aircraft incidents occur during take off or landing, on or close to airport premises or in the surrounding communities. However, crashes can happen anywhere and the range of associated SAR activity is correspondingly wide as the examples below show. High impact events (explosions, mid air collisions, high speed crashes) are generally not survivable but intermediate ones are (ie crash landings, forced landings) and as many of these are close to airport rescue facilities the level of preparedness and response is crucial. Aircraft accidents are invariably major disasters as the following examples show.

- 4 October 1992: An EL Al cargo Boeing 747 crashed into a major apartment building in Amsterdam; the four crew members and 47 people on the ground died.
- 21 December 1988: A Pan-Am Boeing-747 was bombed in mid-air and crashed into the small village of Lockerbie in Scotland, killing all 259 aboard and 11 on the ground.
- 12 August 1985: A Japan Airlines Boeing 747 crashed into Mount Osutaka at an altitude of about 1,200 metres. It took rescue crews nearly 15 hours to reach the site. Out of 520 persons, only four survived. Some had survived, though injured, but were overcome by shock and exposure and later died.
- 31 January 2000: A Kenya Airways Airbus A–310 plunged into the Atlantic off the Ivory Coast in Africa. Ten people were rescued from the cold ocean water, but 169 died.

Low impact crashes **18.46**

The outcome of a low-impact crash is crucially dependent upon rapid response vehicles' (RRVs) support as the following accidents illustrate:

- 22 August 1999: A China Airlines MD–11 crash-landed and burst into flames at Hong Kong International Airport. RRV units were available within minutes. Of the 315 people aboard, three died, and 188 were injured.
- 30 January 1974: A Pan-American Boeing 707 crashed in the jungle less than 900 metres short of the runway at Pago-Pago International Airport. All 101 people aboard survived the impact without serious injuries, but in the subsequent fire and smoke, all but four persons died before they could be rescued.
- 19 November 1996: Two commuter planes collided at Quincy Municipal Airport, Illinois. All passengers survived the initial accident, but the occupants were not able to evacuate. When the local fire department attended 14 minutes later after a 10-mile approach, all 14 persons aboard both aircraft had perished. The airport had no RRV facility.
- 1 March 1978: A Continental Airline's DC–10 crashed during take-off owing to blown tyres and the subsequent collapse of the landing gear. The aircraft carried 198 people and 81,000 gallons of Jet-A fuel. At least 10,000 gallons of kerosene spilled and ignited instantly, engulfing the fuselage in flames and toxic smoke.

Approaching airport fire units encountered people outside the plane on fire and many still trapped inside the burning jet. The first RRV crash truck was on the scene and in foam operation within 90 seconds of the initial alarm. Total extinguishment of the massive fire was accomplished just six minutes after the crash. Three persons died, but 195 others survived, 43 with injuries.

- 22 August 1985: A Boeing 737–200 with 130 passengers plus crew suffered an engine fire at take-off in Manchester. The flight crew aborted but by the time the aircraft turned off the runway, the blaze was already intense. Aviation fuel spilled out of the wing tank and formed a flaming lake. To further hamper the evacuation, wind fanned the flames towards the aircraft, denying two exits. Fire invaded the fuselage within 30 seconds and the rear fuselage collapsed in about a minute. Although RRV arrived at the scene quickly they were unable to save 55 of the passengers, almost all of them were in the rear cabin and many more were injured. Pilot training now includes action to take if a ground fire occurs – turn so the fire is downwind.
- 8 January 1989: A Boeing 737–400 with 126 persons left Heathrow Airport bound for Belfast. 13 minutes into the flight there was an explosion and fire in the left engine due to turbine blade failure. The aircraft crashed just short of East Midlands airport, and the M1 Motorway close to the village of Kegworth. The plane then slid along the motorway and broke up on the embankment. Although the on-airport response was rapid, 45 people died and many of the 79 survivors were injured. Pilot error was blamed for shutting down the wrong engine, causing the aircraft to land short. However, it should be noted that the engines could not be seen from the flight deck.

RRV Requirements **18.47**

The International Civil Aviation Organisation (ICAO) is the worldwide regulatory body for airports and airport emergency services. It recommends that for major airports, the first fire apparatus has to be able to arrive at any point on the operational runways in less than two minutes and all other required apparatus in three minutes or less. Unhappily, these recommendations are not always achieved.

Supporting role of local emergency services **18.48**

The local fire and rescue services play an important support role in the build-up of an incident. They must interact seamlessly with the on-airport RRV, which is a first-response activity and a consideration of response times and level of provision is critical to successful disaster mitigation. It is the local rescuers who will have to deal with casualty handling, prioritisation and management of many injured persons and liaison with hospitals. Command and communication must be prearranged, practiced and adapted as incidents require. Crews must be assigned to begin fire, rescue, medical, and salvage or recovery operations by the designated Incident Commander. This situation can reverse if the accident is off-airport. In this case the local firefighters will make the initial response, but with very limited means. Firefighters are trained to protect the fuselage at all cost because fire can break through in less than two minutes – it is RRV support which will be secondary.

Fuselage **18.49**

However, even if rescuers are able to get to the wreckage, gaining access to and into the fuselage presents a serious problem. Wide body jets are as high as a three-story building. Ladders, turntable ladders and hydraulic platforms are likely to be needed. If doors have not been opened from inside, forcing an entry can be very difficult. The conventional rescue tools carried on frontline appliances may not be very successful, an aircraft's aluminium skin and pressurised windows are nearly unbreakable.

On 19 August 1980, the flight crew of a Saudi Arabian Airlines L–1011 received fire and smoke alerts shortly after take-off from Ryad International Airport. The plane returned safely and stopped on a taxiway. For still unexplained reasons the engines were not shut down for some minutes, the cabin was probably not depressurised, no door was opened from the inside, and no evacuation took place. The Airport Fire Service took nearly 25 minutes to gain access into the fuselage but in the meantime, all 301 people aboard had died. On the other hand, when an Air France Airbus A340, carrying 309 passengers and crew, caught fire after overshooting the runway whilst landing in bad weather at Toronto Airport on 2 August 2005, everyone escaped.

Aircraft crash checklist **18.50**

Fire and Rescue services adjacent to an airport should have an Aircraft Crash Checklist carried on all front-line appliances and contain at least the following information:

- a grid-map of the airport and the designated staging areas;
- the specifically assigned radio frequency;
- priorities, the do's and don'ts at an aircraft accident scene;
- rapid fire control at the fuselage from upwind is essential;
- never approach without proper protective equipment; and
- no freelancing – always work inside established ICS.

Aviation Specific Hazards:

- kerosene fuel can always ignite;
- sharp metal debris can cut;
- engine force can blast objects and persons away;
- damaged aircraft structures can collapse and/or rollover; and
- the unknowns:
  - radioactive materials;
  - explosive devices;
  - chemicals;

- biological samples; and
- other HAZMATS.

Stability of structures 18.51

Assessing stability and structural integrity of buildings and other structures (eg the structure of bridges) is considered in **CHAPTER 3: CONSTRUCTION RELATED DISASTERS**. A cubic metre of concrete has a nominal weight of 2.5 tonnes. A piece of rock the size of a fist falling 10 cm can result in a black eye. Gravity is a potent force and should be respected.

Damage from fire 18.52

Imminent collapse is a major consideration of rescuers. The Fire Service has to consider this aspect in any dynamic risk assessment at a fire ground, usually against the consideration that the building is being progressively weakened by fire. It is normal practice to appoint a safety officer to watch for evidence of collapse and give timely warning to crews in the danger zone and especially the withdrawal of crews from inside.

Fire codes specify that compartments within buildings are designed to standards of structural integrity, should fire occur:

- half an hour for a basic house;
- one hour for a house of multiple occupation;
- two or four hours for high rise flats.

Badly designed structures 18.53

However, structures may not achieve their rated performance due to unknowns such as poor building work, under-specified or poor quality materials. Floors may be overloaded with stock. Premature collapse can also be initiated by inappropriate firefighting methods – cast iron columns will crack, stone piers will spall and maybe shatter, stone staircases may disintegrate suddenly. Sudden cooling can cause uneven contraction and distortion of elements of structure.

Use of high output fire jets which can easily deliver a ton a minute of water on to fire-weakened floors. A ship on fire was capsized in St Catherine's dock in this manner. A fire-weakened bulkhead contributed to the loss of the Titanic, collapsing suddenly due to the effect of cold water on it. And, most notoriously of all, the highly destabilising free surface effect which capsized the *Herald of Free Enterprise* when the inrush of water through the open bow doors all ran to one side. Longitudinal bulkheads would probably have saved her.

Evidence of failure can be observed in cracking of walls (horizontal cracks are generally known to be worse than vertical ones), falling debris and misalignment (any significant out-of-plumb condition is particularly dangerous because there is an overturning moment which increases with the departure from vertical). Levering out of floor joists as floors burn away or steel joists expand, buckle and sag are further indicators.

And a structure is only as good as the foundations it stands on, hence the need for good building design in earthquake zones.

Damage from water **18.54**

Water is a notorious agent for washing out soil and has caused many collapses. Leaks developing in our often-elderly water mains and sewerage system may scour out foundations. The river Plum in Plumstead was discovered 16 feet below a terraced house in Plum Lane when the building developed alarming cracks.

Hidden voids constantly threaten our infrastructure. It has been noted by the Kent Underground Research Group that there has been a great increase in such events in the unusually wet conditions at the turn of this century. These are epitomized by the appearance of dean holes which are ancient chalk mines, dug vertically through the clay soil of the London basin to a depth of anything up to 25 metres and hand mined into trefoil shaped chambers. When worked out, old tree stumps and other agricultural waste were poured into them, jamming at some point in the narrowing funnel shape, then they were subsequently filled in to the surface. Hundreds of years later, the plugs fall out and a large void appears. People have died falling into them, roads and houses have collapsed.

Conclusion – Lessons for Practitioners **18.55**

At the time of writing, the resurgence pool in Porth-yr-Ogof, a well-known tourist spot in the Brecon Beacons National Park close to the village of Ystradfellte, claimed its tenth human life, a 17-year-old army cadet. The dangers of drowning in the deep and cold water there are exceptionally well known, warning posters are displayed and even a memorial bench. All adventure centres have been given due warning and the Cambrian Caving Council has given it great prominence in its literature.

Despite all this, once more mankind dismally shows his lack of ability to learn from experience. We seem unable to prevent disaster and when it does occur, history records either great and heroic success or else disorganisation and tragic failure to respond – as is epitomized in the litany of aircraft disasters recited at **18.45** above. If we are to prevent and mitigate disasters effectively, we must not only learn the lessons they give us, we must put into place an adequate, cost-effective SAR capability and maintain it in a high state of readiness. This is a strict and serious duty on all those in the emergency services and it is invariably treated so. We should remember that 402 rescuers died at the World Trade Centre doing their duty. But it is those who determine and resource the capability who must bear the greatest burden of responsibility, from the very top of government downwards.

19

Terrorism

In this chapter:

- Introduction.
- Definitions of terrorism.
- Target and methodology.
- Chemical, biological, radiological and nuclear (CBRN) attacks.
- Suicide attacks.
- Cyber terrorism.
- Building hardening.
- Information.
- Conclusion.

Introduction 19.1

Despite the fact that there is a low probability of any one individual becoming involved in a terrorist act, the potential for terrorist violence, even mass casualty and mass destruction events, presents a significant risk for virtually everyone in the world.

Governments have a responsibility to protect their populations. Organisations have a duty of care; a responsibility to provide a safe environment, both for those who work for them and for those who may visit their premises or avail themselves of their services in whatever capacity.

Terrorist violence can cause destruction and economic dislocation comparable to that caused by a major natural or technological disaster. The management of terrorist events and the resultant effects are, therefore, of paramount importance.

Definitions of Terrorism 19.2

Those involved in formulating policy for anti-terrorist activities frequently spend a considerable amount of time searching for a definition. Often the definition will vary depending on whether it has been formulated by politicians, law enforcement agencies or the military. The lack of an agreed definition of terrorism has, in the past, been a major obstacle to international counter measures. Indeed, it is often suggested that one man's terrorist is another man's guerilla, is another man's freedom fighter.

Criminal definition 19.3

The first attempt to formulate an internationally accepted definition was made by the League of Nations in 1937 when terrorism was defined as 'all criminal acts directed against a State and intended or calculated to create a state of terror in the minds of particular persons or a group of persons or the general public'. The convention in which this appeared never came into existence and the United Nations still has no agreed definition.

A definition formulated by the Federal Bureau of Investigation in 1988 suggested that terrorism was 'the unlawful use of force or violence against persons or property to intimidate or coerce a government, the civilian population, or any segment therefore, in furtherance of political or social objectives'. However, this definition does not include the threat of force.

Academic definition 19.4

Attempts by academics to define terrorism tend to be somewhat longer. The following definition was proposed by A P Schmid to the United Nations Crime Branch in 1992:

> 'Terrorism is an anxiety-inspiring method of repeated violent action, employed by (semi-) clandestine individual, group or state actors, for idiosyncratic, criminal or political reasons, whereby – in contrast to assassination – the direct targets of violence are not the main targets. The immediate human victims of violence are generally chosen randomly (targets of opportunity) or selectively (representative or symbolic targets) from a target population, and serve as message generators. Threat- and violence-based communication processes between terrorist (organisation), (imperilled) victims, and main targets are used to manipulate the main target (audience(s)), turning it into a target of terror, a target of demands, or a target of attention, depending on whether intimidation, coercion, or propaganda is primarily sought'.

For many years, the United Nations failed to come up with an agreed definition on terrorism but it was referred to in Resolutions. For instance, in 1999 it strongly condemned:

> 'all acts, methods and practices of terrorism as criminal and unjustifiable, wherever and by whomsoever committed.' And, following the original, unadopted League of Nations definition, it suggested 'that criminal acts intended or calculated to provoke a state of terror in the general public, a group of persons or particular persons for political purposes are in any circumstances unjustifiable,

whatever the considerations of a political, philosophical, ideological, racial, ethnic, religious or other nature that my be invoked to justify them' (GA Res 51/210 'Measures to Eliminate International Terrorism').

Finally, in November 2004, a United Nations' panel described terrorism as any act 'intended to cause death or serious bodily harm to civilians or non-combatants with the purpose of intimidating a population or compelling a government or an international organization to do or abstain from doing any act'. However, there has been considerable criticism of this definition and it has not been universally agreed.

Distinguishing criteria 19.5

The definition contained in Title 22 of the United States Code, Section 2656f(d) defines terrorism as 'premeditated, politically motivated violence perpetrated against non-combatant targets by sub national groups or clandestine agents, usually intended to influence an audience.'

This definition has three key criteria that distinguishes it from other forms of violence:

- first, the act must be politically motivated;
- second, it must be directed against non-combatants; and
- third, sub-national groups or clandestine agents commit the acts.

The term non-combatant obviously includes civilians but it also includes military personnel, who, at the time of the incident, were unarmed and/or not on duty.

Insurance industry 19.6

Increasingly, acts of terrorism are being excluded from property insurance policies but there is no generally accepted definition within the insurance industry. At the 2002 Conference of Asia-Pacific Risk and Insurance Association (APRIA) in Shanghai, the following definition was put forward by Harold Skipper, Chairman of the Department of Risk Management at Georgia State University, and Janet Lambert, from the UK firm of Barlow Lyde and Gilbert.

> 'An act, including, but not limited to, the use of force or violence, committed by any person or persons acting on behalf of or in connection with any organisation creating serious violence against a person or serious damage to property or a serious risk to the health or safety of the public undertaken to influence a government for the purpose of advancing a political, religious or ideological cause.'

The statement went on to suggest that such an act shall be certified as an 'Act of Terrorism' by the senior judicial or administrative official designated by the adopting government and should not be subject to appeal.

Use or threat of action 19.7

By virtue of the *Terrorism Act 2001*, terrorism is now defined in the UK as the 'use or threat of action' where:

- the use or threat is designed to influence the government or to intimidate the public or a section of the public; and
- the use or threat is made for the purpose of advancing a political, religious or ideological cause'.

This definition is controversial in that it has been suggested that what has, in the past, been regarded as legitimate political activity could, under certain circumstances, now be regarded as a terrorist act. However, for the purpose of this chapter, terrorism is taken to be the use and/or threat of repeated violence in support of, or in opposition to, some lawful authority, where the intention is to induce fear of similar attacks in as many non-immediate victims as possible with the intention that those threatened will accept and comply with the demands of the perpetrators. To come within this definition any threat of violence must be credible; in other words, there must be a belief amongst the victims, or within the lawful authority, that the perpetrators have the ability to carry out the threat.

Model 19.8

The management of terrorism conforms to one of the standard cycles of disaster events. Figure 1 illustrates the cycle of terrorism management.

Figure 1: Cycle of terrorism management

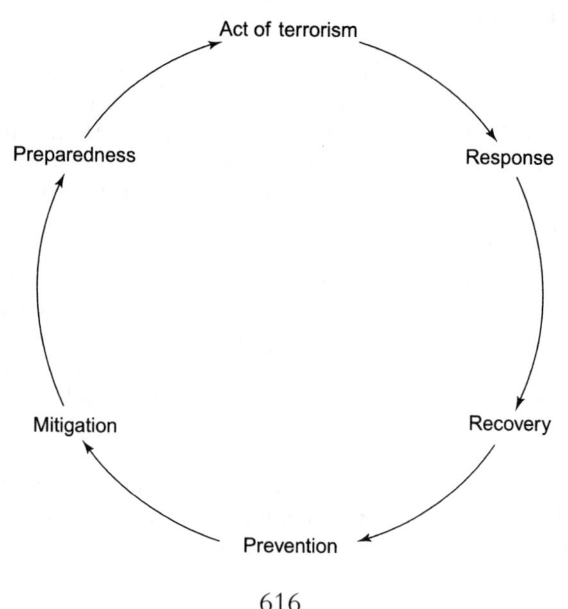

Prevention **19.9**

As the title suggests, action within this phase of the cycle is designed to prevent terrorism from occurring. The following could be classified as preventative measures:

- *Elimination of the grievances of the terrorists*, eg eliminating the sources of the frustration – this has been done in the past in countries such as Kenya, Palestine, Malaya and Cyprus, generally by giving them independence from colonial rule.
- *The elimination of the terrorist* – this is extremely difficult, as many countries have found. In some cases, acts carried out by governments to eliminate the terrorists merely result in more recruits coming forward to aid the terrorists' cause.

Mitigation **19.10**

Action within this phase normally takes the form of specific programmes or actions intended to reduce the effects of terrorism either internationally or on a particular nation, or part of a nation.

- Threat assessment – what are the likely targets? What might the impact be? Accurate threat assessment is based on the collation and analysis of information about the likely intentions of the terrorists but it also includes an assessment of the response capabilities and the likely outcome of any attack. Mathematical modelling, widely used in engineering design, might be of use in calculating the possible outcome of any attack and technology has been used to assess chemical spread arising from accidents for some time.
- Identification of critical infrastructure – what can a country or an organisation do without? What can it not do without? In so far as countries are concerned, it may well be water, electricity, gas, petrol and so on. In so far as organisations are concerned, it may well be the computerised system it has evolved to carry out its business. For the emergency services, communications are likely to be a critical part of their ability to operate. The list can be fairly substantial and each country or organisation might have some critical function that is unique.

  Are all the assets in one place? Where are the alternative supplies? What are the alternative systems? How secure are the critical assets? Such questions need to be answered and acted upon.
- The denial of opportunities to attack the critical infrastructure.
- The construction of buildings likely to be attacked in such a way that damage is minimised – building hardening, as it is known in some quarters, is dealt with in more detail later at **19.20** below.
- Protective measures to reduce the threat to the population in or near areas likely to be affected.

- The cultivation of a climate that is hostile to the terrorists, ie reduces their opportunities to use violence and increases the likelihood that they will be identified and captured.
- Alterations to the terrorists' operational environment, eg reducing their opportunities to launch attacks.
- International treaties that:
  - outlaw support for terrorist organisations;
  - reduce the flow of arms, ammunition and money for terrorist purposes.
- Bilateral and multilateral treaties that:
  - deny terrorists staging areas outside the target nation;
  - provide for the trial or extradition of terrorists.
- The monitoring and controlling of the movement of people across borders, particularly suspected terrorists and other people who are travelling to areas that have been identified as possible targets.
- The introduction of legislation which gives the security forces greater powers, eg in Britain, the *Prevention of Terrorism (Temporary Provisions) Act 1974*, provided for longer periods of detention, the entering and searching of homes and the deportation of suspects. Whilst the introduction of such legislation may be regarded as essential in the fight against terrorism, governments need to exercise extreme care. Terrorists may well seek to provoke an over-reaction to their violence to discredit the actions of the government and/or security forces. If the government and/or security forces are seen to be using violence as an excuse to abridge civil liberties and to reduce legitimate opposition this is likely to work in the terrorists' favour.

Preparedness **19.11**

Preparedness comprises measures taken by the government and the security forces to enable them to respond rapidly and effectively to terrorist acts. Examples of preparedness measures are:

- Continuity of government to ensure consistency in the security effort – this does not mean that the same political party must remain in power; rather that the main political parties should be in broad agreement as to the policy to be followed in the fight against terrorism.
- The organisation and co-ordination of security agencies, including the designation of lead agencies, to provide special units trained to respond to:
  - the taking of hostages by the terrorists;
  - other terrorist activities.

- Operational plans to structure and facilitate the emergency response by the police and/or military and other emergency management personnel – eg fire and medical services.
- The establishment of emergency communications networks to tie together the various agencies involved.
- The provision to the emergency services and other agencies of the equipment necessary to provide an effective response to any terrorist incident – eg protective clothing in the event of an attack using chemical, biological, radiological or nuclear material.
- The provision of an effective information/intelligence gathering system – this is dealt with in some detail at **19.34** below.
- Information programmes to:
  - keep the public informed so that they can effectively avoid the danger areas and prepare for evacuation if warranted;
  - assure the public that the danger is perceived as real and the actions of the authorities are perceived as necessary and credible;
  - maintain majority public support for government action against terrorists.

  Public warning systems need to be developed. Currently, messages are regularly given out to the public that tell them to be vigilant or be more vigilant, or be ever more vigilant. But, in many cases, no one takes any notice. And what does 'be vigilant' mean? A warning system, say, using a scale from 1 to 6, needs to be developed so that the public know more precisely what action they should take at each scale.
- Training and education programmes to ensure that the security forces and other emergency personnel are adequately prepared to respond to acts of terrorism.

  Detection is important in the clinical field. Some kind of biological attack is likely to first appear through general practitioners' surgeries and Hospital Accident and Emergency departments.

  Once detected, there is a need to protect people. Is there an antidote? Is it readily available in sufficient quantities? In the US and the UK particularly, the development of new medicines takes a long time because of the need for extensive health trials – and it is costly.

  In an emergency, will people be able to administer themselves? Have they been trained to administer it themselves if, for instance, it is necessary to do so by injection?
- Exercises and drills so that the plans and organisational arrangements can be tested and changed when necessary – the importance of exercises and drills is described in **CHAPTER 20: TRAINING AND EXERCISING FOR EFFECTIVE PREPAREDNESS AND RESPONSE**.

Response

19.12

Response measures are usually those that are taken immediately prior to and following the act of terrorism. Typical measures include:

- The survey and assessment of the terrorist act.
- The implementation of plans to save life and protect property – these include:
  - evacuation measures;
  - search and rescue;
  - cordoning off the affected area.

  The primary aim is to protect life. But whose life? Problems could arise if it is a serious incident involving a chemical, biological, radiological or nuclear (CBRN) agent. This might be best explained using a hypothetical scenario. Imagine there has been a serious attack by terrorists on a community and a number of people have been contaminated with a nasty biological agent. The Police Service and the Fire Service have thrown a cordon round the area in an attempt to keep the contaminated people from contaminating others. But some of those people refuse to be corralled. They go to leave the area and there are insufficient members of the emergency services to physically stop them.

  What action does the police commander take? Does he/she let them go or, as a last resort, does he/she give an order for lethal force to be used to prevent them from leaving? Those outside the cordon have a right to live; not to be contaminated. But what are the rights of the people who have already been contaminated?
- The gathering of evidence to support a criminal prosecution – September 11th has changed the way the emergency services are likely to respond to terrorism. Whilst the individual services and, indeed, countries will continue to deliver a response on a local basis, counter-terrorism is now international despite the absence of a generally accepted definition. Terrorists detained in one country are more likely to be extradited to stand trial in another country.
- There must be a coordinated response by the emergency services – in the past, it has been suggested that the aims of the emergency services, particularly at terrorist incidents, have, on occasions, been regarded as incompatible. The medical services were keen to save life and additional suffering; the Fire Service were keen to get rid of whatever was causing the problem, be it a fire after an explosion or a chemical release; and the police were keen to preserve as much of the scene as they could in order to accumulate evidence for the prosecution of the perpetrators. There is, hopefully, no incompatibility now. The primary aim is to protect life. And this is right. Whilst the public would like the police to find and bring to justice those who commit acts of terrorism, clearly they would like to live first.

In the emergency services, there is a need for integrated teamwork. The aim of everyone should be to protect the public and pursue the terrorist. Multi-agency training, involving the police, fire and ambulance services, together with hospitals, local government emergency planning units and the military, is essential in order to bring about an efficient and effective response.

Recovery **19.13**

Recovery is the process by which communities and the nation return to their proper level of functioning following an act of terrorism. Typical activities include the following.

- The containment of the terrorist threat, eg the prosecution and conviction of the offenders.
- The restoration of:
  - essential services;
  - repairable homes and other buildings/installations;
  - measures to assist the physical and psychological rehabilitation of those who have suffered from the effects of the terrorist act.
- The provision of assistance to families of victims.
- Long-term measures of reconstruction, including the replacement of buildings and infrastructure which have been destroyed by the act.
- Decontamination of the affected area and/or affected people if it is a CBRN (chemical, biological, radiological and nuclear) incident.

Target and Methodology

The Tokyo subway attack **19.14**

Case study I

Aum Shinrikyo (supreme truth) was a religious sect founded in 1987 and properly registered as such with the authorities. It claimed to have 10,000 members in Japan and another 30,000 in Russia. On Monday, 20 March 1995, operatives from Aum Shinrikyo attacked the Tokyo Subway system using sarin. The sarin, in small plastic bags, was dropped inside five different underground trains approaching the centre of Tokyo during the morning rush hour. As the train pulled into a station, the five operatives pierced the plastic bags with sharpened umbrella tips before fleeing the train. As a result 12 people died and over 5,000 others were afflicted by the release, some suffering severe complications as a result.

During their subsequent investigations, the Japanese police discovered that the group had quantities of peptone, which can be used in the manufacture of biological agents, and found evidence that Aum Shinrikyo had been researching the use of another nerve agent, tabun.

Concerns after Tokyo attack **19.15**

The events in Tokyo were widely viewed as crossing the threshold and evoked a number of serious concerns, amongst which were fears that:

- the effectiveness of improved security measures had prompted terrorists to seek different methods of attack;
- some terrorists considered public indifference to be such that a more spectacular incident involving higher casualty rates was necessary to attract attention. Both the subsequent bombing of the United States Embassies in East Africa in 1999 and the events of September 11th were clearly designed to inflict maximum casualties;
- some groups had now reached the stage where frustration and a sense of impotence had become so powerful that they were prepared to pursue any course of action in an attempt to achieve their goals.

Attacks focused on economic infrastructures can be expected to continue, including those related to energy and the other utilities. Amongst these might be the poisoning of water in reservoirs or the blowing up of a dam, which in addition to causing widespread flooding could mean the loss of hydroelectric power generation. Although, in most cases, security is generally tight, nuclear and chemical plants could be attacked. But nuclear material and chemicals are likely to be more vulnerable to interception whilst being transported. For instance, it is estimated that 2.5 million Americans have commercial driving licenses authorising them to transport fuel and other hazardous materials. So, there is a very real danger of some kind of chemical, biological, radiological or nuclear release.

The threat to civil aviation varies from country to country according to political and other circumstances, eg if terrorists are held in prison, hijacking might take place in an attempt to secure their release.

Incidents of kidnapping and hostage taking have become frequent occurrences in South America, Asia and parts of the former Soviet Union. In Columbia alone, it has been estimated by the US Department of State that there are more than 3,000 kidnappings each year, resulting in US$250 million in ransom payments. Also, there have been random attacks on tourists and the deliberate killing of foreign aid workers. All are likely to continue.

The dangers of some kind of cyber attack, which could affect a wide range of services, are very real. Indeed, in March 2001, the then Foreign Secretary, Robin Cooke, announced to Parliament that computer hacking could cripple the UK quicker than a military strike or terrorist campaign.

Chemical, Biological, Radiological and Nuclear (CBRN) Attacks **19.16**

Despite the number of casualties and the massive damage inflicted in the attack on the US on September 11th, using more conventional means, chemical, biological, radiological and nuclear (CBRN) weapons of some kind still pose the most lethal terrorist threat. Whilst September 11th showed that nothing can be ruled out, the use of a nuclear weapon remains the least likely scenario, given current levels of security and the reluctance of any State to support the use of such a weapon for terrorist purposes. However, the dispersal of a chemical, biological or radioactive substance in a terrorist incident does remain a strong possibility. There is, therefore, a need to develop appropriate strategies and tactics to counter the possibility of such an attack, given the current availability of source materials.

Numerous chemicals have the capacity to cause death or injury. They are sometimes divided into four broad categories:

- choking agents – such as chlorine and phosgene;
- blood agents – including cyanogen chloride and hydrogen cyanide;
- blister agents – eg mustard gas;
- nerve agents – such as sarin, soman, tabun and VX.

Many chemicals are commonly used in industry and agriculture and are, therefore, readily available in large quantities.

Biological agents are arguably even more dangerous for a variety of reasons, not the least of which is the incubation period and, in some cases, likely delays in identification. In a report to The President and Congress in 1999, Governor James S Gilmore identified four primary routes for the acquisition of biological agents. They could be:

- purchased from one of the world's 1,500 germ banks;
- stolen from a hospital, research laboratory or a public health service laboratory where biological agents are cultivated for diagnostic purposes;
- isolated and cultured from natural sources;
- obtained from a rogue State, a disgruntled scientist, or a State sponsor (Gilmore Commission First Annual Report, 1999).

Added to this, radioactive materials are readily available on the 'black market' and can be obtained at the right price. An extensive amount of weapons-making information can be found in open-source literature. Only rudimentary skills and equipment are required to enable a chemical or biological (CB) weapon of some kind to be constructed, either by some individual working alone or by a terrorist organisation. Methods for the dispersal of CB agents are many and varied.

Nonetheless, despite the apparent advantages of using CBRN material as a means of attack, conventional weapons are still favoured by terrorists, principally because of familiarity and the ease of use. The gun and the Improvised Explosive Device (IED) remain the principle methods of terrorist assault. Vehicle bombs have proven a particularly attractive method of attack for terrorists, in part because the trend in recent years has been toward high casualty, indiscriminate targeting, and in part because of ease of manufacture and delivery. The components are widely available and instructions on their manufacture can be obtained from open sources.

The anthrax attacks in the US 19.17

Case study 2

Shortly after the events of September 11th in 2001, envelopes containing anthrax were sent to a number of addresses in the US. The first two envelopes were sent on 18 September to representatives of the media; the second two envelopes were sent on 9 October to Senators Daschle and Leahy. None of the people to whom the letters were addressed became victims. However, as a result, a number of people, particularly postal workers, came into contact with the envelopes and five people subsequently died. Many more became ill.

At the time of writing no person has been convicted of any criminal offence appertaining to these incidents. However, the anthrax is known to have come from the 'Ames Strain' that originated in the US and DNA analysis suggests that it originated from a US Army facility at Fort Detrick, MD.

Effect 19.18

As will be seen, the anthrax attacks caused relatively few deaths. But they did cause:

- widespread apprehension; and
- a large number of false alarms in the US, many of which were deliberately made with the intention of distracting the police.

Whilst the Anthrax problem was on-going in the US, the police in London were receiving 100 calls each day; another 100 calls were being received by police forces elsewhere in the UK. In London, 90% of the calls were made by people who were genuinely concerned. In one case, a travel company sent out sand from a Florida beach, entreating people to come to Florida for their holidays. People receiving this were convinced that they had been in contact with anthrax. Only one out of ten calls was a hoax. At the start, police, fire and ambulance personnel were responding to each call. But, as the number of calls increased, the London Ambulance Service almost collapsed and had to adopt a vigorous risk assessment process to prevent them from doing so. In addition, at the time there was only one laboratory where analysis could take place – thus the response was far too slow. Consequently, at the outset, there were occasions when there was no choice but to decontaminate people who may have been in contact with the substance.

Checklist I **19.19**

The action to be taken if there is a possibility that an organisation or building may be targeted to receive an envelope or package that could contain a biological agent is illustrated in the Checklist below.

Checklist I

Mitigation

- Ensure that people do not have to pass through the mailroom to get to other parts of the building or other facilities.
- Do not install air conditioning; if, because of the heat, air conditioning is necessary, install a system that only supplies the mailroom (but see action to be taken under Response below).
- Sort mail into routine (expected) and suspect (unexpected).
- If, following a risk assessment, the threat of being targeted is considered to be high, consider issuing mailroom staff with protective clothing, eg gloves and face masks.

Preparedness

- Prepare a contingency plan to include the action to be taken in the event of a suspect package being received, the management of any evacuation should it become necessary and follow-up activities to be undertaken.
- Ensure that the staff has been appropriately trained.

Response

- Isolate the area.
- Turn off air-conditioning systems.
- Close all windows and doors to the room in which the envelope or package is located.
- Place the envelope or package in a plastic bag or other container that can be sealed.
- Inform the emergency services.
- Prevent any other people from entering the room pending the arrival of the emergency services.

➡

- Ensure anyone who may have been contaminated by the suspected 'agent' is isolated.
- Evacuate people away from the affected room.
- Be in a position to meet the emergency services on their arrival and brief them.
- Follow the advice of the emergency services.
- Remain calm.

Recovery

- If it is found to be a biological agent, arrange for the affected area to be decontaminated.
- Ensure that those who may have been in contact with the agent are given medical care and, if necessary, psychological debriefing.
- Take necessary action to reduce anxiety amongst both staff and, if appropriate, customers and clients.

Note: The last two points may be necessary even if it turns out not to be a biological agent.

Suicide Attacks 19.20

Suicide attacks can broadly be divided into three main categories. First there are those in which the bomber or bombers seek out a selected VIP to die with. In Asia, the Tamil Tigers frequently used women suicide bombers who belonged to an elite group known as the Black Tamil Tigers. Invariably the explosives are strapped round the waist and they are assumed to be pregnant. In 1991, Rajiv Gandhi, the former Indian prime minister, was assassinated by a Black Tamil Tiger whilst he was electioneering in India. The Tamil girl approached Gandhi with a flower garland and when he touched it she activated the explosives, killing herself, Gandhi and 17 others. An attempt to assassinate the president of Sri Lanka in 1999 only narrowly failed. The president suffered facial injuries and lost her sight in one eye.

The Palestinians, too, have recently started using women as suicide bombers. However, this use of women to undertake terrorist activity is not new because, in the past, they have often used women as part of the team when hijacking aircraft. The most famous of these was Leila Khaled, who was captured when, with a male colleague, she tried to hijack at El Al aeroplane in September 1970. She was subsequently released when the Popular Front for the Liberation of Palestine (PFLP), of which she was a member, hijacked a British aircraft carrying 300 passengers.

Secondly, there are those in which the bomber or bombers seek out a prestigious target. The simultaneous attacks on the United States Embassies in Kenya and Tanzania in August 1998; the attack on the US Naval destroyer, USS Cole, the simultaneous attacks on New York and Washington in 2001, and the simultaneous attacks on the British Consulate and HSBC Bank in Istanbul in November 2003 are typical examples of this type of suicide bombing.

Following the attacks on New York and Washington on 11 September 2001, a newspaper, funded at that time by the Palestinian Authority, suggested that:

> 'the suicide bombers of today are the noble successors of their predecessors. These suicide bombers are... the engines of history... they are the most honourable among us'.

Also, a poll amongst Palestinians in November 2001 found that 73% of Palestinians supported suicide missions against American interests in the Middle East, which is, perhaps, one reason for the high number of attacks in Iraq.

Thirdly, there are those who merely wish to take as many people with them as possible. Whilst such attacks have taken place for some years in places such as the Middle-East, in the conflict between the Palestinians and Israel, and in Sri Lanka, by the Tamil Tigers, they have, since 11 September 2001, become far more widespread. Iraq has seen the most attacks by far but attacks have also taken place, with devastating results, in Bali (in 2002 and 2005), London and Jordan (all in 2005). Each has been carried out by suicide bombers. Such attacks occur without warning, often at the security network's weakest point, e.g. places of public gatherings. One could, of course, argue that a major transport network, which was the venue for the attacks in London in July 2005, is a prestigious target and therefore this attack should be included under type 2. Similarly, the attack in Amman, Jordan, also in 2005, concentrated on three American-owned hotels.

In Russia, however, there have been at least two occasions recently when people, clearly on a suicide mission, have taken hostages. At the Moscow Theatre siege, in October 2003, 41 Chechen terrorists, half of them women, took over 800 people hostage whilst they were watching a play. The Russian security forces brought the siege to an end, killing all 41 terrorists. But, at the same time, 129 hostages also died. Just under a year later, in September 2004, at a school in Beslan, 32 terrorists, many of them women, took up to 1,000 hostages as parents and pupils gathered for the start of the school term. When the siege was brought to an end, 31 of the 32 terrorists were believed to have been killed, along with 344 hostages, 186 of whom were children.

Cyber Terrorism **19.21**

Cyber terrorists have the ability to target, cripple and destroy a wide range of facilities. It follows that the provision of adequate security to protect against hackers and cyber attacks remains a never-ending battle of technology. The advent of the Internet has dramatically changed the types of viruses that are seen today and, more importantly, how quickly they can be spread. Great reliance is placed on the Internet and e-mail in the conduct of daily business and, when viruses strike, the resulting loss of communications could lead to massive economic damage and even death and injuries.

A concerted terrorist assault on computer networks could disrupt communications and power grids in a manner reminiscent of the Canadian winter ice storm of 1998, and could create havoc in the business community. Similarly, the destruction of critical computer networks could seriously cripple key global commercial infrastructures such as air transport, international banking, and stock markets.

As has already been mentioned, one of the most notorious terrorist attacks prior to September 11th 2001 was carried out by Aum Shinriko when it released quantities of sarin into the Tokyo subway system in 1995 (see **19.14** above). Four years later, in 1999, it was revealed that, prior to 1995, Aum had built up a commercial business producing computer systems that had become a major source of their income. Police raids on Aum sites produced masses of evidence showing that the sect had been involved in the installation of over 100 computer systems in government ministries and major companies. It transpired that while Aum had been producing and installing computers, Japanese Government Agencies had been troubled by mysterious hackers.

Building Hardening **19.22**

It is only in the last 20 years or so that commercial buildings have become terrorist targets. As a result, many of the older commercial buildings still in use have not been designed to withstand the effects of an explosion. But, as was discovered when the World Trade Centre was targeted in 1993, even some of the newer buildings have not been designed with sufficient foresight to withstand the attack of a determined terrorist (see further **3.39** above).

World Trade Centre 1993

Case study 3 **19.23**

In 1993, the World Trade Centre complex consisted of five main buildings on a six-acre site. The two principle buildings, the North and South Towers, each had 110 storeys. Over 50,000 people were employed in the 1,200 businesses. In addition, over 80,000 people visited the complex each day. A bomb, the equivalent of 1,000 lbs of TNT, had been driven in a truck into a five-storey underground car park in the South Tower at the World Trade Centre in New York. It exploded at 12.18 pm on 26 February. At least five people were killed and more than 1,000 injured. In addition, there was extensive damage to the tower and commerce was badly disrupted. The cost of the explosion in the first week was estimated at US$330m. Much of this was made up from dislocation and loss of business rather than structural damage or personal injuries. Later, it was estimated that the total cost of the explosion to corporate business, government and insurers was more than US$1 billion.

Effects of bombing **19.24**

Because bombs come in a variety of shapes and sizes it is very difficult to assess the effects precisely. Vehicle bombs, for instance, are likely to be anything from 50 kgs of explosives in a small car up to 2 tons in a lorry. A bomb placed inside a building is likely to cause far greater damage and more injuries than if the same device were placed outside. But it is usually far more difficult and more dangerous for the perpetrators themselves, to plant large bombs in buildings. One of the most notable internal bomb explosions occurred at the Grand Hotel in Brighton in October 1984 during the Conservative Party Conference. On this occasion, the explosion, caused by 12 kgs of Frangex, broke the back of a massive central chimney stack, sending it crashing down through the rooms below. Five people were killed and over 30 injured.

Blast pressure **19.25**

The effects of a terrorist bomb can be massive and at least three major factors need to be considered. First, the blast pressures on both the building and associated services can be considerable. Buildings in urban areas are particularly vulnerable because the buildings on either side of the streets confine the blast wave, which increases pressure on them. Therefore, the pressure caused by the blast drops more slowly than it would if it had been let off in open countryside.

'Cratering' **19.26**

Second, the 'cratering' under the seat of the explosion may seriously affect underground services. At the World Trade Centre, the explosion ripped through three floors of concrete and left a 60 m hole, causing the ceiling to collapse in the Port Authority Trans Hudson (PATH) subway station that was alongside the car park. At least 50 of the injured were on the station platform. The designers of the building had decided that if water pipes, electrical and communications cables, were housed in close proximity to each other in a subterranean area of the building, together with transformers, ventilation controls and back-up generators, maximum protection would be afforded. However, in planning for an underground car park to be next to the utility equipment area, they created a vulnerability that should have been recognised during the design stage.

Immediately following the explosion, most electrical power and virtually all safety systems in the World Trade Centre were crippled. Thick, black caustic smoke billowed up through the ventilation system, lift shafts and fire escapes. Fire alarms, smoke sensors and sprinkler systems were knocked out and the building's public address system apparently failed to work. The carefully prepared Emergency Plan was abandoned because the Emergency Command Post, located in the basement close to the parking levels, was seriously damaged in the explosion.

Fragmentation 19.27

Finally, there is fragmentation. Fragmentation can come from primary and secondary sources. The former includes materials that are immediately around the explosive device. The latter are mainly made up of projectiles caused by the blast wave. By far the most dangerous kind of secondary fragment comes from glass.

The owners of buildings likely to come under attack from a bomb need to define the threat and then determine the level of protection to meet the threat. Many of the problems that arose after the World Trade Centre bombing arose because the designers had failed to consider that terrorists might drive a truck, loaded with explosives, into the underground car park.

Oklahoma City 1995 19.28

Case study 4

On 19 April 1995, there was a massive explosion in Oklahoma City and the front of the Alfred P Murrah building pancaked to the earth, crushing almost everything as it fell. Consisting of nine floors, the building housed a number of federal offices and law enforcement agencies. Built in a U-shape, the front of the building consisted of glazed full-height windows in a reinforced concrete frame. The cause of the explosion was a massive car bomb placed right outside the front of the building. It killed 169 people and injured many more.

Timothy McVeigh, a US citizen, was subsequently arrested and convicted of murder. He has since been executed by lethal injection.

Building design 19.29

It was claimed that the Alfred P. Murrah building had been 'tested to destruction' (see further **3.36** above). But the crater, measuring some ten metres across and two metres deep, was just three metres away from the centre of the north side of the building. More importantly, the explosion occurred in close proximity to one of the four columns, each of which were two floors high and placed 13 metres apart. All four columns collapsed as a result of the explosion. The collapse of the column immediately adjacent to the explosion started a chain reaction, or a progressive collapse, something that should not have happened in a properly designed building.

For existing buildings there is a limit to what can be done to the main structure. It may be possible to create a zone, such as a bomb shelter area, to withstand the effects of most explosions providing all the threats have been realistically assessed. The bomb shelter area is likely to include an Emergency Control Centre with adequate communications that will enable people in the building to be informed of the action to take following an explosion. Many features can be included in a building that will increase its resilience but designing an entire building to withstand a significant explosion is extremely costly. Therefore, in new buildings, it may be better to concentrate on designing certain sections to safeguard key equipment and key areas should there be an explosion. But there are features, such as large areas of glass, which should be excluded from buildings likely to be a target for a bomb.

Even in cases where the main part of the building appears to have survived the explosion, there will be an inevitable delay before organisations can resume normal business. The structure of the building will need to be examined to ensure that it is safe. But, even though the building may be structurally sound, the fabric damage may be severe. The main services, particularly gas and electricity, must be made safe, and glass shards must be removed from window frames. Any delay is likely to cause economic disruption. It has been found that whilst the larger businesses generally survive a bomb explosion, many of the smaller businesses have ceased trading within 12 months.

In addition to public buildings such as those mentioned in the case studies above, shopping centres and underground systems are tempting targets to the terrorist who wishes to instil fear amongst the population. However, every effort must be made to introduce security measures that do not alienate the public.

London 1993 19.30

Case study 5

On 24 April 1993, a two-ton truck bomb exploded in Bishopsgate in the heart of London. The explosion occurred on a Saturday morning when most of Central London's businesses were closed to the public, although staff often worked behind closed doors. A warning was telephoned to the police on this occasion and this enabled people who were on the streets to be evacuated. However, about 100 people were working in the NatWest Tower, a 42-storey building with a large number of windows and, rather than evacuate it, with all its inherent dangers, they took refuge in designated safe areas. When the bomb did explode it blew out every window on the Tower's east face. Also badly hit was the Hong Kong and Shanghai Bank. The only fatality was a press photographer. Clearly, on this occasion, the bomb was designed to cause maximum damage but minimum casualties.

Dangers of glass surfaces 19.31

Dr P D Smith of the Engineering Systems Department at Cranfield University, points out that there are three principle ways of providing protection to windows and other glass surfaces in order to reduce the dangers from sharp-edged fragments or shards, which travel at high speed following an explosion. Firstly, by applying a polyester anti-shatter film (ASF) to the inner surface of the glass – an optional extra in such cases are what is known as bomb blast curtains (BBNC). Secondly, the use of blast-resistant glass; and thirdly, the installation of blast-resistant secondary glass inside the existing exterior glass. Failure to do this can result in people being seriously injured and damage to delicate equipment, such as computers, inside the buildings.

Vehicles **19.32**

In most of the cases involving serious damage to buildings as a result of bomb blast, the bomb had been delivered in a vehicle. Parking was allowed immediately outside the Alfred P Murrah building; indeed there was an indented passenger-loading zone just two metres from the building.

It is essential, therefore, to increase the distance between the building and the bomb if the effects of an explosion are to be reduced. This is done by physically preventing vehicles from approaching or parking close to the buildings. A comprehensive access control system for both vehicles and pedestrians should prevent an explosive device from being brought into a building such as it was at the World Trade Centre.

Checklist 2 **19.33**

Clearly, those responsible for the security of buildings or other enclosed places to which people have regular access need to build in a robust defence strategy. Consideration should be given to the following checklist.

Checklist 2

- A single point of entry and exit allows for greater control. However, it should be borne in mind that check-in procedures that cause lengthy delays can actually increase security risks because people will look for alternative means, particularly of entry. Baggage scanners and the facility to scan identity cards can assist in speeding up the process.
- Security cameras can act as a deterrent. However, they are only as good as the human personnel assigned to monitor them.
- Security personnel must be trained in all aspects of the work they are required to undertake.
- Access to delivery and shipping areas must be tightly controlled. All deliveries should be registered and screened before being accepted.

- If there are street level vents which take in air in the building these should be sealed; new vents should be installed at a level that does not give ready access.
- Waste bins that could be used to hide a bomb should be removed; alternatively, depending on the threat, they should be emptied approximately every 15 to 30 minutes.
- Physical barriers should be erected to keep cars and trucks at a distance from the building.
- High-tech security systems are being developed in an effort to safeguard buildings in the future. Finger, thumb and palm print scanners, together with iris and retinal scans, are being developed.
- The ability to install high-grade filtration systems already exists. But there needs to be a greater emphasis on detection systems. Currently, most buildings have smoke alarms and carbon-monoxide detectors are readily available; so too are detectors that detect the presence of radiation. Chemical detection equipment needs to be installed in places such as the London Underground; it needs to be carried on the vehicles of the emergency services. Effective bio detection equipment that will meet this requirement is still to be developed for civil use although various systems do exist within the military. But are they reliable? Will that equipment function each time it should without false alarms? If the equipment signals a possible chemical or biological release when there is none it will very quickly lose credibility.

Information **19.34**

Hindsight is a specific science. But it is quite clear now that the events of September 11th 2001 was, amongst other things, an intelligence failure of huge magnitude – indeed, in so far as the US is concerned, history might well put it alongside the attack on Pearl Harbour in World War II.

Good quality information is the lifeblood of effective decision-making. Managed properly, it allows for a clear understanding of the threat and assists in the identification of possible courses of action. It enables valuable resources to be targeted effectively against emerging and current trends, ensuring the best opportunities for positive intervention in meeting any terrorist attack.

This is the information age. Almost everything that is done is governed by the availability of more information. By their very nature, terrorist incidents are situations of uncertainty. Therefore, the more information a government or an organisation has about a terrorist group or an incident, preferably before it happens, the better equipped they should be to reduce that uncertainty. Without information the decision-makers cannot hope to make appropriate decisions.

Information Technology 19.35

It has been suggested that 75% of all the information available throughout the history of mankind has been created in the last 20 years. The prime mover in this has been Information Technology (IT). But, at the current rate of technological development, the amount of available information will rapidly double. Therefore, with so much information available, its movement has to be carried out in a recognised format; otherwise it becomes unmanageable. But it is not enough just to possess information. Information is only of value if it is being used. If some vital fact sits gathering dust in someone's in-tray, or in someone's computer, then it is of little or no value. It only realises its value when it is 'moved' and, if necessary, acted upon. In addition, information must be accurate and timely.

Capabilities v Intentions 19.36

But, Hughes-Wilson points out, those charged with the collation and analysis of information confront another, more subtle problem; that of Capabilities v Intentions (Hughes-Wilson, Col J, *Military Intelligence Blunders* (1999) Robinson, p 7). There is a clear distinction between Capability and Intention. Understanding the difference between the capabilities of an individual or group of people, who are a threat to a country or a region or, perhaps, to an organisation, is crucial to understanding what might or might not happen.

For a number of years there has been, perhaps understandably, a concentration on the dangers posed by Chemical, Biological, Radiological and Nuclear (CBRN) terrorism but the attack, when it came, was, in comparison, extremely low-tech. Aircraft hijacking has been going on for well over 30 years and the weapons of the hijackers were knives and bolt-cutters.

The need to distinguish between Capability and Intention is clear. Capability is relatively easy to measure. It was known for a long time that Bin Laden had training camps in Afghanistan and, more precisely, where they were in the country, but determining his intention was far more difficult. As Hughes-Wilson points out 'a person's intentions can change like the weather. Even the most sophisticated information system can break down when confronted with the vagaries of the human mind' (p 7).

Today, Capability can be pretty accurately assessed at all levels. The problem arises with Intention; not if it is possible for the terrorists to strike. Everyone knows they will strike. They have struck before repeatedly. But where will they strike, when will they strike and how will they strike? Answers to questions such as these are rarely provided by technology; they are provided by people.

However, there is an additional problem, particularly in democratic societies. Prior to September 11th, intelligence staffs had been warning for some time that certain terrorist groups had the capability, ie the organisation and the means, and possibly the intention to mount an attack on key targets, such as those in New York and

Washington. Politicians and other decision-makers, however, have to balance the risk of that occurring against the potential disruption to the normal and legitimate activities of the population as a whole.

Conclusion **19.37**

Since September 11th, 2001, there has been a fundamental shift in the way the world looks at terrorism. The events of that day mean that those responsible for counter-terrorism have to think about possibilities that, to-date, have been regarded as unthinkable. September 11th is the benchmark by which future acts of terrorism will be judged. Present and future terrorist groups wanting to make a major impact will be looking to do something spectacular, perhaps by launching simultaneous attacks on major targets using unconventional methods or means.

Fortunately, some of the better-known terrorist groups such as the Provisional IRA in the UK and ETA in Spain have chosen not to use CBRN as a means of attack. They are, after all, not in the business of destroying society; rather they want to acquire things from the society in which they operate. To use CBRN weapons would deny them the popular support they have amongst certain sections within their respective communities. But there are individuals and groups, such as Aum Shinrikyo (see **19.14** above), who have used CBRN material as the means of achieving their revolutionary or personal goals. The determination of such groups should not be underestimated; neither should their technological proficiency.

Future casualty rates associated with domestic and international terrorist incidents will vary. Generally, more frequent, domestic incidents are usually the result of a shooting or bombing attack directed against security forces or specific civilian opponents, such as politicians or key commercial buildings. Such incidents produce a smaller number of casualties. In some countries, when the action is merely to raise money for terrorist purposes, the target may be the employee of a major international company.

Occasionally, the number of casualties are higher, such as when an explosive device is placed in a crowded area, as was the case in the bombing of the Alfred P Murrah Building in Oklahoma City and is all to frequently seen in Israel. International incidents are characterised by large-scale casualties because the incidents, although less frequent, are designed to achieve maximum publicity and shock effect. Some examples of these are to be found in Moscow in 1999 and in the bombing of the US Embassies in Kenya and Tanzania in 1996.

Learning from experience **19.38**

Terrorists learn as much from experience as do the counter-terrorist agencies. For instance, there was an intention to crash an aircraft onto a high-profile building in December 1994 when the Armed Islamic Group (GIA), an Algerian based terrorist group, hijacked an Air France Airbus with 171 passengers on board. On this occasion, the hijackers, dressed as airport security staff, had taken possession of the aircraft whilst it was on the ground in Algiers. Eventually, it took off. The plan was to crash it on the Eiffel Tower in Paris. But the hijackers had a problem – none of them knew how to fly the aircraft. Instead, the plane landed at Marseilles where the French counter-terrorist force (GIGN) stormed it, killing the hijackers and releasing the passengers and flight crew. It was shortly after this that members of Al Qaeda quietly began to enrol at flight schools in Florida.

Countries and organisations must remain alert and cognisant of the threat of terrorism both from the domestic and the global perspective. Until September 11th 2001, there had been a downward trend in the number of international incidents of terrorism in recent years. This had occurred due to a combination of:

- improved international co-operation to combat terrorism, such as the sharing of information;
- a decrease in the level of State sponsorship;
- improved security arrangements in some countries; and
- positive changes within a number of political and economic climates.

However, history has shown that it would be dangerous to make a long-term projection. Terrorism has often been called war on the cheap. For that reason, it will continue to be used by disaffected individuals to attract attention to their cause. The relative ease with which terrorist acts can be carried out and the attraction of terrorism as a political and foreign policy tool is likely to remain.

Collaboration **19.39**

International co-operation is an important avenue to be followed in the battle against terrorism. Its dynamic nature, and the increasing technical competence and practical experience of those who embrace the use of political violence demand a broad and sophisticated response on the part of counter-terrorist forces. Increased collaboration with the intelligence community is an outcome of bilateral and multilateral arrangements. A fostering of such mutually beneficial activities is likely to lead to a more effective response to the diverse threats posed by terrorism.

It was President Reagan who first said of the terrorists, 'You can run but you cannot hide'. President Clinton signed two Directives, one in 1995 and another 1998, that gave authority for 'rapid, effective and co-ordinated response against terrorism'. So, President Bush's commitment following September 11th is not new. What may be new now that such a devastating act of terrorism has been committed in the US is the resolve of the American people to do something about it. But it needs more than that. It needs the resolve of all nations. But, at the end of the day, the risk of capture or even death remains part of the glamour for some terrorists and participation in terrorism is a demonstration of their commitment. Death is, in effect, an opportunity for martyrdom.

20

Training and Exercising for Effective Preparedness and Response

Introduction 20.1

Investigations and inquiries into disasters as disparate as Piper Alpha, Kings Cross, Clapham Junction, Ladbroke Grove, Marchioness, to name a few, have all cited the issue of 'training' and 'exercising' as one of the key failure modes. Normally, it is neither a lack of training arrangements nor the opportunity to attend courses that leads to shortfalls in competence and proficiency; rather it is a lack of implementation and continuous improvement where the dangers lie.

For example, in the Piper Alpha oil platform disaster of 6 July 1988 where 167 persons died, Lord Cullen who headed the inquiry devotes a whole chapter assessing the role of 'training for emergencies'. He notes that Occidental Petroleum had systems and procedures for training of critical emergency staff, however, the *regularity* of training and the lack of regular assessment were key root failures contributing to the disaster.

> 'While the platform management did not exhibit the leadership required in this important area of training, the onshore safety staff did not operate an effective

monitoring system with regard to emergency training.' (The *Public Inquiry into the Piper Alpha Disaster*, Volume 1, Chapter 13, p 218.)

A key recommendation from Cullen was the need for regular, up-to-date and refresher training, drills and exercising, such that it becomes habitual. Lord Cullen is also suggesting the need for training that equips personnel with the competence, confidence and decision-making skill sets, rather than a rote reliance on manuals and procedures.

Ladbroke Grove Rail Inquiry 20.2

Lord Cullen also headed the Ladbroke Grove Public Inquiry relating to the disaster on 5 October 1999, where 31 died and over 400 were injured. Some of the key training/exercising related points in the 'Ladbroke Grove Rail Inquiry Part 1 Report' (HSE, Chapter 9) can be stated as:

- a fragmented approach to driver training;
- a lack of validation of training material;
- 'gaps' existed in the training itself – not comprehensive enough;
- lack of implementation focus in training programmes;
- training and simulation needs to be linked to the management system with audit and review of training arrangements.

This is somewhat a case of Piper Alpha revisited.

In short, there is evidence that within industrial sectors and between them there is a lack of 'isomorphism' – a pattern of learning, sharing experience and best practice to improve competence through training and exercising.

This chapter provides advice and guidelines to directors and management on the key issues to take account of when establishing a training and exercising regime for an organisation.

Terminology 20.3

It is important not to confuse terms. One of the conceptual causes of industrial and technological disasters is when organisations assume that an 'educational course' will yield the same results as 'training', or that 'learning' and 'training' must be the same. Lord Cullen notes some of these confusions in the Piper Alpha Inquiry where 'drills', 'exercises' and 'training' seem to have been confused. Just because off-shore management practice a drill does not mean the employee has been trained in emergency and safety procedure. Note that the page references in Figure 1 refer to the official inquiry.

Figure 1: Key concepts

Term	*Definition*	*Interpretation*
1. Competence	'The ability to do something successfully or efficiently' (p 374).	• A person, whose performance – practical and/or theoretical satisfies a criteria, standard or benchmark set by a body or society. • Competence refers to the 'end' or goal to be achieved.
2. Education	'Give intellectual, moral, and social instruction to (someone, especially a child), typically at a school or university' (p 589); 'The process of receiving or giving systematic instruction, especially at a school or university' (p 589).	• A broad psychosocial and institutional process involving the imparting of 'instructions' (ie theoretical, logical, sequential and systematic details) and interaction with the recipient(s), usually within a formal and structured environment.
3. Training	'The action of teaching a person or animal a particular skill or type of behaviour' (p 1966).	• Training in contrast to education is more skill acquisition focused in relation to a well-defined task, rather than a broad/generic theoretical experience, with the view of solving a problem or acquiring new behaviour that reduces risk and improves efficiency. • Both education and training are 'means' to achieving an 'end' – namely Competence.
4. Exercising	'[2] A task or activity set to practise or test a skill' (p 643).	• Practising key competencies and improving them through rehearsal and repetition. • Both an individual and collective activity

5. Learning	'Gain or acquire knowledge of or skill in (something) by study, experience, or being taught' (p 1048): 'Usage In modern standard English, it is incorrect to use *learn* to mean *teach* ...' (p 1048): 'The acquisition of knowledge or skills through experience, practice, study or by being taught' (p 1048).	• Learning underpins training and education in particular. It is a structured process that is a function of time (hence 'learning curve') of methodically seeing, hearing, writing and vocalisation whether in an institutional or societal context.
6. Simulation	'Imitate the appearance or character of' (p 1737): A derivative noun of 'simulate' (p 1737)	• Simulation is a process of rehearsal, copying or to mimic behaviour and/or action in order to acquire key experiences, that correspond as far as possible to real world occurrences. It aims to bridge the 'actual' from the perceived.

Types of Training, Competence and Exercise 20.4

Lord Cullen also noted in the Piper Alpha Inquiry how training was not clearly distinguished into its various layers. Management (on-shore and off-shore) seems to have assumed that training is a process of accumulating facts, figures and experiences. Cullen overtly noted the failure to understand the refresher, continuous and bespoke nature of training. Figure 2 is a depiction of the various types, range and breadth of training, competence and exercising that the organisation needs to take account of to ensure personnel possess the up-to-date knowledge and key skills.

Figure 2: Generic types of training, competence and exercise

	Training	*Competence*	*Exercising*
1. Continuous	Regular, on-going training at intervals to update the person of latest best practice.	Re-setting and revising key standards, benchmarks and competencies in line with industrial, socio-technical and academic change.	Practicing a range of key motions, actions, drills or processes on a regular basis – set by management or legislation.

2. Specific	Theoretical and practical knowledge in relation to a particular topic or problem, eg first aid or fire.	Testing skills, knowledge and experience in relation to a well defined task, eg extinguishing a fire.	Practicing a particular motion, action, drill or process either to enhance competence or due to it being a newly introduced skill.
3. Professional	An entry-level qualification in order to attain membership, recognition or accreditation.	The standards and ability to be attained before the person is allowed to practice at a defined level; such competencies cover the breadth and range of skills, knowledge and experience.	The minimum required practical and demonstrable skill that is assessed in order to achieve membership or recognition of a professional association.
4. Bespoke	Relevant and very specific to the needs of a particular organisation, eg off-shore first aid training.	Reaching a standard deemed adequate and appropriate by the organisation.	An exercise specific to a particular plant, department, site or country.
5. 'On the job'	Very practical, primary and 'sense-focused' training by touching, seeing, experiencing at first hand the hazards and associated risk controls.	Setting practical standards of 'do' and 'don't'.	Assessing skills with regards to technology or equipment that is regularly used by a person or team, eg a process plant or fire pumps.

6. Induction	For new entrants to brief them of the minimum standard of conduct and safety behaviour expected and showing them the working environment. This may be reinforced by a practical assessment, either on the same day or later, of information retained.	Ensuring the new entrant is aware of the standards and expectations of the organisation and the law.	Testing in practice the response, cognitive and behavioural skills of a new entrant to determine rank, fitness, position or occupation.
7. Academic	That which is regarded as necessary to obtain a certificate or diploma.	The minimum theoretical, conceptual and pedagogic skills needed to satisfy the assessing body of academic ability.	Identifying theoretical principles and methods that can be developed for general use, from the observation of motions, actions, drills and processes.

The Manual Emergency Preparedness, issued by the Civil Contingencies Secretariat in 2005 identifies Training and Exercising as types of testing to ensure that plans, procedures and policies will perform to expectation in practice. In summary, Training is more personnel focused, aiming to assess how much the human resource knows about the Policy and Plan and means of enhancing the knowledge, skill and experience of that resource. Exercising is a broader approach, examining all aspects of the Policy and Plan, not just the human response but also physical and organisational capability to deal with the Major Incident.

Training and Exercising – Key Principles **20.5**

This section looks at the advice and guidance emanating from competent authorities in the United Kingdom on establishing training and exercising arrangements.

Training **20.6**

An illustration of a training cycle is shown in Figure 3 below.

Figure 3: The Training Cycle (HSG65)

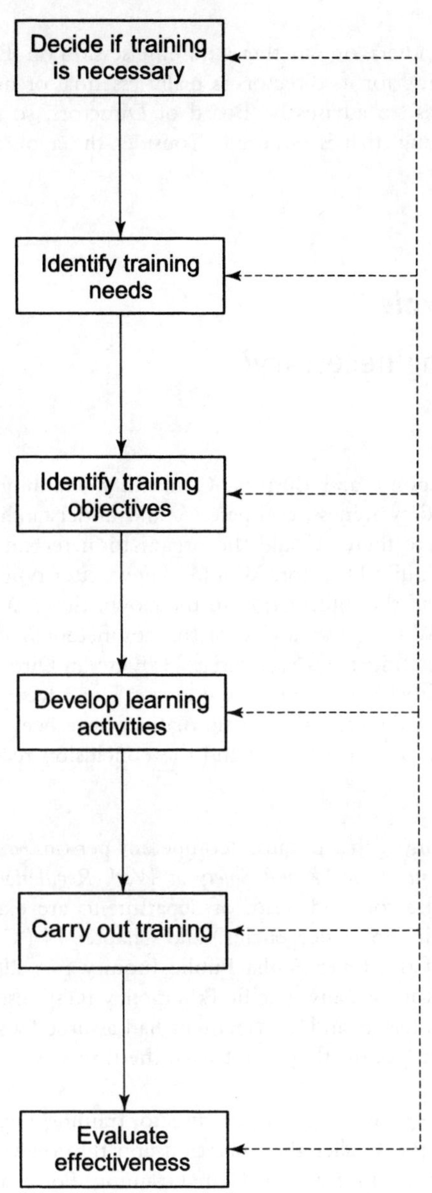

'Successful Health and Safety Management' (HSG 65, HSE) includes a section on the 'Training Cycle' (p 26–32) as in Figure 3, which provides a logical framework to establishing disaster and emergency training programmes. There is no similar guidance as yet from the Civil Contingencies Secretariat on developing disaster and emergency training strategies.

Scenario

20.7

An organisation needs advice on whether a training session on 'Emergency Response and Business Continuity' for its directors is really essential or not. The organisation has 'competent persons' to advise the Board of Directors, so it assumed that it is 'advice' and not 'training' that is required. Consider the application of Figure 3 to provide an outcome.

The training cycle

Step 1: Is training necessary?

20.8

The question means:

- Is Emergency Response and Business Continuity Training the only means of achieving improved awareness, competence and understanding by the directors? What alternatives are there? Could the organisation recruit or appoint a Safety and Business Continuity Director? Would a 'newsletter type' approach be better so that they can read the information in their own time? What do other similar organisations do? What is the advice of the competent bodies such as the Fire Authority, Civil Contingencies Secretariat, HSE etc on Director's minimum level of knowledge of Emergency and related areas? The advisor to the organisation needs to reassure the directors that all alternatives have been considered, research carried out, best practice ascertained and the conclusion reached that training is the best option.

  It must not be assumed that because 'competent persons' exist under *Regulation 7* of the *Management of Health and Safety at Work Regulations 1999* (*as amended 2003*) therefore, directors and heads of departments are exempt from training. Chapter 13 ('Training for Emergencies') and Chapter 14 ('Occidental's Management of Safety') of the Piper Alpha Public Inquiry (see **20.3** above) show no evidence that directors had any specific Emergency Response/Business Continuity Training. The advisors and management had assumed a sort of 'directors are competent and exempt', or 'they don't have the time'.

- With regards to the organisation's size, is director training a realistic option? Step 1 acknowledges that the smaller the operation and the lower the workplace risk, the less may be the need for regular formal training. For example, in the case of a sole director of a small limited company employing only one person, would he/she need a comprehensive one-day course or would being 'kept informed' by suitable sources of information be adequate?

- Will the director training assist in improving the Emergency Response Culture, Business Continuity Management as well as yield insurance/financial savings?

Step 2: Identify training needs **20.9**

Step 1 above asked broader, philosophical and policy questions. Step 2 focuses on specific *needs*:

- What *personal needs* would the director training fulfil? Training can assist the director to:
  - understand the link between disaster prevention, response, business continuity and reputation. Examples, such as Piper Alpha (Occidental Petroleum) and Bhopal (Union Carbide) demonstrate the risks of 'leaving it all to others';
  - improve leadership – individual and collective and reassure personnel and contractors that they are in charge and have oversight. The ValuJet disaster of 11 May 1996, where a DC–9 crashed into the Florida Everglades killing 110, raised issues of pilot and director leadership. Pilots were allegedly turning up to fly planes carrying pizzas and senior officers of ValuJet allegedly did not retain 'strategic risk oversight' of their contractors. The Health and Safety Commission's 'Director's Responsibilities for Health and Safety' now expects individual and collective leadership on safety related matters (see **14.21**);
  - improve the communication skills and enable directors to 'sell' and 'market' the emergency/safety/business continuity message.
- What *job needs* would the director training fulfil? The training will need to update the director on where the organisational risks are and how strategic control can impact on productivity and financial performance. Directors need to understand how Emergency Response and Business Continuity increases corporate success – this key link will enable them in turn to sell the emergency/safety message.
- What *organisational needs* would the director training fulfil? These could include:
  - the need to get individuals to understand the Safety/Business Continuity/Emergency Policy and its implementation;
  - to improve the safety and corporate culture;
  - to reassure industry, customers, suppliers and credit rating agencies that the organisation is following best practice.

Step 3: Identify training objectives **20.10**

Assuming that director training is indeed *necessary* and there is a genuine *need* for training, then objectives will need to be set:

- What are 'training objectives'? This refers to key targets of achievement that the attendees on a training programme need to have realised. These are the key competencies or 'intended learning outcomes'.

- How are the objectives set?
  - Use the Disaster and Emergency Policy or/and the Business Continuity Policy (BCM) as a benchmark. For example, 'at the end of this session, directors will know how they can keep strategic control over key policies and ensure that there is adequate monitoring, implementation and review of key Emergency and BCM arrangements . . . '.
  - Use questionnaires and interviews to ascertain from directors what are the key issues they want addressed and areas of danger they perceive. The training has to be person centred.
  - Enforcement and legal advice – the objectives need to take account of any notices, information, advice or guidance given to the organisation by the enforcing officer, insurer or the courts.
  - Industry standards – key competencies from lead bodies can act as a benchmark, as can manufacturers guidelines.
  - Bodies such as the Institute of Directors (IOD) provide a 'Chartered Director' qualification which covers boardroom responsibility, the findings of Turnbull and other strategic risk control documents. Such a qualification can act as a 'competency checklist' to ensure that any director training provided is at a level, pitch, standard and leads/feeds into such programmes. This reassures directors that the Emergency BCM training has been set at a professional strategic level.
  - Other factors such as corporate needs, political considerations (unions, employers' groups etc), economic competition, resources etc are also important.
- What are the generic objectives of any training programme? Whether the session lasts one hour or longer, there are key underpinning targets that the session needs to achieve, ie:
  - for the attendee to understand the specific theoretical concepts, models or explanations and to be able to understand the practical application to their function and the organisation;
  - to be able to realise the wider corporate and cultural implications of the training topic;
  - to be able to relate it to specific key issues, problems and/or risks;
  - to be able to communicate the key concepts to others;
  - to be able to review, critique and understand the limitations of their role, the organisation and networks to manage disasters/emergencies (knowing the 'systems boundary').

Step 4: Decide on training method 20.11

Some of the key options are summarised in Figure 4:

Figure 4: Training options

	Day release (*a unit of time, eg an hour or day over a given duration*).	**Block** (*eg one week*).	**Distant learning** (*eg at the director's pace*).	**Sessional** (*eg as a one-off session*).
1. In-house/ bespoke.				
2. External (eg at a training centre).				
3. Remote e-learning (eg on-line internet).				

It would be wrong to assume that most directors prefer an in-house sessional approach during, say, a Board Meeting and do not like lengthy programmes. Most directors will want to protect their reputation and preserve their organisation's goodwill and any training that meets that objective is usually welcomed. Each of the above options has advantages and disadvantages that need to be evaluated.

Step 5: Carry out training **20.12**

The key issues to consider are:

- *Where* will the training be conducted from?
  - Is it conveniently located for attendees and the trainer?
  - Is it adequately resourced: flipchart/board, overhead projector, tables, chairs, ventilation, access/egress, a safe learning/working environment etc.
  - Does the venue have welfare facilities such as toilets, drinking water, access for disabled, parking etc?
  - Is access to a practical demonstration/simulation site needed?
- *Who* will do the training?
  - Is the trainer qualified and competent in the topic to be discussed? Has he or she got, and do they understand, the latest legal, technical, managerial and behavioural issues in relation to the topic?
  - Has the trainer a good track record in the topic to be delivered and experience to deliver to directors?

  - Are gender and linguistic issues important? (For instance, what if the audience's English is limited, should someone with knowledge of the relevant language be used?)
  - One should not equate qualifications with good training/teaching skills (there have been many 'highly qualified' people who have been poor at delivery). Therefore, ensure the trainer has the qualities of an effective communicator and listener.
- *When* will the training be conducted?
  - Are the start-finish times realistic and achievable?
  - Is the time sufficient for attendees to reach the destination?
  - Consider time management issues – plan out the training day into sessions or blocks so that the trainer and the attendees know the pace for the training.
- *What* will be covered?
  - Ensure the intended learning objectives are very clear and that they are communicated at the outset to the attendees. Ensure that they also conform to key competencies identified by the HSE, lead bodies or professional bodies. Also, decide what should not be covered and communicate this to attendees so that their expectations are clearly set from the outset.
  - Decide on textbook, lecture notes and cases to be handed out to attendees (relevance, quality of material, user-friendliness, value for money etc).
  - Ensure all documentation is accurate, up to date, genuine and not in breach of copyright. All statements and references cited must be verifiable.
- *Why* is the topic being covered?
  - What reasons are there for its coverage? This can benefit the attendees and the trainer by being clear as to why they are all engaged in the session.

Step 6: Evaluate and feedback **20.13**

In order to evaluate the training, feedback is essential:

- Does the training achieve the objectives set out at the beginning of the session? Were all the topics covered?
- Issue a trainer/tutor 'evaluation form' to seek comments and constructive criticism. Ensure such forms are clear and objectively worded. If need be, explain to the attendees any questions. Such forms provide not only evidence as to the session but invaluable information to improve future training and delivery.
- The trainer needs to be given the feedback also, even if critical.
- Feedback/loop-back to any of the previous steps if the evaluations indicate failings in any areas. Ensure the failing is addressed before the next session.

Exercising **20.14**

The Manual, Emergency Preparedness, issued by the Civil Contingencies Secretariat in 2005 (para 153), describes three specific types of exercising:

- Discussion-based exercises
- Table-top exercises
- Live exercises.

A fourth category combines elements from all three.

These are discussed in more detail below.

Decision-based exercises **20.15**

A decision-based exercise is a broad, brain-storming session, assessing and analysing the efficacy of policies and plans.

Seminars need to be 'inclusive', that is, bringing staff and others together in order to co-ordinate strategic, operational and tactical issues.

Approaches such as 'Technic of Operations Review' (TOR) are useful in drawing out key issues. TOR was developed in the USA and is a qualitative, open-ended but structured method of identifying what each participant regards as the key problem or potential cause(s) of a disaster. TOR would bring together all the key stakeholders and a facilitator would go through five key steps in a given duration of time:

- STEP 1: With regards to each stakeholder, identify and separate the perceptions from the facts. A fact would be that a DC–9 fell over the Florida Everglades, whilst a perception would be that the oxygen canisters were safe and so allowed onto the plane.
- STEP 2: Identify and isolate the basic cause(s) of the disaster or those factors that could cause a disaster. For example, for ValuJet the basic cause was that the canisters were improperly sealed, boxed and stacked. Whilst in the air they released oxygen under air pressure which ignited and in turn was fuelled by the continuous supply of oxygen.
- STEP 3: Identify the root causes. The stakeholders then identify the 'real' causes which range from ineffective control over sub-contractors, poor enforcement by the Federal Aviation Authority and 'burying' a report by one of its officers, the role of the directors and pilots etc.
- STEP 4: Pattern of operational error. This requires those that brainstorm to assess if there is a reoccurring form of decision-making error prevalent in the disaster. For example, identify the pattern of violations, slips, mistakes and lapses. In the case of ValuJet, the mistake was taking the oxygen canisters to a 'COMAT' area instead of a 'HAZMAT' one and ensuring they were properly sealed and disposed off.

- STEP 5: The final step is the 'action' required by stakeholders to ensure the disaster does not happen again. This may require a systemic look at the organisation, focusing on training, investment planning, effective control of contractors and, indeed, director training.

Table-Top exercises **20.16**

The Table Top has been one of the most resilient forms of exercising. It involves using a 2-D map of a site or a generic site and then testing the timing, logistical, cognitive and behavioural skills of participants. Table Tops nowadays also use video images to reinforce the map(s).

Conducting a Table Top does not have to be an expensive exercise. The minimum resource requirements are:

- suitable maps of the site (ordinance survey, GIS), the building (floor plans) and other plans of equipment or machinery;
- clock – to keep to strict time-scales;
- table;
- perhaps toy vehicles, blocks, etc;
- coloured pens/paper;
- documents, eg fire, security, emergency or facilities documentation that could be accessed and used;
- a 'facilitator' to oversee and structure the event;
- 'key note taker' who will make notes on key actions being taken by the players;
- key participants who have pre-defined roles;
- key observers may be invited to make independent assessments. These persons could be from the local authority, external auditors, media etc;
- in some cases a video player is used to record decisions and options. If indoors then the ambience of the room can be changed, eg lighting (high or low or, indeed, modulated during the exercise), the temperature range, ventilation etc;
- the venue could be indoors or at, or near, a potential disaster site.

In short a Table Top is only as good as the planning that goes into it and can never replicate reality, but captures some of the key milestones.

Key issues **20.17**

The key issues being examined by a Table Top exercise include:

Q1. Do the participants know what their roles and functions are?

Q2. Do participants know the limits of their responsibility and how to teamwork with others?

Q3. (a) What human errors are being made and how can they be reduced or eliminated?

(b) Is there a pattern of human error?

(c) Has the Table Top been replicated several times and has the rate of human error changed?

(d) Also, issues of contact between Participants need to be monitored so that decisions are not contaminated, eg opposing team participants discussing decisions whilst leaving the room.

Q4. Are participants aware of the risks to themselves and others at each stage of the exercise? A Table Top, depending on the topic, pace, environment and skill of the facilitator can be stressful. For example, a Table Top of the World Trade Centre for New York firefighters would need to be carefully planned. Likewise, if the exercise is replicating a previous major incident in the organisation's history then preliminary discussion and the consent of participants need to be sought.

Q5. What are the cognitive and behavioural responses to commands given by the facilitator?

Q6. Are participants aware of the gap between reality and the Table Top?

If an organisation is simulating an event that is subject to legal action or proceedings then due diligence needs to be performed and advice sort before embarking on the exercise – certainly if participants are to give evidence at a court or inquiry.

Examples of Table Top exercises **20.18**

An example of two Table Top exercises are shown (see **20.19** and **20.20**) below. In both cases, the following guidelines apply.

- The facilitator will begin by orientating the participants as to the venue, the site and the geography.
- Thereafter, each 'action' is introduced individually, step by step. Participants are only given one action at a time, so they remain focused and do not start concentrating on the other pending ones (but need to anticipate these themselves).
- Participants are given a set time to develop strategies, decisions and responses to the action under discussion. A discussion can ensue as to the output. Participants need to write down their key decisions.
- It is necessary to move to the next action within clear time constraints.
- Facilitators can introduce ad hoc events without pre-warning the participants. For example, a sudden change of the weather, or the emergency services are held in traffic etc. This tests how participants decide under uncertainty.

- The key note taker needs to identify any conflicts, barriers, inter and intra ego problems between participants and/or with the facilitator.
- In addition, Table Top outputs need to be discussed during a de- briefing session and a conclusion and recommendations drawn.

Exercise I: Release of a toxic substance into the ventilation system **20.19**

In this exercise a University's Crisis Management Team (CMT) consists of an Emergency Planning Officer, Vice Chancellor, Facilities Management Head, Heads of key departments, student union representatives and key contractors. In addition, key observers are present. The exercise is to test how the university's management would cope when harmful media once released into the ventilation system.

The facts are as follows:

- 'It is not certain if the release of a dangerous substance which has acute effects into the University's ventilation system was a deliberate act or one of human error', reported a local television company.
- The incident happened at 12.45 hrs on Friday, a week before the exams. The University is at its busiest with many hundreds of students and others on the site. The car park is also full. The substance only requires a minor release but can have effects that can last for 24–48 hours in the ventilation system.
- The weather: It is raining heavily and the temperature is 20 degrees Celsius. The ventilation system is left on at a 'medium' speed in order to generate fresh air.

ACTION 1: A library assistant feels ill and goes to First Aid. It emerges that many members of the students are feeling the same way. The First Aid Room contacts security. The Security Officer contacts the V–C's office. Q. *What is the first action that the CMT needs to take in the first 15 minutes of notification from 12.45–13.00 hrs?*

ACTION 2: 'The unimaginable seems to be happening'; students are coughing and panic is setting in; they are rushing out to get 'fresh air'. Q. *What actions should be taken by the University and what actions should the CMT be taking from 13.00 until 13.45 hrs?*

ACTION 3: 'There seems to be systems failure' – the Security Team also seem to be affected by the release of the substance. Students and employees are also rushing out. Q. *What are the critical issues for the CMT – individually and collectively?*

ACTION 4: You have been under considerable personal, corporate and social pressure as a CMT member. Q. *What are the key legal and behavioural issues that you need to be aware of?*

At any stage the facilitator can introduce an ad hoc event. For example, as the CMT is discussing a response and about to derive an output, the facilitator could announce at Action 3 'a fatality has been announced'. The participants need to include this event into their decision processes, for instance, the event is now also one needing HSE notification under RIDDOR (*Reporting of Injuries, Diseases and Dangerous Occurrences Regulations 1995* (*SI 1995 No 3163*)), the event is now one that becomes a 'high risk' major incident event. Issues of when to shut the university down or keep it open are some of the issues to consider.

Exercise 2: Anti-capitalist rioting **20.20**

In this exercise a large shopping centre is concerned about the risk and threat of rioting. There exists a Crisis Management Team (CMT) consisting of the Managing Director, Communications Manager, Facilities Manager, Administrator, some of the key contractors that have a regular/daily presence at the centre and the Personnel Officer. The CMT are being exposed to a Table Top for the first time.

The facts are as follows:

- 'A group of 200 hard core anti-capitalist demonstrators threw stones and spayed artificial blood on visitors at a fast food site in City X. Two people were seriously hurt', reported a radio station.
- The incident happened on Saturday afternoon at approx 15.00 hrs during summer.
- One of the rioters told us 'our next target is that bourgeois shopping centre'. The rioters were seen running towards the site.
- The shopping centre attracts visitors from across the county and afar and typically gets 40,000 visitors on a Saturday from a cross section of the population (young, old, disabled etc).

ACTION 1: The rioters have approached the main entrance. The security cameras on the perimeter of the site had anticipated the threat and the emergency services had been called and deployed. One of the rioters has a highly poisonous substance with him and releases it into one of the ponds which, it is believed, is on an aquifer. He is not aware of the substance, as he had purchased it from 'a friend' (as he later told the police). Q. *The CMT cannot be contacted at all. What should the shopping centre do?*

NOTE TO CMT: It is not the task of the police to manage neither shopping centres nor the centre's staff; that is management's responsibility.

ACTION 2: One of the rioters was already in the centre car park 'B' and has placed her vehicle that contained a bomb. It has not as yet exploded. Q. *The CMT has now arrived on site. What should you do in light of the above internal and external threats?*

ACTION 3: After an hour, the lady phones in to say, it was all a hoax. There is no bomb. The shoppers are still at the centre and getting anxious to get home. Children are screaming. Q. *What is your next course of action?*

ACTION 4: It has emerged that the substance has changed the colour of the water to a 'purplish' colour. There is serious risk of contamination of an underground aquifer located at point Y. Q. *What is your next course of action?*

ACTION 5: There has been a power failure and the heating is subsequently affected throughout the centre. The shoppers are getting frightened and panic is setting in. There is a serious risk of anger spilling out. Q. *What is your next course of action?*

Again, the facilitator may introduce ad hoc events, eg a fight breaks out between a shop manager and a group of shoppers during Action 3. Or, mass hysteria sets in and the crowd tries to surge out of the shopping centre. Or, 'why are we still in here, surely you should get us to the assembly point?' says a shopper: is she right? The facilitator may wish to suddenly include the role of the emergency services and how they will interface with management. With a scenario such as this, the aim is to convey to management that they need to focus on their skill sets – managing the shopping centre and not presume they have to perform the role of the emergency services. A clear division of labour and interface is a key output being assessed.

For both the above exercises, there needs to be observations about the rate and pattern of interaction between participants, the hierarchy and domination ratio (is one person doing all the leading and decision making or is it more pluralistic?). Other issues include how the participants use any documents provided and is there a dependency on these?

Live exercises 20.21

This is either a simulation of a past Major Incident or rehearsal of a likely one, which actively tests the responses of the individual, organisation and possibly the community (Emergency Services, the media etc). London Underground, the railway sector, COMAH *(Control of Major Accident Hazard Regulations 1999 (as amended 2005))* sites and the civil aviation sector, tend to use such an approach.

The Home Office, when it had overall control of emergency planning and disaster response, produced a document called 'The Exercise Planners Guide' (now under the Civil Contingencies Secretariat). This highlights 28 key issues to address with exercises, in particular, live exercises. Each key point is listed then an interpretation made of what the Home Office/Civil Contingencies Secretariat are implying.

Finance 20.22

As part of the pre-planning, the exercise needs to have a budget which identifies the total direct and indirect costs involved. If resources are limited, consideration can be given to cost reduction through the use of paper-feed rather than telephones or, to use fewer but 'critical' personnel, to reduce the use of expensive features in the scenery etc.

Both a qualitative and quantitative Cost Benefit Analysis (CBA) needs to be performed that highlights personal, job, organisational, social and environmental returns and costs of the live exercise. The larger the organisation, the more detailed this needs to be.

Exercise planners group **20.23**

This is a group of independent observers who can note the happening at the exercise and provide independent testimony as to its occurrence, advice on improving systems or be there to learn for themselves. Such a group could involve the emergency services, enforcers or local authority. For COMAH sites the participation of the community to test its Off-Site Emergency Plans is essential. Larger organisations and exercises involving socially critical systems (a chemical plant near a school for instance) may need to invite the media.

Aim and objectives **20.24**

The aim and objectives need to be set with the participation of key stakeholders, so there is clarity as to purpose of the event.

Q1. What is the overall corporate aim of the exercise?

Q2. What are the intended learning objectives for individuals, teams, departments and sites?

The critical question is, can the aim and the objectives be achieved by another method of exercising, eg Seminar exercise?

Scenario(s) **20.25**

The organisers need to determine the scenario(s) that need to be tested.

Q1. How is a scenario chosen? This could be based on past, Major Incidents, based on public perception of risk or statutory/enforcer requirements.

Q2. How is a scenario structured? At the planning stage of the exercise, the organisers need to produce a 'story board' that depicts a scenario, the key variables and parameters of the scenario and the key competencies/proficiencies being tested. For example:

- Story board: fire spreads in hotel during a busy wedding reception (provide detail as to why, when, where, how, who).
- Variables: number of guests, weather condition, time of day, rate of spread of fire etc.
- Parameters: fire detection systems, access and egress, layout of building and so on.
- Competencies:
  1. To test the response time of hotel managers in activating fire and emergency plans.
  2. To test the verbal and non-verbal communication under pressure of key managers etc.

- Proficiencies: To test how documents are used, which parts are used most, the rate of referral, the problems of use and key problems of understanding of terminology.

Time lapse exercise **20.26**

For each scenario, do the organisers allow it to run in 'real time'? For example, a 'major fire' in a hotel could have been calculated to last × hours. Do the organisers allow the exercise to run for that duration? Again, a careful consideration of costs and benefits need to be made. Alternatively, do the organisers 'speed up' reality and make assumptions, eg 'it is now assumed that four hours have elapsed . . . '. Normally, high risk and socially important sites may need to simulate in real time in order to thoroughly test responses and how personnel react under pressure and frustration, as well as the interaction patterns.

Controlled free play **20.27**

The choice will depend on how much control the organiser wish to exert and what is being tested. Controlled exercises follow a pre-determined script with no digression. 'Free play' is more ad hoc, sudden and allows for uncertainty (similar to the Table Top exercises at **20.18** above). Most exercises use both approaches.

Location **20.28**

Some of the issues associated with location of the exercise relate to:

- permission to carry out the exercise (perhaps planning permission);
- site safety/security and accessibility for participants;
- issues of keeping out trespassers during the exercise;
- the risk created by locating the exercise at that place for traffic, the community, emergency services and participants.
- the Guide points out that visiting the site at a similar time/day when the actual exercise is planned for, enables a visual assessment to be made of how close or far from reality the simulation will be.

Types of exercise **20.29**

Does a Live exercise need to be complimented by a seminar or Table Top? What is the learning strategy? It may be that a shorter and less expensive live event can be performed in conjunction with a range of more economic and bespoke seminar exercises and the results then pooled.

Control post **20.30**

If an organisation has a control centre, then the exercise needs to be close as possible to it or participants need to be able to communicate with it (as in off- or on-shore exercises).

Live **20.31**

The Guide repeats some of its earlier points that a careful selection of venue needs to be made. It also points out the parameters affecting location. For example, a COMAH site or an airport has less choice as the exercise needs to be conducted under realistic conditions. Further, populating a raw site with the equipment and technology of a normal COMAH or airport site may not be feasible.

Exercise base **20.32**

A base near to the exercise is also essential. This could be a nearby building or a tent that will house the key communications equipment, observers and enable the organisers to retain strategic control of the exercise. Where an exercise is testing strategic, operational and tactical decision making, then three bases maybe required. For smaller organisations, fully equipped bases may not be essential and only a small van, table and chairs may be sufficient.

Safety **20.33**

This can be divided into the before, during and after phases.

- Before the exercise, the key issues to address are as follows:
  - Has a Risk Assessment been carried out of the known and foreseeable physical, chemical, biological, ergonomic, behavioural and environmental risks?
  - Issues of Fire Safety need to be considered. Where is the Assembly Point to be sited? Does firefighting equipment exist (or for larger exercises, has the fire brigade been contacted)? Are there trained fire marshals present?
  - Likewise, issues of first aid and medical care need to be accounted for. This ranges form first aid boxes, trained first aiders, rest room, arrangements with a local hospital etc.
  - Furthermore, does the insurance policy extend to exercises? There could be exclusions or conditions that the insurers need to be notified of such events. Does the policy cover observers and trespassers also?
  - Have the organisers been informed of their legal responsibilities towards those attending, the community and the environment?

  - If an exercise involved risk, then issues of safe systems, permits to work and supervision need to be considered.
  - It needs to be noted that even though an exercise is in operation, personnel are still at work and the general duties of the employer will apply.
- During the exercise (but before the actual commencement), all participants and observers need to be told of the safety arrangements in place during the exercise. Second, the key issue is to ensure that a dynamic Risk Assessment is in operation, which identifies occurring risks that were not foreseen during the planning stage. Finally, it should not be forgotten that if a reportable accident does arise during the exercise, this may warrant notification under *RIDDOR (SI 1995 No 3163)*.
- After the exercise, there needs to be monitoring of any injuries, illnesses or diseases. Exercises can also provide invaluable information on the aetiology (causation) and pathology of post traumatic stress – a key cause of time off work for those involved in real incidents as well as the key basis of civil action following disasters.

Welfare **20.34**

Are there adequate welfare facilities in place, eg toilets, rest area, drinking water etc? The psycho-social welfare of the participants is also central. Issues of occupational and disaster stress, anxiety, trauma need to be monitored for. Where exercises involve a close approximation to reality such psychosocial factors need extra attention, eg during oil platform training, working under pressure in a control centre (air traffic controllers) and other high risk situations.

Code words **20.35**

The exercise can be given a code name to ensure that it is noted as an exercise by all stakeholders. The exercise code name can be prefixed with code words to ensure communication is succinct and bespoke to the exercise (consider an emergency service carrying out an exercise without code words – communications could get confused if a real incident arose elsewhere). Below are some of the key words the Guide recommends.

Startex: Start of exercise.
Hold: Suspend exercise for a period.
Resume: Start again after a hold.
Safeguard: Real incident/message outside of exercise.
Abort: Early termination.
Endex: End of exercise.

However, organisations need to use their judgement if such code words need to be used and if they improve response and performance.

Public information **20.36**

If the community needs to be notified of the event (eg off-site COMAH exercises) then the organisers need to do so directly or via the local authority. If the public are not aware, they may enter the exercise area, therefore, 'Exercise in Progress' signs are suggested by the Guide.

Exercise controller **20.37**

The exercise controller will oversee the exercise, make any changes whilst in progress or could terminate the exercise on health and safety grounds. He or she has tactical oversight. Such a person could be an in-house official or externally brought in for the exercise.

Exercise directors **20.38**

The exercise directors' function is to retain strategic, managerial, planning and corporate-legal control over the exercise. They do not actively participate in the exercise but will intervene if chaos, confusion or if a significant decision has to be made. Their primary function is communication – ensuring networks, channels and nodes are spotted, gaps identified and to facilitate where required.

Umpires **20.39**

Umpires are the 'eyes and ears' of the exercise. They can be allotted to each team and will make observations of behaviour, action and reaction of participants. They are indeed akin to cricket umpires, who exist to ensure that the rules of the game are observed but without getting involved in the physical play itself. Umpires are both rule and time keepers.

Observers **20.40**

Observers could be there to either learn, or be a part of a statutory requirement, perhaps to enforce, or indeed as stakeholders (community, media, council etc). Observers need to be clearly identifiable through the vests they are given. How many, who and why are critical issues needed to be answered by the organisers.

Identification **20.41**

For health and safety purposes all parties involved in the exercise need to be identifiable. A system of recording all those involved when they arrive needs to be in place and, indeed, logging them out upon completion. An exercise is akin to a football match with the different teams, referees, the public, media and marshals being distinguishable and located in identifiable areas. Clear identification can also prevent intruders and trespassers.

Communications 20.42

The exercise needs to test the following:

- Verbal communication of each participant, team and group. The clarity of messages sent and retrieved, accents, pitch, tone, modulation, accuracy of statements etc are critical. The Channel Tunnel fire of 18 November 1996 highlighted initial linguistic barriers between the UK and French fire-fighting teams as well as the train driver and the UK rail control centre.
- Non-verbal communication – body language, attitudes and physical reaction rates need to be assessed either descriptively or quantitatively. Do Participants become 'numb', 'freeze' or shocked when they get into the exercise? Do ego problems results and how does body language change?
- Electronic communication – how proficiently do participants utilise equipment and technology? Is there a dependence culture or can they 'survive' or function without it?
- Interface communication – what inferences can be drawn on the communication internally and between strategic, operational and tactical participants?
- Communication nodes and networks – is there a pattern of dominance, control or hierarchy developing? Are rules and orders being obeyed by participants?

There are important 'dramatological' issues of how people react in simulations and rehearsal settings. Participants may adopt 'artificial' behaviours which they would not necessarily use in reality, knowing this is 'just' an exercise. There could be behavioural affects introduced by the presence of observers. Lastly, the organisers need to note any barriers to effective communication and draw lessons form these.

Logging and recording 20.43

Consider the following:

Q1. What do you log and record? All critical actions and reactions commensurate with the scenario and the competencies/proficiencies being tested. The Umpires will fulfil this role. More detailed and larger scale exercises will have 'administrators' who will be the scribes.

Q2. How is the logging and recording carried out? Nowadays the use of video technology, sound recording and transliteration software is quite common and cost effective.

Q3. Why do all this? The exercise evidence could be used for an inquiry, legal action, be required by an insurer or in-house emergency and disaster policy formulation.

Media participation **20.44**

This could refer to both in-house and external media. Organisations have a choice of either an 'open ended' exercise where any bone fide media source can attend or, the organisation's media team is the sole attendee and they put together a press release. If the external media is involved, briefing them on the aim and objectives, ensuring clear lines of communication between them and the organisation (via a media officer) and ensuring their safety are issues to consider. Organisations may produce a 'press pack' beforehand relating to the exercise.

Media coverage **20.45**

If the media are not properly briefed or not involved then there could be an unannounced media attention of what they may genuinely consider to be a real disaster. In such an event, the media officer needs to have arrangements in place to take the media to a 'docking station', provide information and reassure them of the mock nature of the event.

Briefings **20.46**

Before the commencement and during the exercise, team leaders, be they strategic, operational or tactical, need to keep participants fully briefed of the actions, tasks, key issues, risks and reoccurring problems arising from the exercise.

Pre-exercise final arrangements **20.47**

This is a final check to ensure that the 'game' will go ahead smoothly. A checklist approach is useful. The above points hitherto can be included in a checklist. Rushing into an exercise needs to be avoided, irrespective of the time and money that goes into it. Lessons can be learned from the Challenger disaster of 1986, when pressures to launch the Shuttle overlooked the safety concerns over the performance of the 'O-rings' under cold temperature conditions raised by the contractors to NASA (see **5.15** above). Not only was the Shuttle lost, but also more importantly, seven astronauts lost their lives. The final test of whether an exercise goes ahead is that of health and safety.

De-briefing **20.48**

De-briefing needs to be performed at various levels – individual, group, team, department, site and inter-agency. Briefings can be 'hot' (immediately after the exercise) or 'cold' (sometime afterwards) the Guide says. In any case, de-briefing allows key milestones, performance issues and lessons to be identified.

The Guide recommends the appointment of a 'neutral de-brief co-ordinator' who has not necessarily been involved in the exercise who can structure the responses, identify gaps and raise key strategic, operational and tactical issues.

De-briefing needs to be relevant to the participants and can be an interesting/learning experience. Organisations with digital video cameras and basic media software can, in a matter of hours, construct CD-ROMs that can be used as a part of the de-briefing.

Finally, the Guide points out that all those involved in the exercise need to be thanked for their input and support.

Exercise report 20.49

The exercise report can be in four forms as listed below:

- A short one or two-page executive summary for senior corporate officers of outcomes and issues that they need to be aware of.
- Summaries for participants and other key staff (enabling them to communicate to their departments).
- A short one or two-page summary for other stakeholders, eg council, media etc.
- A full report, professionally structured that can form a part of 'corporate memory' and be used for in-house training and educational programmes.

Issues of confidentiality, data protection and legality of the report needs to be taken into account and be shown to suitable legal officers before being released.

Other types of exercises 20.50

In addition to the above, there are also two other types of exercising that need to be noted. These are Control Post exercises and Synthetic Simulation.

Control Post exercises 20.51

In 'The Exercise Planners Guide' the Home Office/Civil Contingencies Secretariat mentions this fourth type of exercising which is concerned with testing the information flow and communications skills between control post holders. For example, a radio control centre of a railway company, its nodes near key railway signals, platform managers and head office strategic controllers may be tested to assess whether they communicate in unison. This form of exercise is 'testing without the troops'.

Synthetic simulation 20.52

Whilst this is not mentioned in Home Office or other emergency service guidance, this fifth approach is becoming a popular option. At one end it used basic Windows or Lotus programs. On the other end it applies the techniques used in flight and military simulation. Thus, using computers, one can model the site and in 3-D format move around the screen. This then enables various eye-points around the site, achieved by moving the mouse. Some advanced systems enable 'computer generated entities' to be included into the data set. For example, a collision can be simulated outside the production plant or the rate of noxious substance release modelled and calculated across the local community.

Three computer-related software packages are commonly used by the Emergency Services in the UK. They are Hydra, Minerva and Vector Command.

Hydra **20.53**

Hydra is a syndicate-based command simulation system used to train members of the emergency services in the tactical and strategic management of large-scale incidents. It allows command officers who would typically be responsible for managing major emergencies and disasters to work together in syndicates on common problems within a safe training environment where they can share best practice and exercise command skills. It is particularly useful in simulating protracted incidents which may take days or even weeks to resolve.

Minerva **20.54**

Minerva has been designed as a real-time command simulation system allowing simulated emergencies or, indeed, serious outbreaks of public disorder, to unfold uninterrupted and in real-time from the point of first approach to the emergency to the point at which effective incident management has been established. Minerva is a team-based simulation system allowing for interaction and problem-solving between members of a command team, each of whom has different roles and responsibilities at a real incident.

Emergency Command System **20.55**

Drawing on its many years of experience creating advanced training systems, particularly for the UK Fire Service, Vector Command has now produced the Emergency Command System, an advanced planning, training and command system developed specifically for use by emergency managers at all levels. Run over selected time periods, the Emergency Command System provides managers with a realistic and complex series of tasks and challenges. Exercise administrators provide 'injects' into the exercises. Often these are unexpected events that test managers' decision-making abilities under pressure.

Conclusion **20.56**

Training and exercising are opposite sides of the same coin. To improve the competence and enhance the learning of personnel, both training and exercising need to be used. There is a powerful argument as continuously stated by Lord Cullen in all his disaster-related reports that regular and up-to-date training/exercising for emergencies and disasters is better than a one-off event. It could be added that smaller scale events, but held on a more frequent basis which form a part of the corporate diary, will embed training and exercising into the corporate habit and psyche. Finally, one must not forget the purpose of all this effort, namely to equip individuals and groups with capabilities to save lives and reduce destruction.

Index